T0213343

MATHEMATICS IN INDUSTRY   29

More information about this series at http://www.springer.com/series/4650

E. Jan W. ter Maten • Hans-Georg Brachtendorf
Roland Pulch • Wim Schoenmaker
Herbert De Gersem

Editors

Nanoelectronic Coupled Problems Solutions

Editors
E. Jan W. ter Maten
Applied Mathematics & Numerical Analysis
Bergische Universität Wuppertal
Wuppertal, Germany

Roland Pulch
Institute of Mathematics and Computer Science
Ernst-Moritz Universität Greifswald
Greifswald, Germany

Herbert De Gersem
TEMF
TU Darmstadt
Darmstadt, Germany

Hans-Georg Brachtendorf
Fachhochschule Oberösterreich
Wels, Austria

Wim Schoenmaker
MAGWEL NV
Leuven, Belgium

ISSN 1612-3956   ISSN 2198-3283   (electronic)
Mathematics in Industry
ISBN 978-3-030-30728-8   ISBN 978-3-030-30726-4   (eBook)
https://doi.org/10.1007/978-3-030-30726-4

Mathematics Subject Classification (2010): 65M20, 65N06, 65N08, 65N30, 65L80, 78M31, 78M34, 93A30, 35Q61, 74F15

Foreword

This book represents an important outcome from the nanoCOPS project, a very timely and quite successful R&D project, partly funded by the EU under the 7th Framework Program (FP7).

I had the opportunity to closely follow the project throughout its three year existence and to witness first hand the progress made. I was invited and accepted to become a project reviewer, together with my esteemed and sorely missed colleague Prof. Tom Brazil from University College in Dublin, Ireland, and Dr. Murat Eskiyerli, from Revolution Semiconductor, in Ankara, Turkey.

Like many other FP7 funded projects, nanoCOPS boasted a select set of partners from academia and industry looking for solutions to problems of common interest. The industrial partners brought to the table their considerable design, characterization and fabrication expertise as well as some of the challenges they faced in validating complex design under variable conditions exposed to electromagnetic, mechanical and thermal stresses. The academic partners were tasked with developing the algorithms required to addresses such problems in an efficient manner. One of the industrial partners, MAG-WEL, was instrumental in ensuring that such algorithms were made available as ready to use tools such that validation on a set of relevant test cases was performed.

The results obtained during the nanoCOPS project are far reaching and many of the ideas, concepts and algorithms developed have since found their way onto other applications and fields. Originally the project dealt with modeling, simulation and validation of strongly coupled power MOS devices and RF circuitry by exploiting the inherent characteristics and different dynamics of its sub-systems with the goal of achieving robust variability-resilient designs.

The project relied on top notch complementary expertise from top to bottom. Experienced electronic engineers conveyed the main issues stemming from the difficulty in validating complex strongly coupled systems subject to multi-domain stresses in a world of uncertainty. Renowned numerical analysis experts ensured that the computational models being developed satisfied requirements for efficient and robust simulation. The development of techniques and algorithms to co-simulate such systems under these challenging conditions led to the development of novel algorithms and techniques, later translated into tools allowing for the validation of electronic systems accounting for electromagnetic coupling and heat transfer. Efficiency requirements and the need to account for design or operating parameter variations led to the development of novel parametric model order reduction algorithms. The project was also one of the first to weigh in heavily on techniques for uncertain quantification, accounting for parameter variations and their effect on designs.

In this book, many of the new algorithms and innovative techniques developed are described in detail with accompanying data resulting from test case analysis and validation. The book is therefore of interest to both practitioners and researchers with interests ranging from scientific computing to modeling and simulation of complex coupled multi-domain systems under varying conditions.

I hope you will all enjoy it as much as I did throughout the run of the project while many of the results here presented were still being imagined, implemented, tested, validated and made into useful tools that will enabling the designs of tomorrow.

Lisbon, Portugal, July 2019

Luis Miguel Silveira

IEEE Fellow
Professor of Electrical and Computer Engineering
INESC ID and IST Tecnico Lisboa

Preface

This book results from cooperations between mathematicians and electronic engineers from academia and industry It collects backgrounds and outcomes with their further perspectives of the FP7 project *nanoCOPS*, *"Nanoelectronic Coupled Problems Solutions"* (Nov. 2013-Oct. 2016), with partners from academia: Bergische Universität Wuppertal, Technische Universität Darmstadt, Humboldt Universität zu Berlin, Universität Greifswald, Max Planck Institute for Dynamics of Complex Technical Systems in Magdeburg (all from Germany), Brno University of Technology (Czech Republic) and Fachhochschule Oberösterreich (Hagenberg im Mühlkreis, Austria) and partners from industry: ON Semiconductor (Oudenaarde, Belgium), ACCO Semiconductor (Louveciennes, France), NXP Semiconductors (Eindhoven, the Netherlands) and MAGWEL NV (Leuven, Belgium). Here the tool provider MAGWEL was the key interface between modelling, development, implementation and validation of algorithms and the industrial end-users.

The project did focus on developing algorithms for coupled problems, in which electromagnetics, electronic circuits and heat transfer were considered, with further steps to implications for mechanical stress.

The algorithms did concentrate on efficient simulation in the time domain covering holistic / monolithic full system time integration and co-simulation both. Dedicated focus was on multirate time integration to deal with waveforms with strong differences of dynamical behaviour (f.i. due to different behaviour on different parts of the geometry, or due to being different physical quantities) or even having Fourier components with a large gap in frequencies (important for communication and mobility).

Uncertainty Quantification studied the effects due to parameter variations for robust optimization of designs of coupled problems as well as for topology optimization and shape optimization and yield estimation and to minimize the RF coupling from digital parts (aggressors) to analogue parts (victims). Estimation of failure probabilities was obtained using Large Deviation Theory and adjoint techniques. More general density estimation of parametrized waveforms was considered using techniques from calibration to determine industrially relevant quantities like capability index and by inverse modelling. Efficient sparse Model Order Reduction for coupled problems made the simulation of large systems possible and thus enhanced uncertainty quantification by providing reduced models as response surface model. During the project here also post error control for the Krylov-based tensor methods was developed.

Special electromagnetic-thermal modelling was made on bond wires that couple a chip to the surrounding package. This included a study on stresses to prevent damages. Uncertainty Quantification did cover the effects due to geometrical variations. Another topic was the study of ageing and reliability prediction, especially for bond wire ageing.

In all modelling, simulation and validation a unique opportunity for mathematicians was given by the excellent expertise in measurements by the industrial partners and the Brno University of Technology.

The book concludes with extensive chapters on test cases for Power-MOS Devices and RF-circuitry, on Measurements, on Validation and on methodology and best-practices for optimized driver design.

A project like nanoCOPS did start from successful cooperations between several partners made in the past within various projects. New proposals start from needs, from being timely, and are based on proven experience in cooperating, that provide short communication lines and good understanding. New algorithms should be implemented, be validated by other means of simulations or tests, and, when possible, compared to measurements. In all steps direct contact between developers, tool provider and end-user is important.

From international projects we mention here

- ICESTARS: "Integrated Circuit/EM Simulation and Design Technologies for Advanced Radio Systems-on-chip" [EU FP7, 2007-2010]
- MEDEA+ 2t204, ELIAS, "End-of-life investigations for automotive systems" [2007-2010].
- Artemos: "Agile Rf Transceivers and front-Ends for future smart Multi-Standard cOmmunications applicationS". [ENIAC Joint Undertaking Project, 2011-2014]

Parallel with nanoCOPS there also did run CORTIF: "Coexistence Of Radiofrequency Transmission In the Future" [CATRENE - EUREKA Cluster for Application and Technoogy Research in Europe on NanoElectronics Project, 2014-2017].

An interested reader may notice that some contributions reveal exchange between the nanoCOPS, Artemos and CORTIF projects.

Apart from these projects there were several national projects between various consortium partners.

The book is written to an audience of people interested in mathematics, electronics and physics at the interplay of modelling, scientific computing, optimizing, computer science and validation. In the book we collected the contributions in six parts: (1) Equations and Discretizations, (2) Time Integration for Coupled Problems, (3) Uncertainty Quantification, (4) Model Order Reduction, (5) Robustness, Reliability and Ageing, and (6) Test Cases, Measurements, Validation and Best Practices. In principle, all chapters can be read independently from each other. They show a nice 'final' cooperation between academia and industry.

We thank all colleagues for the cooperation during the nanoCOPS project. Especially we want to thank Prof.Dr. Caren Tischendorf (Humboldt Universität zu Berlin) for organizing several minisymposia where we could expose our results and ideas and for organizing the videos to highlight different achievements. We also thank here Dr. H.H.J.M. (Rick) Janssen (NXP Semiconductors), who structured the validation process for our algorithms.

We thank our EU Project Officer, Dr. Henri Rajbenbach (Bruxelles, Belgium), and our EU Project Reviewers, Prof. Thomas J. Brazil (University College Dublin, Ireland), Prof. Luis Miguel Silveira (INESC-ID and University of Lisbon, Portugal), and Dr. Murat Eskiyerli (Revolution Semiconductor, Ankara, Turkey). Their deep interest and experienced comments inspired us to further improve and extend algorithms, with more validations and to broaden exposure of our highlights to a larger community. This book is one of the outcomes of this recommendation. During the preparation of this book we were deeply shocked by the unexpected passing of Prof. Brazil in April 2018.

The editors thank all our authors for their contributions and for their patience in finalizing the book. We also thank our respective universities, as well as Eindhoven University of Technology, in providing excellent environments for preparing the book.

Finally we thank Springer Verlag to include the book in the book series Mathematics in Industry, that was set up jointly with ECMI, the European Consortium for Mathematics in Industry.

June 2019

Wuppertal (Germany)/Eindhoven (the Netherlands),	*E. Jan W. ter Maten*
Hagenberg im Mühlkreis (Austria),	*Hans-Georg Brachtendorf*
Greifswald (Germany),	*Roland Pulch*
Leuven (Belgium),	*Wim Schoenmaker*
Darmstadt (Germany),	*Herbert De Gersem*

Acknowledgements

The scientific outcomes in this book result from the Collaborative Project *nanoCOPS*, "Nanoelectronic COupled Problems Solutions", which was supported by the European Union in the FP7-ICT-2013-11 Program under Grant Agreement Number 619166, November 2013 – October 2016.
See also the website `http://www.fp7-nanoCOPS.eu/`.

Contents

Part VI   Test Cases, Measurements, Validation and Best Practices

20   Test Cases for Power-MOS Devices and RF-Circuitry 459

Rick Janssen, Renaud Gillon, Aarnout Wieers, Frederik Deleu,
Hervé Guegnaud, Pascal Reynier, Wim Schoenmaker,
E. Jan W. ter Maten

List of Contributors

Nicodemus Banagaaya
Max Planck Institut für Dynamik komplexer technischer Systeme, Sandtorstr. 1, Magdeburg, 39106, Germany, e-mail: Banagaaya@mpi-magdeburg.mpg.de;Banagaaya@gmail.com

Patrice Barroul
ACCO Semiconductor, 36-38 Rue de la Princesse, Louveciennes, 78430, France, e-mail: Patrice.Barroul@acco-semi.com

Theo G.J. Beelen
Eindhoven University of Technology, Department of Mathematics and Computer Science, PostBox 513, 5600 MB Eindhoven, the Netherlands, e-mail: T.G.J.Beelen@tue.nl;Th.Beelen@gmail.com

Peter Benner
Max Planck Institut für Dynamik komplexer technischer Systeme, Sandtorstr. 1, Magdeburg, 39106, Germany, e-mail: Benner@mpi-magdeburg.mpg.de

Kai Bittner
Fachhochschule Oberösterreich, Softwarepark 11, Hagenberg im Mühlkreis, 4232, Austria, e-mail: Kai.Bittner@fh-hagenberg.at

Hans-Georg Brachtendorf
Fachhochschule Oberösterreich, Softwarepark 11, Hagenberg im Mühlkreis, 4232, Austria, e-mail: Hans-Georg.Brachtendorf@fh-hagenberg.at

Thorben Casper
Technische Universität Darmstadt, Dolivostraße 15, Darmstadt, 64293, Germany, e-mail: Casper@gsc.tu-darmstadt.de

Herbert De Gersem
Technische Universität Darmstadt, Schloßgartenstraße 8, Darmstadt, 64293, Germany, e-mail: DeGersem@temf.tu-darmstadt.de

Bart De Smedt
MAGWEL NV, Vital Decosterstraat 44 – bus 27, Leuven, 3000, Belgium,
e-mail: Bart.DeSmedt@magwel.com

Frederik Deleu
ON Semiconductor Belgium BVBA, Westerring 15, Oudenaarde, 9700,
Belgium, e-mail: Frederik.Deleu@onsemi.com

Alessandro Di Bucchianico
Eindhoven University of Technology, Department of Mathematics and
Computer Science, PostBox 513, 5600 MB Eindhoven, the Netherlands,
e-mail: A.D.Bucchianico@tue.nl

Jos J. Dohmen
NXP Semiconductors, High Tech Campus 46, 5656 AE Eindhoven, the
Netherlands, e-mail: Jos.J.Dohmen@nxp.com

Jiri Drinovsky, Jiří Dřínovský
Brno University of Technology, Technicka 3082/12, Brno, 61600, Czech
Republic, e-mail: Drino@feec.vutbr.cz

David J. Duque Guerra
Technische Universität Darmstadt, Dolivostraße 15, Darmstadt, 64293,
Germany, e-mail: Duque@gsc.tu-darmstadt.de

Oryna Dvortsova
Eindhoven University of Technology, PostBox 513, 5600 MB Eindhoven, the
Netherlands, e-mail: Oryna.Dvortsova@gmail.com

Lihong Feng
Max Planck Institut für Dynamik komplexer technischer Systeme, Sandtorstr. 1, Magdeburg, 39106, Germany, e-mail: Feng@mpi-magdeburg.mpg.de

Renaud Gillon
ON Semiconductor Belgium BVBA, Westerring 15, Oudenaarde, 9700,
Belgium, e-mail: Renaud.Gillon@onsemi.com

Tomas Gotthans, Tomáš Götthans
Brno University of Technology, Technicka 3082/12, Brno, 61600, Czech
Republic, e-mail: Gotthans@feec.vutbr.cz

Hervé Guegnaud
ACCO Semiconductor, 36-38 Rue de la Princesse, Louveciennes, 78430,
France, e-mail: Herve.Guegnaud@acco-semi.com

Michael Günther
Bergische Universität Wuppertal, Gauß-Straße 20, Wuppertal, 42119,
Germany, e-mail: Guenther@math.uni-wuppertal.de

Rick Janssen
NXP Semiconductors, High Tech Campus 46, 5656 AE Eindhoven, the

Netherlands, e-mail: Rick.Janssen@nxp.com

Tomas Kratochvil, Tomáš Kratochvíl
Brno University of Technology, Technicka 3082/12, Brno, 61600, Czech
Republic, e-mail: Kratot@feec.vutbr.cz

E. Jan W. ter Maten
Bergische Universität Wuppertal, Gauß-Straße 20, Wuppertal, 42119,
Germany, e-mail: Jan.ter.Maten@math.uni-wuppertal.de
Eindhoven University of Technology, Department of Mathematics and
Computer Science, PostBox 513, 5600 MB Eindhoven, the Netherlands,
e-mail: E.J.W.ter.Maten@tue.nl

Peter Meuris
MAGWEL NV, Vital Decosterstraat 44 – bus 27, Leuven, 3000, Belgium,
e-mail: Peter.Meuris@magwel.com

Jiri Petrzela, Jiří Petřzela
Brno University of Technology, Technicka 3082/12, Brno, 61600, Czech
Republic, e-mail: Petrzelj@feec.vutbr.cz

Yannick Poupin
ACCO Semiconductor, 36-38 Rue de la Princesse, Louveciennes, 78430,
France, e-mail: Yannick.Poupin@acco-semi.com

Roland Pulch
Universität Greifswald, Walther-Rathenau-Straße 47, Greifswald, 17487,
Germany. e-mail: Roland.Pulch@uni-greifswald.de

Piotr Putek
Bergische Universität Wuppertal, Gauß-Straße 20, Wuppertal, 42119, Germany, e-mail: Putek@math.uni-wuppertal.de;Piotr.Putek@gmail.com

Pascal Reynier
ACCO Semiconductor, 36-38 Rue de la Princesse, Louveciennes, 78430,
France, e-mail: Pascal.Reynier@yahoo.fr

Ulrich Römer
Technische Universität Braunschweig, Schleinitzstr. 20, 2. OG, Braunschweig, 38106, Germany, e-mail: U.Roemer@tu-braunschweig.de

Wim Schoenmaker
MAGWEL NV, Vital Decosterstraat 44 – bus 27, Leuven, 3000, Belgium,
e-mail: Wim.Schoenmaker@magwel.com

Sebastian Schöps
Technische Universität Darmstadt, Dolivostraße 15, Darmstadt, 64293,
Germany, e-mail: Schoeps@gsc.tu-darmstadt.de

Roman Sotner, Roman Šotner
Brno University of Technology, Technicka 3082/12, Brno, 61600, Czech

Republic, e-mail: Sotner@feec.vutbr.cz

Christian Strom
Humboldt Universität zu Berlin, Unter den Linden 6, Berlin, 10099,
Germany, e-mail: StrohmCh@math.hu-berlin.de

Bratislav Tasić
NXP Semiconductors, High Tech Campus 46, 5656 AE Eindhoven, the
Netherlands, e-mail: Bratislav.Tasic@nxp.com

Caren Tischendorf
Humboldt Universität zu Berlin, Unter den Linden 6, Berlin, 10099,
Germany, e-mail: Caren.Tischendorf@math.hu-berlin.de

Aarnout Wieers
ON Semiconductor Belgium BVBA, Westerring 15, Oudenaarde, 9700,
Belgium, e-mail: Aarnout.Wieers@onsemi.com

Yao Yue
Max Planck Institut für Dynamik komplexer technischer Systeme, Sandtorstr. 1, Magdeburg, 39106, Germany, e-mail: Yue@mpi-magdeburg.mpg.de

Chapter 1
Nanoelectronic Coupled Problems Solutions – Highlights from the nanoCOPS Project

Caren Tischendorf, E. Jan W. ter Maten, Wim Schoenmaker

Abstract   This introductory chapter summarizes the objectives and the highlights of the FP7-ICT project nanoCOPS from which the outcomes of this book emerged. It is based on the Brochure that was made at the end of the project to broadly advertise the achievements of the project. We identify the role of mathematics to solve the coupled problems that were posed by the partners from semiconductor industry. The size of the problems asked for developing reduction techniques that allowed for parameterization. Applications needed uncertainty quantification for reliability purposes and optimization. All methods developed have been tested and validated on industrial test examples.

1.1 FP7 Project nanoCOPS - Overview

The increasing performance of electronic devices as mobile phones, laptops, control units for automotive applications, medical diagnostic equipments and lots more is connected with fundamental innovations in semiconductor design and manufacturing. New materials and structures in the nanometer regime allow the excess of existing limits, but also induce ever-increasing process variations and reliability degradation. Particular problems are overheating in

Caren Tischendorf
Humboldt Universität zu Berlin, Germany,
e-mail: `Caren.Tischendorf@math.hu-berlin.de`

E. Jan W. ter Maten
Bergische Universität Wuppertal, Germany,
e-mail: `Jan.ter.Maten@math.uni-wuppertal.de`

Wim Schoenmaker
MAGWEL N.V., Leuven, Belgium,
e-mail: `Wim.Schoenmaker@magwel.com`

© Springer Nature Switzerland AG 2019   1
E. J. W. ter Maten et al. (eds.), *Nanoelectronic Coupled Problems Solutions*,
Mathematics in Industry 29, https://doi.org/10.1007/978-3-030-30726-4_1

integrated circuits and signal interferences depending on the operating conditions and (3D) layout configurations. The project *nanoCOPS* develops algorithms and software tools to quantify such undesirable effects for reliable validations of new nanoelectronic designs. It focusses on coupled simulations of electromagnetic, heat and stress phenomena for designs with power-transistor devices and high frequency circuitry for wireless communication. Due to the market demands, the scientific challenges are to

- create efficient and robust simulation techniques for strongly coupled systems, that exploit the different dynamics of sub-systems and that can deal with signals that differ strongly in the frequency range,
- include variability such that robust design, worst case analysis, and yield estimation with tiny failures are possible,
- reduce complexity.

Predecessor Project ICESTARS

Already in 2009 nine partners started working together on the FP7 project ICESTARS (**I**ntegrated **C**ircuit/**EM** **S**imulation and Design **T**echnologies for **A**dvanced **R**adio **S**ystems-on-Chip). They developed new tools and methods for the functional design development of more powerful chips for faster data transfer in mobile communications. This joint project was funded by the European Union and included among its participants not only private companies but also universities. In 2012, European funding ran out for the highly successful project ICESTARS. Outcomes can be found on http://www.icestars.eu/. In the course of further developments, however, new problems emerged, and to address them a new research project was conceived.

Fig. 1.1 Left:
ICESTARS
http://www.icestars.eu/.
Right: nanoCOPS
http://www.fp7-
nanocops.eu.

Project nanoCOPS

This new project was been funded by the European Union since 2013 and carried the name *nanoCOPS* (**Nano**electronic **Co**upled **P**roblems **S**olutions). The consortium contained several partners from the earlier ICESTARS project. Coordinator was the University of Wuppertal. The consortium includes the semiconductor companies NXP Semiconductors (the Netherlands), ACCO Semiconductor (France), ON Semiconductor (Belgium) and the electronic design tool provider MAGWEL NV (Belgium), as well as the academic institutions University of Wuppertal, Humboldt University of Berlin, Ernst-Moritz-Arndt University of Greifswald, Technical University of Darmstadt, Max Planck Institute for Dynamics of Complex Technical Systems in Magdeburg (all Germany), Brno University of Technology (Czech Republic) and University of Applied Sciences of Upper Austria in Hagenberg im Mühlkreis (Austria).

Fig. 1.2: Logos of the participating project members.

1.1.1 Objectives

nanoCOPS was concerned with semiconductor devices and the development of new chips. One of the objectives was a meaningful and efficient circuit design. It is particularly important to accommodate more and more devices on the ever smaller carrier surface so that more information can be transmitted without interference. The underlying problem is that physical limits are reached, and thus the individual circuits and components are getting closer and closer to each other, increasing the risk of the so-called crosstalk effect, which leads to unwanted electromagnetic interaction between the individual components. This can no longer be ignored in the current development of ever more powerful chips.

This is why in the project *nanoCOPS* tools were being developed to explore these physical limits. Models and simulations were created to see how far one can reduce the areas while increasing the number of components, and without risking the negative effects of adverse interactions.

Moreover, *nanoCOPS* was pursuing a number of superordinate research tasks. In particular, new tools for simulations were being developed that go beyond

Fig. 1.3 Layout of a power transistor (stretched vertical direction.

the existing standardized simulations. Electromagnetic field simulations can be combined with thermal simulations and are integrated directly into the chip simulation. In this way, highly sensitive, interference-prone circuit components can be identified during development and can be designed more robustly.

Focus was on power MOS transistors, see Fig. 1.3, which can reduce the "power hungry" electrical appliances. Problems arise when electrical and thermal interactions affect the component and have an impact on material stress and the lifespan of the transistors. It is therefore important to correctly assess heat generation, in order to be able to develop more efficient and more durable transistors. The *nanoCOPS* project has enabled the design of such modern transistors. At the same time, a long lifespan is guaranteed and a more accurate prediction of the yield of the product is achieved.

1.1.2 The Role of Mathematics

As with almost all developments in more effective modelling and simulation programs, new mathematical methods were required. In order to ensure sufficient accuracy of the simulations, an extremely high geometric resolution was needed. This resulted in high-dimensional model equations that can only be solved in a reasonable time by means of a mathematical model reduction. However, the existing model reduction methods do not work for all model systems. A particular challenge was represented by nonlinear parameter-dependent systems. In *nanoCOPS*, new model reduction techniques were being developed with which non-linear temperature-dependent circuit behaviour can be studied efficiently.

Another key area was the quantification of possible uncertainties arising from

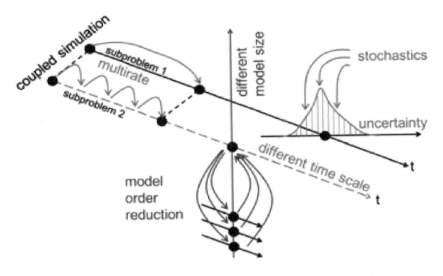

Fig. 1.4: The mathematics has to deal with varying dynamics between different subproblems or with large frequency differences in components of signals: both effects are called multirate and lead to different time scales that have to be properly tracked in time. This makes that Model Order Reduction has to deal with very different model sizes as well as with large different time scales. Uncertainty in parameters, or in layout, results in uncertainty of output signals. Performance requirements lead to optimization for signals with tight bounds on an amount of variation around the (time-varying) mean (up to 6-*sigma*) [8, 11].

the chip production process. The used semiconductor material never shows a perfectly consistent accuracy. The resulting defects are investigated mathematically with stochastic methods and the impact of these disturbances is described. Fig. 1.4 graphically summarizes the various aspects in mathematics. In [8, 11] (open access) some further details are given.

Research on semiconductor materials predominantly focuses on conventional silicon wafers. Nevertheless, the developed tools are not limited to this material. A number of different parameters that result from other wafer materials were included in the investigations.

1.1.3 Applications

To minimise signal interference in mobile devices such as telephones, mobile communications, mobile TV, navigation and also in pacemakers, high-frequency signals of different frequency have to be simulated efficiently. There is also an interaction between electronic circuits, electromagnetic fields and heat generation. These factors, as well as the ageing of the components, cause changes that cannot be accurately predicted. One of the objectives of *nanoCOPS* was therefore to construct probability distributions that can lead to reliable predictions on the operation of the entire integrated circuit over a long period.

1.2 Successes - Innovative nanoCOPS Developments

The problems to be solved have ever been linked to each other. There are two main themes: *electro-thermal simulations* and *circuit simulation*. They were the common working basis for all the *nanoCOPS* partners. As a result, partial but promising answers have been found to the most pressing issues.

A central role is frequently played by the large number of data that goes far beyond the simulation of a single transistor finger or a relatively simple integrated back-end structure. The claim has always been to solve the electro-thermal problem at full-chip level or at least to include a significant portion of the chip. In particular, it should be noted here that there is a strong temperature dependence in power semiconductor applications. On the other hand, the temperature depends on internal and external heat sources and the available heat sinks. In extreme cases, high temperature can lead to power concentration and a further increase in temperature. This is because of the temperature-dependent thermal conductivity that cannot be neglected, e.g., in silicon. It gives an undesirable feedback leading to less thermal conduc-

Fig. 1.5 Brochure of the FP7 project nanoCOPS, available on http://www.fp7-nanocops.eu in addition to reports and videos.

tivity with increasing temperature. Unfortunately, existing circuit simulators do not have the ability to represent the thermal properties in enough detail. Tools that are normally suitable for simulating thermal behaviour are not yet sufficiently developed to predict a system's relationship between temperature and power.

Thanks to *nanoCOPS*, the company MAGWEL can now offer a solution to solve electrical and thermal behaviour in a deterministic manner. Their PTM-ET simulator models connect the metal heating as a function of the applied current. They understand in detail how the different parameters contribute to heat flow.

In the next sections we highlight various innovative *nanoCOPS* developments. They offer a number of absolutely new and groundbreaking developments for nanoelectronic design. Fig. 1.5 refers to the Brochure that can be downloaded from the project website, `http://www.fp7-nanocops.eu`, in addition to reports and videos.

1.2.1 Coupled Electro-Thermal Simulation

There is a strong dependence between temperature and power in circuits. High temperatures and extreme power levels are large sources of error. Until now, two separate operations were needed for the simulation of these processes. For the first time, *nanoCOPS* has successfully combined the two steps into one, speeding up the simulation significantly and delivering much better results. Among other things, the product PTM-ET from MAGWEL [10] calculates the computed dynamic currents of Joule self-heating in metal and active area, heat flow models in chip and package, and supports thermal planning and non-linear temperature-dependent models. It provides self-consistent solution of electrical and thermal equations in a 3D solver [11]. New, detailed modeling and simulation of bond wires have been added [4]. Thanks to *nanoCOPS*, company MAGWEL now provides a field-solver based with a specification to microelectronics. The uniqueness of PTM-ET is the faithful handling of nonlinearities that are present in the physical properties of the material and equations. Fig. 1.6 shows the temperature distribution in a plane for field effect transistors.

1.2.2 Parametric Model Order Reduction

The Max Planck Institute Magdeburg has developed sparse Parametric Model Order Reduction (PMOR) methods for electro-thermal simulation (ET) simulation, especially for being used together with the simulation tool PTM-ET [6]. For the Krylov-based methods a new posteriori error estimate was developed [5]. These methods simulate a real and original circuit in a smaller model, which makes much more extensive simulations possible. The accuracy and efficiency of the proposed PMOR methods are validated by both simulation and uncertainty quantification based on reduced-order models. A unique feature is that even temperature-dependent parameters and non-linear, quadratic functions are integrated into a common simulation.

Fig. 1.7 shows a layout view of a tested power cell. Fig. 1.8 lists the reduction rates for several test cases. They are all above 90% - see also [6].

1.2.3 EM Field Simulations

Another theme is the coupling of electromagnetic fields with circuits. This field is becoming more and more complex. Unlike all previous suppliers of simulators, the tool developed by *nanoCOPS* can now simulate not just individual components in 3D, but also the circuits [7,18]. An additional difference is that non-linear materials such as semiconductors can be integrated directly into the simulation. With the MAGWEL program Device-Electro-Magnetic Modeller (devEM) [10], all kinds of disturbance can be simulated [11]. To date, this graphical approach is unique and allows a much faster detection of errors.

Fig. 1.9 gives the geometrical structure of an inductor device from ACCO Semiconductor. This device was used in coupling to an circuit with oscillating behaviour. Fig. 1.10 presents the magnetic field of the inductor at two different frequencies.

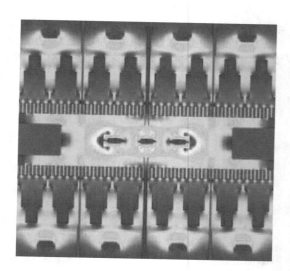

Fig. 1.6 Temperature distribution for field effect transistors.

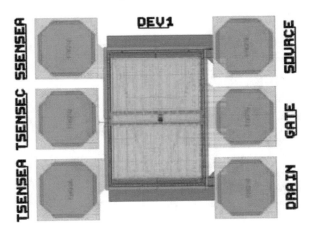

Fig. 1.7: Layout view of the tested power cell.

1.2.4 Test Time Reduction with new Library for Uncertainty Quantification

The UQ-library provides Uncertainty Quantification (UQ) by creating a response surface model based on an expansion in orthogonal polynominals. NXP, the University of Wuppertal and TU Eindhoven have developed a special algorithm for fast fault simulation in NXP's in-house circuit simulator [11, 20]. For this functionality, this simulator is the current market leader. NXP can identify locations on a chip that are probably affected by limited manufacturing accuracies that cause faulty behaviour at predefined time points for measurements. Inclusion of sensitivity analysis brought accel-

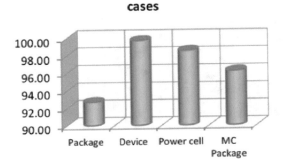

Fig. 1.8: Reduction rates for several test cases.

Fig. 1.9: Geometrical structure of a new inductor device from ACCO Semi-conductor.

erations in CPU time of a factor 20 or more.

Modelling interaction between digital and analogue parts of an RFIC, see Fig. 1.11, led to identify the dominant parts and provided ways to reduce the unwanted impact on the analogue part in an optimal way [13].

The same techniques were applied to minimize loss in a permanent magnet machine [12, 14, 17]

1.2.5 Robust Topology Optimization

In the manufacture of metal parts, there is inevitably uncertainty and error involved. In this field, research is about how strongly and in which way the

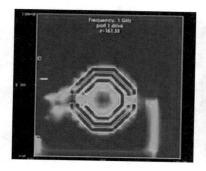
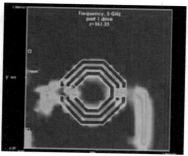

Fig. 1.10: Magnetic field at 1 GHz (left) and at 5 GHz (right).

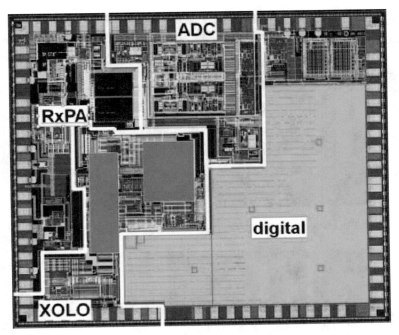

Fig. 1.11: Integrated automotive transceiver design from NXP, used in a study to quantify impact of radio frequency signals from the digital part (agressor) to analogue blocks (victims) [13].

performance of a transistor is dependent on and affected by this uncertainty. Different voltages cause different structures. This results in very high temperatures which can lead to the failure of the transistor. For the first time, the work of *nanoCOPS* has provided a model that reveals the dangers and also leads to an improvement of the metal carrier. This allows for a significant reduction of heat generation. Fig. 1.12 reveals details of a power transistor design for which Fig. 1.13 gives the heat flux for the initial and for the optimised topology design [15, 16].

1.2.6 Ageing and Failure Analysis

In order to analyse the ageing process, circuit devices and boards are tested in a temperature cabin at different temperatures ranging up to 180°C. In order to avoid a destruction of thousands of circuits for testing, new models have been developed by TU Darmstadt in cooperation with Brno University of Technology, ON Semiconductor and MAGWEL for test bond wires. The measurements show that a fused bond wire does not mean an immediate

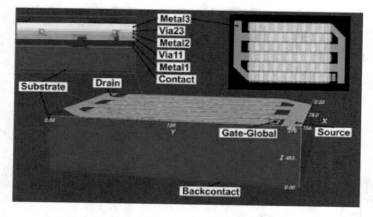

Fig. 1.12: Power transistor design including contacts [15].

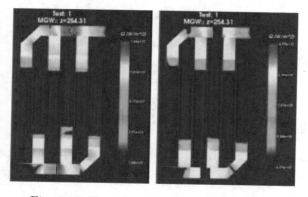

Fig. 1.13: Heat flux. Left: initial design. Right: optimised design [15].

open loop. Various degradation processes have been intercepted by the delayed time base feature of the oscilloscopes.

Fig. 1.14 gives an impression of the experimental testboards with the measurement devices. Fig. 1.15 depicts X-rays of very thin bond wires to connect contacts with parts of the chip [4].

1.2.7 Bond Wire Model in Electro-Thermal Field Simulation

When designing bond wires for the packaging of integrated circuits as shown at the left in Fig. 1.16, engineers are left with various design parameters. There is a trade-off between minimal cost and maximum performance. More-

Fig. 1.14: Experimental testboards with the measurement devices.

Fig. 1.15: X-rays of bond wires.

over, the thinner the wire, the higher the probability of failure ('breaking') during operation.

Bond wire calculators allow to estimate parameters based on simplified models. In *nanoCOPS*, an improved bond wire model was developed and included in MAGWEL's electro-thermal simulations field simulator PTM-ET [10]. By this, the overall behaviour of chip and bonding wires can be reliably analysed in the design phase. It was mathematically shown that the coupled model preserves important principles such as energy conservation.

Finally, in combination with the tools from uncertainty quantification, failure probabilities, as shown at the right in Fig. 1.16, can be precisely estimated. Eventually, cheaper and more robust integrated circuits can be designed.

1.2.8 Electromagnetic and Thermal Netlist Extraction from Nonlinear Field Problems

Many field simulations can be carried out efficiently using dedicated tools, like MAGWEL's PTM-ET simulator for electro-thermal problems [10]. However, some problems require an export of the model, e.g., for additional postpro-

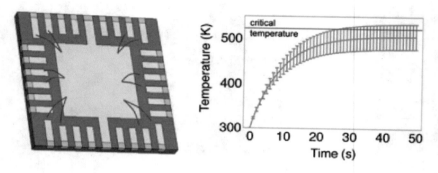

Fig. 1.16: Package (left) and critical temperature of bond wires with six-sigma interval (right) [4].

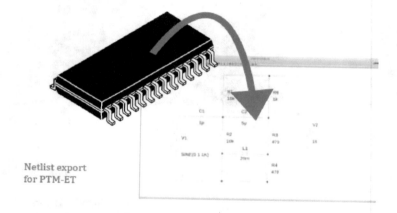

Fig. 1.17: Netlist export from PTM-ET for allowing coupling to a circuit simulator [10].

cessing or coupling to surrounding circuitry, see Fig. 1.17.

For those cases, *nanoCOPS* developed an algorithm for the lossless extraction of nonlinear equivalent electro-thermal and electromagnetic netlists from 3D field models as constructed by PTM-ET.

In this algorithm, apart from the spatial discretisation of the fields, neither simplifications are applied nor it is limited to particular circuit simulators.

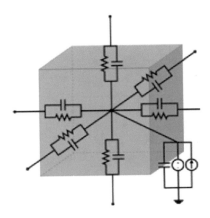

Fig. 1.18: Mapping of integrals to circuit elements.

1.2.9 Towards an Integrated Workflow for Uncertainty Quantification in Semiconductor Simulation

Within the *nanoCOPS* project, a beta version of a graphical interface was designed, implemented and added to MAGWEL's software suite. In Fig. 1.19 the interface is seen to be built upon Sandia's Dakota library [1] for Uncertainty Quantification (UQ) and allows an unique user experience: relevant statistical data is gathered and the impact of tolerances and random distributions of design parameters on quantities of interest are computed without further user interaction, see also Fig. 1.20.

To speed up simulations, the hybrid algorithm by Li, Li and Xiu [9] for rare failure probabilities was implemented and made available to the industrial partners [4]. The required number of simulations for the quantification is thereby reduced from several 100,000 to several 100.

1.2.10 Holistic Coupled Circuit/Electromagnetic Field Simulation

The Python integrated Circuit/EM environment PyCEM [19], see Fig. 1.21, provides the opportunity to solve coupled circuit and EM equation systems in one holistic framework. The essential advantage is the robustness of the holistic approach. Fig. 1.22 provides results obtained with this environment. The violation of certain contraction conditions may lead to serious convergence problems in co-simulation approaches. The holistic approach usually

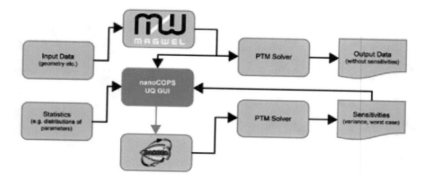

Fig. 1.19: Integrated Simulation Workflow for Uncertainty Quantification.

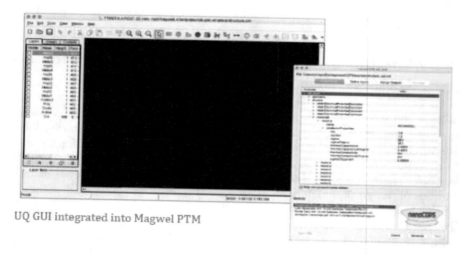

UQ GUI integrated into Magwel PTM

Fig. 1.20: Graphical User Interface for Uncertainty Quantification that offers comparison to Monte Carlo Simulation as well as to Worst Case Corner Analysis.

leads to reliable solutions [11]. The holistic approach with a unified time step control is new and unique in the world for circuit/EM couplings [19].

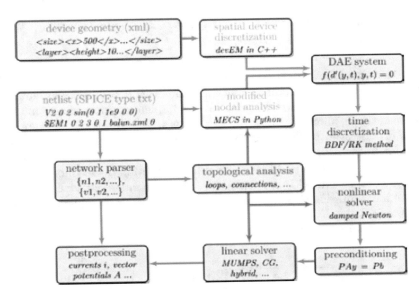

Fig. 1.21: Modular structure of the holistic circuit/electromagnetic field solver PyCEM.

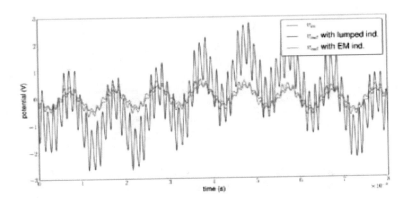

Fig. 1.22: Comparison of simulation results obtained with lumped inductors and with full EM inductor models using the holistic circuit/electromagnetic field solver PyCEM.

1.2.11 Multirate Circuit Simulation Incorporating Reduced Order Models

The in-house simulator LinzFrame of the Fachhochschule Oberösterreich, see Fig. 1.23, follows a strict modular concept. Moreover, it comprises an automatic differentiation suite which simplifies the implementation of new models significantly. It allows that partial derivatives with respect to the state variables do not need to be coded, but are computed automatically. Model libraries for linear devices, SPICE transistor models and a stimulus library including modulated sources, libraries to industry relevant device models and the Simkit library from NXP are available. A Laplace model interface, which allows the incorporation of compressed models from MOR, complements the model libraries.

The analysis tool comprises standard methods such as DC, AC and tran-

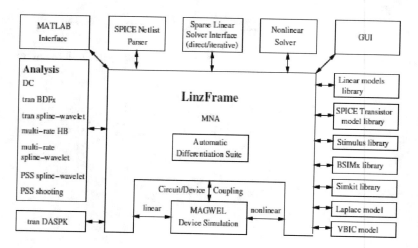

Fig. 1.23: Modular concept of the in-house simulator LinzFrame of Fachhochschule Oberösterreich used for multirate circuit simulation.

sient analysis with polynomial and trigonometric BDFx (MBDFx) methods, an interface to the DASPK simulator for solving higher index Differential-Algebraic Equations (DAEs), and a spline-wavelet transient simulator. Several tools for multirate simulation Harmonic Balance (HB), BDF and spline-wavelet techniques are incorporated. Periodic Steady-State (PSS) methods for driven and autonomous circuits such as oscillators complete the tool [2, 3, 11]. A holistic coupling with an EM/device simulator enables mixed-level circuit/ electromagnetic simulation.

1.3 Conclusion

Designs in nanoelectronics often lead to large-size simulation problems and include strong feedback couplings. Industry demands the provisions of variability to guarantee quality and yield. It also requires the incorporation of higher abstraction levels to allow for system simulation in order to shorten the design cycles, while at the same time preserving accuracy. The scientific challenges that are reported in this book are: (1) to create efficient and robust simulation techniques for strongly-coupled systems, that exploit the different dynamics of sub-systems within multi-physics problems and that allow designers to predict reliability and ageing; (2) to include a variability capability such that robust design and optimization, worst case analysis, and yield estimation with tiny failure probabilities are possible (including large deviations like 6-sigma); (3) to reduce the complexity of the sub-systems while ensuring that the operational and coupling parameters can still be varied and that the reduced models offer higher abstraction models that are efficient to simulate. The methods developed for this advance a methodology for circuit-and-system-level modelling and simulation based on best practice rules to deal with coupled electromagnetic field-circuit-heat problems as well as coupled electro-thermal-stress problems that emerge in nanoelectronic designs. The new methods are robust and allow for strong feedback coupling when integrating systems to increase the performance of both existing devices and when integrating systems to produce new devices. With the new techniques it is possible to efficiently analyze the effects due to variability. Our methods are designed to solve reliability questions resulting from manufacturability. They facilitate robust design as well as enable worst case analysis. They can also be used to study effects due to ageing. Ageing causes variations in parameters over a long-term period, which cannot be predicted exactly and thus are typically uncertain. Novel Model Order Reduction (MOR) techniques, developed here for the fast repeated simulation of the coupled problems under consideration, are applicable to both coupled systems and parametrized sub-systems. As such they are an essential ingredient for the Uncertainty Quantification. In summary, our solutions are:

- advanced co-simulation/multirate/monolithic techniques, combined with envelope/wavelet approaches;
- new generalized techniques in Uncertainty Quantification (UQ) for coupled problems, tuned to the statistical demands from manufacturability;
- enhanced Parametric Model Order Reduction (PMOR) techniques for coupled problems and for UQ.

All the new algorithms produced were implemented, transferred and tested by the EDA-vendor MAGWEL. Validation was conducted on industrial designs provided by partners from semiconductor industry. These industrial end-users gave feedback during the project life-time, contributed to measurements and supplied material data as well as process data. A thorough comparison to measurements on real devices was being made to demonstrate the industrial applicability.

References

1. ADAMS, B.M., ET AL: *Dakota, A Multilevel Parallel Object-Oriented Framework for Design Optimization, Parameter Estimation, Uncertainty Quantification, and Sensitivity Analysis*, Version 6.0 User's Manual, Sandia Technical Report SAND2014-4633, July 2014. Updated November 2015 (Version 6.3). Sandia National Laboratories, 2017, https://dakota.sandia.gov/.
2. BITTNER, K., AND BRACHTENDORF, H.-G.: *Optimal frequency sweep method in multi-rate circuit simulation.* COMPEL, 33(4):1189–1197, 2014.
3. BITTNER, K., AND BRACHTENDORF, H.-G.: *Fast algorithms for grid adaptation using non-uniform biorthogonal spline wavelets.* SIAM J. Scient. Computing, 37(2):283–304, 2015.
4. CASPER, T., DE GERSEM, H., GILLON, R., GÖTTHANS, T., KRATOCHVÍL, T., MEURIS, P., AND SCHÖPS, S.: *Electrothermal simulation of bonding wire degradation under uncertain geometries.* In: L. FANUCCI, AND J. TEICH (EDS.): *Proceedings of the 2016 Design, Automation & Test in Europe Conference & Exhibition (DATE)*, pp. 1297–1302. IEEE (2016). http://www.date-conference.com
5. FENG, L., ANTOULAS, A.C., AND BENNER, P.: *Some a posteriori error bounds for reduced order modelling of (non-)parametrized linear systems.* ESAIM: Mathematical Modelling and Numerical Analysis (ESAIM:M2AN), 51:6, pp. 2127–2158, 2017. http://dx.doi.org/10.1051/m2an/2017014.
6. FENG, L., YUE, Y., BANAGAAYA, N., MEURIS, P., SCHOENMAKER, W., AND BENNER, P.: *Parametric Modeling and Model Order Reduction for (Electro-)Thermal Analysis of Nanoelectronic Structures.* Journal for Mathematics in Industry, 6:10, Springer, 2016. Open access: http://dx.doi.org/10.1186/s13362-016-0030-8.
7. GALY, P., AND SCHOENMAKER, W.: *In-depth electromagnetic analysis of ESD protection for advanced CMOS technology during fast transient and high-current surge.* IEEE Transactions on Electron Devices, 61(6):1900–1906, June 2014.
8. JANSSEN, R., TER MATEN, J., TISCHENDORF, C., BRACHTENDORF, H.-G., BITTNER, K., SCHOENMAKER, W., BENNER, P., FENG, L., PULCH, R., DELEU, F., AND WIEERS, A.: *The nanoCOPS Project on Algorithms for Nanoelectronic Coupled Problems Solutions.* In: Schrefler, B., Oñate, E., and Papadrakakis, M. (Eds): *Coupled Problems in Science and Engineering VI – COUPLED PROBLEMS 2015.* Proceedings of the VI International Conference on Coupled Problems in Science and Engineering, May 18-20, 2015, San Servolo Island, Venice, Italy. Publ.: CIMNE - International Center for Numerical Methods in Engineering, Barcelona, Spain, 2015, ISBN 978-84-943928-3-2, pp. 1029–1036, 2015. E-book: http://congress.cimne.com/coupled2015/frontal/doc/Ebook_COUPLED_15.pdf (108Mb).
9. LI, J., LI, J., AND XIU, D.: *An efficient surrogate-based method for computing rare failure probability.* J. Comp. Physics 230, pp. 8683–8697, 2011.
10. MAGWEL NV: *Device-Electro-Magnetic Modeler (devEM)* and *Power Transistor Modeler: PTM*, MAGWEL NV, Leuven, Belgium, 2018, http://www.magwel.com/.
11. TER MATEN, E.J.W., PUTEK, P., GÜNTHER, M., PULCH, R., TISCHENDORF, C., STROHM, C., SCHOENMAKER, W., MEURIS, P., DE SMEDT, B., BENNER, P., FENG, L., BANAGAAYA, N., YUE, Y., JANSSEN, R., DOHMEN, J.J., TASIĆ, B., DELEU, F., GILLON, R., WIEERS, A., BRACHTENDORF, H.-G., BITTNER, K., KRATOCHVÍL, T., PETŘZELA, J., ŠOTNER, R., GÖTTHANS, T., DŘÍNOVSKÝ, J., SCHÖPS, S., DUQUE GUERRA, D.J., CASPER, T., DE GERSEM, H., RÖMER, U., REYNIER, P.,

BARROUL, P., MASLIAH, D., AND ROUSSEAU, B.: *Nanoelectronic COupled Problems Solutions – nanoCOPS: Modelling, Multirate, Model Order Reduction, Uncertainty Quantification, Fast Fault Simulation.* Journal Mathematics in Industry 7:2, 2016. Open access: http://dx.doi.org/10.1186/s13362-016-0025-5.

12. PUTEK, P.: *Mitigation of the cogging torque and loss minimization in a permanent magnet machine using shape and topology optimization.* Engineering Computation **33**, 831–854, 2016.

13. PUTEK, P., JANSSEN, R., NIEHOF, J., TER MATEN, E.J.W., PULCH, R., TASIĆ, B., AND GÜNTHER, M.: *Nanoelectronic COupled Problems Solutions: Uncertainty Quantification for Analysis and Optimization of an RFIC Interference Problem.* Journal Mathematics in Industry 8:12, 2018. Open access: http://dx.doi.org/10.1186/s13362-018-0054-3.

14. PUTEK, P., TER MATEN, E.J.W., GÜNTHER M., AND SYKULSKI, J.K.: *Variance-Based Robust Optimization of a Permanent Magnet Synchronous Machine.* IEEE Transactions on Magnetics, **54:3**, 2018 (article 8102504, 2017). http://dx.doi.org/10.1109/TMAG.2017.2750485.

15. PUTEK, P., MEURIS, P., PULCH, R., TER MATEN, E.J.W., GÜNTHER, M., SCHOENMAKER, W., DELEU, F., AND WIEERS, A.: *Shape optimization of a power MOS device under uncertainties.* In: Proceedings of 2016 Design, Automation & Test in Europe Conference & Exhibition (DATE), Dresden, 319–324, 2016. http://dx.doi.org/10.3850/9783981537079_0998.

16. PUTEK, P., MEURIS, P., PULCH, R., TER MATEN, E.J.W., SCHOENMAKER, W., AND GÜNTHER, M.: *Uncertainty Quantification for Robust Topology Optimization of Power Transistor Devices.* IEEE Transactions on Magnetics, **52** 1–4, 2016.

17. PUTEK, P., PULCH, R., BARTEL, A., TER MATEN, E.J.W., AND GÜNTHER, M., AND GAWRYLCZYK, K.M.: *Shape and topology optimization of a permanent-magnet machine under uncertainties.* Journal of Mathematics in Industry, **6:11**, 2016. Open access: http://dx.doi.org/10.1186/s13362-016-0032-6.

18. SCHOENMAKER, W., CHEN, Q., AND GALY, P.: *Computation of self-induced magnetic field effects including the Lorentz force for fast-transient phenomena in integrated-circuit devices.* IEEE Transactions on Computer-Aided Design of Integrated Circuits and Systems, 33(6):893–902, June 2014.

19. SCHOENMAKER, W., MEURIS, P., STROHM, C., AND TISCHENDORF, C.: *Holistic coupled field and circuit simulation.* In: *2016 Design, Automation Test in Europe Conference Exhibition (DATE)*, pp. 307–312, March 2016.

20. TASIĆ, B., DOHMEN, J.J., JANSSEN, R., TER MATEN, E.J.W., BEELEN, T.G.J., AND PULCH, R.: *Fast Time-Domain Simulation for Reliable Fault Detection.* Proceedings of Design, Automation and Test in Europe (DATE) 2016, Paper 0994, pp. 301–306, 2016.

Part I
Equations, Discretizations

Part I covers the following topics.

- **EM-Equations, Coupling to Heat and to Circuits**
  Authors:
  Wim Schoenmaker (MAGWEL, Leuven, Belgium),
  Hans-Georg Brachtendorf, Kai Bittner (FH Oberösterreich, Hagenberg, Austria),
  Caren Tischendorf, Christian Strohm (Humboldt Universität zu Berlin, Germany).
- **Bond Wire Models**
  Authors:
  David J. Duque Guerra, Thorben Casper, Sebastian Schöps, Herbert De Gersem (Technische Universität Darmstadt, Germany),
  Ulrich Römer (Technische Universität Braunschweig, Germany),
  Renaud Gillon, Aarnout Wieers (ON Semiconductor, Oudenaarde, Belgium),
  Tomas Kratochvil, Tomas Gotthans (Brno University of Technology, Czech Republic),
  Peter Meuris (MAGWEL, Leuven, Belgium).
- **Discretizations**
  Authors:
  Wim Schoenmaker (MAGWEL, Leuven, Belgium),
  Hans-Georg Brachtendorf, Kai Bittner (FH Oberösterreich, Hagenberg, Austria),
  Caren Tischendorf, Christian Strohm (Humboldt Universität zu Berlin, Germany).
- **Automated Generation of Netlists from Electrothermal Field Models**
  Authors:
  Thorben Casper, David J. Duque Guerra, Sebastian Schöps,
  Herbert De Gersem (Technische Universität Darmstadt, Germany).

Chapter 2
EM-Equations, Coupling to Heat and to Circuits

Wim Schoenmaker, Hans-Georg Brachtendorf, Kai Bittner,
Caren Tischendorf, Christian Strohm

Abstract In this chapter we present the underlying physical equations of the systems that are addressed in the remainder of this book. Two kind of couplings are considered, in particular: (1) the coupling of electromagnetic fields with circuits and (2) the electro-thermal coupling in power transistors. These cases are not exhaustive but are used as typical for identifying generic guidelines for addressing coupled problems.

2.1 EM-Equations – Metals, Insulators, Semiconductors

The mathematical modelling of electromagnetic problems starts from the Maxwell equations and the associated constitutive relations. The Maxwell's equations are a set of four partial differential equations that allows the evaluation of the electric and magnetic fields for prescribed sources. In general, any electromagnetic field can be described on a microscopic scale by two vectors fields $\mathbf{E}(\mathbf{r}, t)$ and $\mathbf{B}(\mathbf{r}, t)$ specifying the electric field and the magnetic induction in an arbitrary point $\mathbf{r} = (x, y, z)$ in space at an arbitrary time t respectively. The Maxwell equations are:

Wim Schoenmaker
MAGWEL NV, Leuven, Belgium, e-mail: `Wim.Schoenmaker@magwel.com`

Hans-Georg Brachtendorf, Kai Bittner
Fachhochschule Oberösterreich, Hagenberg im Mühlkreis, Austria, e-mail: `{Hans-Georg.Brachtendorf,Kai.Bittner}@fh-hagenberg.at`

Caren Tischendorf, Christian Strohm
Humboldt Universität zu Berlin, Germany, e-mail: `{Caren.Tischendorf,StrohmCh}@math.hu-berlin.de`

© Springer Nature Switzerland AG 2019
E. J. W. ter Maten et al. (eds.), *Nanoelectronic Coupled Problems Solutions*,
Mathematics in Industry 29, https://doi.org/10.1007/978-3-030-30726-4_2

$$\nabla.\mathbf{D} = \rho \,, \tag{2.1}$$

$$\nabla.\mathbf{B} = 0 \,, \tag{2.2}$$

$$\nabla \times \mathbf{E} = -\frac{\partial \mathbf{B}}{\partial t} \,, \tag{2.3}$$

$$\nabla \times \mathbf{H} = \mathbf{J} + \frac{\partial \mathbf{D}}{\partial t} \,. \tag{2.4}$$

Equations (2.1) and (2.2) are the Gauss' laws for the electric and magnetic fields, whereas equation (2.3) is the Maxwell-Faraday's law for induction, and equation (2.4) is the Maxwell-Ampère's law.

The quantities $\mathbf{E}$ and $\mathbf{H}$ are the electric and magnetic field intensities and are measured in units of [volt/m] and [ampere/m], respectively. The quantities, $\mathbf{D}$ and $\mathbf{B}$ are the electric and magnetic flux densities and are expressed in units of $[\mathrm{coulomb/m^2}]$ and [tesla]. The quantities ρ and $\mathbf{J}$ are the volume electric charge density and electric current density, which are the sources of the electromagnetic fields and are measured in units of $[\mathrm{coulomb/m^3}]$ and $[\mathrm{ampere/m^2}]$. For a full treatment see, see also [16], [22].

2.1.1 Charge Conservation

In order to guarantee charge conservation, Maxwell added the displacement current term to Ampere's law in equation (2.4). Indeed, taking the divergence of both sides of Ampere's law and using Gauss's law $\nabla.\mathbf{D} = \rho$, we obtain :

$$\begin{aligned}
\nabla \cdot (\nabla \times \mathbf{H}) &= \nabla.\mathbf{J} + \nabla.\frac{\partial \mathbf{D}}{\partial t} \\
&= \nabla.\mathbf{J} + \frac{\partial}{\partial t}\nabla.\mathbf{D} \\
&= \nabla.\mathbf{J} + \frac{\partial \rho}{\partial t} = 0.
\end{aligned} \tag{2.5}$$

Hence the charge conservation law reads :

$$\nabla.\mathbf{J} + \frac{\partial \rho}{\partial t} = 0 \,. \tag{2.6}$$

Equation (2.6) is the current continuity or conservation of charge equation. We note that equation (2.6) always needs to be fulfilled in order to have a valid solutions of the Maxwell equations (2.1). Integrating both sides over a closed and a static. e.g. constant in time, volume V surrounded by the surface S, and using the divergence theorem, we obtain the integrated form of equation (2.6) :

$$\oint_S \mathbf{J}.\mathrm{d}S = -\frac{\mathrm{d}}{\mathrm{d}t}\int_V \rho \,\mathrm{d}V. \tag{2.7}$$

The left-hand side represents the total amount of charge flowing in/out the volume enclosed by the surface S per unit time and the right-hand side represents the amount by change of the charge inside the volume V per unit time.

2.1.2 Constitutive Relations

The electric and magnetic flux densities $\mathbf{D}$, $\mathbf{B}$ are related to the electric and magnetic field intensities $\mathbf{E}$, $\mathbf{H}$ via the so-called the constitutive relations, which depends on the materials in which the fields exist. Various scenarios are listed below :

- In the absence of magnetic and dielectric materials, the relations are simply:

$$\mathbf{D} = \epsilon_0 \mathbf{E} , \tag{2.8}$$
$$\mathbf{H} = \frac{1}{\mu_0} \mathbf{B} , \tag{2.9}$$

  where ϵ_0 and μ_0 are permittivity and permeability of vacuum.
- In case of homogeneous isotropic dielectric and magnetic materials:

$$\mathbf{D} = \epsilon \mathbf{E} , \tag{2.10}$$
$$\mathbf{H} = \frac{1}{\mu} \mathbf{B} , \tag{2.11}$$

  where ϵ and μ are the permittivity and permeability respectively and they are related to the electric and magnetic susceptibilities of the materials as follows:

$$\epsilon = \epsilon_0 (1 + \chi) , \tag{2.12}$$
$$\mu = \mu_0 (1 + \chi_m) . \tag{2.13}$$

  The susceptibilities χ, χ_m are the measures of the electric and magnetic polarization properties of the material.
- In inhomogeneous materials or compositions of materials the permittivity ϵ depends on the location within the material:

$$\mathbf{D}(\mathbf{r},t) = \epsilon(\mathbf{r}) \mathbf{E}(\mathbf{r},t) , \quad \mathbf{r} = (x,y,z). \tag{2.14}$$

In general, constitutive relations maybe inhomogeneous, anisotropic, nonlinear, and frequency dependent (dispersive).

The constitutive relation that links the current $\mathbf{J}$ to the electric field or to the carrier densities is determined by the medium under consideration. For conductors the current $\mathbf{J}$ is given by Ohm's law,

$$\mathbf{J} = \sigma \mathbf{E}, \tag{2.15}$$

where $\sigma > 0$ is the electrical conductivity of the conductor and $\sigma = 0$ for insulators. In the case of semiconductors, $\mathbf{J} = \mathbf{J}_p + \mathbf{J}_n$ consists of positively and negatively charged carriers, e.g., holes and electrons, obeying the current-continuity equations:

$$\nabla.\mathbf{J}_p + q\frac{\partial p}{\partial t} = -qU(n,p) , \qquad (2.16)$$

$$\nabla.\mathbf{J}_n - q\frac{\partial n}{\partial t} = qU(n,p) . \qquad (2.17)$$

The charge and current densities are :

$$\rho = q(p - n + N_D - N_A) , \qquad (2.18)$$
$$\mathbf{J_n} = q\mu_n n\mathbf{E} + k_B T \mu_n \nabla n , \qquad (2.19)$$
$$\mathbf{J_p} = q\mu_p p\mathbf{E} - k_B T \mu_p \nabla p , \qquad (2.20)$$

where $U(n,p)$ is the generation or recombination of the charge carriers, $D = N_D - N_A$ is the net doping determined by the donor and acceptor concentration. Furthermore, n is the electron concentration, p is the hole concentration, q is the elementary charge, T is the Kelvin temperature and μ_n, μ_p are the mobilities of electron and holes respectively. The recombination/generation rate depends on different physical mechanisms and there is an extended literature on the subject. A generally accepted model is due to Shockley, Read and Hall, e.g. the SRH model [4] [2]. For simulation purposes it is simplified to an expression as given below:

$$U(p,n) = \frac{pn - n_i^2}{\tau_n(p + p_1) + \tau_p(n + n_1)} . \qquad (2.21)$$

Here τ_p, τ_n, p_1 and n_1 are effective parameters representing carrier life times and effective carrier densities. The intrinsic carrier concentration is n_i. For silicon the value of the intrinsic concentration at room temperature 300 °K is 1.5×10^{10} cm^{-3}.

2.2 Potentials

The Maxwell equations can be written in an alternative form that involves the scalar potential $\mathbf{V}$ and vector potential $\mathbf{A}$. Consequently, we obtain the reformulated Maxwell's equations which we call the $\mathbf{A}$-V formulation. From equation (2.2) and Helmholtz' theorem we find that there exists a vector field $\mathbf{A}$ (vector potential)such that:

$$\mathbf{B} = \nabla \times \mathbf{A} . \qquad (2.22)$$

Using equation 2.3 we obtain :

$$\nabla \times \mathbf{E} + \frac{\partial}{\partial t}(\nabla \times \mathbf{A}) = 0 . \tag{2.23}$$

We may rewrite this equation as:

$$\nabla \times \left(\mathbf{E} + \frac{\partial \mathbf{A}}{\partial t}\right) = 0 . \tag{2.24}$$

Therefore, there exists a scalar potential V, such that

$$\mathbf{E} + \frac{\partial \mathbf{A}}{\partial t} = -\nabla V . \tag{2.25}$$

The electric field $\mathbf{E}$ and magnetic field $\mathbf{B}$ can be written using a scalar potential $\mathbf{V}$ and vector potential $\mathbf{A}$:

$$\mathbf{B} = \nabla \times \mathbf{A} , \tag{2.26}$$

$$\mathbf{E} = -\nabla V - \frac{\partial \mathbf{A}}{\partial t} . \tag{2.27}$$

These results can be inserted in the Maxwell equations (2.1) and (2.4) such that the Maxwell's equations are now exclusively expressed in potentials as given below:

$$\nabla.(\epsilon(-\nabla V - \partial_t \mathbf{A})) = \rho , \tag{2.28}$$

$$\nabla \times \left(\frac{1}{\mu}\nabla \times \mathbf{A}\right) = \mathbf{J} + \partial_t \left(\epsilon(-\nabla V - \partial_t \mathbf{A})\right) . \tag{2.29}$$

2.2.1 Gauge Invariance and Gauge Fixing

If we set $\mathbf{A} \to \mathbf{A}' = \mathbf{A} + \nabla \psi$ and $V \to V' = V - \frac{\partial \psi}{\partial t}$, where $\psi = \psi(\mathbf{r}, t)$ is some continuous function, then

$$\mathbf{E}' \equiv -\nabla V' - \frac{\partial \mathbf{A}'}{\partial t} = -\nabla\left(V - \frac{\partial \psi}{\partial t}\right) - \frac{\partial}{\partial t}(\mathbf{A} + \nabla \psi) = \mathbf{E}, \tag{2.30}$$

and also $\mathbf{B}' \equiv \nabla \times (\mathbf{A} + \nabla \psi) = \mathbf{B}$.

We observe that the Maxwell equations do not change in their appearance. Another way of looking at this is the following: there is an underlying symmetry or invariance in the system of equations. This symmetry is named a 'gauge' symmetry. From Noether's theorem we know that if there is some symmetry in the equations then there must be some conserved quantity. Indeed, as the corresponding conserved quantity is the total charge, that does not change under the time evolution of the dynamical system. A consequence of the symmetry is that the scalar and vector potentials are not uniquely defined, since $\mathbf{B}' = \mathbf{B}$ and $\mathbf{E}' = \mathbf{E}$. In order to eliminate this non uniqueness, we need to do a gauge fixing or choosing a gauge. Whatever, restriction or gauge choice one makes, it should should not alter the physical content of the equations. Furthermore, the choice should not trigger constraints that are not present in the original formulation. Whereas in electrodynamics these issues are rather harmless, in quantum chromodynamics which is also a theory based on gauge invariance the consequences of gauge fixing are dramatic but fortunately surmountable. A peculiar aspect of gauge fixing is that one needs to fix the gauge to eliminate the non-uniqueness of the potentials thereby distroying the symmetry, but since there are many ways in doing so, there is still a symmetry! This is the famous Becchi-Rouet-Stora-Tyutin or BRST symmetry. An illustration of the possibility to constrict the solution space by gauge fixing was given by Gribov, who showed that the complete solution space of the non-gauge fixed formulation consists of a number of copies of the gauge-fixed formulation, As said, in electrodynamics these concerns have not been obstructing the use of potentials (so far) but it is recommended to be aware of these subtleties. We will proceed with the potential formulation of the Maxwell equations and consider some gauge choices.

A particular choice of the scalar and vector potentials is a gauge, and a scalar function ψ used to change the gauge is called gauge function. Gauge fixing can be done in many ways but in our discussion the two below:

- The **Coulomb gauge** is defined by the gauge condition

$$\nabla . \mathbf{A} = 0. \tag{2.31}$$

- The **Lorenz gauge** is defined by the gauge condition

$$\nabla . \mathbf{A} + \frac{1}{c^2} \frac{\partial V}{\partial t} = 0, \tag{2.32}$$

where $c = \frac{1}{\sqrt{\mu\epsilon}}$ is the speed of light. A more extended discussion of gauge condition can be found in [22].

2.3 Maxwell Equations and Material Interfaces

The conditions for the Maxwell equations (2.1) - (2.4) across materials boundaries are given below in terms of the tangential and normal components:

$$E_{1t} - E_{2t} = 0, \tag{2.33}$$
$$H_{1t} - H_{2t} = (\mathbf{J}_s \times \hat{\mathbf{n}}) \cdot \hat{\mathbf{t}}, \tag{2.34}$$
$$D_{1n} - D_{2n} = \rho_s, \quad B_{1n} - B_{2n} = 0, \tag{2.35}$$

where $\hat{\mathbf{n}}$ is a unit vector perpendicular to the interface of the two materials and $\hat{\mathbf{t}}$ is a unit vector tangential to the material interface. The quantities ρ_s, $\mathbf{J}_s$ are the surface charge and surface-current densities on the boundary surface. Thus, the tangential component of the $\mathbf{E}$-field is continuous across the interface, the difference of the tangential components of the $\mathbf{H}$-field is equal to the surface-current density, the difference of the normal components of the flux density $\mathbf{D}$ is equal to the surface charge density and the normal components of the magnetic flux density $\mathbf{B}$ is again continuous across the interface. Very often there are no surface charges or current densities on the interface. In such cases, the conditions may be stated as:

$$E_{1t} = E_{2t},   \tag{2.36}$$
$$H_{1t} = H_{2t},   \tag{2.37}$$
$$D_{1n} = D_{2n},   \tag{2.38}$$
$$B_{1n} = B_{2n}.   \tag{2.39}$$

These boundary conditions will act as a guideline when constructing interface conditions for V and $\mathbf{A}$.

2.4 Lumped Circuit Modelling

The common approach for simulating circuits is the Modified Nodal Analysis (MNA). It bases on the Kirchhoff's laws described by

$$A\mathbf{i} = 0,   \tag{2.40}$$
$$\mathbf{v} = A^\top \mathbf{e},   \tag{2.41}$$

with the incidence matrix A (with only suitable choices of elements 0, 1 and -1 in each row) mapping branches to nodes of the circuit. The circuit variables are the vector $\mathbf{i}$ of all branch currents, the vector $\mathbf{v}$ of all branch voltages and the vector $\mathbf{e}$ of all nodal potentials. They are completed by the local constitutive element equations

$$\mathbf{i}_1 = \tfrac{\mathrm{d}}{\mathrm{d}t} q(\mathbf{v}_1, t) + g(\mathbf{v}_1, t),   \tag{2.42}$$
$$\mathbf{v}_2 = \tfrac{\mathrm{d}}{\mathrm{d}t} \phi(\mathbf{i}_2, t) + r(\mathbf{i}_2, t),   \tag{2.43}$$

for lumped current and voltage-controlling elements, respectively. Notice, all basic types as capacitances, inductances, resistances and sources are covered by a suitable choice of the functions q, g, ϕ and r. Performing the modified nodal analysis, we get the following reduced equation system having only the nodal potentials e and the currents i_2 of the voltage controlling elements, see [9]:

$$A_1 \tfrac{\mathrm{d}}{\mathrm{d}t} q(A_1^\top \mathbf{e}, t) + A_1 g(A_1^\top \mathbf{e}, t) + A_2 \mathbf{i}_2 = 0, \tag{2.44}$$

$$\tfrac{\mathrm{d}}{\mathrm{d}t} \phi(\mathbf{i}_2, t) + r(\mathbf{i}_2, t) - A_2^\top \mathbf{e} = 0, \tag{2.45}$$

where the incidence matrix $A = (A_1, A_2)$ is split with respect to the current and voltage controlling elements. The equations (2.44)-(2.45) are generated automatically from net lists providing the node to branch element relation for the entries of A_1 and A_2 as well as the element related functions q, g, ϕ and r.

2.5 Coupled Modelling of EM fields and Circuits

We assume the interface between the electromagnetic field model and the lumped circuit model to be perfectly electric conducting such that $\mathbf{B} \cdot \hat{\mathbf{n}} = 0$ and $\mathbf{E} \cdot \hat{\mathbf{t}} = 0$ with $\hat{\mathbf{n}}$ and $\hat{\mathbf{t}}$ being the outer unit normal vectors transversal and parallel to the contact boundary. It is convenient, also for later use in the temporal discretization to introduce the quasi-canonical momentum field $\boldsymbol{\Pi} = \partial_t \mathbf{A}$, such that $\mathbf{E} = -\nabla V + \boldsymbol{\Pi}$. This motivates the boundary conditions, [3],

$$(\nabla \times \mathbf{A}) \cdot \hat{\mathbf{n}} = 0 \,,$$

$$(\boldsymbol{\Pi} + \nabla V) \cdot \hat{\mathbf{t}} = 0 \,. \tag{2.46}$$

Denoting by Γ_k the k-th contact of the electromagnetic field model element with Γ_0 being the reference contact we get the current through Γ_k as

$$\mathbf{i}_k = \int_{\Gamma_k} [\mathbf{J} - \partial_t(\varepsilon(\nabla V + \boldsymbol{\Pi}))] \cdot \hat{\mathbf{n}} \, \mathrm{d}\sigma \,. \tag{2.47}$$

Note that equations (2.6) and the boundary condition (2.46) guarantee that the sum of all contact currents equals zero, that means

$$\sum_k \mathbf{i}_k = 0.$$

This is necessary requirement for all lumped element descriptions in order to preserve the Kirchhoff's current law. To reveal the relation to the voltages $\mathbf{v}_k$ between the contact Γ_k and the reference contact Γ_0, we express the potential V as

$$V(x, t) = V_{bi}(x) + V_c(x, t) \tag{2.48}$$

with the contact potential

$$V_c(x, t) = \begin{cases} \mathbf{v}_k & \text{if } x \in \Gamma_k \\ 0 & \text{if } x \notin \Gamma_k. \end{cases}$$

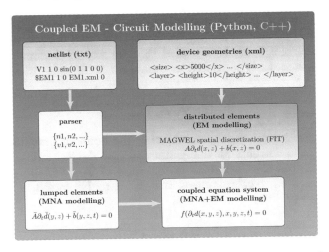

Fig. 2.1: Holistic modelling of field and lumped elements. The netlist contains all elements.

Here, we assumed the reference contact Γ_0 to be the mass node for simplicity. The potential V_{bi} describes the position dependent built-in potential arising by varying doping concentrations and bonding different materials.

2.5.1 Code Implementation Aspects

The MAGWEL simulator devEM provides the spatially discretized field equations as a DAE system (A) that may be written as

$$\mathbf{F}(\mathbf{X}) = 0 \,, \tag{2.49}$$

where $\mathbf{X}$ is the collection of unknowns generated by the discretizion of the field equations and $\mathbf{F}$ is the collection of equations determining the solution for the discretized field variables. Strictly speaking, we should formulate the system of equations as follows :

$$\mathbf{F}(\mathbf{X}, \mathbf{X}_{bc}) = 0 \,, \tag{2.50}$$

where $\mathbf{X}_{bc}$ are the variables that serve as Dirichlet's boundary condition values. The lumped elements and the connections of all are modelled by MNA resulting in DAE system (B). The latter can be denoted in a condensed way as :

$$\mathbf{G}(\mathbf{Y}, \mathbf{Y}_{bc}) = 0 \,, \tag{2.51}$$

where $\mathbf{Y}$ are the collection of nodal and branch variables and $\mathbf{Y}_{bc}$ represents the voltage and current boundary conditions which in lumped modelling are voltage and current sources. It means that we have a combined, but so far uncoupled, system as is presented in equation (2.52) :

$$\begin{bmatrix} \mathbf{F}(\mathbf{X},\mathbf{X}_{bc}) \\ \mathbf{G}(\mathbf{Y},\mathbf{Y}_{bc}) \end{bmatrix} = 0 \, . \tag{2.52}$$

The coupling is introduced by considering the boundary conditions for the voltages and the currents at the contacts of the devices submitted to field solving. It implies that a subset of variables of $\mathbf{X}_{bc}$ will be promoted to degrees of freedom, e.g. unknowns that need to be solved. Therefore, we write equation (2.50) as :

$$\mathbf{F}(\mathbf{X},\mathbf{X}',\mathbf{X}''_{bc}) = 0 \, , \tag{2.53}$$

where $\mathbf{X}'$ are the 'would-be' boundary conditions variables that become degrees of freedom in the coupled system and $\mathbf{X}''_{bc}$ still remain regular Dirichlet type boundary conditions. In a similar way, we may split the boundary conditions for the lumped circuit modelling equations (2.51).

$$\mathbf{G}(\mathbf{Y},\mathbf{Y}'\mathbf{Y}''_{bc}) = 0 \, , \tag{2.54}$$

Since there are now new degrees of freedom, additional equations needed to be provided in order to arrive at a solvable complete system. These equations are provided by the conditions as formulated in equations (2.47) and (2.48) and can be written in condensed notation as :

$$\mathbf{H}(\mathbf{X},\mathbf{X}',\mathbf{Y},\mathbf{Y}') = 0 \, , \tag{2.55}$$

or by denoting the primed variables using $\mathbf{Z} = \left(\mathbf{X}',\mathbf{Y}' \right)$:

$$\mathbf{H}(\mathbf{X},\mathbf{Y},\mathbf{Z}) = 0 \, . \tag{2.56}$$

The DAE system of the circuit and the field model DAE are coupled in the following combined system not explicltly denoting the dependency on $\mathbf{X}''_{bc}$ and $\mathbf{Y}''_{bc}$:

$$\begin{bmatrix} \mathbf{F}(\mathbf{X},\mathbf{Z}) \\ \mathbf{G}(\mathbf{Y},\mathbf{Z}) \\ \mathbf{H}(\mathbf{X},\mathbf{Y},\mathbf{Z}) \end{bmatrix} = 0 \, . \tag{2.57}$$

Holistic solving means that the system (2.57) is solved using a simultaneous Newton-Raphson solver. Thus the Newton-Raphson update equations are :

$$
\begin{bmatrix} \Delta X \\ \Delta Y \\ \Delta Z \end{bmatrix} = - \begin{bmatrix} \frac{\partial F_i}{\partial X_j} & 0 & \frac{\partial F_i}{\partial Z_j} \\ 0 & \frac{\partial G_i}{\partial Y_j} & \frac{\partial G_i}{\partial Z_j} \\ \frac{\partial H_i}{\partial X_j} & \frac{\partial H_i}{\partial Y_j} & \frac{\partial H_i}{\partial Z_j} \end{bmatrix}^{-1} * \begin{bmatrix} \mathbf{F(X,Z)} \\ \mathbf{G(Y,Z)} \\ \mathbf{H(X,Y,Z)} \end{bmatrix} . \tag{2.58}
$$

From equation 2.58) it follows that a code implementation of the full coupling requires the evaluation of all entries of the Newton-Raphson matrix. In particular, the entries $\frac{\partial F_i}{\partial Z_j}$ and $\frac{\partial H_i}{\partial X_j}$ are generated by the field solver, whereas the circuit simulator needs to generate the entries $\frac{\partial G_i}{\partial Z_j}$ and $\frac{\partial H_i}{\partial Y_j}$. Finally, the entries $\frac{\partial H_i}{\partial Z_j}$ can be constructed using data both from the field solver and the circuit simulator. The construction of the coupled simulation is shown in Fig. 2.1. It is realized in a Python framework including C++ implementations of the MAGWEL's field solver devEM. The netlist contains all elements. Elements that shall be modelled by EM fields are marked with a \$ in the first position. The geometric and material structure is given in the corresponding xml file listed in the same line. The data flow is shown in 2.2. From the

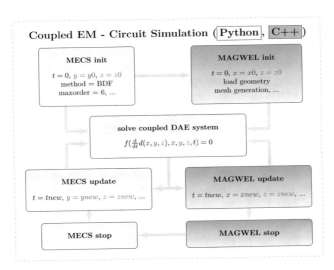

Fig. 2.2: Flow diagram for the coupled field circuit simulation. It is realized in a Python framework including C++ implementations of the field solver devEM.

circuit simulator's perspective a new element is defined that triggers the use of the field solver for the generation of the field equations on a computational grid. The syntax is illustrated in Fig. 2.3 and must be read as follows: There is net list element with name EM1 whose field solver details can be found in the file structure_balun.xml. The contacts of this element are con-

```
Rs1 0 1 50
Vs2 0 2 sin(0 1 1e9 0 0)
Vs3 0 3 sin(0 1 -1e9 0 0)
$EM1 0 2 3 0 1 structure_balun.xml 0
```

Fig. 2.3: Net list with a \$-line that triggers a field solving approach for a netlist element. The example line is read as follows: There is a net list element with name EM1 whose field solver details can be found in the file structure_balun.xml. The five outer contacts of this element are connected to the nodes 0, 2, 3, 0 and 1. The last number in the line refers to the reference contact (usually the mass node).

nected to the nodes 0, 2, 3, 0 and 1. The last number in the line refers to the reference contact (usually the mass node).

2.6 Electrothermal Coupling

With the increased need of smart grids for a sustainable energy infrastructure, power devices play an important role in, both, energy harvesting and distribution and thus in controlling overall energy efficiency. Automotive applications are also an important field that requires the handling of demanding electro-thermal operational constraints to the design of both components and systems. Power devices consist of several thousands of parallel channel devices to deliver high throughput of current and/or current control in both CMOS and bipolar technology. The layout of metal interconnects, bond pads, and bondwires (or bumps/balls) of large-area power semiconductor devices has a profound effect on metal de-biasing, device ON-resistance (Rdson, or RDS(on)1) and reliability (electro-migration, thermal-induced stress, device & material life times, etc.). Metal-interconnects resistance is especially critical in the context of large-area devices that are designed to have a very low (up to a few tens of milliohms) Rdson value [7].

In order to get a glance of a power device, Fig. 2.4 shows the complex geometrical finger structure in a typical device design of a power transistor, using six layers of metal. The source drain contacts are located at the top of the design. A series of metal stripes and via patterns will ultimately transport the current to the drains and away from the sources of the individual channels. Having found the electro-thermal solution, hot spot detectors (design-rule checking) can be activated to post-check the full power transistor array. The current flow is really multi-dimensional.

For being able to model and simulate such devices we first need to accurately capture the complicated current-flow patterns in such devices with

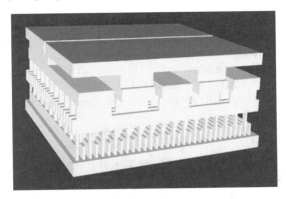

Fig. 2.4: Typical layout of the power transistor (stretched vertical direction) showing its complex geometry.

complicated top metal layouts constrained by the package, wire bonding, or ball array requirements. The design challenge is further increased by the need to separate out the consequences of the competing physical effects, which are strongly coupled, and moreover depend on the time-dependent input stimuli.

Several approaches for electro-thermal simulations of power DMOS transistors have been presented, e.g. [5, 6, 8, 11, 13, 15, 19, 21, 24, 25]. In most approaches, the DMOS is divided into several parts to account for nonuniform temperatures and different power densities. Different thermal simulator strategies are proposed, e.g., in [6, 13, 19, 21, 25], the thermal characteristics of an integrated circuit is analyzed by discretizing the layout information in a layered fashion. The thermal equations are solved on the generated mesh. This analysis method starts from an electrically characterized structure: electrical power dissipation is known on beforehand and used as an input parameter in the simulation.

The metallization stack is a multi layered structure consisting of dielectric, vias, contacts, and metal interconnect structures which are meshed in 3D. Actually, although electrical and thermal models are in essence described by linear (Poisson) models, non linear elements enter the simulation in two ways:

- Joule self heating and heat flow in metal.
- Electrical and thermal conductivities of both metal and substrate are temperature dependent. This is necessary for accurate simulations of large temperature swings for instance in failure conditions like short circuits as well as non linear temperature dependent electrical and thermal resistivities.

The thermal boundary conditions are modelled by thermal contacts (2D rectangular shapes), which can be placed anywhere on the surface of the die or anywhere inside the chip. Thermal contacts have a constant temperature over the surface of the contact. A fixed or time variable temperature can be applied directly to the contact, or indirectly through a thermal resistance that is attached to the contact. A back contact under the die can, for example, be put at the room temperature to model a heat sink. Die attach, mold, lead frame, leads and other elements in the thermal path can be modelled with 3D structures. The bond wires are modelled as discrete elements with a thermal resistance. The heat Q_{out} removed from the device is an area integral over the thermal contacts :

$$Q_{out}(t) = \int \kappa \nabla T \cdot \mathrm{d}S \qquad (2.59)$$

where κ is the thermal conductivity and $T(\mathbf{r},t)$ the temperature, respectively. Thermal contacts can also be used as fixed or variable (time dependent) 2D heat generators or cooling elements, which provide a uniform heat flux (positive or negative) through its surface but where the temperature distribution over the contact surface needs not to be uniform.

2.6.1 Electric Field Solver Aspects

The electrical part of the integrated field solver [17, 20] addresses the current continuity equation:

$$\nabla \cdot \boldsymbol{J} + \partial \rho = 0 \ , \ \boldsymbol{J} = \sigma \boldsymbol{E} \ , \ \boldsymbol{E} = -\nabla V \ , \ \rho = -\nabla(\varepsilon \nabla V) \ , \qquad (2.60)$$

where $\boldsymbol{J}$ is the current density, $\boldsymbol{E}$ the electrical field, V the electrical potential, σ the electrical conductivity, ε the permittivity, and ρ the charge density. Since we do not consider local charging effects, i.e. $\rho = 0$, the current continuity equation, being the last one in (2.60) reduces to a Poisson problem for conductive domains, consisting of interconnects and active devices. The power dissipation is located mainly in the active material constituting the transistor fingers, and to a lesser degree in the interconnect metallization.

2.6.2 Thermal Field Solver Aspects

The thermal part of the solver addresses the heat e quation

$$\nabla \Phi_{\mathrm{q}} + \partial_t w(T) = Q \ , \ \Phi_q = -\kappa \nabla T \ , \ w(T) = C_{\mathrm{T}}(T - T_{ref}) \ . \qquad (2.61)$$

Here, Φ_{q} is the heat flux and w the local energy storage, characterized by the thermal capacitance C_{T}. Here we ignore (nonlinear) effects due to radiation [14]. Q represents heat sources or sinks. The solution of this equation provides the desired temperature information to feed into (2.60). However, the solution is only computable, provided that the heat source is known. The source may consist of several contributors. The boundaries of the simulation domain may contain heat injecting or extracting properties. Besides these sources, the Joule self heating is of particular interest:

$$Q_{\mathrm{SH}} = \boldsymbol{E} \cdot \boldsymbol{J} \qquad (2.62)$$

Note that this nonlinear term is determined by (2.60), and therefore it is mandatory to solve (2.60) and (2.61) simultaneously.

Just as for the active devices in the electrical part of the system, we also apply a compact model for the self-heating of the devices. For transistor structures, the self-heating is determined as a function of the source-drain voltage, the gate-source voltage and the local temperature. Here we assume that to each gate finger fragment we may assign a unique temperature value, which may vary in going from one fragment or finger to another.

$$\Sigma_{SH} = V_{DS} I_{DS}(V_{DS}, V_{GS}, T) \,. \qquad (2.63)$$

This almost completes the definition of our holistic electro-thermal approach. In the present stage we have implemented two kinds of boundary conditions: at metallic contacts we can select electrical voltage boundary conditions or current boundary conditions or a (primitive) circuit may be selected from a build-in library of circuits. The metallic contacts also serve as heat sinks / sources meaning that fixed temperature boundary conditions can be selected. Alternatively, one may opt for thermal-flux boundary conditions. The side walls of the simulation domain are dealt with by using Neumann boundary conditions. This corresponds to ideal thermally and electrically insulating walls, which is a valid assumption for the applications considered here. Finally we note that the boundary conditions can be time-dependent. Even when the voltages adapt instantaneously to the time-dependent contact voltage, i.e., no charge effects are considered in equation (2.60), we still deal with a transient problem because of the thermal capacitive term in equation (2.61). Thus our solver will be able to explore in a fully self-consistent way the occurrence of thermal runaway. Of course, predictive simulation requires the availability of accurate compact models over a sufficiently wide temperature range.

2.6.3 Compact Device Representation

Despite the fact that we address the electrical and thermal variables from a field perspective we do not sustain this practice all the way down into the active devices. Doing so, would mean to imply that the field solving approach not only must be applied at a much smaller length scale (sub micron) but, moreover that new degrees of freedom (the electron and hole Fermi levels) must be considered. The purpose of the underlying scheme is not to contribute to progress in process and device technology but to provide an EDA tool suitable to optimize designs. Just because the active devices are very limited in size, we may replace them by entities with negligible volume. The active devices are then only "visible" through their contacts to which we may assign compact models.

References

1. ALÌ, G., CULPO, M., PULCH, R., ROMANO, V., AND SCHÖPS, S.: *PDEA Modeling and Discretization*. In: Günther, M. (Edt.): *Coupled Multiscale Simulation and Optimization*, Mathematics in Industry 21, Springer, pp. 15–102, 2015.
2. ANILE, A.M., NIKIFORAKIS, N., ROMANO, V., AND RUSSO, G.: *Discretization of Semiconductor Device Problems (II)*. In: SCHILDERS, W.H.A., AND TER MATEN, E.J.W. (EDS.): *Numerical Methods in Electromagnetism, Special Volume XIII – Handbook of Numerical Analysis*, Elsevier B.V., pp. 443-522, 2005.
3. BAUMANNS, S.: *Coupled Electromagnetic Field/Circuit Simulation. Modeling and Numerical Analysis*. PhD-Thesis, Universität zu Köln, 2012.
4. BREZZI, B.F., MARINI, L.D., MICHELETTI, S., PIETRA, P., SACCO, R., AND WANG, S: *Discretization of Semiconductor Device Problems (I)*. In: SCHILDERS, W.H.A., AND TER MATEN, E.J.W. (EDS.): *Numerical Methods in Electromagnetism, Special Volume XIII – Handbook of Numerical Analysis*, Elsevier B.V., pp. 317–441, 2005.
5. CHANDRA, R.: *Semiconductor Chip Design Having Thermal Awareness Across Multiple Sub-system Domains*. US patent no. US 7,472,363, 2008.
6. CHAUFFLEUR, X., TOUNSI, P., DORKEL, J.-M., DUPUY, P., AND FRADIN, J.-P.: *Non linear 3D electrothermal investigation on power MOS chips*. In: Proc. IEEE 2004 BCTM, Montreal, Canada, pp. 156–159, 2004.
7. DENISON, M., PFOST, M., PIEPER, K.-W., MÄRKL, S., METZNER, D., AND STECHER, M.: *Influence of inhomogeneous current distribution on the thermal SOA of integrated DMOS transistors*. In: Proc. ISPSD, Kitakyushu, Japan, May 2004.
8. DIGELE, G., LINDENKREUZ, S., AND KASPER, E.: *Fully coupled dynamic electrothermal simulation*. IEEE Trans on VLSI Systems, 5(3) pp. 250–257, Sep. 1997.
9. ESTÉVEZ SCHWARZ, D., AND TISCHENDORF, C.: *Structural analysis of electric circuits and consequences for MNA*. Internat. J. Circ. Theor. Appl., 28(2) pp. 131–162, 2000.
10. GALY, P., AND SCHOENMAKER, W.: *In-depth electromagnetic analysis of esd protection for advanced CMOS technology during fast transient and high-current surge*. IEEE Transactions on Electron Devices, 61(6), pp. 1900–1906, 2014.
11. GILLON, R., JORIS, P., OPRINS, H., VANDEVELDE, B., SRINIVASAN, A., AND CHANDRA, R.: *Practical chip-centric electro-thermal simulations*. In: Proc. THERMINIC Workshop, Rome, Italy, pp. 220–223, Sep. 2008.
12. GÜNTHER, M., FELDMANN, U., AND TER MATEN, J.: *Modelling and Discretization of Circuit Problems*. In: SCHILDERS, W.H.A., AND TER MATEN, E.J.W. (EDS.): *Numerical Methods in Electromagnetism, Special Volume XIII – Handbook of Numerical Analysis*, Elsevier B.V., pp. 523–659, 2005.
13. IRACE, A., BREGLIO, G., AND SPIRITO, P.: *New developments of THERMOS3, a tool for 3D electro-thermal simulation of smart power MOSFETs*. Microelectronics Reliability, 47, pp. 1696–1700, 2007.
14. JANSSEN, H.H.J.M., TER MATEN, E.J.W., AND VAN HOUWELINGEN, D.: *Simulation of coupled electromagnetic and heat dissipation problems*. IEEE Transactions on Magnetics, 30-5, pp. 3331–3334, 1994.

15. KRABBENBORG, B., BOSMAN, A., DE GRAAFF, H.C., AND MOUTHAAN, A.J.: *Layout to circuit extraction for three-dimensional thermal-electrical circuit simulation of device structures*. IEEE Transactions on Computer-Aided Design of Integrated Circuits and Systems, vol. 15, pp. 765-774, 1996.
16. MAGNUS, W., AND SCHOENMAKER, W.: *Introduction to Electromagnetism*. In: SCHILDERS, W.H.A., AND TER MATEN, E.J.W. (EDS.): *Numerical Methods in Electromagnetism, Special Volume XIII – Handbook of Numerical Analysis*, Elsevier B.V., pp. 3–103, 2005.
17. MAGWEL NV: *An Electro-thermal module of a power transistor modeler: PTM-ET*. http://www.magwel.com/, 2014.
18. MARKOWICH, P.: *The Stationary Semiconductor Device Equations*. Springer, Wien, 1986.
19. PFOST, M., BOIANCEANU, C., LOHMEYER, H., AND STECHER, M.: *Electro-Thermal Simulation of Self-Heating in DMOS Transistors up to Thermal Runaway*. IEEE Trans. on Electron Devices, 60(2), pp. 699–707, 2013.
20. PFOST, M., COSTACHESCU, D., MAYERHOFER, A., STECHER, M., BYCHIKHIN, S., POGANY, D., AND GORNIK, E.: *Accurate temperature measurements of DMOS power transistors up to thermal runaway by small embedded sensors*. IEEE Transactions on Semiconductor Manufacturing, 25 (3), pp. 294–302, 2012.
21. PFOST, M., JOOS, J., AND STECHER, M.: *Measurement and simulation of selfheating in DMOS transistors up to very high temperatures*. In: Proceedings of ISPSD 2008, Orlando, FL, pp. 209–212, 2008.
22. SCHOENMAKER, W.: *Computational Electrodynamics – A Gauge Approach With Application in Microelectronics*. The River Publishers Series in Electronic Materials and Devices. ISBN: 9788793519848. River Publishers, Gistrup, Denmark, 2017.
23. SCHOENMAKER, W., MEURIS, P., AND MAGNUS, W.: *Strategy for electromagnetic interconnect modeling*. IEEE Transactions on Computer-Aided Design of Integrated Circuits and Systems, 20(2), pp 753–762, 2001.
24. TEICHMANN, J., TASCHNER, G., AND DIETZ, F.: *Electrothermal Simulation and Measurement of SOI Smart Power Transistors for short pulses*. In: Proc. of the European Solid State Device Research Conference, pp. 440–443. 2009.
25. YU, Z., YERGEAU, D., DUTTON, R.W., NAKAGAWA, S., CHANG, N., LIN, S., AND XIE, W.: *Full-chip thermal simulation*. In: Proceedings of ISQED, pp. 145–149, 2000.

Chapter 3
Bond Wire Models

David J. Duque Guerra, Thorben Casper, Sebastian Schöps,
Herbert De Gersem, Ulrich Römer, Renaud Gillon, Aarnout Wieers,
Tomas Kratochvil, Tomas Gotthans, Peter Meuris

Abstract In this chapter, models for the heating in bond wires are presented. First, an analytic model is developed, which allows a fast characterisation of the bond wire properties, e.g., its time to failure. Secondly, a more accurate model using the finite integration technique for spatial discretisation is constructed. This model is also used to study the impact of manufacturing uncertainties, e.g., variable bond wire lengths, on the performance of bond wires.

3.1 Introduction

One of the cheapest and mostly used electric connections between chips and their lead frames or pins are bond wires [14–18, 20, 22]. The bond wire technique enables fast and fully automated assembly manufacturing steps. Mostly, fine aluminium, copper and gold wires are used. By means of bonding, elec-

David J. Duque Guerra, Thorben Casper, Sebastian Schöps, Herbert De Gersem
Technische Universität Darmstadt, Germany, e-mail: `{Duque,Casper,Schoeps}@gsc.tu-darmstadt.de;DeGersem@temf.tu-darmstadt.de`

Ulrich Römer
Technische Universität Braunschweig, Germany,
e-mail: `U.Roemer@tu-braunschweig.de`

Renaud Gillon, Aarnout Wieers
ON Semiconductor BVBA, Oudenaarde, Belgium,
e-mail: `{Renaud.Gillon,Aarnout.Wieers}@onsemi.com`

Tomas Kratochvil (Tomáš Kratochvíl), Tomas Gotthans (Tomáš Götthans)
Brno University of Technology, Brno, Czech Republic,
e-mail: `{KratoT,Gotthans}@feec.vutbr.cz`

Peter Meuris
MAGWEL NV, Leuven, Belgium, e-mail: `Peter.Meuris@magwel.com`

© Springer Nature Switzerland AG 2019
E. J. W. ter Maten et al. (eds.), *Nanoelectronic Coupled Problems Solutions*,
Mathematics in Industry 29, https://doi.org/10.1007/978-3-030-30726-4_3

43

tric paths between the chip and its package are established. Bond wires have diminished in size together with the downsizing in electronics during the last decades. Not only the bond wire diameters have decreased, also ever tighter connections have been used. On the contrary, the current densities passing through bond wires have been continuously increasing, together with the increasing number of transistors per chip. If the bond wires cannot properly dissipate their Joule losses to the terminals and the surrounding material, irreversible damage occurs. Today, an important source of failure in integrated circuit (IC) devices is identified as *bond wire breakage* [17].

The technical relevance of bond wire failure motivates the development of bond wire heating models of different fidelities that are applicable at all phases of an electronic design process. For fast evaluation of bond wire performance, especially during the first design steps, one wants to have a set of simple formulae available that allows to predict a safe operation range of particular bond wire configurations. The analytical model should include the major effects such as the heating due to the current, the according temperature rise, the increase of the bond wire resistance with increasing temperature, the heat flux through the electric connections and the heat flux towards the surrounding material. Such a model should predict failure times within 10 °C. In later design steps and for accurate studies, a model that attains accuracies comparable to the accuracies of the input data is desired. For that purpose, only 3D electrothermally coupled field models are applicable.

This chapter develops an analytical model for predicting bond wire failure. Furthermore, a 3D electrothermal formulation is discretised by the finite-integration technique and used to study the variance of the bond wire failure time because of uncertainties of the geometry caused by manufacturing.

3.2 Analytical Model

In the literature, one finds several analytic and semi-analytic models for estimating the performance of bond wires [14–17, 20]. Each of them introduces particular assumptions in order to come up with closed-form formulae describing the bond wire properties. In particular, the underlying partial differential equation (PDE) is drastically simplified, which causes the influence of the package geometry to be neglected. In the model developed here, the moulding compound material, the package dimensions, the bond wire extent and the wire material are taken into account by appropriate material laws, well defined field boundary conditions and a set of well motivated assumptions. The study aims to answer the question how much current a bond wire can carry without failure.

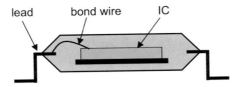

Fig. 3.1: IC lead-frame package (a model for a single bond wire is shown in Fig. 3.2).

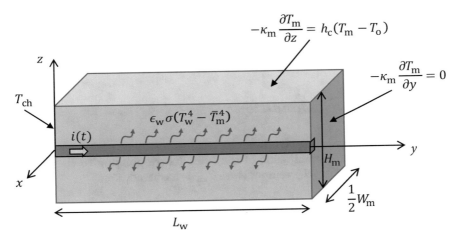

Fig. 3.2: Bond-wire heat transfer problem.

3.2.1 Problem Setup

A conceptual figure of a classic IC lead-frame package is shown in Fig. 3.1. There are several bond wire connections between chip terminals and leads. However, for the sake of simplicity, it is assumed that each of them interacts only thermally with the surrounding material. Moreover, it is assumed that only the length and not the shape of the bond wire determines its behaviour. With these assumptions, the problem can be reduced to that of a single bond wire surrounded by a moulding compound as shown in Fig. 3.2. This simplified model consists of a brick of height H_m, width W_m and length L_m encapsulating the bond wire. The compound material (m) has a thermal conductivity κ_m, specific heat $c_{e;m}$, and mass density ρ_m. Centred within the compound brick, a bond wire (w) of length $L_w = L_m$ with specific heat $c_{e;w}$ and electric resistivity $\rho_{e;w}$ is considered. The thermal conductivity and electrical resistivity of the bond wire are linearised, i.e.,

$$\kappa_{\mathrm{w}}(\widetilde{T}_{\mathrm{w}}) := \kappa_{\mathrm{o}}\left(1+\alpha_{\kappa}\widetilde{T}_{\mathrm{w}}\right), \tag{3.1}$$

$$\rho_{\mathrm{e;w}}(\widetilde{T}_{\mathrm{w}}) := \rho_{\mathrm{e;o}}\left(1+\alpha_{\rho}\widetilde{T}_{\mathrm{w}}\right), \tag{3.2}$$

with $\widetilde{T}_{\mathrm{w}} = T_{\mathrm{w}} - T_{\mathrm{o}}$, where T_{w} is the temperature of the bond wire, T_{o} is the reference (ambient) temperature, κ_{o} is the reference thermal conductivity, $\rho_{\mathrm{e;o}}$ is the reference resistivity, α_{κ} is the thermal conductivity temperature coefficient and α_{ρ} is the electrical resistivity temperature coefficient. All other material parameters are assumed to be constant. A constant electric current $i(t) = I_0$ heats up the bond wire during a time interval t_{p}. The analytic model should be able to determine the temperature T_{w} along the wire as a function of time.

3.2.2 Boundary Conditions

The chip is expected to have a large thermal capacitance with respect to the wire. Therefore, at the chip side of the brick (corresponding to the leftmost wall), a constant chip (ch) temperature T_{ch} is assumed, i.e.,

$$T_{\mathrm{m}}(x,0,z,t) = T_{\mathrm{ch}}, \tag{3.3}$$

$$T_{\mathrm{w}}(x,0,z,t) = T_{\mathrm{ch}}. \tag{3.4}$$

The lead side (corresponding to the rightmost wall) is considered as a thermal insulator, except for the leads (ld) themselves, which carry a prescribed temperature T_{ld}, i.e.,

$$-\kappa_{\mathrm{m}}\frac{\partial}{\partial y}T_{\mathrm{m}}(x,L_{\mathrm{w}},z,t) = 0, \tag{3.5}$$

$$T_{\mathrm{w}}(x,L_{\mathrm{w}},z,t) = T_{\mathrm{ld}}, \tag{3.6}$$

On the lateral walls of the compound, convective heat transfer is modelled, i.e.,

$$-\kappa_{\mathrm{m}}\frac{\partial}{\partial z}T_{\mathrm{m}}\left(x,y,\pm\frac{H_{\mathrm{m}}}{2},t\right) = h_{\mathrm{c}}\left(T_{\mathrm{m}}(x,y,\pm\frac{H_{\mathrm{m}}}{2},t)-T_{\mathrm{o}})\right), \tag{3.7}$$

$$-\kappa_{\mathrm{m}}\frac{\partial}{\partial x}T_{\mathrm{m}}\left(\pm\frac{W_{\mathrm{m}}}{2},y,z,t\right) = h_{\mathrm{c}}\left(T_{\mathrm{m}}(\pm\frac{W_{\mathrm{m}}}{2},y,z,t)-T_{\mathrm{o}})\right), \tag{3.8}$$

where h_{c} is the scalar convective heat transfer coefficient [2]. At the wire surface, thermal conduction and radiation occur, i.e.,

$$-\int_{S_{\mathrm{w}}}\kappa_{\mathrm{w}}\nabla_{\mathrm{T}}T_{\mathrm{w}}\cdot\mathrm{d}\boldsymbol{S} = \mathrm{TH_{c}} + \int_{S_{\mathrm{w}}}\varepsilon_{\mathrm{w}}\sigma_{\mathrm{SB}}\left(T_{\mathrm{w}}^4 - T_{\mathrm{o}}^4\right)\mathrm{d}S, \tag{3.9}$$

where $\nabla_T T_w$ denotes the gradient in the transverse plane, i.e., the plane (x, y, z) with $y = ct$, TH_c is the conductive part of the thermal flux occurring on the wire surface, S_w is the bond wire surface, ε_w is the wire emissivity and σ_{SB} is the Stefan-Boltzmann constant [2].

3.2.3 Thermal Field in the Bond Wire

The bond wire is considered an incompressible piece of metal. Thereby, the heat equation for the wire reads

$$\rho_{e;w} c_{e;w} \frac{\partial T_w}{\partial t} = \nabla \cdot (\kappa_w \nabla T_w) + \dot{q}_i, \tag{3.10}$$

where $\dot{q}_i$ is the volumetric thermal power density related to the Ohmic losses given by

$$\dot{q}_i = \rho_{e;o} \left(1 + \alpha_\rho \tilde{T}_w\right) \left(\frac{I_0}{A_w}\right)^2, \tag{3.11}$$

where A_w is the cross-sectional area of the bond wire. The major heat flux in the transversal plane is caused by the thermal radiation. This fact is characterised by a transversal Biot number being small [2]. Then, the nabla operator can be approximated by $\nabla = \partial_y + \nabla_T \approx \partial_y$ and, after substituting (3.9), (3.10) becomes

$$\rho_{e;w} c_{e;w} \frac{\partial T_w}{\partial t} = \frac{\partial}{\partial y} \left(\kappa_w \frac{\partial T_w}{\partial y}\right) - \varepsilon_w \sigma_{SB} \left(T_w^4 - T_o^4\right) \frac{C_w}{A_w} + \dot{q}_i, \tag{3.12}$$

where $C_w(y)$ is the cross-sectional perimeter of the bond wire. The result is a one-dimensional nonlinear partial differential equation as a function of the temperature $T_w(y, t)$ which now only depends on y and t.

To provoke an expedient solution, the radiation term is linearised, i.e.,

$$T_w^4 - T_o^4 = \left(T_w^3 + T_w^2 T_o + T_w T_o^2 + T_o^3\right) (T_w - T_o) \cong \chi_w(y, t) (T_w - T_o). \tag{3.13}$$

Here, $\chi_w(y, t)$ is a slowly-varying coefficient function. To obtain a simple model, we even assume $\chi_w(y, t) \equiv \chi_w$ to be constant in space and time. This is acceptable as long as the heat transfer by conduction along the bond wire is dominating all other heat transfer mechanisms.

To deal with the temperature-dependent thermal conductivity in (3.12), the temperature is transformed as

$$\tilde{\theta}_w \left(\tilde{T}_w\right) := \frac{1}{\kappa_o} \int_0^{\tilde{T}_w} \kappa_w(s) \, ds, \tag{3.14}$$

from which together with (3.1) follows that $\widetilde{\theta}_\mathrm{w} = \widetilde{T}_\mathrm{w} + \frac{1}{2}\alpha_\kappa \widetilde{T}_\mathrm{w}^2$, and $\partial_y \widetilde{\theta}_\mathrm{w} = \frac{\kappa_\mathrm{w}}{\kappa_\mathrm{o}}\partial_y \widetilde{T}_\mathrm{w}$. For the transient term in (3.12), we apply the approximation $\partial_t \widetilde{\theta}_\mathrm{w} = \partial_t \widetilde{T}_\mathrm{w}$. Altogether, we obtain

$$\rho_{\mathrm{e};\mathrm{w}} c_{\mathrm{e};\mathrm{w}} \frac{\partial \widetilde{\theta}_\mathrm{w}}{\partial t} = \kappa_\mathrm{o} \frac{\partial^2 \widetilde{\theta}_\mathrm{w}}{\partial y^2} - F_{\mathrm{o};\mathrm{w};\mathrm{r}} \widetilde{\theta}_\mathrm{w} + G_{\mathrm{o};\mathrm{w}} + \frac{1}{2} H_{\mathrm{o};\mathrm{w};\mathrm{r}}, \tag{3.15}$$

where

$$G_{\mathrm{o};\mathrm{w}} = \frac{I_0^2 \rho_{\mathrm{e};\mathrm{o}}}{A_\mathrm{w}^2}, \tag{3.16}$$

$$F_{\mathrm{o};\mathrm{w};\mathrm{r}} = \varepsilon_\mathrm{w} \sigma_{\mathrm{SB}} \chi_\mathrm{w} \frac{C_\mathrm{w}}{A_\mathrm{w}}, \tag{3.17}$$

$$H_{\mathrm{o};\mathrm{w};\mathrm{r}} = \frac{2 I_0^2 \rho_{\mathrm{e};\mathrm{o}} \alpha_\rho \widetilde{T}_{\mathrm{w};\mathrm{e}}}{A_\mathrm{w}^2} + \varepsilon_\mathrm{w} \sigma_{\mathrm{SB}} \chi_\mathrm{w} \frac{C_\mathrm{w}}{A_\mathrm{w}} \alpha_\kappa \widetilde{T}_{\mathrm{w};\mathrm{e}}^2, \tag{3.18}$$

where $\widetilde{T}_{\mathrm{w};\mathrm{e}}$ is the bond wire *effective* temperature. This is defined as an auxiliary source term that stems from the employed linearisation. From the expression of $H_{\mathrm{o};\mathrm{w};\mathrm{r}}$ we realise that $\widetilde{T}_{\mathrm{w};\mathrm{e}}$ must be such that

$$\widetilde{T}_{\mathrm{w};\mathrm{e}} < \frac{2 I_0^2 \rho_{\mathrm{e};\mathrm{o}} \alpha_\rho}{\varepsilon_\mathrm{w} \sigma_{\mathrm{SB}} \chi_\mathrm{w}^{(i)} A_\mathrm{w} C_\mathrm{w} |\alpha_\kappa|}. \tag{3.19}$$

To solve (3.15) we express $\widetilde{\theta}_\mathrm{w}(y,t) = \widetilde{\theta}_{\mathrm{w};1}(y,t) + \widetilde{\theta}_{\mathrm{w};2}(y)$ and get

$$\widetilde{\theta}_\mathrm{w}(y,t) = \sum_k C_{\mathrm{w};k;\mathrm{r}}^\mathrm{t} e^{-\frac{\kappa_\mathrm{o}}{\rho_{\mathrm{e};\mathrm{w}} c_{\mathrm{e};\mathrm{w}}} \lambda_{y;\mathrm{w},k}^2 t} e^{-\frac{F_{\mathrm{o};\mathrm{w};\mathrm{r}}}{\rho_{\mathrm{e};\mathrm{w}} c_{\mathrm{e};\mathrm{w}}} t} \sin\left(\lambda_{y;\mathrm{w},k} y\right)$$

$$+ C_{1;y;\mathrm{w};\mathrm{r}}^\mathrm{s} \cosh\left(\sqrt{\frac{F_{\mathrm{o};\mathrm{w};\mathrm{r}}}{\kappa_\mathrm{o}}} y\right) + C_{2;y;\mathrm{w};\mathrm{r}}^\mathrm{s} \sinh\left(\sqrt{\frac{F_{\mathrm{o};\mathrm{w};\mathrm{r}}}{\kappa_\mathrm{o}}} y\right) \tag{3.20}$$

$$+ \frac{1}{2} \frac{H_{\mathrm{o};\mathrm{w};\mathrm{r}}}{F_{\mathrm{o};\mathrm{w};\mathrm{r}}} + \frac{G_{\mathrm{o};\mathrm{w}}}{F_{\mathrm{o};\mathrm{w};\mathrm{r}}}; \quad \lambda_{y;\mathrm{w},k} = \frac{k\pi}{L_\mathrm{w}}, k > 0.$$

The coefficients $\{C_{\mathrm{w};k;\mathrm{r}}^\mathrm{t}, C_{1;y;\mathrm{w};\mathrm{r}}^\mathrm{s}, C_{2;y;\mathrm{w};\mathrm{r}}^\mathrm{s}\}$ follow from the initial condition $T_\mathrm{w} = T_\mathrm{o}$ at $t = 0$, and from the BCs (3.5) and (3.3). The final result is then

$$\widetilde{T}_\mathrm{w}(y,t) \cong \frac{\sqrt{2\alpha_\kappa \widetilde{\theta}_\mathrm{w}(y,t) + 1}}{\alpha_\kappa} - \frac{1}{\alpha_\kappa}, \tag{3.21}$$

provided that the radicand in (3.21) is positive. Moreover, when $|\alpha_\kappa| \ll 1$, we find that $\widetilde{T}_\mathrm{w} \cong \widetilde{\theta}_\mathrm{w}$. In the following, we still need to find expressions for $\widetilde{T}_{\mathrm{w};\mathrm{e}}$ and χ_w.

3.2.4 Thermal Field in the Moulding Compound

In the moulding compound, the temperature is described by

$$\rho_{\mathrm{m}} c_{\mathrm{e;m}} \frac{\partial T_{\mathrm{m}}}{\partial t} = \nabla \cdot (\kappa_{\mathrm{m}} \nabla T_{\mathrm{m}}) + \varepsilon_{\mathrm{w}} \sigma_{\mathrm{SB}} \chi_{\mathrm{w}} \widetilde{T}_{\mathrm{w}} C_{\mathrm{w}} \delta(x)\delta(z), \qquad (3.22)$$

where the connection with the bond wire is established along the Dirac-delta functions $\delta(\cdot)$. The temperature in the moulding compound rises because of the chip temperature exerted at the leftmost wall and because of the heat flux coming from the bond wire. Equation (3.22) is the heat equation in a homogeneous and linear medium. This allows to invoke superposition and Green's functions to solve the equation. The effect of the chip temperature is found by solving for $\widetilde{T}_{\mathrm{m}} \equiv T_{\mathrm{m}} - T_{\mathrm{o}}$ with the BCs (3.5)–(3.8). The effect of the bond wire heating is found by searching Green's function $\widetilde{G}_{\mathrm{m}} \equiv G_{\mathrm{m}} - T_{\mathrm{o}}$ enforcing the same BCs. The overall solution for the temperature reads

$$T_{\mathrm{m}}(x,y,z,t) = T_{\mathrm{o}} + \sum_{n,p} C^{\mathrm{s}}_{\mathrm{m};n,p} e^{\lambda_{y;\mathrm{m};n,p} y} \left(1 + e^{2\lambda_{y;\mathrm{m};n,p}(L_{\mathrm{w}}-y)} \right) \cos\left(\lambda_{x;\mathrm{m},n} x\right)$$

$$\cos\left(\lambda_{z;\mathrm{m},p} z\right) + \sum_{n,m,p} C^{\mathrm{t}}_{\mathrm{m};n,m,p} e^{-\frac{\kappa_{\mathrm{m}}}{\rho_{\mathrm{m}} c_{\mathrm{e;m}}} (\lambda^{2}_{x;\mathrm{m},n} + \lambda^{2}_{y;\mathrm{m},m} + \lambda^{2}_{z;\mathrm{m},p}) t}$$

$$\cos\left(\lambda_{x;\mathrm{m},n} x\right) \sin\left(\lambda_{y;\mathrm{m},m} y\right) \cos\left(\lambda_{z;\mathrm{m},p} z\right)$$

$$+ \varepsilon_{\mathrm{w}} \sigma_{\mathrm{SB}} \chi_{\mathrm{w}} C_{\mathrm{w}} \int_{0}^{t} \int_{y'} G_{\mathrm{m}}(x,y,z,t-\tau,y') \widetilde{T}_{\mathrm{w}}(y',\tau)\, \mathrm{d}y'\mathrm{d}\tau,$$

$$(3.23)$$

where

$$\lambda_{x;\mathrm{m},n} \tan\left(\lambda_{x;\mathrm{m},n} W_{\mathrm{m}}/2\right) := h_{\mathrm{c}}/\kappa_{\mathrm{m}}, \qquad (3.24)$$

$$\lambda_{z;\mathrm{m},p} \tan\left(\lambda_{z;\mathrm{m},p} H_{\mathrm{m}}/2\right) := h_{\mathrm{c}}/\kappa_{\mathrm{m}}, \qquad (3.25)$$

$$\lambda_{y;\mathrm{m},m} := (2m+1)\pi/2L_{\mathrm{w}}, , \quad m \geq 0, \qquad (3.26)$$

$$\lambda^{2}_{y;\mathrm{m};n,p} := \lambda^{2}_{x;\mathrm{m},n} + \lambda^{2}_{z;\mathrm{m},p} \qquad (3.27)$$

and

$$G_{\mathrm{m}}(x,y,z,t,y') = T_{\mathrm{o}} + \sum_{n}\sum_{p} C^{\mathrm{s}}_{\mathrm{g};n,p} e^{\lambda_{y;\mathrm{m};n,p} y} \left(1 + e^{2\lambda_{y;\mathrm{m};n,p}(L_{\mathrm{w}}-y)} \right)$$

$$\cos\left(\lambda_{x;\mathrm{m},n} x\right) \cos\left(\lambda_{z;\mathrm{m},p} z\right)$$

$$+ \sum_{n}\sum_{m}\sum_{p} C^{\mathrm{t}}_{\mathrm{g};n,m,p}(y') e^{-\frac{\kappa_{\mathrm{m}}}{\rho_{\mathrm{m}} c_{\mathrm{e;m}}} (\lambda^{2}_{x;\mathrm{m},n} + \lambda^{2}_{y;\mathrm{m},m} + \lambda^{2}_{z;\mathrm{m},p}) t}$$

$$\cos\left(\lambda_{x;\mathrm{m},n} x\right) \sin\left(\lambda_{y;\mathrm{m},m} y\right) \cos\left(\lambda_{z;\mathrm{m},p} z\right). \qquad (3.28)$$

Now, it is possible to determine the constant function χ_w by enforcing the moulding and wire temperature to be the same at the common interface point y_o, averaged along the wire length and averaged in time. This leads to the expression

$$\int_0^{t_p} \int_0^{L_w} \lim_{z \to 0} \lim_{x \to 0} \widetilde{T}_m \left(x, y_o, z, t \right) dy dt = \int_0^{t_p} \int_0^{L_w} \widetilde{T}_w \left(y_o, t \right) dy dt, \; y_o \in [0, L_w].$$

(3.29)

The assumption of a constant function χ_w allows to incorporate the influence of the compound on the wire temperature. However, this assumption makes it impossible to enforce rigorously the full continuity of the temperature at the moulding-wire interface. This problem is alleviated by imposing this continuity in a weak way as suggested by (3.29). The factor χ_w is not unique because it depends on the choice of the collocation point y_o [12]. A maximum/minimum value for χ_w is attained in the point where T_w is maximal/minimal. We fix a proper value for χ_w using (3.23) and (3.29) and approximating $\widetilde{T}_w$ by $\theta_{w;0}$, which can be seen as the solution of the *linearised* wire heat transfer equation. Here, the subscript $_0$ refers to (3.20) for which $\chi_{w;e,0} = \left(\bar{T}_{w;e}^3 + \bar{T}_{w;e}^2 T_o + \bar{T}_{w;e} T_o^2 + T_o^3 \right)$, and $\bar{T}_{w;e} = \widetilde{T}_{w;e} + T_o$.

3.2.5 Results for the Analytical Bond Wire Model

The analytical bond wire model has been implemented in Mathematica[TM]. Numerical tests have been carried out for bond wires of gold (Au), copper (Cu) and aluminium (Al). Bond wires with length $L_w = 2.5\,\mathrm{mm}$ and with diameters $D_w \in \{0.8, 1.0, \ldots, 1.8, 2.0\}\,\mathrm{mil}$ are considered. A diameter of $1\,\mathrm{mil}$ equals $10^{-3}\,\mathrm{in} = 25.4\,\mu\mathrm{m}$. Current pulses with time span $t_p = 50\,\mathrm{ms}$ and with amplitudes $I_0 \in \{0, \ldots, 6\}\,\mathrm{A}$ are selected. The extent of the moulding compound is $W_m = 4.45\,\mathrm{mm}$ by $H_m = 1.48\,\mathrm{mm}$. The moulding compound consists of epoxy resin with $\kappa_m = 0.870\,\mathrm{W/(m \cdot K)}$, $c_{e;m} = 882\,\mathrm{J/(Kg \cdot K)}$, and $\rho_m = 1860\,\mathrm{kg/m^3}$.

To verify the analytical model, (3.23) has been computed at $z = 0$ for an Au bond wire of diameter $2\,\mathrm{mil}$ carrying a current of $3.7\,\mathrm{A}$ during $50\,\mathrm{ms}$. The environmental conditions have been set to $T_{ch} = 80\,^\circ\mathrm{C}$, $T_{ld} = 40\,^\circ\mathrm{C}$, $T_o = 20\,^\circ\mathrm{C}$ and $h_c = 25\,\mathrm{W/(m^2 \cdot K)}$.

In Fig. 3.3, each component present in (3.23) is shown at $t_o = 50\,\mathrm{ms}$. Fig. 3.3a and Fig. 3.3b illustrate the steady-state and transient component of (3.23). The temperatures values along the x-axis, i.e., at $(x, 0, 0)$, $\widetilde{T}_{m;2} = 60\,^\circ\mathrm{C}$ and $\widetilde{T}_{m;1} = 0\,^\circ\mathrm{C}$ indicate that the prescribed BCs are satisfied. Fig. 3.3c shows the term of (3.23) with the Green's function, which has been computed with a constant factor χ_w resulting from imposing (3.29) at $y_o = L_w$, where $T_w = T_{ld}$ is minimum. This amounts to underestimating the thermal radiation. Hence, the maximum heat flux occurs at the wire mid-point and decreases towards

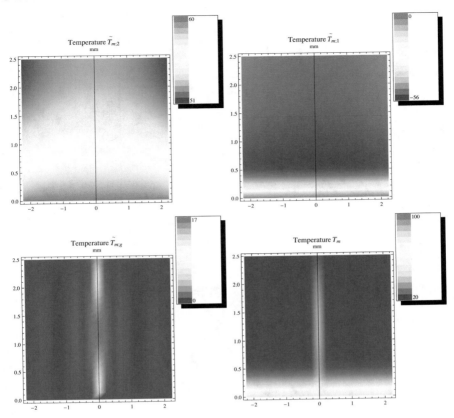

Fig. 3.3: Top view (at the $z = 0$-plane) of the steady temperature component $\widetilde{T}_{\mathrm{m};2}$ (a, Top-Left), the transient temperature component $\widetilde{T}_{\mathrm{m};1}$ (b, Top-Right), the heat kernel temperature component $\widetilde{T}_{\mathrm{m};g}$ (c, Bottom-left) and the moulding compound temperature T_{m} (d, Bottom-Right).

the extremes. Fig. 3.3d shows the temperature T_{m} attaining $T_{\mathrm{m}} = T_{\mathrm{ch}} = 80\,^{\circ}\mathrm{C}$ along the x-axis and a minimum value $T_{\mathrm{o}} = 20\,^{\circ}\mathrm{C}$. Notice that T_{m} in Fig. 3.3d serves as a figure of merit accounting for the effect of the compound on the wire temperature and not as the *real* compound temperature.

The major design question is up to which excitations a bond wire configuration can be operated without failure. For the above described setting and for a variable current pulse amplitude I_0, the temperatures at the wire mid-points are plotted in Fig. 3.4 for Al, Au, and Cu bond wires. The critical temperatures are indicated by red lines. The results confirm those from literature [12, 20] under similar settings. The results also indicate that the assumptions necessary to obtain closed-form formulae are acceptable. The analytic model allows a fast evaluation of the bond wire performance under

David J. Duque Guerra et al.

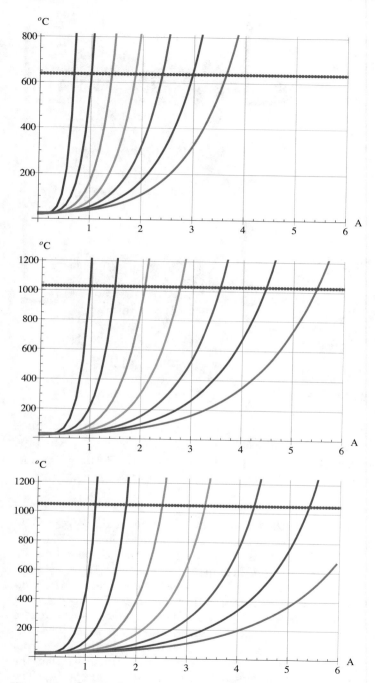

Fig. 3.4: Bond-wire current capacities for diameters $D_{\mathrm{w}} \in \{0.8, 1.0, \ldots, 1.8, 2.0\}$ mil (corresponding to the up-going curves from left to right) and $L_{\mathrm{w}} = 2.5$ mm: a) Al-wire; b) Au-wire; c) Cu-wire. The red lines indicate the critical temperatures.

all relevant environment and excitation conditions. The model can be used for parameter studies and within optimisation procedures. From the model, look-up tables can be derived such that the analytic bond wire model can be inserted as nonlinear circuit elements in larger models.

3.3 Electrothermal Field Models for Bond Wires

Analytic bond wire models are convenient and fast but certainly incorporate many assumptions. On the contrary, field numerical models can be conceived to incorporate all relevant physical effects, but may become large and lead to long computation times. Nonetheless, electrothermal field numerical models of bond wire configuration are absolutely essential for detailed studies, for innovative developments and for calibrating analytical and semi-analytical tools. Following [6], we present a method to incorporate bond wires in electrothermal field simulation.

3.3.1 Electrothermal Field Problem

The current applied to the bond wire causes Ohmic losses, which on their turn cause a temperature increase, causing an increase of the electric resistivity of the bond wire material. This problem setting is coupled, highly nonlinear and includes effects with spatial and temporal dependencies.

3.3.1.1 Electroquasistatic Sub-Problem

When neglecting magnetically induced effects, the electric scalar potential $\varphi(\mathbf{r},t)$ is governed by

$$-\nabla \cdot (\sigma(T)\nabla\varphi) - \nabla \cdot \left(\varepsilon\nabla\frac{\partial\varphi}{\partial t}\right) = 0, \qquad (3.30)$$

where the electric conductivity $\sigma(T)$ depends on the temperature $T(\mathbf{r},t)$, whereas the permittivity ε does not depend on the temperature. Eq. (3.30) is called the *electroquasistatic formulation* [9,10] and is completed by boundary conditions at the terminals of the bond wire and the outside boundaries of the moulding compound. When capacitive effects can be neglected, i.e., when the occurring frequencies are comparably low and/or the conductivities are comparably large, (3.30) can be approximated by the electrokinetic formulation

$$-\nabla \cdot (\sigma(T)\nabla\varphi) = 0. \qquad (3.31)$$

For studying the major effects of bond wire heating, (3.31) is sufficiently accurate.

3.3.1.2 Thermal Sub-Problem

In the thermal sub-problem, heat conduction, heat convection and heat radiation needs to be considered. The according partial differential equation reads

$$\rho c \frac{\partial T}{\partial t} - \nabla \cdot (\lambda(T)\nabla T) = Q(T,\varphi), \tag{3.32}$$

where the thermal conductivity $\lambda(T)$ depends on the temperature, whereas the volumetric heat capacity ρc is assumed to be constant, according to the assumption that the temperature increase does not invoke a significant deformation. The power density is modelled by

$$Q(T,\varphi) = Q_{\mathrm{m}}(T,\varphi) + Q_{\mathrm{bnd}}(T) + Q_{\mathrm{w}}(T,\varphi). \tag{3.33}$$

Here, Q_{m} accounts for the Joule heating in the chip, outer contacts and in the mold compound, Q_{bnd} models the heat exchange with the environment at the boundary and Q_{w} models the Joule heat generated in the bond wire. The latter is considered as an external heat source because the bond wire is not resolved by the grid of the field model as it will become clear below.

The term $Q_{\mathrm{bnd}} = Q_{\mathrm{conv}} + Q_{\mathrm{rad}}$ models the heat exchange through the boundary. The contribution of the adiabatic boundaries (the $y = L_{\mathrm{w}}$-plane in Fig. 3.2) is zero. At convective and radiative boundaries, the heat fluxes are

$$Q_{\mathrm{conv}} = -h_{\mathrm{c}}\left[T(t) - T_{\mathrm{o}}\right], \qquad \text{at the convective boundary,} \tag{3.34}$$

$$Q_{\mathrm{rad}} = -\varepsilon_{\mathrm{w}}\sigma_{\mathrm{SB}}\left[T^4(t) - T_{\mathrm{o}}^4\right], \qquad \text{at the radiative boundary.} \tag{3.35}$$

3.3.1.3 Coupled Electrothermal Problem

The combination of the transient heat equation (3.32) and the static electrokinetic formulation (3.31) leads to the nonlinear electrothermally coupled system of PDEs

$$-\nabla \cdot (\sigma(T)\nabla\varphi) = 0, \tag{3.36}$$

$$\rho c \frac{\partial T}{\partial t} - \nabla \cdot (\lambda(T)\nabla T) = Q(T,\varphi), \tag{3.37}$$

to be completed with appropriate boundary and initial conditions for both φ and T. Field coupling occurs in two directions, i.e., the Joule heating due to the bond wire currents is modelled by

$$Q_{\mathrm{m}} = \sigma(T)\nabla\varphi \cdot \nabla\varphi, \tag{3.38}$$

whereas the temperature effect on the electric and thermal conductivities are represented by $\sigma(T)$ and $\lambda(T)$, to be further specified below.

This coupling can be illustrated by a circuit consisting of an electric and a thermal part as shown in Fig. 3.5. At the electrical side, a voltage source V_0 generates a current in the nonlinear electrical conductance $\sigma(T)$, whereas, at the thermal side, the heat source Q_m caused by the current forces a heat flux through the nonlinear thermal conductance $\lambda(T)$. This generated heat flux leads to changes in temperatures and thus influences $\lambda(T)$ as well as $\sigma(T)$. The electrical side is connected to the thermal side via the power density Q_m. In the other direction, the coupling is given by the influence of the temperature T on the electric conductivity σ.

The electrothermal coupling can also be interpreted in terms of the relevant field quantities arranged in electromagnetic and thermal *houses*, also called *Tonti diagrams* [1, 13, 23] (Fig. 3.6). Equations (3.36) and (3.37) are combinations of black arrows, whereas the coupling effects are represented by red arrows.

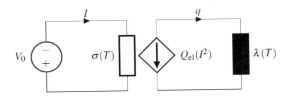

Fig. 3.5: Electrothermal coupling illustrated by the usage of circuit elements.

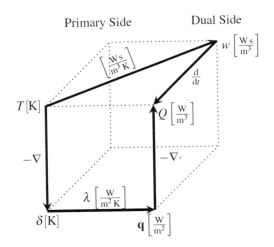

Fig. 3.6: Continuous electrothermal house.

3.3.2 Field Discretization by the Finite Integration Technique

Only volume discretisation methods are applicable to discretise the nonlinear coupled electrothermal problem. Most frequently, a finite-element (FE) method is used [11]. Here, the coupled electrothermal problem is discretized in space by the finite-integration technique (FIT) [24, 25], which allows a convenient consideration of semiconductors (see also [21]).

Field Allocation A staggered 3D hexahedral grid pair consisting of a primary and a dual grid is used (Fig. 3.7). The material distribution is assumed to be conform to the primary grid. The relevant field quantities are allocated at the nodes, edges, faces and volumes of the grid. In particular, the electric potentials φ as well as the temperatures $\mathbf{T}$ are allocated at the nodes of the primary grid, the voltages and the temperature drops are allocated at the edges of the primary grid, the electric currents and the heat fluxes are associated with the faces of the dual grid and the current and heat sources are defined in the cells of the dual grid [8]. The discrete counterpart of the gradient operator at the primary grid is an incidence matrix containing the values 0, 1 and -1, is denoted by $\mathbf{G}$ and allows to express the voltages and temperature drops by $\widehat{\mathbf{e}} = -\mathbf{G}\varphi$ and $\widehat{\mathbf{t}} = -\mathbf{G}\mathbf{T}$. The discrete counterpart of the divergence operator defined at the dual grid is denoted by $\widetilde{\mathbf{S}}$ and allows to express the currents $\widehat{\widehat{\mathbf{j}}}$ and heat fluxes $\widehat{\widehat{\mathbf{q}}}$ accumulating at the dual cells by $\widetilde{\mathbf{S}}\widehat{\widehat{\mathbf{j}}}$ and $\widetilde{\mathbf{S}}\widehat{\widehat{\mathbf{q}}}$. The staggering of both grids results in the important property $\mathbf{G} = -\widetilde{\mathbf{S}}^{\mathsf{T}}$. Notice that this reasoning is also valid for an unstructured FE mesh.

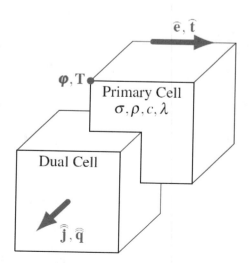

Fig. 3.7 Primary and dual cells organised in a staggered FIT grid.

Material Matrices The currents $\widehat{\widehat{\mathbf{j}}}$ and the heat fluxes $\widehat{\widehat{\mathbf{q}}}$ allocated at the dual faces need to be related to the voltages $\widehat{\mathbf{e}}$ and temperature drops $\widehat{\mathbf{t}}$ associated with the corresponding primary edges. This is established by the so-called *electrical* and *thermal conductance matrices* $\mathbf{M}_\sigma$ and $\mathbf{M}_\lambda$, i.e., $\widehat{\widehat{\mathbf{j}}} = \mathbf{M}_\sigma \widehat{\mathbf{e}}$ and $\widehat{\widehat{\mathbf{q}}} = \mathbf{M}_\lambda \widehat{\mathbf{t}}$. Moreover, the heat power $\mathbf{Q}_{\dot{\mathbf{T}}}$ in the dual cells needs to be related to the temperature change $\frac{\partial \mathbf{T}}{\partial t}$, as organised by the *thermal capacitance matrix* $\mathbf{M}_{\rho c}$ in $\mathbf{Q}_{\dot{\mathbf{T}}} = \mathbf{M}_{\rho c} \frac{\partial \mathbf{T}}{\partial t}$. In a FE setting, these matrices are symmetric and positive definite and are expressed by

$$M_{\sigma,i,j} = \int_V \sigma \mathbf{w}_j \cdot \mathbf{w}_i \, \mathrm{d}V; \tag{3.39}$$

$$M_{\lambda,i,j} = \int_V \lambda \mathbf{w}_j \cdot \mathbf{w}_i \, \mathrm{d}V; \tag{3.40}$$

$$M_{\rho c,i,j} = \int_V \rho c N_j N_i \, \mathrm{d}V, \tag{3.41}$$

where $N_j(\mathbf{r})$ and $\mathbf{w}_j(\mathbf{r})$ are nodal and edge shape functions respectively [4,19] and V is the computational domain. In FIT and in case of a pair of mutually orthogonal grids, these matrices are diagonal and positive definite and have the entries

$$M_{\sigma,i,i} = \frac{\overline{\sigma}_i \tilde{A}_i}{\ell_i}; \tag{3.42}$$

$$M_{\lambda,i,i} = \frac{\overline{\lambda}_i \tilde{A}_i}{\ell_i}; \tag{3.43}$$

$$M_{\rho c,j,j} = \overline{\rho c}_j \tilde{V}_j, \tag{3.44}$$

where ℓ_i is the length of primary edge i, $\tilde{A}_i$ is the area of dual facet i, $\tilde{V}_j$ is the volume of dual cell j, $\overline{\sigma}$ and $\overline{\lambda}$ are found by averaging σ and λ over the primary cells touching the considered primary edge and $\overline{\rho c}_j$ is obtained by averaging the volumetric heat capacity over the primary cells touching the dual cell j. The convection boundary conditions give rise to an additional term $\mathbf{K}_{\text{conv}} (\mathbf{T} - \mathbf{T}_\text{o})$ where

$$K_{\text{conv},i,i} = h_{\text{c},i} \tilde{A}_i, \tag{3.45}$$

and $\tilde{A}_i$ is the area of the dual facet at the boundary around node i and $h_{\text{c},i}$ is the thermal convection coefficient. One can show that the FIT matrices can be obtained by approximating the FE integrals (3.42)-(3.44) by the midpoint rule [3].

Electrothermal Coupling In the FE setting, the Joule heating is discretised by

$$Q_i = \sum_p \sum_q \widehat{e}_p \widehat{e}_q \int_V \sigma \mathbf{w}_q \cdot \mathbf{w}_p N_i \, \mathrm{d}V. \tag{3.46}$$

In FIT, the Joule heating is discretised in two steps. At first, the electric field vectors $\mathbf{E}_k$ at the midpoints of the primary cells numbered by k are found by interpolation from the voltages at the surrounding primary edges. The power density follows from $Q_{\text{el},k} = \sigma_k \mathbf{E}_k \cdot \mathbf{E}_k$. Finally, the power associated with the dual cells is found to be

$$\mathbf{Q}_{\text{el}}(T, \varphi) = \widetilde{\mathbf{D}}_V \overline{\mathbf{q}} \qquad (3.47)$$

where $\widetilde{\mathbf{D}}_V$ is a diagonal matrix containing the volumes of the dual cells and $\overline{\mathbf{q}}$ results from averaging the powers from the primary cells to the dual cells.

Discrete Electrothermal Problem Putting together the topological operators $\mathbf{G}$ and $\widetilde{\mathbf{S}}^\top$ and the material matrices $\mathbf{M}_\sigma$, $\mathbf{M}_\lambda$ and $\mathbf{M}_{\rho c}$, the discrete counterpart to (3.36) and (3.37) is found, i.e.,

$$-\widetilde{\mathbf{S}}\mathbf{M}_\sigma(\mathbf{T})\mathbf{G}\varphi = \mathbf{0}, \qquad (3.48)$$

$$\mathbf{M}_{\rho c}\dot{\mathbf{T}} - \widetilde{\mathbf{S}}\mathbf{M}_\lambda(\mathbf{T})\mathbf{G}\mathbf{T} = \mathbf{Q}(\mathbf{T}, \varphi). \qquad (3.49)$$

The discrete temperatures $\mathbf{T} = \mathbf{T}(t)$ and the electrical potentials $\varphi = \varphi(t)$ serve as degrees of freedom. The power $\mathbf{Q} = \mathbf{Q}_m + \mathbf{Q}_{\text{bnd}} + \mathbf{Q}_w$ is the discrete representative of Q and includes the Joule heating in the field model, the heat loss at the convective and radiative boundaries as well as a contribution from the bond wire as will be developed below. The discrete quantities and their relations are brought together in a *discrete electrothermal house* [1, 7] (Fig. 3.8). The figure illustrates a one-by-one equivalence between the continuous and the discrete setting. The representation is valid for both the FIT and FE cases. The difference between both is limited to the construction of the material matrices and coupling operators as developed above.

3.3.3 Inserting a Bond Wire Model in the Field Model

The cross section of a bond wire is typically much smaller than its length and than the dimensions of the moulding compound. For that reason, it is recommended not to resolve the bond wire cross-section by the grid. Instead, a bond wire is modelled by lumped elements [6]. We assume that the electric conduction dominates any capacitive effects. Moreover, we neglect the thermal heat capacity of a bond wire with respect to the thermal heat capacity of the moulding compound. A bond wire is inserted in the field model by inserting its conductance matrix

$$\mathbf{G}_w = \begin{bmatrix} G_w & -G_w \\ -G_w & G_w \end{bmatrix}, \qquad (3.50)$$

where G_{bw} stands for either the electrical conductance $G_{\mathrm{w}}^{\mathrm{el}}$ or the thermal conductance $G_{\mathrm{w}}^{\mathrm{th}}$, into the overall system matrices. The end points of a bond wire are assumed to match points of the computational grid (Fig. 3.9). These connections are formalised in the connecting matrix $\mathbf{P}_j$ which contains a 1 for connecting the entrance of bond wire j to one of the grid nodes and a -1 for connecting the exit of bond wire j to one of the grid nodes. With N_{bw} bond wires, each with its connecting matrix $\mathbf{P}_j$, the extended discrete systems of equations read

$$\widetilde{\mathbf{S}}\mathbf{M}_\sigma(\mathbf{T})\widetilde{\mathbf{S}}^\top\boldsymbol{\varphi} + \sum_{j=1}^{N_{\mathrm{bw}}} \mathbf{P}_j G_{\mathrm{bw},j}^{\mathrm{el}}(T_{\mathrm{bw},j})\mathbf{P}_j^\top \boldsymbol{\varphi} = \mathbf{0}, \tag{3.51}$$

$$\mathbf{M}_{\rho c}\dot{\mathbf{T}} + \widetilde{\mathbf{S}}\mathbf{M}_\lambda(\mathbf{T})\widetilde{\mathbf{S}}^\top\mathbf{T} + \sum_{j=1}^{N_{\mathrm{bw}}} \mathbf{P}_j G_{\mathrm{bw},j}^{\mathrm{th}}(T_{\mathrm{bw},j})\mathbf{P}_j^\top \mathbf{T} = \mathbf{Q}(\mathbf{T}). \tag{3.52}$$

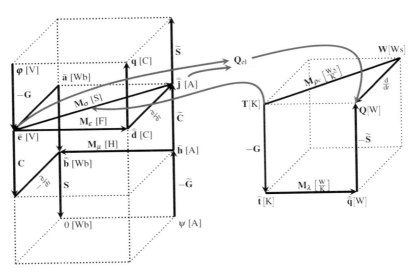

Fig. 3.8: Discrete electrothermal house.

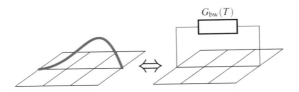

Fig. 3.9: Bonding wire modelling by a lumped element approach.

Each bond wire is modelled as a single lumped element. Hence, a linear temperature distribution is assumed along the bond wire. The average bond wire temperature then follows by averaging the temperatures at its ends as formalised by

$$T_{\mathrm{bw},j} = \mathbf{X}_j^\top \mathbf{T}, \tag{3.53}$$

where $\mathbf{X}_j$ contains only two non-zero entries equal to $1/2$ at the indices representing the connecting nodes. The bond wire model can be improved by subdividing into several sections. The Joule heating of a single bond wire is

$$Q_{\mathrm{bw},j} = \boldsymbol{\varphi}^\top \mathbf{P}_j G_{\mathrm{bw},j}^{\mathrm{el}}(T_{\mathrm{bw},j}) \mathbf{P}_j^\top \boldsymbol{\varphi}, \tag{3.54}$$

which is combined by

$$\mathbf{Q}_{\mathrm{w}} = \sum_{j=1}^{N_{\mathrm{bw}}} \mathbf{X}_j Q_{\mathrm{bw,j}}, \tag{3.55}$$

in order to obtain the contribution of the bond wires to the right hand side of (3.49).

3.4 Bond Wire Degradation under Uncertain Geometries

The manufacturing of bond wires comes with unavoidable imperfections, e.g., in the material properties, the size of the bond wires and the position of the connections. In this section, following [6], the above developed electrothermal field model is used to study the effect of manufacturing tolerances on the thermal behaviour of a package. In particular, uncertainties on the bond wire lengths will be considered.

3.4.1 Chip Package with Bond Wires

A considered chip is connected to a subset of 28 contacts by 12 bond wires. The bond wires become visible in X-ray pictures (Fig. 3.10). Because only the contacts are accessible, experiments necessarily apply a voltage over two neighbouring contact pads, thereby exciting two adjacent bond wires (e.g. wires 3 and 4 or wires 7 and 8).

Because the material properties and the cross-section of bond wires are accurately known, the major uncertainty is expected to be the bond wire length. The length of each of the bond wires heavily depends on the bonding process. Using the available X-ray pictures in Fig. 3.10, the truly fabricated lengths have been measured.

3.4.2 Modeling of the Bond Wire Length

The length of a bond wire depends on the direct distance d between contact pad and chip, the connection point on the contact pad and the bending of the wire. When exactly manufactured, the connection point is equally spaced with a distance a to three of the contact pad edges (Fig. 3.11, Top-Left). A deviation of that perfect position causes an elongation Δs yielding a longer distance $D = d + \Delta s$ (Fig. 3.11, Top-Right). The bond wires follow a bended line through the moulding compound. Hence, an additional elongation Δh as illustrated in Fig. 3.11 (Bottom) has to be accounted for. In concrete, the total bond wire length is given by $L = d + \Delta s + \Delta h$.

The bond wire lengths have been measured for 12 bond wires connected to a chip by X-ray imaging (Fig. 3.10). Because of the perspective of the camera, the elongation Δh could only be measured for 6 wires. For the other wires, the average of the obtained 6 values is used. Instead of using the total length of the bond wires as reference quantity, the relative elongation $\delta = \frac{L-d}{L}$ is selected. The random elongations for all bond wires, independent of their lengths, are determined by the probability density function for δ. The histogram depicted in Fig. 3.12 indicates a normal distribution with expectation value $\mu_\mathrm{w} = 0.17$ and standard deviation $\sigma_\mathrm{w} = 0.048$. The statistics are here hampered by the rather small number of samples. More accurate statistics would necessitate the fabrication and measurement of additional chips.

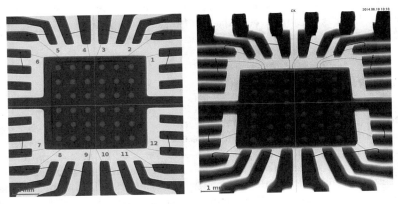

Fig. 3.10: X-ray photos (Top View and Perspective View) of the investigated chip where the considered bond wires are marked with red numbers. The wires short-circuiting some of the leads are not considered.

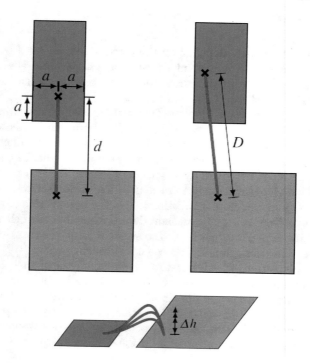

Fig. 3.11: Variability of the bond wire length due to construction tolerances. Top-Left: Exact position on contact pad (in grey). Top-Right: Elongation due to misplacement. Bottom: Elongation due to bending.

3.4.3 Monte Carlo Simulation

When a high number of random input parameters needs to be considered, the Monte Carlo (MC) method is a suitable choice [5]. The MC method solves a large number of instantiations of the model for random sets of parameters. Here, the model is described by (3.48)–(3.49) and involves the solution of an electrothermal field model for every instantiation, which can be done in parallel. The root-mean-square error originating from the finite number of samples M decreases with increasing M, i.e.,

$$\text{error}_{\text{MC}} = \frac{\sigma}{\sqrt{M}}, \qquad (3.56)$$

where σ is the standard deviation of the quantity to be computed. According to this estimate, one has to expect a rather slow convergence, as indicated by the $\sqrt{M}$-term in the denominator.

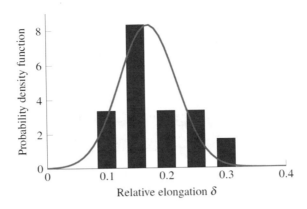

Fig. 3.12: Probability density function for the relative elongation δ.

3.4.4 Chip Model

The considered model is shown in Fig. 3.13. For all contact pads, we assume a uniform width, here valued $w_{\text{pad}} = 0.311\,\text{mm}$. The majority, i.e. 24 contact paths have the length $\ell_{\text{pad}} = 1.01\,\text{mm}$, whereas 4 contact paths have the length $L_{\text{pad}} = 1.261\,\text{mm}$. The structure is easily represented by a hexahedral mesh and hence, a 3D FIT model can readily be applied. The bond wires and contact pads are made of copper. The mold compound is made of epoxy resin. The chip is assumed to be a good electrical and thermal conductor and therefore gets the same material data as copper. The material data (for a reference temperature of $T = 300\,\text{K}$) are gathered in Table 3.1.

3.4.5 Boundary, Excitations and Initial Conditions

The boundaries of the contact pads are modelled by perfect electric conducting (PEC) material. The remaining boundary parts are electrical insulators. A voltage drop of $V_{\text{w}} = 40\,\text{mV}$ is exerted on the six pairs of bond wires by

Table 3.1: Material Properties @ $T = 300\,\text{K}$

Region	Material	λ [W/K/m]	σ [S/m]
Compound	Epoxy resin	0.87	$1 \cdot 10^{-6}$
Contact pad	Copper	398	$5.80 \cdot 10^{7}$
Chip	Copper	398	$5.80 \cdot 10^{7}$
Bonding wire	Copper	398	$5.80 \cdot 10^{7}$

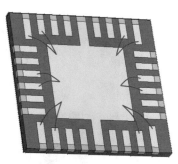

Fig. 3.13: Model of the investigated chip.

applying a potential of $V_{\mathrm{dc}} = \pm 20\,\mathrm{mV}$ to each pair of PEC boundaries. Except for the excited PEC parts, the initial potential is $V_{\mathrm{init}} = 0\,\mathrm{V}$ at time $t = 0\,\mathrm{s}$. As thermal boundary conditions, convection and radiation conditions with a heat transfer coefficient of $h = 25\,\mathrm{W/m^2/K}$ and an emissivity of $\varepsilon = 0.2475$ are applied. At $t = 0$, the entire chip is assumed to be at the ambient temperature $T_\infty = 300\,\mathrm{K}$.

3.4.6 Failure Probability

The major source of failure is the temperature of the bond wire. Because the bond wires themselves are not resolved by the grid, the temperatures at their end points are considered as representative. The bond wire temperatures then follow from (3.53). The expectation value $E_j(t)$ for the temperature of each of the bond wires is found by averaging over the samples, i.e.,

$$E_j(t) = \frac{1}{M} \sum_{m=1}^{M} T_{\mathrm{bw},j}^{(m)}(t). \tag{3.57}$$

A failure occurs when any bond wire exceeds a critical value. For that reason, only the maximum of the expectation values is relevant, i.e.,

$$E_{\max}(t) = \max_j[E_j(t)], \text{ for } j \in \{1, ..., N_{\mathrm{bw}}\}. \tag{3.58}$$

The derivation of other stochastic moments, e.g., the variance or standard deviation, is similar.

Table 3.2: Simulation Parameters.

Parameter	Value
Bonding wire voltage V_w	40 mV
End time	50 s
No. of time steps	51
No. of MC samples	1000
Wires' diameter	25.4 µm
Average wires' length $\overline{L}$	1.55 mm
Ambient temperature	300 K
Heat transfer coefficient	25 W/m^2/K
Emissivity	0.2475

3.4.7 Results

The expected value and the standard deviation of the bond wire temperature was calculated by Monte Carlo simulation with $M = 1000$ samples and for the field simulation parameters collected in Table 3.2. The probability density function from Fig. 3.12 was used. According to (3.56), the Monte Carlo error is found to be error$_{MC} = 0.147$ K. The expectation value $E(t)$ for the temperature of the hottest wire is plotted over time in Fig. 3.14. The error bars indicate the uncertainty on the temperature due to the variability of the bond wire lengths as discussed in Section 3.4.2. Eventually, a failure mainly occurs due to the damage of the surrounding mold material. For that reason, the critical temperature related to the degradation of epoxy resin, i.e., $T_{critical} = 523$ K ≈ 250 °C, is selected as a threshold for failure. This critical temperature is shown as a red line in Fig. 3.14.

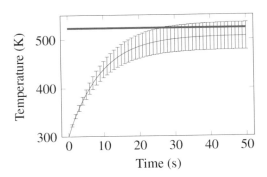

Fig. 3.14: Expected temperature of the hottest bond wire with plotted $6\sigma_{MC}$-deviation over time. In red, the critical temperature of the wire's material is indicated.

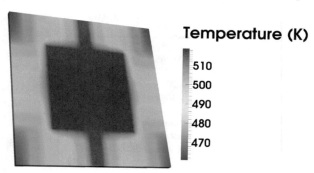

Fig. 3.15: Spatial temperature distribution at $t = 50\,\text{s}$.

Directly after starting the excitation, the chip and bond wire temperatures increase drastically. Thanks to the convection and radiation heat fluxes at the package boundaries, a stationary situation settles in after $t \approx 50\,\text{s}$. The expected value for the temperature of the hottest bond wire is then still below the critical temperature T_{critical}. From a designer's perspective, all bond wires are thus expected to survive without failure. However, the uncertainty in the bond wire lengths is responsible for a significant variation of the temperature, i.e., a standard deviation of $\sigma_{\text{MC}} = 4.65\,\text{K}$ is found. The 6σ-deviation is visualised by the error bars in Fig. 3.14. When a reliability with 6σ is required, the design needs to be adapted, i.e., by increasing the bond wire diameter.

The distribution of the temperature at $t = 50\,\text{s}$ is shown in Fig. 3.15. As expected, the regions where the contacts are closer to the chip and carry the shortest bond wires, exhibit the largest temperature increase. Hence, it is also not surprising that one of these bond wires reaches the highest temperature. From this study, it becomes clear that hot spots in the bond wires are reduced when the variation on the lengths of the bond wires is reduced.

3.5 Conclusions

In this chapter, transient electrothermal models for bond wires have been developed. A first simple analytic model allows to predict the temperature and the time-to-failure by a set of closed-form formulae and are of great help in fast design procedures. The second one uses a full 3D transient electrothermal field simulation and is therefore very accurate as long as accurate data are provided. The field model has been used to compute moments of the temperature in the bond wires due to uncertainties on the bond wire lengths related to manufacturing tolerances.

Acknowledgements This work is a part of the project 'Nanoelectronic Coupled Problems Solutions' (nanoCOPS) funded by the European Union within FP7-ICT-2013 (grant no. 619166). The second author is also supported by the 'Excellence Initiative' of the German Federal and State Governments and the Graduate School of Computational Engineering at Technische Universität Darmstadt. The authors would like to thank Jiří Petřzela, Roman Sotner and Jiří Dřínovský for the bond wire measurements done at Brno University of Technology.

References

1. ALOTTO, P., FRESCHI, F., AND REPETTO, M.: *Multiphysics problems via the cell method: The role of Tonti diagrams*. IEEE Trans. Magn. **46**(8), 2959–2962 (2010). doi:10.1109/TMAG.2010.2044487
2. BERGMAN, T.L., LAVINE, A.S., INCROPERA, F.P., AND DEWITT, D.P.: *Introduction to Heat Transfer, 6 edn*. John Wiley & Sons, Ltd. (2011)
3. BONDESON, A., RYLANDER, T., AND INGELSTRÖM, P.: *Computational Electromagnetics*. Texts in Applied Mathematics. Springer (2005). doi:10.1007/b136922
4. BOSSAVIT, A.: *Computational Electromagnetism: Variational Formulations, Complementarity, Edge Elements*. Academic Press, San Diego (1998). http://natrium.em.tut.fi/~bossavit/
5. CAFLISCH, R.E.: *Monte Carlo and Quasi-Monte Carlo Methods*. Acta. Num. **7**, 1–49 (1998). doi:10.1017/S0962492900002804
6. CASPER, T., DE GERSEM, H., GILLON, R., GÖTTHANS, T., KRATOCHVÍL, T., MEURIS, P., AND SCHÖPS, S.: *Electrothermal simulation of bonding wire degradation under uncertain geometries*. In: L. FANUCCI, AND J. TEICH (EDS.): *Proceedings of the 2016 Design, Automation & Test in Europe Conference & Exhibition (DATE)*, pp. 1297–1302. IEEE (2016). http://www.date-conference.com
7. CLEMENS, M.: *Large systems of equations in a discrete electromagnetism: formulations and numerical algorithms*. IEE. Proc. Sci. Meas. Tech. **152**(2), 50–72 (2005). doi:10.1049/ip-smt:20050849
8. CLEMENS, M., GJONAJ, E., PINDER, P., AND WEILAND, T.: *Self-consistent simulations of transient heating effects in electrical devices using the finite integration technique*. IEEE Trans. Magn. **37**(5), 3375–3379 (2001). doi:10.1109/20.952617
9. CLEMENS, M., WILKE, M., BENDERSKAYA, G., DE GERSEM, H., KOCH, W., AND WEILAND, T.: *Transient electro-quasistatic adaptive simulation schemes*. IEEE Trans. Magn. **40**(2), 1294–1297 (2004). doi:10.1109/TMAG.2004.824582
10. DIRKS, H.K.: *Quasi-stationary fields for microelectronic applications*. Electr. Eng. **79**(2), 145–155 (1996). doi:10.1007/BF01232924
11. DRIESEN, J., BELMANS, R.J.M., AND HAMEYER, K.: *Methodologies for coupled transient electromagnetic-thermal finite-element modeling of electrical energy transducers*. IEEE Trans. Ind. Appl. **38**(5), 1244–1250 (2002). doi:10.1109/TIA.2002.803024
12. DUQUE, D.J., SCHÖPS, S., AND WIEERS, A.: *Fast and reliable simulations of the heating of bond wires*. In: G. RUSSO, V. CAPASSO, G. NICOSIA, AND V. ROMANO (EDS.): *Progress in Industrial Mathematics at ECMI 2014*. Series Mathematics in Industry, Vol. 22. pp. 819–827, Springer, Berlin (2017). doi:10.1007/978-3-319-23413-7_114
13. FRESCHI, F., AND REPETTO, M.: *Tonti diagrams and algebraic methods for the solution of coupled problems*. In: B. MICHIELSEN, AND J.R. POIRIER (EDS.): *Scientific Computing in Electrical Engineering SCEE 2010*. Series Mathematics in Industry, Vol. 16, pp. 195–203. Springer, Berlin (2012). doi:10.1007/978-3-642-22453-9_21

14. LOH, E.: *Physical analysis of data on fused-open bond wires.* IEEE Trans. Compon., Hybrids, and Manuf. Technol. **6**(2), 209–217 (1983). doi:10.1109/tchmt.1983.1136162
15. LOH, E.: *Heat transfer of fine-wire fuse.* IEEE Trans. Compon. Packag. Manuf. **7**(3), 264–267 (1984)
16. MALLIK, A., AND STOUT, R.: *Simulation methods for predicting fusing current and time for encapsulated wire bonds.* IEEE Trans. Electron. Packag. Manuf. **33**(4), 255–264 (2010). doi:10.1109/TEPM.2010.2055568
17. MERTOL, A.: *Estimation of aluminum and gold bond wire fusing current and fusing time.* IEEE Trans. Compon. Packag. Manuf. B **18**(1), 210–214 (1995)
18. MOUTHAAN, K.: *Modelling of RF high power bipolar transistors.* PhD-Thesis, Technical University Delft, The Netherlands (2001)
19. NÉDÉLEC, J.C.: *Mixed finite elements in* $\mathbb{R}^3$. Numerische Mathematik **35**(3), 315–341 (1980). doi:10.1007/BF01396415
20. NÖBAUER, G.T., AND MOSER, H.: *Analytical approach to temperature evaluation in bonding wires and calculation of allowable current.* IEEE Transactions on Advanced Packaging **23**(3), 426—435 (2000)
21. SCHOENMAKER, W., BRACHTENDORF, H.G., BITTNER, K., TISCHENDORF, C., AND STROHM, C.: *EM-equations, coupling to heat and to circuits.* Chapter in this Book, 2019.
22. SCHUSTER, C., LEONHARDT, G., AND FICHTNER, W.: *Electromagnetic simulation of bonding wires and comparison with wide band measurements.* IEEE Trans. Adv. Packag. **23**(1), 69–79 (2000). doi:10.1109/6040.826764
23. TONTI, E.: *On the formal structure of physical theories.* Tech. Rep., Politecnico di Milano, Milano, Italy (1975)
24. WEILAND, T.: *A discretization method for the solution of Maxwell's equations for six-component fields.* AEÜ **31**, 116–120 (1977)
25. WEILAND, T.: *Time domain electromagnetic field computation with finite difference methods.* Int. J. Numer. Model. Electron. Network. Dev. Field **9**(4), 295–319 (1996). doi:10.1002/(SICI)1099-1204(199607)9:4<295::AID-JNM240>3.0.CO;2-8

Chapter 4
Discretizations

Wim Schoenmaker, Hans-Georg Brachtendorf, Kai Bittner,
Caren Tischendorf, Christian Strohm

Abstract This chapter discusses the techniques for converting continous systems to discrete systems. Specific attention is paid to the mimetic discretization approach. Furthermore the set up of coupled problems is addressed. Finally, it is shown that discretization requires a careful handling of the various terms in the discretization procedure, otherwise numerical unstable equations will appear.

4.1 Spatial Discretization

The continuous field equations as described in [3] need to be converted to approximated equations that can be solved using numerical analysis. For that purpose a discretization grid is computed and a transition recipe is given that maps the continuous equations on discrete equations. In a nut shell this defines the discretization procedure. It should be emphasized that each stage in procedure is highly non-trivial and it has taken the research community decades to find adequate methods that are capable of addressing industrial relevant design challenges and up to today the improvements of all stages are active research fields. The first stage, e.g. mesh generation is subjected to a number constraints: (1) the mesh should be economical, meaning that it size of number of mesh nodes should not be unnecessarily large and (2) the mesh

Wim Schoenmaker
MAGWEL NV, Leuven, Belgium, e-mail: `Wim.Schoenmaker@magwel.com`

Hans-Georg Brachtendorf, Kai Bittner
Fachhochschule Oberösterreich, Hagenberg im Mühlkreis, Austria, e-mail: {`Hans-Georg.Brachtendorf,Kai.Bittner`}`@fh-hagenberg.at`

Caren Tischendorf, Christian Strohm
Humboldt Universität zu Berlin, Germany, e-mail: {`Caren.Tischendorf,StrohmCh`}`@math.hu-berlin.de`

© Springer Nature Switzerland AG 2019
E. J. W. ter Maten et al. (eds.), *Nanoelectronic Coupled Problems Solutions*,
Mathematics in Industry 29, https://doi.org/10.1007/978-3-030-30726-4_4

element should be preferably acute, meaning that the center of the mesh cell should be a point inside the mesh cell.

The transition from continuous equations to discrete equations is also a delicate step. A careless translation, for instance replacing brute-force all derivatives by finite differences will lead to inconsistencies or conflicts with basic physical requirement such as charge conservation. This is extensively discussed in [2]. Again, at this stage one must be already have in mind the next stage being the solving of the equations. A smart discretization method will lead to discretized equations that can be more easily solved. The finite-volume method (FVM) and the finite-surface method (FSM) have the property that basic physical requirements are respected. The FVM and FSM are mimetic and relax the demand that for finite grid sizes, every variable should be calculable at every space-time point. In other words: only in the limit of zero grid-node distance ($\lim_{h \to 0}$) such detailed information may be extracted.

Finally, the solving stage is based on combining Newton-Raphson iterative solving for non-linear systems and either direct or indirect solving of the linear systems that are generated in each Newton-Raphson step. The algorithms for solving the systems of linear equations, depend crucially on the properties of the system matrix. If this matrix has a high condition number the linear solving is hampered. Therefore, the prior stages, meshing and discretization, must already anticipate avoidance of ill-defined system matrices. In this chapter we summarize the essential discretization steps that are applied to the continuous equations. For in-depth details we refer to [2].

4.1.1 Semiconductor Equations

In this section we consider the discretization of Gauss' law. It is one of the equations that constitute the drift-diffusion model for semiconductors. For insulators it suffices to solve Gauss' law. In what follows we introduce the following notation. For numbering the space grid points of adjacent grid nodes we use i and j. Moreover, let Δw_i be a finite volume element, associated with a grid node i, and $\sigma_{ij} = \pm 1$ the orientation of a link connecting node i and node j. It is set positive when oriented from inside to outside of the volume. The links between nodes i and j are denoted with $\langle ij \rangle$ and have an associated length h_{ij} and an area ΔS_{ij} for the dual surface. Every link has an *intrinsic* orientation vector of length 1 and is denoted by $\mathbf{e}_{ij}$. The projection of the vector potential onto a link $\langle ij \rangle$ is marked with index i and j, e.g. $\mathbf{e}_{ij} \cdot \mathbf{A} = A_{ij}$. When applying the finite-volume method, each node generates a balance equation corresponding to elaborating the divergence of a flux over the surface of the dual volume element of each node. Each surface element naturally gets a normal vector pointing away from the node under consideration. This vector $\mathbf{n}$, that is also found on a each link, can be parallel or anti-parallel to $\mathbf{e}$. The resulting sign is denoted as s, e.g. $s_{ij} = \mathbf{n} \cdot \mathbf{e} = \pm 1$.

The discretization of Gauss' law is done using the usual finite-volume method or finite-integration technique. Starting from

$$\epsilon \nabla \cdot (\nabla V + \partial_t \mathbf{A}) + \rho = 0,$$

it leads by integration over a volume element ΔV to

$$\epsilon \oint_{\partial \Delta V} (\nabla V + \Pi) \cdot d\mathbf{S} + \int_{\Delta V} \rho \, dV = 0.$$

A summation over all nodes j, incident with i via a branch $\langle ij \rangle$ will replace the continuous expressions resulting into

$$\sum_j \epsilon \frac{d_{ij}}{h_{ij}} (V_i - V_j - \sigma_{ij} \Pi_{ij} \, h_{ij}) - \varrho_i \, \Delta w_i = 0, \qquad (4.1)$$

where Δw_i is a volume element. The electric field strength along a link $\langle ij \rangle$ - positive when directed from inside to outside of a volume - is obtained by

$$-E_{ij} = \left(\nabla V + \frac{\partial}{\partial t} \mathbf{A} \right) \cdot \mathbf{e}_{ij} = (\nabla V + \Pi) \cdot \mathbf{e}_{ij} = \frac{V_j - V_i + \sigma_{ij} \Pi_{ij} h_{ij}}{h_{ij}}.$$

4.1.2 Discretization of the Maxwell-Ampere System

The four Maxwell's equations read

$$\nabla \times \mathbf{E} = -\partial_t \mathbf{B}, \quad \nabla \times \mathbf{H} = \mathbf{J} + \partial_t \mathbf{D}, \quad \nabla \cdot \mathbf{D} = \rho, \quad \nabla \cdot \mathbf{B} = 0, \qquad (4.2)$$

where $\mathbf{E}, \mathbf{D}$ are the electric field strength and the displacement, and $\mathbf{H}, \mathbf{B}$ the magnetic field strength and induction, respectively. Moreover ρ and $\mathbf{J}$ are the electric charge density and current density, respectively. In what follows we assume linear isotropic materials, i.e.

$$\mathbf{D} = \epsilon \mathbf{E}, \quad \mathbf{B} = \mu \mathbf{H}. \quad \mathbf{E} = -(\nabla V + \partial_t \mathbf{A}), \quad \mathbf{B} = \nabla \times \mathbf{A}, \qquad (4.3)$$

where ϵ is the dielectric constant and μ the permeability. It should be noted that materials of different types can be stacked or blocks of different materials can be placed next to each other. This results into abrupt jumps in the overall permittivity ϵ and permeability μ. However, generally we assume that the parameters depend on the space coordinate. Furthermore, we rewrite the Maxwell equations using the scalar potential V and vector potential $\mathbf{A}$.

The Maxwell-Ampère's law is addressed in a slightly different way. Let $\mathbf{J}_c$ be the conduction current. It reads

$$\frac{1}{\mu}\nabla\times\nabla\times\mathbf{A} = \mathbf{J}_c - \epsilon\frac{\partial}{\partial t}\left(\nabla V + \Pi\right). \tag{4.4}$$

In order to obtain unique solution we must impose a gauge condition. Here we use

$$\frac{1}{\mu}\nabla\left(\nabla\cdot\mathbf{A}\right) + \xi\varepsilon\nabla\left(\frac{\partial}{\partial t}V\right) = 0, \tag{4.5}$$

where $0 \leq \xi \leq 1$ is a free parameter.

Remark 4.1. The Coulomb gauge is obtained with $\xi = 0$ and the Lorenz gauge with $\xi = 1$ as special cases.

Adding the gauge condition (i.e. zero to the right-hand side of (4.4), performing an integration over a dual-surface area ΔS of a link $\langle ij \rangle$ and multiplying the result with the length $L = h_{ij}$ of the link under consideration gives with $\mathbf{J}_c = \sigma\mathbf{E}$

$$\epsilon L\frac{\partial}{\partial t}\int_{\Delta S}d\mathbf{S}\cdot\Pi = -L\int_{\Delta S}d\mathbf{S}\cdot\nabla\times\left(\frac{1}{\mu}\nabla\times\mathbf{A}\right) + L\int_{\Delta S}d\mathbf{S}\cdot\frac{1}{\mu}\nabla\left(\nabla\cdot\mathbf{A}\right)$$

$$-L\int_{\Delta S}d\mathbf{S}\cdot\sigma\nabla V - L\int_{\Delta S}d\mathbf{S}\cdot\sigma\Pi$$

$$-\epsilon L\int_{\Delta S}d\mathbf{S}\cdot\frac{\partial}{\partial t}\left(\nabla V\right) + \xi\epsilon L\int_{\Delta S}d\mathbf{S}\cdot\nabla\left(\frac{\partial}{\partial t}V\right). \tag{4.6}$$

The discretization of each term will now be discussed. Starting at the left-hand side, we define a link variable Π_{ij} for the link going from node i to node j. The surface integral is approximated by taking Π constant over the dual area. Thus

$$\epsilon L\frac{\partial}{\partial t}\int_{\Delta S}d\mathbf{S}\cdot\Pi \simeq \epsilon L\,\Delta S_{ij}\frac{d\Pi_{ij}}{dt}. \tag{4.7}$$

We can assign to each link a volume being $\Delta v_{ij} = L\,\Delta S_{ij}$.

Remark 4.2. Note that $\Delta v_{ij} \neq \Delta w_{ij}$, since Δv_{ij} is the volume corresponding to the area of a dual surface multiplied with the length of a primary-mesh link whereas Δw_{ij} is a dual volume of a primary-mesh node.

The first term on the right-hand side is dealt with using Stoke's theorem twice in order to evaluate the circulations

$$-L\int_{\Delta S}d\mathbf{S}\cdot\nabla\times\left(\frac{1}{\mu}\nabla\times\mathbf{A}\right) = -L\oint_{\partial(\Delta S)}d\mathbf{l}\cdot\left(\frac{1}{\mu}\nabla\times\mathbf{A}\right). \tag{4.8}$$

The circumference $\partial(\Delta S)$ consists of N segments. Each segment corresponds to a dual link that pierces through a *primary*-mesh surface. Therefore, we may approximate the right-hand side of (4.8) as

$$-L\oint_{\partial(\Delta S)}d\mathbf{l}\cdot\left(\frac{1}{\mu}\nabla\times\mathbf{A}\right) = -L\sum_{k=1}^{N}\Delta l_k\frac{1}{\mu_k}\left(\nabla\times\mathbf{A}\right)_k. \tag{4.9}$$

where the sum goes over all primary-mesh surfaces that were identified above as belonging to the circulation around the starting link. Note that we also attached an index on μ. This will guarantee that the correct value is taken depending in which material the segment Δl_k is located. Next we must obtain an appropriate expression for $(\nabla \times \mathbf{A})_k$. For that purpose, we consider the primary-mesh surfaces. In particular, an approximation for this expression is found by using

$$(\nabla \times \mathbf{A})_k \simeq \frac{1}{\Delta S_k} \int_{\Delta S_k} d\mathbf{S} \cdot \nabla \times \mathbf{A} = \frac{1}{\Delta S_k} \oint_{\partial(\Delta S_k)} d\mathbf{l} \cdot \mathbf{A}. \qquad (4.10)$$

The last contour integral is evidently replaced by the collection of primary-mesh links variables around the primary-mesh surface. As a consequence, the first term at the right-hand side of (4.6) becomes

$$-L \sum_{k=1}^{N} \Delta l_k \frac{1}{\mu_k} \frac{1}{\Delta S_k} \left(\sum_{l=1}^{N'} \Delta l_{\langle kl \rangle} A_{\langle kl \rangle} \right), \qquad (4.11)$$

where we distinguished the link labeling from node labeling (ij) to surface labeling $\langle kl \rangle$.

Next we consider the second term of (4.6). Now we use the fact that each link has a specific *intrinsic* orientation from 'front' to 'back' that was earlier set equal to $\mathbf{e}$,

$$L \int_{\Delta S} d\mathbf{S} \cdot \frac{1}{\mu} \nabla (\nabla \cdot \mathbf{A}) \simeq \int_{\Delta S} d\mathbf{S} \cdot \frac{1}{\mu} (\nabla \cdot \mathbf{A})_{back} - \int_{\Delta S} d\mathbf{S} \cdot \frac{1}{\mu} (\nabla \cdot \mathbf{A})_{front}. \qquad (4.12)$$

The two terms in (4.12) are now discretized as

$$\int_{\Delta S} d\mathbf{S} \cdot \frac{1}{\mu} (\nabla \cdot \mathbf{A}) = \frac{\Delta S}{\mu \Delta v} \int_{\Delta v} dv \nabla \cdot \mathbf{A}$$
$$= \frac{\Delta S}{\mu \Delta v} \oint_{\partial(\Delta v)} d\mathbf{S} \cdot \mathbf{A}$$
$$= \frac{\Delta S}{\mu \Delta v} \sum_{j}^{n} \Delta S_{ij} A_{ij},$$

where the sum is now from the front or back node to their corresponding neighbour nodes. The boundary conditions enter this analysis in a specific way. Suppose the front or back node is on the surface of the simulation domain. Then the closed surface integral around such a node will require a dual area contribution from a dual area outside the simulation domain. These surfaces are by definition not considered. However, we can return to the gauge condition and use

$$\int_{\Delta S} d\mathbf{S} \cdot \frac{1}{\mu} (\nabla \cdot \mathbf{A}) = -\xi \Delta S \, \epsilon \frac{\partial V}{\partial t}. \tag{4.13}$$

At first sight this looks weird: First we insert the gauge condition to get rid of the singular character of the curl-curl operation and now we 'undo' this for nodes at the surface. This is however fine because for the Dirichlet boundary conditions for $\mathbf{A}$ there are no closed circulations around primary surfaces and there is no uniqueness problem and therefore the double circulation operator is well defined.

The next two terms are rather straightforward. For the third term we consider ∇V constant over the dual surface. Thus we obtain

$$-L \int_{\Delta S} d\mathbf{S} \cdot \sigma \nabla V = (V_{front} - V_{back}) \left(\sum \Delta S_i \sigma_i \right). \tag{4.14}$$

The variation of σ is taken into account by looking at each volume contribution separately. The fourth term can be dealt with in a similar manner

$$-L \int_{\Delta S} d\mathbf{S} \cdot \sigma \Pi = L \, \Pi_{ij} \left(\sum \Delta S_i \sigma_i \right). \tag{4.15}$$

4.2 Time Discretization

There exists several views to address above system of equations in the temporal regime. We can distinguish between the linear case (no semiconductors present) and the non-linear case (semiconductors included). In the latter case it is not useful to isolate the time differentiations from the equations. Therefore, we write the system of equations as a system of Differential-Algebraic Equations (DAE):

$$M * \frac{d}{dt} \mathbf{X}(t) + H(\mathbf{X}(t), t) + F * \mathbf{X}_{bc}(t) = 0, \tag{4.16a}$$

$$G(\mathbf{X}) = 0. \tag{4.16b}$$

Here, $\mathbf{G}$ is determined by the Gauss' equation. The other equations are collected in (4.16a). $\mathbf{X}$ is the vector of unknowns $(V, \mathbf{A}, \Pi, n, p)$ and $\mathbf{X}_{bc}(t)$ is the vector of boundary-condition values, which are coupled into the system via the operator F. We assumed that the boundary condition can be linked into the system using a linear operator. This is definitely the case for linear materials (insulators and metals) but this assumption must be revised for semiconductors.

The discretized Maxwell equations for conductor / insulator systems lead to a linear system

$$M * \frac{\mathrm{d}}{\mathrm{d}t}\mathbf{X}(t) + H * \mathbf{X}(t) + F * \mathbf{X}_{bc}(t) = 0, \tag{4.17a}$$

$$G(\mathbf{X}) = 0. \tag{4.17b}$$

This DAE will be coupled to the equations of the Modified Nodal Analysis (MNA) that describe the circuit's behaviour. This coupling is given in two ways: on one hand, the currents at the contacts of the elements described by the Maxwell equations are added to the Kirchoff's current law equation in the network equations, and on the other hand, the boundary conditions for these elements depend on the node potentials of the circuit. The coupled system could have the following form

$$E * \frac{\mathrm{d}}{\mathrm{d}t}d(\mathbf{Y},t) + b(\mathbf{Y},t) + c(\mathbf{Y}_{ext},t) = 0, \tag{4.18a}$$

$$g(\mathbf{X},\mathbf{Y},\frac{\mathrm{d}}{\mathrm{d}t}\mathbf{X},t) = 0, \tag{4.18b}$$

$$M * \frac{\mathrm{d}}{\mathrm{d}t}\mathbf{X} + H(\mathbf{X},t) + F * \mathbf{Y} = 0, \tag{4.18c}$$

$$\mathbf{G}(\mathbf{X},\mathbf{Y}) = 0. \tag{4.18d}$$

with $\mathbf{Y} = (e, j_L, j_V, j_M)$ being the network variables: all node potentials e, the currents j_L through inductances, the currents j_V through voltages sources and the currents j_M through the elements that are represented in detail by the discretized field equations. We refer to these elements as 'electromagnetic' or EM elements. Note that the list *includes* all currents entering and leaving the EM elements. As argued before, each contact of the EM element induces a current variable j_M^i.

Equation (4.18a) describes the circuit equations. We did put the external current and voltage sources into a separate function $c(\mathbf{Y}_{ext},t)$. These variables are the true external boundary conditions of the coupled system contrary to the variables j_M^i that are contributing to the set of unknowns.

The equations (4.18c) and (4.18d) describe the Maxwell equations corresponding to the field-solver (FS) problem. Here, we made the important assertion that (a sub set) of the MNA variables act as boundary-condition variables, i.e. $\mathbf{X}_{bc}(t) \subset \mathbf{Y}(t)$ The coupling between both is given by (4.18b). In more detail, equation system (4.18a) has the form

$$A_C \frac{\mathrm{d}}{\mathrm{d}t}q_C(A_C^{\mathrm{T}}e,t) + A_R g_R(A_R^{\mathrm{T}}e,t) + A_L j_L + A_V j_V + A_I i_s(t) = 0, \tag{4.19a}$$

$$\frac{\mathrm{d}}{\mathrm{d}t}\phi(j_L,t) - A_L^{\mathrm{T}}e = 0, \tag{4.19b}$$

$$A_V^{\mathrm{T}}e - v_s(t) = 0, \tag{4.19c}$$

where A_C, A_R, A_L, A_V and A_I are the element related reduced incidence matrices. They describe the branch-node relationships for capacitors, resistors, inductors, voltage sources and current sources respectively. The independent variable $t \in [t_a, t_b]$ represents the time.

The unknowns are the node potentials, excepting the mass node $e(t) : \mathbb{R} \to \mathbb{R}^{n_N - 1}$ and the currents through inductors and voltage sources $j_L(t) : \mathbb{R} \to \mathbb{R}^{n_L}$ and $j_V(t) : \mathbb{R} \to \mathbb{R}^{n_V}$ respectively, where $n_N - 1$ represents the number of nodes (excepting the mass node) in the directed graph associated to the circuit, n_L is the number of capacitive branches in this graph and n_V, the number of branches that correspond to voltage sources.

Equations (4.19a) are the Kirchoff's current law equations for the circuit and Equations (4.19b)-(4.19c) describe its inductors and voltage sources. In particular, Equations (4.19c) are the Kirchoff's voltage law equations for the voltage sources in the circuit.

An initial value $(t_a, e_a, j_{La}, j_{Va})$ for (4.19) is called *consistent* if there is a solution of (4.19) that fulfils $e(t_a) = e_a$, $j_L(t_a) = j_{La}$, $j_V(t_a) = j_{Va}$.

In order to bridge the communication between the world of circuit simulation, we explain the terminology that is used in both environments. The voltage, e, are known in the field solver as V provided that they are located at metal or insulator field nodes. If the the latter nodes are attached to semiconductor nodes, i.e. the node is intrinsic semi conductor or on a semiconductor/insulator interface, it is identified with the Fermi potential. For nodes that are located at the semiconductor/metal interface, the voltage e is identified with the voltage $V|_{metal}$. Note that the potential makes a discrete jump when moving from the metal into the semiconductor.

The coupling equations (4.18b) look like

$$g_{\text{FS} \to \text{MNA}} \left(A_M^{\text{T}} e, j_M, \mathbf{X}, \frac{\mathrm{d}}{\mathrm{d}t} \mathbf{X} \right) = 0. \tag{4.20}$$

The applied potential at the EM elements is given by $A_M^{\text{T}} e$, which constitute a subset of the variables $\mathbf{Y}$.

Furthermore , there exists a second coupling that informs us how the MNA system impacts the solutions from the field solver. In a first approach, we will use Dirichlet's boundary conditions. This implies that the applied voltages at the contacts are the linked to the field-solver degrees of freedom as the contact voltages

$$g_{\text{MNA} \to \text{FS}} (A_M^{\text{T}} e, \mathbf{X}) = 0. \tag{4.21}$$

This coupling is already accounted for by the third term in (4.18c).

In order to gain a more pictorial understanding of the coupled problem, we consider the combined vector $\mathbf{Z}$ of $\mathbf{X}$ and $\mathbf{Y}$.

$$\mathbf{Z} = \begin{bmatrix} \mathbf{X} \\ \mathbf{Y} \end{bmatrix}. \tag{4.22}$$

In this vector, $\mathbf{X} = \{V_1, V_2, ..., A_1, A_2..., \Pi_1, \Pi_2, ..., p_1, p_2, ..., n_1, n_2...\}$ represents all field degrees of freedom and $\mathbf{Y} = \{e_1, e_2, ..., j_1, j_2, , , , , , j_M^1, j_M^2, ...\}$ represents all MNA variables as identified above.

The integrated circuit-field-solver will address the following equation:

$$\frac{d}{dt}\mathcal{U}(\mathbf{Z},t) + \mathcal{V}(\mathbf{Z},t) + \mathcal{W}(\mathbf{Z}_{ext}) = 0, \tag{4.23}$$

$$\mathcal{U} = \begin{bmatrix} \mathcal{U}_{11} & \mathcal{U}_{12} \\ \mathcal{U}_{21} & \mathcal{U}_{22} \end{bmatrix}, \qquad \mathcal{V} = \begin{bmatrix} \mathcal{V}_{11} & \mathcal{V}_{12} \\ \mathcal{V}_{21} & \mathcal{V}_{22} \end{bmatrix}, \qquad \mathcal{W} = \begin{bmatrix} \mathcal{W}_{11} & \mathcal{W}_{12} \\ \mathcal{W}_{21} & \mathcal{W}_{22} \end{bmatrix}. \tag{4.24}$$

The entries of the $\mathcal{U}, \mathcal{V}$ and $\mathcal{W}$ are all functions of $\mathbf{Z}$ and t. At this stage, we would like to introduce a new notation. Whereas a matrix-vector multiplication by default refers to a linear operation, we are in need for allowing non-linear functions in given parts of the assemble of matrices.

Consider a collection of functions $\{F_{ij} | i = 1, ...n, j = 1, ...m\}$. The ij-th function has as input a vector $\mathbf{X}_j$ with dimension D_j and generates an output vector $\mathbf{Y}_i$ of dimension D_i. Thus

$$\mathbf{Y}_i = F_{ij}(\mathbf{X}_j). \tag{4.25}$$

The namings $\mathbf{X}$ and $\mathbf{Y}$ are *not* referring to the previous use of these vector names! The collection of functions can be grouped in a matrix.

$$\begin{bmatrix} \mathbf{Y}_1 \\ \mathbf{Y}_2 \\ \vdots \\ \mathbf{Y}_n \end{bmatrix} = \begin{bmatrix} F_{11}(\mathbf{X}_1) & F_{12}(\mathbf{X}_2) & \cdots & F_{1m}(\mathbf{X}_m) \\ F_{21}(\mathbf{X}_1) & F_{12}(\mathbf{X}_2) & \cdots & F_{1m}(\mathbf{X}_m) \\ \vdots & \vdots & & \vdots \\ F_{n1}(\mathbf{X}_1) & F_{n2}(\mathbf{X}_2) & \cdots & F_{nm}(\mathbf{X}_m) \end{bmatrix}. \tag{4.26}$$

We will now use an underlining of the indices if the function is non-linear.

Definition:

$$F_{\underline{ij}} * \mathbf{X}_j = F_{ij}(\mathbf{X}_j), \qquad \text{for a non-linear function.} \tag{4.27}$$

Using this notation equation (4.23) becomes (now $\mathbf{X}$ and $\mathbf{Y}$ refer the field variables and MNA variables)

$$\frac{d}{dt}\begin{bmatrix} \mathcal{U}_{11} & 0 \\ 0 & \mathcal{U}_{\underline{22}}(t) \end{bmatrix} * \begin{bmatrix} \mathbf{X} \\ \mathbf{Y} \end{bmatrix} + \begin{bmatrix} \mathcal{V}_{11} & \mathcal{V}_{12} \\ \mathcal{V}_{21} & \mathcal{V}_{22} \end{bmatrix} * \begin{bmatrix} \mathbf{X} \\ \mathbf{Y} \end{bmatrix} + \begin{bmatrix} 0 & 0 \\ 0 & \mathcal{W}_{\underline{22}} \end{bmatrix} * \begin{bmatrix} 0 \\ \mathbf{Y}_{ext} \end{bmatrix} = 0. \tag{4.28}$$

The entries in (4.28) were written down earlier as

$$\frac{d}{dt}\begin{bmatrix} M & 0 \\ 0 & E \cdot d \end{bmatrix} * \begin{bmatrix} \mathbf{X} \\ \mathbf{Y} \end{bmatrix} + \begin{bmatrix} H & F \\ g_{FS \to MNA} & b(t) \end{bmatrix} * \begin{bmatrix} \mathbf{X} \\ \mathbf{Y} \end{bmatrix} + \begin{bmatrix} 0 & 0 \\ 0 & c \end{bmatrix} * \begin{bmatrix} 0 \\ \mathbf{Y}_{ext} \end{bmatrix} = 0. \tag{4.29}$$

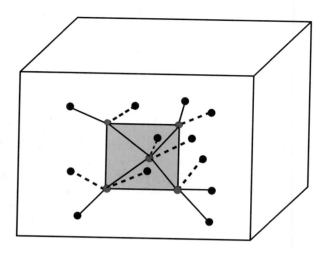

Contact plane (green) with contact nodes and connected nodes (black)
All the links to the contact nodes are also shown.

Fig. 4.1: Illustration of the contact nodes and links participating in the coupling.

4.2.1 How does the Function $g_{FS \to MNA}$ looks like?

In order to answer the question, we will consider a simple example and attempt to submit it to a combined circuit-field solving procedure. As a starting point we will will consider a simple one-winding prototype inductor and attach a small circuit to it. The inductor has two contacts (it is not considered as a multiple-port system but as an EM element that requires voltages as input and gives a current response. There are two contacts (one is visible as a red plate) to which we can attach a circuit. In the next section we will consider the equation construction in detail. The coupling that we want to describe is the between a contact current, j_M, a contact voltage e and a set of field degrees of freedom $\mathbf{X}$. Current continuity at the contact plane provides a connection between these variables. This is illustrated in Fig. 4.1.

The sum of the current into the contact plane must be adding up to zero. From the "inside" , i.e. from the field solvers perspective a collection of currents arrives at the contact plane. This current is a sum of all contributions from the links attached to the contact. Each link contributes the current $J_{ij}^{\text{cond}} + J_{ij}^{\text{displ}}$. Since the displacement current also contributes to the current, we obtain that the "inside" current is

$$I_{\text{inside}} = \sum_k w_k^1 \left(e_{\text{contact}} - V_k\right) + \sum_l \left(w_l^2 A_l + w_l^3 \Pi_l\right). \qquad (4.30)$$

where the factors w^i are determined by material parameters ϵ and σ and geometrical details and the sums go over the nodes and links that connected to the nodes of the contact planes.

From the MNA perspective we deal with a current variable j_M. The function g just expresses the continuity of the current

$$j_M + I_{\text{inside}} = 0. \qquad (4.31)$$

4.2.2 A Modified-Nodal Analysis Approach for Setting Up the Coupled System

The foregoing section was clearly composed starting from a field-solver 'language' and adding MNA equations to it. Of course there is no superiority in this approach. One can just as well start from an MNA perspective and add the field equation. This will be done in this section. Note that we list the variable: first MNA and next FS.

In order to treat the whole coupled system (4.18) by time integration, we rewrite it as a DAE with the following structure

$$A \frac{\mathrm{d}}{\mathrm{d}t} d(Y,t) + b(Y,t) = 0, \qquad (4.32)$$

where Y is the vector of unknowns $Y = (e, j_L, j_V, j_M, X)$. The matrix A is usually a singular but constant matrix. As shown in the DAE should have a properly stated term in order to avoid time stepsize reductions because of certain stability problems. It means, we should formulate A and d such that

$$\ker A \cap \operatorname{im} D = \{0\}.$$

with $D(Y,t) := \partial_Y d(Y,t)$.

For example, the DAE (4.19) can be written as a DAE with the form (4.32) with

$$A = \begin{pmatrix} A_C & 0 \\ 0 & I \\ 0 & 0 \end{pmatrix}, \qquad d(e, j_L, j_V) = \begin{pmatrix} q_C(A_C^{\mathrm{T}} e, t) \\ \phi(j_L, t) \end{pmatrix},$$

and

$$b(e, j_L, j_V) = \begin{pmatrix} A_R g_R(A_R^{\mathrm{T}} e, t) + A_L j_L + A_V j_V + A_I i_s(t) \\ -A_L^{\mathrm{T}} e \\ A_V^{\mathrm{T}} e - v_s(t) \end{pmatrix}.$$

The coupled system (4.18) can be written as (4.32) with

$$A = \begin{pmatrix} A_C & 0 & 0 & 0 \\ 0 & I & 0 & 0 \\ 0 & 0 & 0 & 0 \\ 0 & 0 & I & 0 \\ 0 & 0 & 0 & \mathbf{M} \\ 0 & 0 & 0 & 0 \end{pmatrix}, \qquad d = \begin{pmatrix} q_C(A_C^T e, t) \\ \phi(j_L, t) \\ q_M(\mathbf{X}, A_M^T e, t) \\ \mathbf{X} \end{pmatrix},$$

and

$$b = \begin{pmatrix} A_R g_R(A_R^T e, t) + A_L j_L + A_V j_V + A_M j_M + A_I i_s(t) \\ -A_L^T e \\ A_V^T e - v_s(t) \\ g_M(A_M^T e, j_M, \mathbf{X}, t) \\ \mathbf{H}(t, \mathbf{X}, A_M^T e, j_M) \\ \mathbf{G}(\mathbf{X}, A_M^T e, j_M) \end{pmatrix}.$$

For the transient simulation of DAEs (4.32), we need an implicit time integration. The first method of choice are the Backward Differentiation Formulas (BDF). They do not produce discretization errors in the algebraic equations. Furthermore, one can get higher order of convergence without enlarging the system of equations. Note, the implicit Euler method is a special case of BDF having order 1.

4.2.3 Example of MNA of a Simple Circuit

In this paragraph we will illustrate the model analysis with a simple example. The purpose of this paragraph is to show the MNA 'in action'. In Fig. 4.2, we show the variables of a simple RC network.

The circuit has two nodes which are marked as red spots. Each nodes introduces a nodal voltage e_i into the nodal analysis. There a three currents in the circuit. Each branch generates its own current. The currents are i_{VS}, i_R, i_C. At the nodes we apply the Kirchhoff current laws (KCL):

- At node #1: $i_{VS} + i_R = 0$.
- At node #2: $-i_R + i_C = 0$.

Besides the KCL we have to set up the branch-constituent equations (BCE). There are three branches, leading to

- At branch for i_{VS}: $V_S = e_1$.
- At branch for i_R: $V_R = e_1 - e_2$.
- At branch for i_C: $V_C = e_2$.

Next we will need the voltage-current characteristics of the elements:

- For the resistor: $V_R = R\, i_R$.
- For the capacitance: $i_C = C \frac{dV_C}{dt}$.

The modified nodal analysis collects all equations for $\{i_i, e_i\}$. Here we have 5 unknowns. The full problem can be casted into the following form. Let $\mathbf{x}(t)$ be the collection of unknowns. Then the system is fully described by the following differential-algebraic equation:

$$\mathbf{A} * \frac{\mathrm{d}}{\mathrm{d}t}\mathbf{x}(t) + \mathbf{B} * \mathbf{x}(t) + \mathbf{f} = 0. \tag{4.33}$$

The entries in equation (4.33) are:

$$\mathbf{x}(t) = \begin{bmatrix} i_{VS} \\ i_R \\ i_C \\ e_1 \\ e_2 \end{bmatrix}, \quad \mathbf{f} = \begin{bmatrix} 0 \\ 0 \\ 0 \\ 0 \\ -V_S \end{bmatrix}, \quad \mathbf{A} = \begin{bmatrix} 0 & 0 & 0 & 0 & 0 \\ 0 & 0 & 0 & C & 0 \\ 0 & 0 & 0 & 0 & 0 \\ 0 & 0 & 0 & 0 & 0 \\ 0 & 0 & 0 & 0 & 0 \end{bmatrix}, \quad \mathbf{B} = \begin{bmatrix} 0 & -R & 0 & 1 & -1 \\ 0 & 0 & -1 & 0 & 0 \\ 1 & -1 & 0 & 0 & 0 \\ 0 & -1 & 1 & 0 & 0 \\ 0 & 0 & 0 & 1 & 0 \end{bmatrix}.$$
$$\tag{4.34}$$

We can easily eliminate the variables i_R and i_C leading to a system for 3 unknowns. On the contrary, we want to modify the problem formulation and *extend* the number of unknowns. For that purpose, we review Fig. 4.2. The current through the resistor will be described once from the left-node perspective and once from the right-node perspective. This is illustrated in

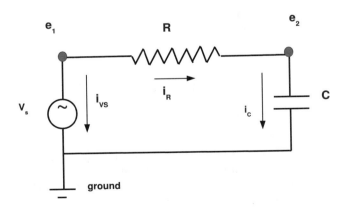

Fig. 4.2: Elementary circuit for illustrating MNA.

Fig. 4.3. The equation $i_R R + e_1 - e_2 = 0$, is complementary to the equation $i_L R + e_2 - e_1 = 0$. Of course, all that is added is a new trivial variable $i_L = -i_R$. However, although this knowledge is known for this simple example, before doing any computation, we can anticipate circumstances where such knowledge can only be obtained after an elaborate computation. For example, $i_L = i_R$ may be the result of a complicated resistive network representing R. The extension shows that part of the MNA can be an impedance matrix at some location in the net list

$$\begin{bmatrix} i_R \\ i_L \end{bmatrix} = \begin{bmatrix} 1 & -1 \\ -1 & 1 \end{bmatrix} * \begin{bmatrix} e_1 \\ e_2 \end{bmatrix} = \mathbf{Y} * \begin{bmatrix} e_1 \\ e_2 \end{bmatrix} \quad (4.35)$$

and where $\mathbf{Y}$ is the admittance matrix.

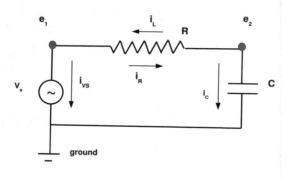

Fig. 4.3: Elementary circuit for illustrating MNA with nodal-perspective currents.

Thus from this elementary example we can already extract quite some insight how to couple circuit simulators and field solvers. For each contact pad of the field solver we assign a voltage and current variable. All these voltage and current variables will be loaded into the state vector of the modified-nodal analysis. Suppose that there are n contact pads. (They may be organised in $n/2$ ports.) Then there are $2n$ variables entering the MNA. We should not a priori demand that the total current is zero. This is obtained as a consequence of solution from the field solver.

4.2.4 BDF for DAEs

First, we explain the standard way of using k-step BDF methods when integrating DAEs in standard form

$$f(x',x,t) = 0, \quad t \in [t_0, t_F].$$

Suppose that the approximations $x_{n-j} \approx x(t_{n-j})$, $j = 1,2,\ldots,k$ have already been calculated. We denote $\tau_m = t_m - t_{m-1}$, $m = 1,2,\ldots$ as the m-th time step. An approximation x_n to $x(t_n)$ will be obtained by solving the nonlinear equation

$$f\left(\frac{1}{\tau_n} \sum_{j=0}^{k} \alpha_{j,n} x_{n-j}, x_n, t_n\right) = 0.$$

The BDF coefficients $\alpha_{0,n}, \alpha_{1,n}, \ldots, \alpha_{k,n}$ depend on the stepsizes $\tau_n, \tau_{n-1}, \ldots,$ τ_{n-k+1} (if the time stepsize is not constant). They are determined such that the derivative $x'(t_n)$ is approximated by

$$x'(t_n) \approx \frac{1}{\tau_n} \sum_{j=0}^{k} \alpha_{j,n} x_{n-j}.$$

with order k. In case of index-1 DAEs, the order of convergence of the BDF method equals k. In case of index-2 DAEs, one may lose one order of τ. In case of index-3 DAEs, the BDF method may fail completely. Therefore, we should check whether our coupled system has at most index 2 as the circuit equations have.

In order to obtain an approximation Y_n to the value of $Y(t_n)$, where $Y(t)$ is the exact solution of (4.32), we proceed in a similar way. In this case, the derivative $d'(Y(t_n), t_n)$ is replaced by a sum approximating it, Y_n is then the solution of the nonlinear equation

$$A\left(\frac{1}{\tau_n} \sum_{j=0}^{k} \alpha_{j,n} d(Y_{n-j}, t_{n-j})\right) + b(Y_n, t_n) = 0.$$

If e.g. the MNA equations are solved with BDF methods, the approximations $e_n, j_{L,n}$ and $j_{V,n}$ to $e(t_n), j_L(t_n), j_V(t_n)$, $n = 1,2,\ldots$ are the solution of the nonlinear system of equations

$$A_C \frac{1}{\tau_n} \left(\sum_{j=0}^{k} \alpha_{j,n} q_C(A_C^{\mathrm{T}} e_{n-j}, t_{n-j}) \right) + A_R g_R(A_R^{\mathrm{T}} e_n, t_n) +$$

$$A_L j_{L,n} + A_V j_{V,n} + A_I i_s(t_n) = 0,$$

$$\frac{1}{\tau_n} \left(\sum_{j=0}^{k} \alpha_{j,n} \phi(j_{L,n-j}, t_{n-j}) \right) - A_L^{\mathrm{T}} e_n = 0,$$

$$A_V^{\mathrm{T}} e_n - v_s(t_n) = 0.$$

In the special case of the implicit Euler method we have $k = 1$, $\alpha_{0,n} = 1$, $\alpha_{1,n} = 1$. The values $e_n, j_{L,n}$ and $j_{V,n}$ are then obtained by solving the following system of equations

$$A_C \frac{1}{\tau_n} \left(q_C(A_C^{\mathrm{T}} e_n, t_n) - q_C(A_C^{\mathrm{T}} e_{n-1}, t_{n-1}) \right) +$$

$$A_R g_R(A_R^{\mathrm{T}} e_n, t_n) + A_L j_{L,n} + A_V j_{V,n} + A_I i_s(t_n) = 0,$$

$$\frac{1}{\tau_n} \left(\phi(j_{L,n}, t_n) - \phi(j_{L,n-1}, t_{n-1}) \right) - A_L^{\mathrm{T}} e_n = 0,$$

$$A_V^{\mathrm{T}} e_n - v_s(t_n) = 0.$$

4.2.5 Coupled System Modified-Nodal Analysis and Discretized Drift-Diffusion Equations

Suppose the Drift-Diffusion (DD) equations for describing the semiconductor devices in an electrical circuit are discretized in space with e.g., the Scharfetter-Gummel discretization method.

The DAE associated to the coupled system of discretized Drift-Diffusion equations and MNA equations has the general form

$$A_C \frac{\mathrm{d}}{\mathrm{d}t} q_C(A_C^{\mathrm{T}} e, t) + A_R g_R(A_R^{\mathrm{T}} e, t) +$$

$$A_L j_L + A_V j_V + A_S j_S + A_I i_s(t) = 0, \tag{4.36a}$$

$$\frac{\mathrm{d}}{\mathrm{d}t} \phi(j_L, t) - A_L^{\mathrm{T}} e = 0, \tag{4.36b}$$

$$A_V^{\mathrm{T}} e - v_s(t) = 0, \tag{4.36c}$$

$$j_S + g_S(A_S^{\mathrm{T}} e, \Psi, N, P) + \frac{\mathrm{d}}{\mathrm{d}t} q_S(\Psi, A_S^{\mathrm{T}} e, t) = 0, \tag{4.36d}$$

$$T\Psi + D(C - N + P) = 0, \tag{4.36e}$$

$$DN' + j_N(A_S^{\mathrm{T}} e, \Psi, N) + R(N, P) = 0, \tag{4.36f}$$

$$DP' - j_P(A_S^{\mathrm{T}} e, \Psi, P) + R(N, P) = 0. \tag{4.36g}$$

The matrix A_S is an incidence matrix, j_S denotes the currents at the semiconductor's contacts. The potentials applied at the semiconductor's contacts depend on $A_S^T e$ (that's why (4.36d)-(4.36g) depend on it). The matrices T and D are non-singular and constant in time. They depend on the spatial mesh. $\Psi(t), N(t)$ and $P(t)$ are for each t approximations to the values of the electrostatic potential and the electrons and holes densities respectively on the mesh points. Equation (4.36e) is the discretized Poisson-equation, while (4.36f)-(4.36g) are the discretized continuity equations for the electrons and holes densities. In equation (4.36d) the current at the semiconductor's contacts is calculated, it is the sum of the current caused by electrons and holes $g_S(A_S^T e, \Psi, N, P)$ and the displacement current $\frac{d}{dt} q_S(\Psi, A_S^T e, t)$. This DAE can be written as a DAE with properly stated leading term with

$$A = \begin{pmatrix} A_C & 0 & 0 & 0 & 0 \\ 0 & I & 0 & 0 & 0 \\ 0 & 0 & 0 & 0 & 0 \\ 0 & 0 & I & 0 & 0 \\ 0 & 0 & 0 & 0 & 0 \\ 0 & 0 & 0 & I & 0 \\ 0 & 0 & 0 & 0 & I \end{pmatrix}, \quad d = \begin{pmatrix} q_C(A_C^T e, t) \\ \phi(j_L, t) \\ q_S(\Psi, A_S^T e, t) \\ DN \\ DP \end{pmatrix}$$

and

$$b = \begin{pmatrix} A_R g_R(A_R^T e, t) + A_L j_L + A_V j_V + A_S j_S + A_I i_s(t) \\ -A_L^T e \\ A_V^T e - v_s(t) \\ j_S + g_S(A_S^T e, \Psi, N, P) \\ T\Psi + D(C - N + P) \\ j_N(A_S^T e, \Psi, N) + R(N, P) \\ -j_P(A_S^T e, \Psi, P) + R(N, P) \end{pmatrix}$$

The index of this DAE is always less or equal to two. If it is solved with the k-steps BDF method, approximations $e_n, j_{L,n}, j_{V,n}, j_{S,n}, \Psi_n, N_n$ and P_n to $e(t_n), j_L(t_n), j_V(t_n), j_S(t_n), \Psi(t_n), N(t_n)$ and $P(t_n)$ are obtained by solving the following system of nonlinear equations

$$A_C \frac{1}{\tau_n} \left(\sum_{j=0}^{k} \alpha_{j,n} q_C(A_C^{\mathrm{T}} e_{n-j}, t_{n-j}) \right) + A_R g_R(A_R^{\mathrm{T}} e_n, t_n) +$$

$$A_L j_{L,n} + A_V j_{V,n} + A_S j_{S,n} + A_I i_s(t_n) = 0,$$

$$\frac{1}{\tau_n} \left(\sum_{j=0}^{k} \alpha_{j,n} \phi(j_{L,n-j}, t_{n-j}) \right) - A_L^{\mathrm{T}} e_n = 0,$$

$$A_V^{\mathrm{T}} e_n - v_s(t_n) = 0,$$

$$j_{S,n} + g_S(A_S^{\mathrm{T}} e_n, \Psi_n, N_n, P_n) + \frac{1}{\tau_n} \sum_{j=0}^{k} \alpha_{j,n} q_S(\Psi_{n-j}, A_S^{\mathrm{T}} e_{n-j}, t_{n-j}) = 0,$$

$$T\Psi_n + D(C - N_n + P_n) = 0,$$

$$D \frac{1}{\tau_n} \sum_{j=0}^{k} \alpha_{j,n} N_{n-j} + j_N(A_S^{\mathrm{T}} e_n, \Psi_n, N_n) + R(N_n, P_n) = 0,$$

$$D \frac{1}{\tau_n} \sum_{j=0}^{k} \alpha_{j,n} P_{n-j} - j_P(A_S^{\mathrm{T}} e_n, \Psi_n, P_n) + R(N_n, P_n) = 0.$$

4.3 Stability Analysis

It is a well-known fact that electrical systems containing resistors will respond to transient signals in such a way that when the stimulus stops at some time instant the electromagnetic fields will gradually decay due to two physical mechanisms. First of all the resistances convert electrical energy into heat such that the electric energy decreases. Secondly for open systems there is radiation loss which also results into the situation that the electrical energy decays when time proceeds.

When constructing a transient simulation tool of electrical systems it is required that the basic fact of above energy decay mechanism is mimicked by the simulator. For circuit simulation tools this fact is easily reproduced because the circuit equations that contain resistors are in general stable. We can identify stability as a property of the circuit equations in the following way.

Let $\mathbf{X}(t)$ be the collection of all system or circuit variables. The complete system of circuit equations is given by the state-space equations

$$E \frac{d}{dt} \mathbf{X} + A\mathbf{X} = 0. \tag{4.37}$$

If E is a non-singular matrix we may rewrite (4.37) as

$$\frac{d}{dt}\mathbf{X} + J\mathbf{X} = 0, \quad J = E^{-1}A. \tag{4.38}$$

Stability corresponds to the property of J that all its eigenvalues have real parts larger than or equal zero. If some eigenvalues have real part less than zero, the system has modes that explode when time proceeds which conflicts the energy conservation law and the system is therefore unphysical. Of course if it is impossible to create initial conditions such that when decomposed into the eigenvector base there are no components corresponding to negative real-part eigenvalues one may conclude that these modes will neither develop in the future and therefore the formulation is physical acceptable. Unfortunately this does not mean that if such a formulation of the system equations exists, e.g. J has negative real-part eigenvalues but the initial condition projected onto the negative real-part eigenvectors is empty, the simulation set up is physically save. While the transient time steps accumulate, numerical noise can mix into the transient solution and after some time leap the solution can still explode and yet becomes physically unacceptable. This was nicely demonstrated in [1].

As is seen from (4.38), the stability criterion could be straightforwardly connected to the formal solution

$$\mathbf{X}(t) = \mathbf{X}(0)\mathrm{e}^{-Jt}. \tag{4.39}$$

This solution is easily obtained because the system equations are *first* order in time. When the Maxwell-Ampere equations are considered we must account for the wave-like solutions and these equations are *second* order in time. The stability analysis must be revised. The Maxwell-Ampère equations in the potential formulation are given in [4]. By introducing the variable $\Pi = \partial_t \mathbf{A}$ the second-order system of equations is converted to first-order. Of course this step does not change the characteristic features of the solution, but it makes the system accessible to regular stability analysis.

Of critical importance are the details of the implementation of the gauge condition. In order to make the double-curl operator more Laplacian-like we added the gauge condition to this equation. There are two terms that contain a mixture of a spatial and a time derivative, i.e.

$$-\frac{\partial}{\partial t}(\epsilon \nabla V) \quad \text{and} \quad \xi \nabla \left(\epsilon \frac{\partial V}{\partial t} \right). \tag{4.40}$$

It turns out that these terms need a different discretization based on the origin of appearance in the Maxwell-Ampère equation. The first term in (4.40) needs to be discretized as is done for the term $\frac{\partial}{\partial t}(-\mu_0 \epsilon \Pi)$ The discretization is based on the finite-*surface* integration, whereas the second term in (4.40) needs to be discretized as is done for the term $\nabla(\nabla \cdot \mathbf{A})$. The latter is discretized using the finite-*volume* discretization. We observed that dealing with both terms using the finite-volume discretization leads to an unstable

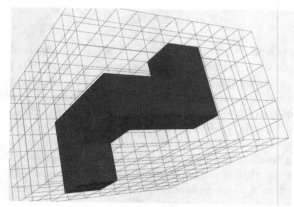

Fig. 4.4: Twisted bar used for computing the spectrum of the matrix that determines the system stability.

discretized formulation of the Maxwell-Ampère system. This is demonstrated in the following numerical example. In Fig. 4.4, a twisted bar is shown and a coarse mesh is used. This allows us to do a detailed eigenvalue analysis of the discretized system. In Fig. 4.5 left panel, the spectrum of the matrix J is shown based on a finite-volume implementation of both terms in (4.6). In Fig. 4.5 right panel, the spectrum of J is shown where J results from a discretization of (4.40) using the finite-*surface* integration method for the left term and keeping the discretization of the right term unaltered.

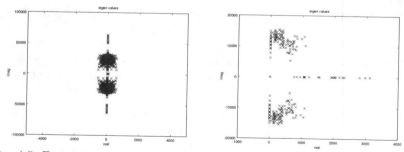

Fig. 4.5: Zoom-in to the eigenvalue spectrum around the real axis using exclusively finite-volume discretization for terms containing a mixed temporal and spatial differentiation (left). Zoom-in to the eigenvalue spectrum around the real axis using finite-volume discretization and finite-surface integration for terms containing a mixed temporal and spatial differentiation (right).

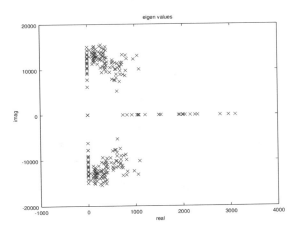

Fig. 4.6: Zoom-in to the eigenvalue spectrum around the real axis using finite-volume discretization and finite-surface integration for terms containing a mixed temporal and spatial differentiation.

We demonstrated that the conversion of continuous terms to discrete representative terms must account for the original motivation behind their presence.

4.3.0.1 Some Theoretical Considerations

Let us return to the Maxwell-Ampère equation (4.4). Since $\mathbf{J}_c = \sigma\left(-\nabla V - \Pi\right)$ and $\Pi = \partial_t \mathbf{A}$ we obtain

$$\epsilon \frac{\partial^2}{\partial t^2}\mathbf{A} + \sigma \frac{\partial}{\partial t}\mathbf{A} = M_{op}\mathbf{A} + \mathbf{J}_s, \qquad (4.41)$$

$$M_{op} = \nabla\left(\nabla \cdot \left(\frac{1}{\mu}\right)\right) - \nabla \times \left(\frac{1}{\mu}\nabla\times\right) \simeq \left(\frac{1}{\mu}\right)\nabla^2. \qquad (4.42)$$

Here M_{op} is a spatial differential operator and $\mathbf{J}_s$ is a source term. For a planar structure, the component A_z decouples from the equations system. Moreover the source term for this component is zero. The second order spatial derivative will lead to wave-like solutions. Consider the very simple one-DOF equation and k^2 is the result of the Laplace operator :

$$\epsilon \frac{\partial^2 x}{\partial t^2} + \sigma \frac{\partial x}{\partial t} + k^2 x = 0. \qquad (4.43)$$

There are solutions of the type $x(t) = x_0 \exp(\lambda t)$. Inserting this solution gives :

$$\epsilon\lambda^2 + \sigma\lambda + k^2 = 0. \tag{4.44}$$

The solutions of this equation for $\sigma > 0$ are:

$$\lambda_{1,2} = -\frac{\sigma}{2\epsilon}\left(1 \pm \sqrt{1 - \frac{4\epsilon k^2}{\sigma^2}}\right). \tag{4.45}$$

This can not lead to an unstable eigenvalue since the argument of the square root is a number less than one. If the argument is less than zero we get wave-like solutions. Our observation critically depends on the assumption that the 'Laplace operator' M_{op} gives rise to $k^2 \geq 0$. Unstable eigenvalues can arise if M_{op} gives rise to negative eigenvalues for the imposed boundary conditions. It should also be noted that if $\sigma = 0$ then the eigenvalues become purely imaginary.

4.4 The Impact of Meshing

In this section we consider a structure with contacts at the edge of the simulation domain. The structure and its mesh are shown in Fig. 4.7. The Manhat-

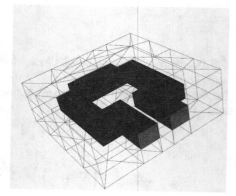

Fig. 4.7: Test structure: 2D view (left) and 3D view (right).

tan meshing gives rise to to an eigenvalue spectrum, which has no negative real-parts. However, when using $2D$ - Delaunay meshing, we find that the spectrum has severe negative real-part eigenvalues: (-1.92 10^{10} + 0.0i), (-4.92 + 4.26 10^3i), (-4.92 - 4.26 10^3i). As is seen in Fig. 4.7 left panel, some cells have obtuse angles. This will lead to negative dual areas. One may modify the meshing algorithm by assigning a dual volume to each node in each cell by starting from the center of gravity for the surfaces of the cell and the cell volume. Using this modified method of obtaining dual volumes and dual areas, the negative real-part eigenvalues are removed again.

Converting the Maxwell-Ampère equations that are second-order in time differentiation into equations that are first-order in time differentiation, the standard techniques for stability consideration become applicable. We found that the discretization of each term must be done in accordance with the original motivation of the appearance of the term. Carelessly swapping temporal and spatial differential operators may quickly lead to erroneous discretization set up. We also noted that a stable implementation requires that the discrete Laplace operator must be implemented such that its continuous spectrum property, i.e. semi-definiteness must be preserved.

References

1. BAUMANNS, S., CLEMENS, M., AND SCHÖPS, S.: *Structural Aspects of Regularized Full Maxwell Electrodynamic Potential Formulations Using FIT*. In: Proceedings of the 2013 International Symposium on Electromagnetic Theory, ISET, 24PM1C-01 pp. 1007–1010, 2014.
2. SCHOENMAKER, W.: *Computational Electrodynamics – A Gauge Approach with Applications in Microelectronics*. The River Publishers Series in Electronic Materials and Devices. ISBN: 9788793519848. River Publishers, Gistrup, Denmark, 2017.
3. SCHOENMAKER, W., BRACHTENDORF, H.-G., BITTNER, K., TISCHENDORF, C., AND STROHM, C.: *EM-Equations, Coupling to Heat and to Circuits*. Chapter in this Book, 2019.
4. SCHOENMAKER, W., CHEN, Q., AND GALY, P.: *Computation of Self-Induced Magnetic Field Effects Including the Lorentz Force for Fast-Transient Phenomena in Integrated-Circuit Devices*. IEEE Transactions on Computer-Aided Design of Integrated Circuits and Systems, **33**, pp. 893–902, 2014.

Chapter 5
Automated Generation of Netlists from Electrothermal Field Models

Thorben Casper, David J. Duque Guerra, Sebastian Schöps,
Herbert De Gersem

Abstract The equivalence between field and circuit models is shown and formalised. Accordingly, an algorithm for the automated generation of circuit netlists out of field models is presented. Such netlists can be used by standard circuit simulators. The field-to-netlist extraction is organised for coupled electrothermal field models. The resulting circuit model is an exact representation of the semi-discrete field model. A 3D electrothermal field model of a microelectronic chip package serves as an illustration of the overall procedure.

5.1 Introduction

Contemporary electronic designs heavily rely upon simulation for predicting device performance. Closed-form models are only applicable in very rare situations. The majority of electronic designs is based on circuit simulation, thereby incorporating the individual parts of the device by lumped parameters and groups of lumped parameters in an overall network. Standard tools such as PSPICE [33], Xyce [35] and Cadence® Spectre® [9] have revolutionised the electronics industry. The accuracy of circuit simulations depends on the accuracy of the lumped-parameter representation of the device components. Here, one has to rely upon hand calculations or fitting which are not capable of resolving nonlinear effects and field inhomogeneities. Many manufacturers provide models for their devices, albeit for a limited operation range, e.g., under the assumption of a constant operation temperature.

When field effects come into play, it may make more sense to use a field solver which solves the Maxwell equations and the heat equation on a com-

Thorben Casper, David J. Duque Guerra, Sebastian Schöps, Herbert De Gersem
Technische Universität Darmstadt, Germany, e-mail: {Casper,Duque,Schoeps}@gsc.
tu-darmstadt.de;DeGersem@temf.tu-darmstadt.de

© Springer Nature Switzerland AG 2019
E. J. W. ter Maten et al. (eds.), *Nanoelectronic Coupled Problems Solutions*,
Mathematics in Industry 29, https://doi.org/10.1007/978-3-030-30726-4_5

putational grid. For that purpose, the Finite-Difference (FD), Finite-Element (FE) and Finite-Volume (FV) methods have become standard. Also for field simulation, powerful tools such as, e.g., CST Design Studio^{TM} [18], ANSYS HFSS [2] and EMPro [27] are available on the market.

In the beginning era of circuit simulatio, the Simulation Program with Integrated Circuit Emphasis (SPICE) soon arose as the standard tool for describing and solving circuits [32]. The description of circuits was and is organised in so-called *netlists*. Electrothermal effects were, however, not considered. Merely a constant operating temperature could be imposed. Especially with the emergence of power electronic applications, strongly temperature dependent materials and components needed to be considered [28, 31]. Self-heating effects and thermal couplings between the devices became relevant and the thermal load for different operation modes became a design issue. Therefore, a SPICE extension enabling a simultaneous simulation of electrical and thermal circuits was introduced [44]. The coupling was organised along an extra temperature node [23, 30]. Hybrid approaches exist where electric circuit simulation is accomplished with 3D thermal field simulation [12, 34, 48]. Alternatively, mesh-based equivalent thermal circuits [40] are constructed, such that electric and thermal circuits can be coupled without interaction along a 3D field solver.

There is a growing interest in coupling field and circuit solvers. Circuit simulators typically allow embedding look-up tables with data obtained by field solvers in an a priori off-line step. Circuit simulators also provide the possibility to insert state-space models which have been obtained on beforehand, e.g., with model order reduction techniques [4, 20, 47]. More and more circuit simulators also allow a connection to external routines that describe the behaviour of a black box part of the circuit. With that methodology, a field solver can be called from the circuit solver in a master-slave relation [37]. The other way around, the above mentioned field simulators allow to couple a few circuit elements into the field model and thereby enable the description of device parts for which a field model would be prohibitively complicated [5, 39]. Although apparently easy, field-circuit coupling comes with a price. Algebraic system solvers encounter problems with the resulting hybrid systems of equations. Largely different time scales can only be overcome with multi-rate time integrators [6, 21]. Some field-circuit couplings are unsolvable unless part of the models are preprocessed using model order reduction. These problems further explode when field-circuit couplings have to be organised for an electrothermal problem, especially when the electrothermal coupling has to be expressed both at the field as well as at the circuit level.

Following [11], this chapter shows the equivalence between field and circuit models. Accordingly, we provide a recipe to translate field models into the circuit formalism. More precisely, we offer a procedure to extract a netlist from a nonlinear field model. At first, this seems to provide no more than a translation. However, expressing field models by netlists allows to embed them in the most powerful circuit simulators available at the time. The field solver is exploited to discretize the problem in space. After netlist generation, the circuit solver deals with the electrothermal coupling, the nonlinearities and the time integration. Hence, this procedure exploits the spatial accuracy of the electrothermal field model and enables its representation at the circuit level after an automated and loss-free extraction of an equivalent electrothermal circuit model (Fig. 5.1). The approach keeps the possibility to couple with additional (external) SPICE circuitries and preserves the possibility to apply standard techniques for model order reduction to reduce the eventual networks [24].

5.2 Electrothermal Field Problem

In the considered models, electric and thermal conductive as well as capacitive effects are relevant. Moreover, the electric current causes a temperature increase due to Joule heating, whereas increasing temperatures cause the electric conductivity to decrease. This behaviour is described by the electro-quasistatic approximation to Maxwell's equations in combination with the heat equation, i.e. [16, 19],

$$-\nabla \cdot \left(\varepsilon \nabla \frac{\partial \varphi}{\partial t}(t) \right) - \nabla \cdot (\sigma(T)\nabla\varphi(t)) = 0, \tag{5.1}$$

$$\rho c \frac{\partial T}{\partial t} - \nabla \cdot (\lambda(T)\nabla T(t)) = Q_{\mathrm{el}}(\varphi), \tag{5.2}$$

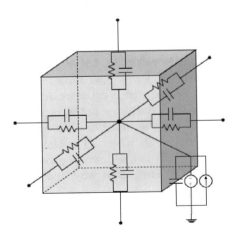

Fig. 5.1: Visualisation of the approach to extract lumped elements from a 3D field model.

where φ is the electric scalar potential, T is the temperature, $Q_{\mathrm{el}}(\varphi) = \sigma(T)\nabla\varphi \cdot \nabla\varphi$ is the Joule heating term, ε is the permittivity, $\sigma(T)$ is the electric conductivity depending on the temperature, ρ is the volumetric mass density, c is the specific heat capacity, $\lambda(T)$ is the temperature dependent thermal conductivity and t denotes the time. The materials may be non-linear, inhomogeneous or anisotropic. Equations (5.1) and (5.2) need to be accompanied by appropriate boundary and initial conditions for $\varphi(\mathbf{r},t)$ and $T(\mathbf{r},t)$.

5.3 Finite Integration Technique

5.3.1 Grid Topology

The formulation expressed by (5.1) and (5.2) is treated on a mutually orthogonal grid pair. For conciseness, the method is sketched for a 3D hexahedral but not necessarily equidistant primary grid and a corresponding dual grid obtained by connecting the midpoints of the primary cells. The Finite Integration Technique (FIT) uses a canonical indexing of the n primary nodes which is carried over to the incident primary edges, primary facets and primary cells and to the associated dual cells, dual facets, dual edges and dual nodes [13, 45, 46]. It is assumed that the material distribution coincides with the primary cells, up to a staircase discretisation error, i.e., the geometry error occurring at curved material interfaces when using a hexahedral grid. The electric potentials $\boldsymbol{\varphi} \in \mathbb{R}^n$ and the temperatures $\mathbf{T} \in \mathbb{R}^n$ are associated with the primary nodes (Fig. 5.2). By taking differences between adjacent nodes, we find the voltages and the temperature drops along the primary edges. This is formalised into $\widehat{\mathbf{e}} = -\mathbf{G}\boldsymbol{\varphi}$ and $\widehat{\mathbf{t}} = -\mathbf{G}\mathbf{T}$, where $\mathbf{G} \in \{-1,0,1\}^{3n \times n}$ is the discrete gradient matrix reflecting the topology of the primary grid. We associate electrical currents $\widehat{\widehat{\mathbf{j}}}$ and heat fluxes $\widehat{\widehat{\mathbf{q}}}$ to dual facets, each of which corresponds to the primary edge orthogonally crossing them. At the dual grid, one can express the divergence operator to enforce the continuity of the electric current and the thermal flux. This gives rise to expressions of the form $\widetilde{\mathbf{S}}\widehat{\widehat{\mathbf{j}}} = \mathbf{0}$ and $\widetilde{\mathbf{S}}\widehat{\widehat{\mathbf{q}}} = \mathbf{Q}$, where $\mathbf{Q}$ is the heat production in the dual cells and $\widetilde{\mathbf{S}} \in \{-1,0,1\}^{n \times 3n}$ is the discrete divergence matrix related to the topology of the dual grid. For a mutually dual grid pair, it holds that $\mathbf{G} = -\widetilde{\mathbf{S}}^\top$ [13, 45, 46].

For a tensor-product Cartesian grid as used here, the discrete gradient operator is found by $\mathbf{G} = \begin{bmatrix} \mathbf{P}_x & \mathbf{P}_y & \mathbf{P}_z \end{bmatrix}^\top$, where the matrices $\mathbf{P}_x$, $\mathbf{P}_y$ and $\mathbf{P}_z$ mimic the spatial derivatives in x-, y- and z-direction. When defining the primary edges to be oriented in positive x-, y- and z-direction, and assuming the canonical index of the primary edges to be carried over from the downstream primary nodes, one find that the $n \times n$ matrices $\mathbf{P}_\xi$ with $\xi \in \{x,y,z\}$ contain

−1 entries at their main diagonals and +1 entries at one of the above diagonals. The discrete gradient operator and the discrete divergence operator have many similarities to a non-reduced circuit incidence matrix and its negative transpose, i.e., they only have 0, 1 and −1 as entries and their column sums (or for the transposed matrices, their row sums) are zero [26,38,43].

5.3.2 Discrete Hodge Operators

In FIT, the spatial discretisation amounts to relating field quantities allocated at the primary grid to field quantities at the dual grid by the material parameters. In the differential-forms framework, one speaks about *discrete Hodge operators* or *constitutive equations* [1,3,8,42]. In an algebraic setting, we speak about material matrices. The relevant relations are here

$$\widehat{\widehat{\mathbf{j}}} = \mathbf{M}_\sigma \widehat{\mathbf{e}}, \tag{5.3}$$

$$\widehat{\widehat{\mathbf{d}}} = \mathbf{M}_\varepsilon \widehat{\mathbf{e}}, \tag{5.4}$$

$$\widehat{\widehat{\mathbf{q}}} = \mathbf{M}_\lambda \widehat{\mathbf{t}}, \tag{5.5}$$

$$\mathbf{Q} = \mathbf{M}_{\rho c} \frac{\partial \mathbf{T}}{\partial t}, \tag{5.6}$$

where $\widehat{\widehat{\mathbf{d}}}$ are the electric fluxes associated with the dual facets, $\mathbf{M}_\sigma$ is the electric conductance matrix, $\mathbf{M}_\varepsilon$ is the electric capacitance matrix, $\mathbf{M}_\lambda$ is the thermal conductance matrix and $\mathbf{M}_{\rho c}$ is the thermal capacitance matrix. Because of the one-to-one correspondence of primary edge (nodes) and dual facets (cells) and because of the mutual orthogonality of the grid pair, the material matrices are diagonal and have the entries

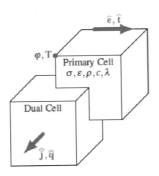

Fig. 5.2: Allocation of electrical and thermal quantities at the primary and dual grid.

$$\mathbf{M}_{\sigma;j,j} = \frac{\overline{\sigma}_j|\widetilde{A}_j|}{|L_j|}, \tag{5.7}$$

$$\mathbf{M}_{\varepsilon;j,j} = \frac{\overline{\varepsilon}_j|\widetilde{A}_j|}{|L_j|}, \tag{5.8}$$

$$\mathbf{M}_{\lambda;j,j} = \frac{\overline{\lambda}_j|\widetilde{A}_j|}{|L_j|}, \tag{5.9}$$

$$\mathbf{M}_{\rho c;i,i} = \overline{\rho c}_i|\widetilde{V}_i|, \tag{5.10}$$

where i indexes the primary nodes and dual cells, j indexes the primary edges and dual facets, $|L_j|$ is the length of the primary edge L_j, $|\widetilde{A}_j|$ is the area of the dual facet $\widetilde{A}_j$ and $|\widetilde{V}_i|$ is the volume of dual cell $\widetilde{V}_i$. The material parameters assumed to coincide with the primary cells are combined in order to obtain $\overline{\sigma}_j$, $\overline{\varepsilon}_j$, $\overline{\lambda}_j$ and $\overline{\rho c}_i$ averaged over the dual facets and dual volumes, respectively [10, 14]. As an example, consider a primary edge L_j surrounded by four primary cells with conductivities σ_1, σ_2, σ_3 and σ_4. For an equidistant grid, one then finds

$$\overline{\sigma}_j = \frac{1}{4}\sum_{p=1}^{4}\sigma_p. \tag{5.11}$$

Averaging in case of a non-equidistant grid would involve coefficient factors related to the cell sizes.

The FIT as explained here suffers from the staircase error introduced by associating the materials with the hexahedral primary grid cells. Many approaches have been proposed to at least partially alleviate the staircase error. Among them, partially filled cells [36], conforming techniques [15], subgridding [41] and cell methods [17] are the most popular. A mathematically precise description of these methods requires a lengthy development for which we refer to the cited literature. In particular, FIT can be understood as a finite element approach with simplified quadrature or mass lumping [7].

5.3.3 Electrothermal Coupling

The voltages $\widehat{\mathbf{e}}$ at the primary edges together with the electric currents at the associated dual facets are the origin of Joule heating. The heat powers are found by

$$\widehat{\mathbf{Q}}_{\mathrm{el}} = \widehat{\mathbf{e}} \odot \widehat{\widetilde{\mathbf{j}}}, \tag{5.12}$$

where $\odot$ denotes the component-wise Hadamard product. The vector $\widehat{\mathbf{Q}}_{\mathrm{el}} \in \mathbb{R}^{3n}$ can be interpreted as heat powers allocated at the *shifted cells* $\widehat{V}_j$ with volume $|\widehat{V}_j| = |\widetilde{A}_j||L_j|$ (Fig. 5.3). The shifted cells combine two half dual cells sharing a common dual facet and follow the indexing of these dual facets.

The heat powers at the dual cells are found by interpolation, i.e.,

$$Q_{\text{el},i} = \sum_{p=1}^{6} \frac{|\widetilde{V}_i|}{2|\widehat{V}_p|} \widehat{Q}_{\text{el},p}, \tag{5.13}$$

where p iterates over all shifted cells that intersect the dual cell i. The relations (5.13) are collected into the incidence matrix $\mathbf{P}_Q \in \{0,1\}^{n \times 3n}$. Then, the Joule losses $\mathbf{Q}_{\text{el}} \in \mathbb{R}^n$ read

$$\mathbf{Q}_{\text{el}} = \frac{1}{2} \widetilde{\mathbf{D}}_V \mathbf{P}_Q \widehat{\mathbf{D}}_V^{-1} \widehat{\mathbf{Q}}_{\text{el}}, \tag{5.14}$$

where $\widetilde{\mathbf{D}}_V \in \mathbb{R}^{n \times n}$ and $\widehat{\mathbf{D}}_V \in \mathbb{R}^{3n \times 3n}$ are diagonal matrices containing the dual volumes $|\widetilde{V}_i|$ and the shifted volumes $|\widehat{V}_j|$, respectively.

5.3.4 Discrete Electrothermal Formulation

Gathering all matrix operators, one finds the discrete counterpart of (5.1) and (5.2) to be

$$\widetilde{\mathbf{S}} \mathbf{M}_\varepsilon \widetilde{\mathbf{S}}^\top \frac{\mathrm{d}\boldsymbol{\varphi}}{\mathrm{d}t} + \widetilde{\mathbf{S}} \mathbf{M}_\sigma(\mathbf{T}) \widetilde{\mathbf{S}}^\top \boldsymbol{\varphi} = \mathbf{0}, \tag{5.15}$$

$$\mathbf{M}_{\rho c} \dot{\mathbf{T}} + \widetilde{\mathbf{S}} \mathbf{M}_\lambda(\mathbf{T}) \widetilde{\mathbf{S}}^\top \mathbf{T} = \mathbf{Q}_{\text{el}}(\boldsymbol{\varphi}). \tag{5.16}$$

The degrees of freedom are the electric scalar potentials $\boldsymbol{\varphi}(t)$ and the temperatures $\mathbf{T}(t)$ at the primary nodes. The boundary conditions still need to be inserted in (5.15) and (5.16). After spatial discretisation by FIT, we are left with a system of ordinary differential equations, which is further solved by an appropriate implicit time integration scheme.

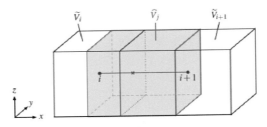

Fig. 5.3: Shifted volume $\widehat{V}_j$.

5.4 Circuit Theory as a Discretisation of Maxwell's Equations

The idea of this section is to derive the Kirchhoff Voltage Law (KVL), the Kirchhoff Current Law (KCL), the branch relations and finally the Modified Nodal Analysis (MNA) [22, 25] along the same lines as in the previous section about the Finite Integration Technique (FIT). This will illustrate the close correspondence between discrete field formulations and circuit analysis. The equivalences will then be exploited in the next section to turn a discrete field model into a network description.

5.4.1 Kirchhoff's Voltage Law (KVL) from Faraday's Law

When magnetic effects can be neglected, Faraday's law is reduced to

$$\oint_{\partial A} \mathbf{E} \cdot \mathrm{d}\mathbf{s} = 0. \tag{5.17}$$

Here, A is a surface whose contour consists of b branches L_j, i.e., $\partial A = \cup_{j=1}^{b} L_j$. When loops in an electric circuit are thought to coincide with loops in a primary grid, a direct resemblance is found. The integral form of Faraday's law corresponds to the sum of voltage drops V_j along the branches L_j giving

$$\oint_{\partial A} \mathbf{E} \cdot \mathrm{d}\mathbf{s} = \sum_{j=1}^{b} \int_{L_j} \mathbf{E} \cdot \mathrm{d}\mathbf{s} = \sum_{j=1}^{b} V_j = 0. \tag{5.18}$$

The key point in MNA is to enforce (5.18) by defining n nodal voltages v_i and expressing the branch voltages V_j by

$$\mathbf{V} = \mathbf{A}^{\top} \mathbf{v}, \tag{5.19}$$

with the circuit incidence matrix $\mathbf{A} \in \{-1, 0, 1\}^{n \times b}$. The rule that $a_{ij} = +1$ if branch L_j is directed away from node i and $a_{ij} = -1$ if branch L_j is directed towards node i is in direct correspondence to the definition of the FIT matrix $-\mathbf{G}$. The nodal voltages $\mathbf{v}$ are the mirror images of the electric potentials $\boldsymbol{\varphi}$ in the FIT formulation. In both cases, reference nodes need to be selected at which reference potentials are assigned to ensure a unique solution.

5.4.2 Kirchhoff's Current Law (KCL) from the Continuity Equation

By applying the divergence operator to Ampère's law, one finds the continuity equation

$$\int_{\partial V} \mathbf{J} \cdot d\mathbf{A} + \int_V \frac{\partial \varrho}{\partial t} \, dV = 0, \tag{5.20}$$

where $\varrho(\mathbf{r},t)$ is here the charge density and V an arbitrary volume. If we choose a volume $\widetilde{V}_i$ around a node i of the circuit and assume the total charge in $\widetilde{V}_i$ to be zero (which forces capacitive effects to be located at the circuit branches and thus outside $\widetilde{V}_i$), we find

$$\int_{\partial \widetilde{V}_i} \mathbf{J} \cdot d\mathbf{A} = 0. \tag{5.21}$$

When currents I_j are leaving $\widetilde{V}_i$ at a finite number s of conductors with cross-sectional areas $\widetilde{A}_j$, we come to the KCL given by

$$\sum_{j=1}^{s} I_j = \sum_{j=1}^{s} \int_{\widetilde{A}_j} \mathbf{J} \cdot d\mathbf{A} = 0. \tag{5.22}$$

In the MNA context, KCL is expressed using the circuit incidence matrix, i.e.,

$$\mathbf{A}\mathbf{I} = \mathbf{0}. \tag{5.23}$$

5.4.3 Modified Nodal Analysis

In an electrothermal setting, only electric resistances, electric capacitances and current sources, as well as thermal resistances, thermal capacitances and thermal heat sources are relevant. Moreover, here, we make a generic development valid for both the electric and the thermal case. The circuit consists of b_R resistive branches, b_C capacitive branches and b_I current sources. The circuit incidence matrix $\mathbf{A}$ is partitioned accordingly [22], i.e.,

$$\mathbf{A} = [\mathbf{A}_R \quad \mathbf{A}_C \quad \mathbf{A}_I], \tag{5.24}$$

where $\mathbf{A}_R \in \{-1,0,1\}^{n \times b_R}$, $\mathbf{A}_C \in \{-1,0,1\}^{n \times b_C}$ and $\mathbf{A}_I \in \{-1,0,1\}^{n \times b_I}$. The branch currents and the branch voltages are expressed similarly, i.e.,

$$\mathbf{I}^\top = \begin{bmatrix} \mathbf{I}_R^\top & \mathbf{I}_C^\top & \mathbf{I}_I^\top \end{bmatrix}, \tag{5.25}$$

$$\mathbf{V}^\top = \begin{bmatrix} \mathbf{V}_R^\top & \mathbf{V}_C^\top & \mathbf{V}_I^\top \end{bmatrix}. \tag{5.26}$$

where $\mathbf{I}_R, \mathbf{V}_R \in \mathbb{R}^{b_R}$, $\mathbf{V}_C, \mathbf{I}_C \in \mathbb{R}^{b_C}$ and $\mathbf{V}_I, \mathbf{I}_I \in \mathbb{R}^{b_I}$. The KVL, the KCL and the branch relations are now further specified by

$$\mathbf{V}_R = \mathbf{A}_R^\top \mathbf{v}, \quad \mathbf{V}_C = \mathbf{A}_C^\top \mathbf{v}, \quad \mathbf{V}_I = \mathbf{A}_I^\top \mathbf{v}, \tag{5.27}$$

$$\mathbf{A}_R \mathbf{I}_R + \mathbf{A}_C \mathbf{I}_C + \mathbf{A}_I \mathbf{I}_I = \mathbf{0}, \tag{5.28}$$

$$\mathbf{I}_R = \mathbf{G}\mathbf{V}_R, \quad \mathbf{I}_C = \mathbf{C}\dot{\mathbf{V}}_C \quad \text{and} \quad \mathbf{I}_I = \mathbf{I}_s(t), \tag{5.29}$$

where $\mathbf{G} \in \mathbb{R}^{b_R \times b_R}$ is a diagonal conductance matrix, $\mathbf{C} \in \mathbb{R}^{b_C \times b_C}$ is a diagonal capacitance matrix and $\mathbf{I}_I = \mathbf{I}_s(t)$ is a vector of source currents. Putting (5.27), (5.28) and (5.29) together, we find the system of equations according to the MNA, viz.

$$\mathbf{A}_C \mathbf{C} \mathbf{A}_C^\top \dot{\mathbf{v}} + \mathbf{A}_R \mathbf{G} \mathbf{A}_R^\top \mathbf{v} = -\mathbf{A}_I \mathbf{I}_s(t). \tag{5.30}$$

5.4.4 Equivalence between Electroquasistatic Field and Electric Circuit

The MNA formulation for the electric circuit is equivalent to the discrete electroquasistatic FIT formulation. This is directly obvious when comparing (5.15) and (5.16) with (5.30). Due to the matching conventions on the orientation of the edges, one can express the equivalence formally by

$$\mathbf{A}_C^{\text{el}} = \mathbf{A}_R^{\text{el}} = \widetilde{\mathbf{S}}, \quad \mathbf{C}^{\text{el}} = \mathbf{M}_\varepsilon, \quad \mathbf{G}^{\text{el}} = \mathbf{M}_\sigma,$$
$$\mathbf{I}_s^{\text{el}} = \mathbf{0} \text{ and } \mathbf{v}^{\text{el}} = \boldsymbol{\varphi}, \tag{5.31}$$

where the superscript ^{el} indicates the circuit quantities for the electroquasistatic case.

5.4.5 Equivalence between Thermal Field and Thermal Circuit

The Partial Differential Equation (PDE) (5.2) for the temperature differs from the PDE (5.1) for the electric scalar potential. In (5.2), the time derivative for the temperature appears as a mass term, whereas in (5.1), the time derivative for the electric scalar potential appears together with a Laplace operator. This structures is preserved in the discrete setting, i.e., compare $\mathbf{M}_{\rho c}\frac{\mathrm{d}\mathbf{T}}{\mathrm{d}t}$ in (5.16) to $\widetilde{\mathbf{S}}\mathbf{M}_\varepsilon\widetilde{\mathbf{S}}^\top\frac{\mathrm{d}\boldsymbol{\varphi}}{\mathrm{d}t}$ in (5.15). The underlying physical reasoning is that the amount of heat stored in a material depends on the difference between the temperature and a reference temperature and not on the temperature difference between two adjacent nodes. In thermal networks, this is expressed by connecting thermal capacitances to a common thermal ground node put at the reference temperature, here considered as node $n+1$. For that reason, the thermal MNA formulation looks slightly different from the electric one, i.e.,

$$\widehat{\mathbf{A}}_C \mathbf{C} \widehat{\mathbf{A}}_C^\top \dot{\widehat{\mathbf{v}}} + \widehat{\mathbf{A}}_R \mathbf{G} \widehat{\mathbf{A}}_R^\top \widehat{\mathbf{v}} = -\widehat{\mathbf{A}}_I \mathbf{I}_s(t) \tag{5.32}$$

where $\widehat{\mathbf{A}}_C \in \{-1,0,1\}^{(n+1)\times b_C}$, $\widehat{\mathbf{A}}_R \in \{-1,0,1\}^{(n+1)\times b_R}$, $\widehat{\mathbf{A}}_I \in \{-1,0,1\}^{(n+1)\times b_I}$ and $\widehat{\mathbf{v}} \in \mathbb{R}^{n+1}$ result from extending the corresponding quantities to account for the additional node, i.e.,

$$\begin{aligned}
\widehat{\mathbf{A}}_C^\top &= [\mathbb{I} \quad -\mathbb{1}], & \widehat{\mathbf{A}}_R^\top &= [\mathbf{A}_R^\top \quad \mathbf{0}], \\
\widehat{\mathbf{A}}_I^\top &= [\mathbb{I} \quad -\mathbb{1}] \quad \text{and} & \widehat{\mathbf{v}}^\top &= [\mathbf{v}^\top \quad v_{\text{gnd}}],
\end{aligned} \tag{5.33}$$

where $\mathbb{I}$ is the $n \times n$ identity matrix, $\mathbb{1}$ is a vector of ones with dimension n and v_{gnd} is the potential of the thermal ground node.

The equivalence between (5.32) and (5.16) is found by discarding line $n+1$ in (5.32) in order to apply a reference temperature, and setting

$$\begin{aligned}
\mathbf{A}_C^{\text{th}} &= \mathbf{A}_I^{\text{th}} = \mathbb{I}, \ \mathbf{A}_R^{\text{th}} = \widetilde{\mathbf{S}}, \ \mathbf{C}^{\text{th}} = \mathbf{M}_{\rho c}, \\
\mathbf{G}^{\text{th}} &= \mathbf{M}_\lambda, \ \mathbf{I}_s^{\text{th}} = -\mathbf{Q}_{\text{el}} \ \text{and} \ \mathbf{v}^{\text{th}} = \mathbf{T},
\end{aligned} \tag{5.34}$$

where the superscript th indicates the quantities for the thermal problem.

The fact that the FIT formulation (5.15) and (5.16) and the circuit formulation (5.30) are equivalent follows from the basic laws of physics. In this section, this equivalence is made explicit by choosing a matching notation for both cases and clearly defining the topology of the grid used for field simulation. This equivalence motivates to organise a translation from field to circuit models. The relations (5.31) and (5.34) indicate how a field model can be brought into an equivalent circuit netlist, independent from the geometry of the 3D problem behind. This will offer a powerful tool to include field effects in an overall circuit model.

5.5 SPICE Netlist Generation

In this section, the translation from an electrothermal field model into a combination of an electric and a thermal circuit is carried out in two steps. First, the topology of the electric circuit is determined, also accounting for the extraction of the Joule loss term. Then, the topology of the thermal circuit is fixed such that the Joule loss term is easily inserted. Furthermore, the treatment of nonlinear materials is discussed.

5.5.1 Electric Circuit Stamp

There is a direct correspondence between the electric scalar potentials φ allocated at the primary nodes of the FIT field model and the potentials at the nodes of the MNA circuit model. In the previous section, it has been shown that the branch voltages operate on the electric capacitances as well as on the electric resistances, see (5.27) and (5.31), which indicates a parallel connection. The branch voltages are $\mathbf{V}_R^{el} = \mathbf{V}_C^{el} = \widetilde{\mathbf{S}}^\top \boldsymbol{\varphi}$ and are obtained as the differences of the potentials at the adjacent nodes. Hence, the diagonal entries of the FIT conductance matrix $\mathbf{M}_\sigma$ and the FIT capacitance matrix $\mathbf{M}_\varepsilon$, i.e., $M_{\sigma;j,j}$ and $M_{\varepsilon;j,j}$, should be incorporated as lumped conductances and capacitances on two parallel circuit branches connected to the end points of edge L_j. Dirichlet boundary conditions in the field model are modelled by voltage sources connected to the ground node. In Fig. 5.4a, the part of the electric circuit stamp corresponding to one primary edge L_j with its incident nodes i and $i+1$ is shown.

Thermal losses occur at every primary edge and accordingly every branch of the exported circuit. At branch L_j, the thermal loss is

$$\widehat{Q}_{el,j} = U_j I_j, \tag{5.35}$$

where U_j is the branch voltage and I_j is the branch current. This thermal loss corresponds to the thermal loss derived for the field model in Section 5.2. All thermal losses are collected in a vector $\widehat{\mathbf{Q}}_{el}$ from which the thermal losses $\mathbf{Q}_{el}$ for the dual volumes can be computed according to (5.14).

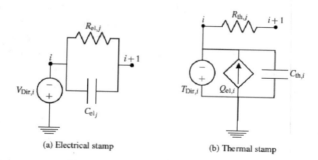

(a) Electrical stamp   (b) Thermal stamp

Fig. 5.4: Equivalent circuit stamps for the primary edge L_j between the primary nodes i and $i+1$: (a) Electrical stamp, (b) Thermal stamp.

5.5.2 Thermal Circuit Stamp

The thermal conductances are fully equivalent to the electric conductances and are organised as branches between the thermal circuit nodes. The thermal capacitances are, however, different, as already discussed above. The temperature drops relevant for the thermal heat capacity are not the temperature drops between adjacent nodes but the temperature drops between the nodes and the thermal reference node. This is expressed by the corresponding part of the incidence matrix $\mathbf{A}_C^{\mathrm{th}} = \mathbb{I}$ and by the temperature differences to the thermal reference node $\mathbf{V}_C = \mathbb{I}\mathbf{T}$. The thermal capacitances are taken from the diagonal of $\mathbf{M}_{\rho c}$ and are organised as connections between each of the nodes and the thermal reference node as shown in Fig. 5.4b. Dirichlet conditions in the field model correspond to temperature sources connecting the corresponding nodes to the thermal ground. The heat powers $\mathbf{Q}_{\mathrm{el}}$ coming from the electric circuit are introduced as thermal flux sources between the corresponding nodes and the thermal ground node by the incidence matrix $\mathbf{A}_I^{\mathrm{th}} = \mathbb{I}$. The thermal circuit is shown for an exemplary primary edge in Fig. 5.4b.

5.5.3 Accounting for Nonlinear Material Properties

The electric conductivity $\sigma(T)$ and the thermal conductivity $\lambda(T)$ depend on the temperature T. The introduction of this nonlinearity to the representing circuit is exemplarily shown for the electric conductivity. The nonlinearity immediately causes the material properties to become inhomogeneous, even in regions with the same material. The material parameters associated with the primary grid cells need to be updated between the successive steps of the circuit solver. The averaging procedures, e.g., (5.11) for the electric conductivity, is needed to map the material properties to the branch relations. This can be organised by a netlist with nonlinear entries.

In more detail, the nonlinear update procedure is a follows. On a primary edge L_j, the average temperature $\overline{T}_j$ is calculated by averaging the temperatures at the nodes that the edge is connected to. This temperature is used to evaluate the electric resistivity of the primary volumes V_p that the edge is embedded in, i.e.,

$$\rho_p(\overline{T}_j) = \frac{1}{\sigma_p(\overline{T}_j)} = \rho_{0,p}\left(1 + \alpha_p\left(\overline{T}_j - T_0\right)\right), \tag{5.36}$$

with $\rho_{0,p}$ the resistivity at the reference temperature T_0 and $\alpha_p \in \mathbb{R}$ the temperature coefficient for primary cell V_p. From these obtained values for σ_p, the average conductivity $\overline{\sigma}_j$ of edge L_j is calculated using (5.11). Finally, the nonlinear conductance $R_j(\overline{T}_j)$ is found from

$$R_j(\overline{T}_j) = \frac{1}{\overline{\sigma}_j(\overline{T}_j)} \frac{|\widetilde{A}_j|}{|L_j|}, \tag{5.37}$$

where $|L_j|$ is the length of primary edge L_j and $|\widetilde{A}_j|$ is the area of dual facet $\widetilde{A}_j$. The evaluation of nonlinear thermal conductivities is analogous. To incorporate more complicated material laws does not cause additional issues for what the spatial discretisation and the translation to the circuit level is concerned.

5.5.4 Extraction Algorithm

The extraction of a netlist from a FIT field model can be organised in a fully automated procedure. After constructing the hexahedral grid, assembling the topological operators and calculating the entries of the material matrices, a routine is called which writes the necessary information to a file. The pseudocode for the definition of an electrothermal netlists including nonlinearities is listed in Algorithm 5.1.

Algorithm 5.1 Electrothermal SPICE netlist generation

1: **for** edge L_j between primary nodes i and k **do**
2:      write R_el(j) node(i) node(k) $R_j(\overline{T}_j)$
3:      write C_el(j) node(i) node(k) $\mathbf{M}_\varepsilon(j,j)$
4:      write R_th(j) node(i) node(k) $\mathbf{M}_\lambda^{-1}(j,j)$
5:      write C_th(i) gnd node(i) $\mathbf{M}_{\rho c}(i,i)$
6:      write I_Loss gnd node(i) $Q_{\mathrm{el},i}$
7:      **if** i is electric Dirichlet node **then**
8:          write Vdir_el(i) node(i) gnd $V_{\mathrm{Dir},i}$
9:      **end if**
10:     **if** i is thermal Dirichlet node **then**
11:         write Vdir_th(i) node(i) gnd $T_{\mathrm{Dir},i}$
12:     **end if**
13: **end for**

5.6 Validation and Example

For validation of the presented methodology, this section first uses a simple benchmark geometry on which the netlist generation algorithm is applied. To validate also on a 3D problem, a microelectronic chip package is used in the second part of the section.

Table 5.1: Material properties of the benchmark example.

Property	$0 < x < \ell$	$\ell < x < \ell + d$
σ (S/m)	3	0
ε_{r}	1	$1.13 \cdot 10^5$
λ (W/K/m)	400	400
ρc $\left(\mathrm{J/K/m^3}\right)$	8000	8000

5.6.1 Benchmark

The benchmark example is deliberately kept very simple. It is a brick with dimensions $4\,\mathrm{mm} \times 1\,\mathrm{mm} \times 1\,\mathrm{mm}$ consisting of two regions with different materials (Fig. 5.5). The left region has a length $\ell = 3\,\mathrm{mm}$ and is mainly resistive, whereas the right region has a length $d = 1\,\mathrm{mm}$ and is mainly capacitive (see also Table 5.1). A voltage drop along the x-direction is applied to the series connection of both regions by applying $0\,\mathrm{V}$ to the right boundary and a sinusoidal potential with an amplitude of $1\,\mathrm{kV}$ and a frequency of $76.9\,\mathrm{kHz}$ at the left boundary. The remaining boundary conditions are electrically insulating. For the thermal field, all boundaries are adiabatic. As initial conditions, the electric potential and temperatures are set to zero.

The applied voltage drop is modelled by a voltage source V_{Dir} connecting the nodes at the left and right sides. The resistive part is modelled by a single resistance and the capacitive part by a single capacitor. Both parts share the same thermal properties. After setting up the FIT field model, a replacement circuit is extracted using Algorithm 5.1. Both the FIT field model and the extracted MNA circuit model have the same spatial resolution. Hence, identical results are expected if the time stepping errors are neglible. Both models are simulated in their own framework. The electric potential at $x = 0$ and $x = \ell$ are plotted in Fig. 5.6. From the electric solution, the thermal

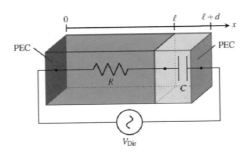

Fig. 5.5: Geometry of the benchmark example. A resistive part of length ℓ and a capacitive part of length d are modelled and excited with a sinusoidal voltage source imposed as Dirichlet conditions.

loss term $\mathbf{Q}_{el}$ is computed and inserted in the heat equation. The results for the temperature at $x = 0$ and $x = \ell$ are plotted in Fig. 5.7.

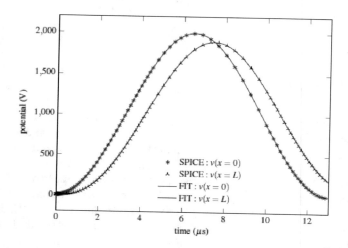

Fig. 5.6: Electric potential at selected points of the benchmark example calculated by the SPICE (crosses) and FIT (solid line) simulations.

Both figures show a very good agreement. The relative error between the solutions is calculated by

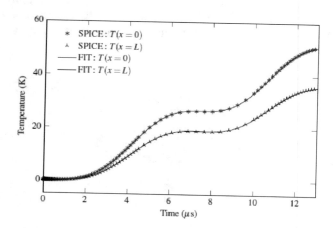

Fig. 5.7: Temperature at selected points of the benchmark example as the result of the SPICE (crosses) and FIT (solid line) simulation.

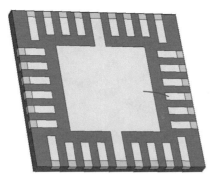

Fig. 5.8: Chip package with one bond wire.

$$\text{error} = \frac{\max_i \|\mathbf{T}_{\text{SPICE}}(t_i) - \mathbf{T}_{\text{FIT}}(t_i)\|_2}{\max_i \|\mathbf{T}_{\text{FIT}}(t_i)\|_2} \approx 0.52\,\%, \qquad (5.38)$$

where $\mathbf{T}_{\text{SPICE}}$ and $\mathbf{T}_{\text{FIT}}$ are n-by-n_t arrays with n the number of grid points and n_t the number of time steps. Because of the matching spatial discretisations, the error is fully attributed to the different time integrators.

5.6.2 Chip Package

The possibilities of electrothermal circuit extraction become more pronounced for a more complex example which is here chosen to be a microelectronic chip package (Fig. 5.8) [10]. Besides the chip and the moulding compound, one bond wire connecting one of the contacts of the chip to a contact pad is considered. The bond wire has an electrical conductance $G_{\text{bw}}^{\text{el}} = 1\,\text{S}$ and a thermal conductance $G_{\text{bw}}^{\text{th}} = 1\,\text{kW/K}$. The relative permittivity of all materials is 1. The boundaries are adiabatic. The system is excited by a voltage drop $V_0(t) = 10\,\text{V}\,[1 - \exp(-t)]$ exerted on the bond wire. With Algorithm 5.1, a netlist is extracted from the FIT field model and the single bond wire is added as a lumped element represented by an additional line in the netlist. This electrothermal circuit model is simulated by SPICE. The temperature at the hottest node is compared to the one obtained by field simulation in Fig. 5.9. The error following from (5.38) is 0.07 %.

5.7 Conclusions

The electrothermal circuit extraction procedure developed in this chapter enables to translate an electrothermal field model with all spatial details into

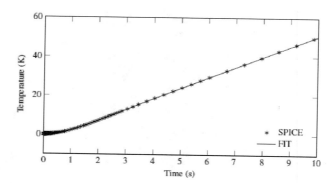

Fig. 5.9: Temperature at the hottest point of the microelectronic chip package obtained by SPICE simulation.

a circuit framework. The extraction procedure relies upon the equivalent between field and circuit formulation as is apparent when the field discretisation is interpreted in a mimetic way. The extraction method incorporates the nonlinear coupling between the electric and thermal fields along the Joule losses and the temperature-dependent electric conductivities. The extraction procedure does not introduce any further simplifications, except for the spatial discretisation which already has been carried out a priori. However, it remains possible to apply a model reduction after circuit extraction, possibly also including any circuitry outside the field model.

Acknowledgements This work is a part of the project 'Nanoelectronic Coupled Problems Solutions' (nanoCOPS) funded by the European Union within FP7-ICT-2013 (grant no. 619166). The work is also supported by the 'Excellence Initiative' of the German Federal and State Governments and the Graduate School of Computational Engineering at Technische Universität Darmstadt. The authors would like to thank Abdul Moiz for his work spent on the implementation of the automated electrothermal netlist generation.

References

1. ALOTTO, P., FRESCHI, F., AND REPETTO, M.: *Multiphysics problems via the cell method: The role of Tonti diagrams.* IEEE Trans. Magn. **46**(8), 2959–2962 (2010). doi:10.1109/TMAG.2010.2044487
2. ANSYS: *HFSS: High frequency electromagnetic field simulation.* Manual (2018). https://www.ansys.com
3. AUCHMANN, B., AND KURZ, S.: *A geometrically defined discrete Hodge operator on simplicial cells.* IEEE Trans. Magn. **42**(4), 643–646 (2006). doi:10.1109/TMAG.2006.870932

4. BANAGAAYA, N., FENG, L., AND BENNER, P.: *Sparse (P)MOR for Electro-Thermal Coupled Problems with Many Inputs.* Chapter in this Book, 2019.
5. BENDERSKAYA, G., DE GERSEM, H., WEILAND, T., AND CLEMENS, M.: *Transient field-circuit coupled formulation based on the finite integration technique and a mixed circuit formulation.* COMPEL **23**(4), 968–976 (2004). doi:10.1108/03321640410553391
6. BITTNER, K., AND BRACHTENDORF, H.G.: *Multirate Circuit - EM - Device Simulation.* Chapter in this Book, 2019.
7. BONDESON, A., RYLANDER, T., AND INGELSTRÖM, P.: *Computational Electromagnetic.* Texts in Applied Mathematics. Springer (2005). doi:10.1007/b136922
8. BOSSAVIT, A.: *Computational Electromagnetism: Variational Formulations, Complementarity, Edge Elements.* Academic Press, San Diego (1998). http://natrium. em.tut.fi/~bossavit/
9. CADENCE: *Spectre circuit simulator* (2018). https://www.cadence.com/content/ cadence-www/global/en_US/home/tools/custom-ic-analog-rf-design/ circuit-simulation/spectre-circuit-simulator.html
10. CASPER, T., DE GERSEM, H., GILLON, R., GÖTTHANS, T., KRATOCHVÍL, T., MEURIS, P., AND SCHÖPS, S.: *Electrothermal simulation of bonding wire degradation under uncertain geometries.* In: L. FANUCCI, AND J. TEICH (EDS.): *Proceedings of the 2016 Design, Automation & Test in Europe Conference & Exhibition (DATE)*, pp. 1297–1302. IEEE (2016). http://www.date-conference.com
11. CASPER, T., DE GERSEM, H., AND SCHÖPS, S.: *Automatic generation of equivalent electrothermal SPICE netlists from 3D electrothermal field models.* In: W. VAN DRIEL, AND P. RODGERS (EDS.): *17th International Conference on Thermal, Mechanical and Multi-Physics Simulation and Experiments in Microelectronics and Microsystems (EuroSimE 2016)* (2016). doi:10.1109/EuroSimE.2016.7463329
12. CHVALA, A., DONOVAL, D., MAREK, J., PRIBYTNY, P., MOLNAR, M., AND MIKOLASEK, M.: *Fast 3-D electrothermal device/circuit simulation of power superjunction MOSFET based on SDevice and HSPICE interaction.* IEEE Trans. Electron. Dev. **61**(4), 1116–1122 (2014). doi:10.1109/TED.2014.2305848
13. CLEMENS, M., GJONAJ, E., PINDER, P., AND WEILAND, T.: *Self-consistent simulations of transient heating effects in electrical devices using the finite integration technique.* IEEE Trans. Magn. **37**(5), 3375–3379 (2001). doi:10.1109/20.952617
14. CLEMENS, M., AND WEILAND, T.: *Discrete electromagnetism with the finite integration technique.* PIER **32**, 65–87 (2001). doi:10.2528/PIER00080103. http: //ceta.mit.edu/PIER/pier32/03.00080103.clemens.pdf
15. CLEMENS, M., AND WEILAND, T.: *Magnetic field simulation using conformal FIT formulations.* IEEE Trans. Magn. **38**(2), 389–392 (2002). doi:10.1109/20.996104
16. CLEMENS, M., WILKE, M., BENDERSKAYA, G., DE GERSEM, H., KOCH, W., AND WEILAND, T.: *Transient electro-quasistatic adaptive simulation schemes.* IEEE Trans. Magn. **40**(2), 1294–1297 (2004). doi:10.1109/TMAG.2004.824582
17. CODECASA, L., SPECOGNA, R., AND TREVISAN, F.: *Symmetric positive-definite constitutive matrices for discrete eddy-current problems.* IEEE Trans. Magn. **43**(2), 510–515 (2007). doi:10.1109/TMAG.2006.887065
18. CST AG: *CST STUDIO SUITE 2016* (2017). https://www.cst.com
19. DIRKS, H.K.: *Quasi-stationary fields for microelectronic applications.* Electr. Eng. **79**(2), 145–155 (1996). doi:10.1007/BF01232924
20. FENG, L., AND BENNER, P.: *Parametric Model Order Reduction for Electro-Thermal Coupled Problems.* Chapter in this Book, 2019.
21. GÜNTHER, M. (ED.): *Coupled Multiscale Simulation and Optimization in Nanoelectronics*, Series Mathematics in Industry, Vol. 21. Springer, Berlin (2015). doi:10.1007/978-3-662-46672-8
22. GÜNTHER, M., FELDMANN, U., AND TER MATEN, J.: *Modelling and Discretization of Circuit Problems.* In: W.H.A. SCHILDERS AND E.J.W. TER MATEN: *Handbook of Numerical Analysis*, Vol. 13, pp. 523–659. Elsevier BV, Amsterdam (2005). doi:10.1016/s1570-8659(04)13006-8

23. HEFNER, A., AND BLACKBURN, D.: *Thermal component models for electrothermal network simulation.* IEEE Trans. Compon. Packag. Manuf. A **17**(3), 413–424 (1994). doi:10.1109/95.311751

24. HINZE, M., KUNKEL, M., AND MATTHES, U.: *POD model order reduction of electrical networks with semiconductors modeled by the transient drift-diffusion equations.* In: M. GÜNTHER, A. BARTEL, M. BRUNK, S. SCHÖPS, AND M. STRIEBEL (EDS.): *Progress in Industrial Mathematics at ECMI 2010*, Series Mathematics in Industry, Vol. 17. Springer, Berlin (2012). doi:10.1007/978-3-642-25100-9

25. HO, C.W., RUEHLI, A.E., AND BRENNAN, P.A.: *The modified nodal approach to network analysi.* IEEE Trans. Circ. Syst. **22**(6), 504–509 (1975). doi:10.1109/TCS.1975.1084079

26. KETTUNEN, L.: *Fields and circuits in computational electromagnetism.* IEEE Trans. Magn. **37**(5), 3393–3396 (2001). doi:10.1109/20.952621

27. KEYSIGHT TECHNOLOGIES: *EMPro 3d EM simulation software.* Manual (2018). https://www.keysight.com

28. KOŠEL, V., ILLING, R., GLAVANOVICS, M., AND ŠATKA, A.: *Non-linear thermal modeling of DMOS transistor and validation using electrical measurements and FEM simulations.* Microelectron. J. **41**(12), 889–896 (2010). doi:10.1016/j.mejo.2010.07.016

29. TER MATEN, E.J.W., BRACHTENDORF, H.G., PULCH, R., SCHOENMAKER, W., AND DE GERSEM, H. (EDS.): *Nanoelectronic Coupled Problems Solutions.* Series Mathematics in Industry, Vol. NN, Springer, 2019.

30. MAWBY, P., IGIC, P., AND TOWERS, M.: *New physics-based compact electro-thermal model of power diode dedicated to circuit simulation.* In: Proceedings IEEE Circuits and Systems ISCAS, vol. 3, pp. 401–404 (2001). doi:10.1109/ISCAS.2001.921332

31. MÄRZ, M., AND NANCE, P.: *Thermal modeling of power-electronic systems* (2000)

32. NAGEL, L.W.: *SPICE2: a computer program to simulate semiconductor circuits.* Tech. Rep. No. UCB/ERL M520 1975, University of Berkeley (1975)

33. ORCAD: *PSpice User's Guide, 17.2 edn.* (2016). http://www.orcad.com/products/orcad-pspice-designer/overview

34. VAN PETEGEM, W., GEERAERTS, B., SANSEN, W., AND GRAINDOURZE, B.: *Electrothermal simulation and design of integrated circuits.* IEEE J. Solid. State. Circ. **29**(2), 143–146 (1994). doi:10.1109/4.272120

35. SANDIA NATIONAL LABORATORIES: *Xyce: Parallel Electronic Simulator* (2018). https://xyce.sandia.gov

36. SCHAUER, M., HAMMES, P., THOMA, P., AND WEILAND, T.: *Nonlinear update scheme for partially filled cells.* IEEE Trans. Magn. **39**(3), 1421–1423 (2003). doi:10.1109/TMAG.2003.810415

37. SCHOENMAKER, W., BRACHTENDORF, H.G., BITTNER, K., TISCHENDORF, C., AND STROHM, C.: *EM-Equations, Coupling to Heat and to Circuits.* Chapter in this Book, 2019.

38. SCHUHMANN, R., AND WEILAND, T.: *Conservation of discrete energy and related laws in the finite integration technique.* PIER **32**, 301–316 (2001). doi:10.2528/PIER00080112. http://ceta.mit.edu/PIER/pier32/12.00080112.shuhmann.pdf

39. SCHÖPS, S., DE GERSEM, H., AND WEILAND, T.: *Winding functions in transient magnetoquasistatic field-circuit coupled simulations.* COMPEL **32**(6), 2063–2083 (2013). doi:10.1108/COMPEL-01-2013-0004

40. SIMPSON, N., WROBEL, R., AND MELLOR, P.: *An accurate mesh-based equivalent circuit approach to thermal modeling.* IEEE Trans. Magn. **50**(2), 269–272 (2014). doi:10.1109/TMAG.2013.2282047

41. THOMA, P., AND WEILAND, T.: *A consistent subgridding scheme for the finite difference time domain method.* Int. J. Numer. Model. Electron. Network. Dev. Field **9**(5), 359–374 (1996). doi:10.1002/(SICI)1099-1204(199609)9:5<359::AID-JNM245>3.0.CO;2-A

42. TREVISAN, F.,AND KETTUNEN, L.: *Geometric interpretation of discrete approaches to solving magnetostatic problems.* IEEE Trans. Magn. **40**(2), 361–365 (2004). doi:10.1109/TMAG.2004.824107

43. VANDEKERCKHOVE, S., VANDEWOESTYNE, B., DE GERSEM, H., VAN DEN ABEELE, K., AND VANDEWALLE, S.: *Mimetic discretisation and higher order time integration for acoustic, electromagnetic and elastodynamic wave propagation.* J. Comput. Appl. Math. **259**, 65–76 (2014). doi:10.1016/j.cam.2013.02.027

44. VOGELSONG, R., AND BRZEZINSKI, C.: *Extending SPICE for electro-thermal simulation.* In: R. MILANO, M. HARTRANFT, AND D. BROWN (EDS.): Proceedings of the IEEE 1989 Custom Integrated Circuits Conference, pp. 21.4/1–21.4/4 (1989). doi:10.1109/CICC.1989.56803

45. WEILAND, T.: *A discretization method for the solution of Maxwell's equations for six-component fields.* AEÜ **31**, 116–120 (1977)

46. WEILAND, T.: *Time domain electromagnetic field computation with finite difference methods.* Int. J. Numer. Model. Electron. Network. Dev. Field **9**(4), 295–319 (1996). doi:10.1002/(SICI)1099-1204(199607)9:4<295::AID-JNM240>3.0.CO;2-8

47. WITTIG, T., MUNTEANU, I., SCHUHMANN, R., AND WEILAND, T.: *Model order reduction and equivalent circuit extraction for FIT discretized electromagnetic systems.* Int. J. Numer. Model. Electron. Network. Dev. Field **15**(5-6), 517–533 (2002)

48. WÜNSCHE, S., CLAUSS, C., SCHWARZ, P., AND WINKLER, F.: *Electro-thermal circuit simulation using simulator coupling.* IEEE Trans. Very Large Scale Integr. (VLSI) Syst. **5**(3), 277–282 (1997). doi:10.1109/92.609870

Part II
Time Integration for Coupled Problems

Part II covers the following topics.

- **Holistic / Monolithic Time Integration**
  Authors:
  Wim Schoenmaker (MAGWEL, Leuven, Belgium),
  Hans-Georg Brachtendorf, Kai Bittner (FH Oberösterreich, Hagenberg, Austria),
  Caren Tischendorf, Christian Strohm (Humboldt Universität zu Berlin, Germany).
- **Non-Intrusive Methods for the Cosimulation of Coupled Problems**
  Authors:
  Sebastian Schöps, David J. Duque Guerra, Herbert De Gersem (Technische Universität Darmstadt, Germany),
  Andreas Bartel, Michael Günther (Bergische Universität Wuppertal, Germany),
  Roland Pulch (Universität Greifswald, Germany).
- **Multirate Circuit - EM - Device Simulation**
  Authors:
  Kai Bittner, Hans-Georg Brachtendorf (FH Oberösterreich, Hagenberg, Austria).

Chapter 6
Holistic / Monolithic Time Integration

Wim Schoenmaker, Hans-Georg Brachtendorf, Kai Bittner,
Caren Tischendorf, Christian Strohm

Abstract This chapter discusses a number of techniques to solve the discrete systems of equations that result from the coupling of the circuit and electromagnetic field equations. We briefly summarize the modified nodal analysis for the lumped modeling of circuits. Then we discuss the coupling relations for including EM models into the circuit simulation systems and combine them with the spatially discretized Maxwell equations. Finally, we apply an adaptive time stepping scheme to the resulting coupled differential-algebraic equation system and in particular focus on a number of technical steps to find the solutions of the discretized coupled equations.

6.1 Lumped Circuit Modeling

The common approach for simulating circuits is the Modified Nodal Analysis (MNA). It bases on Kirchhoff's laws described by

$$Ai = 0, \qquad\qquad v = A^\top e \qquad\qquad (6.1)$$

with the incidence matrix A, mapping branches to nodes of the circuit. The circuit variables consists of all branch currents i, of all branch voltages v and of all nodal potentials e. They are completed by the constitutive element

Wim Schoenmaker
MAGWEL NV, Leuven, Belgium, e-mail: `Wim.Schoenmaker@magwel.com`

Hans-Georg Brachtendorf, Kai Bittner
Fachhochschule Oberösterreich, Hagenberg im Mühlkreis, Austria, e-mail: {`Hans-Georg.Brachtendorf,Kai.Bittner`}`@fh-hagenberg.at`

Caren Tischendorf, Christian Strohm
Humboldt Universität zu Berlin, Germany, e-mail: {`Caren.Tischendorf,StrohmCh`}`@math.hu-berlin.de`

© Springer Nature Switzerland AG 2019
E. J. W. ter Maten et al. (eds.), *Nanoelectronic Coupled Problems Solutions*,
Mathematics in Industry 29, https://doi.org/10.1007/978-3-030-30726-4_6

equations

$$i_{[1]} = \tfrac{d}{dt}q(v_{[1]}) + g(v_{[1]}, t), \qquad v_{[2]} = \tfrac{d}{dt}\phi(i_{[2]}) + r(i_{[2]}, t) \qquad (6.2)$$

for lumped current and voltage controlling elements, respectively. Notice, all basic types as capacitances, inductances, resistances and sources are covered by a suitable choice of the functions q, g, ϕ and r. Performing the modified nodal analysis, we get the following reduced equation system having only the nodal potentials e and the currents i_2 of the voltage controlling elements, see [9]:

$$A_{[1]}\tfrac{d}{dt}q(A_{[1]}^\top e) + A_{[1]}g(A_{[1]}^\top e, t) + A_{[2]}i_{[2]} = 0, \qquad (6.3)$$

$$\tfrac{d}{dt}\phi(i_{[2]}) + r(i_{[2]}, t) - A_{[2]}^\top e = 0, \qquad (6.4)$$

where the incidence matrix $A = (A_{[1]}, A_{[2]})$ is split with respect to the current and voltage controlling elements. The equations (6.3)-(6.4) are generated automatically from netlists providing the node to branch element relation (for entries of $A_{[1]}$ and $A_{[2]}$) as well as the element related functions $\tilde{q}$, g, ϕ and r.

6.2 Modeling the Coupling

We assume that the contacts between the electromagnetic field model and the lumped circuit model to be perfectly electric conducting such that $\mathbf{B} \cdot n_\perp = 0$ and $e \cdot n_\| = 0$ with $n_\perp$ and $n_\|$ being the outer unit normal vectors transversal and parallel to the contact boundary. This motivates the boundary conditions, cf. [3],

$$(\nabla \times \mathbf{A}) \cdot n_\perp = 0, \qquad (\nabla V + \boldsymbol{\Pi}) \cdot n_\| = 0. \qquad (6.5)$$

Denoting by Γ_k the k-th contact of the electromagnetic field model element with Γ_0 being the reference contact we get the current through Γ_k as

$$i_k = \int_{\Gamma_k} [\boldsymbol{J} - \partial_t(\epsilon(\nabla V + \boldsymbol{\Pi}))] \cdot n_\perp \, d\sigma \qquad (6.6)$$

with $\boldsymbol{\Pi} = \partial_t \mathbf{A}$ the canonical momentum introduced in Chapter [3, Section 2.5]. Note that the equation for $\boldsymbol{J}$ in Chapter [3, Section 2.5] and the boundary condition (6.5) guarantee that the sum of all contact currents equals zero, that means

$$\sum_k i_k = 0.$$

This model property is necessary for all lumped element descriptions in order to preserve the Kirchhoff's current law. In order to reveal the relation to the voltages v_k between the contact Γ_k and the reference contact Γ_0, we express the scalar potential V as

$$V(x,t) = V_{bi}(x) + V_c(x,t) \tag{6.7}$$

with the contact potential

$$V_c(x,t) = \begin{cases} v_k & \text{if } x \in \Gamma_k \\ 0 & \text{else.} \end{cases}$$

Here, we assumed the reference contact Γ_0 to be the ground node for simplicity. The potential V_{bi} describes the position dependent built-in potential arising by varying doping concentrations and bonding different materials.

6.3 Holistic DAE System

The circuit equation system (6.3) - (6.4) is extended by currents i_M through the contact of the EM-element and summarized as a DAE of the form

$$A_C \frac{\mathrm{d}}{\mathrm{d}t} d_C(y_C) + b_C(y_C, t) + B_{CM} i_M = 0, \tag{6.8}$$

where $y_C = (e, i_{[2]})^\top$ collects the circuit variables. The spatially discretized Maxwell system (4.1), (2.15)-(2.20), and (4.6), together with the coupling boundary condition (6.5), are summarized into the system (see also Section 2.5)

$$A_M \frac{\mathrm{d}}{\mathrm{d}t} d_M(y_M, y_C) + b_M(y_M, y_C, t) = 0, \tag{6.9}$$

where $y_M = (V, n, p, \mathbf{A}, \mathbf{\Pi})^\top$ are the Maxwell system variables and

$$d_M(y_M, y_C) = \begin{pmatrix} y_M \\ (B_{CM}^\top, 0) y_C \end{pmatrix} = \begin{pmatrix} y_M \\ B_{CM}^\top e \end{pmatrix} = \begin{pmatrix} y_M \\ v_{app} \end{pmatrix}$$
$$b_M(y_M, y_C, t) = \beta_M(y_M, v_{app}, t)$$

with v_{app} describing the voltages at the EM device contacts. Analogously to the discretization of the Maxwell-Ampère equation which results in (4.6), we discretize the coupling condition (6.6) and summarize it to

$$A_{MC} \frac{\mathrm{d}}{\mathrm{d}t} d_{MC}(y_M, y_C) + b_{MC}(y_M, y_C, t) = i_M \tag{6.10}$$

with $d_{MC}(y_M, y_C) = d_M(y_M, y_C)$ and $b_{MC}(y_M, y_C, t) = \beta_{MC}(y_M, v_{app}, t)$.

Collecting (6.8), (6.9) and (6.10), we obtain the coupled DAE system

$$\underbrace{\begin{bmatrix} A_M & 0 & 0 \\ 0 & A_{MC} & 0 \\ 0 & 0 & A_C \end{bmatrix}}_{=:A} \frac{d}{dt} \underbrace{\begin{pmatrix} d_M(y_M, y_C) \\ d_{MC}(y_M, y_C) \\ d_C(y_C) \end{pmatrix}}_{=:d} + \underbrace{\begin{pmatrix} b_M(y_M, y_C, t) \\ b_{MC}(y_M, y_C, t) - i_M \\ b_C(y_C, t) - B_{CM} i_M \end{pmatrix}}_{=:b} = 0.$$

$$(6.11)$$

For also including Temperature in the full system we refer to [4].

6.4 Time Discretization

The resulting holistic, monolithic, DAE system (6.11)

$$A \frac{d}{dt} d(x) + b(x, t) = 0$$

is solved by the BDF methods (cf. [1]), i.e. we solve the nonlinear systems

$$\frac{1}{h_n} \sum_{i=0}^{k} \alpha_{ni} A d(x_{n-i}, t_{n-i}) + b(x_n, t_n) = 0 \qquad (6.12)$$

at each time point t_n. Here, $h_n := t_n - t_{n-1}$ is the time stepsize,

$$\alpha_{ni} = \frac{t_n - t_{n-1}}{t_n - t_{n-i}} \prod_{j=1, j \neq i}^{k} \frac{t_n - t_{n-j}}{t_{n-i} - t_{n-j}}, \quad i = 1, \ldots, k,$$

$$\alpha_{n0} = -\sum_{i=1}^{k} \alpha_{ni},$$

are the BDF coefficients and x_{n-i} are the numerical approximations of the exact solution $x(t_{n-i})$ at the time points t_{n-i}. The implemented time integration scheme has the option to switch to a Runge-Kutta scheme (Radau IIa), for starting and restarting the integration.

The adaptive time step control estimates the error for the dynamic components $d(x_n, t_n)$. Such a control is more stable and reliable for higher index DAE systems. The error is estimated by the difference of $d(x_n, t_n)$ and $d(x_n^p, t_n)$ with x_n^p being the predictor

$$x_n^p := \sum_{i=1}^{k+1} \gamma_{ni} x_{n-i}$$

and

$$\gamma_{ni} := \prod_{j=1, j \neq i}^{k+1} \frac{t_n - t_{n-j}}{t_{n-i} - t_{n-j}}, \quad i = 1, \ldots, k+1.$$

6.5 Implementation Issues

The construction of the coupled simulation is shown in Fig. 6.1. It is realized in a Python framework combining a Python implementation of the circuit solver MECS of the Humboldt University in Berlin [5] with a C++ implementation of the MAGWEL's field solver devEM.

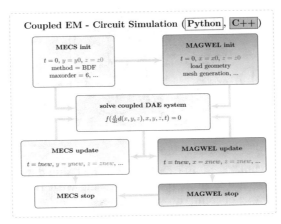

Fig. 6.1: Flow diagram for the coupled field circuit simulation. It is realized in a Python framework including C++ implementations of the field solver devEM.

```
Rs1 0 1 50
Vs2 0 2 sin(0 1 1e9 0 0)
Vs3 0 3 sin(0 1 -1e9 0 0)
$EM1 0 2 3 0 1 structure_balun.xml 0
```

Fig. 6.2: netlist with a $-line that triggers a field solving approach for a netlist element. The example line is read as follows: There is a netlist element with name EM1 whose field solver details can be found in the file structure_balun.xml. The five outer contacts of this element are connected to the nodes 0, 2, 3, 0 and 1. The last number in the line refers to the reference contact (usually the ground node).

From the circuit simulator's perspective a new element is defined that triggers the use of the field solver for the generation of the field equations on a computational grid. The syntax is illustrated in Fig. 6.2 and must be read as follows:

There is netlist element with name EM1 whose field solver details can be found in the file structure_balun.xml. The contacts of this element are connected to the nodes 0, 2, 3, 0 and 1. The last number in the line refers to the reference contact (usually the ground node).

A complete model can be illustrated as shown in Fig. 6.3.

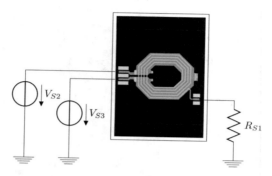

Fig. 6.3: Test Benchmark: Simple circuit with the balun device. The RF input signals V_{S2} and V_{S3} are sinusoidal ones operating with 1GHz frequency. The RF output is given as current through the resistance R_{S1}.

6.5.1 Difficulties with Iterative Solving

Depending on the spatial discretization and its resulting grid refinement we have to deal with a system size far above 100.000 degrees of freedom (DOFs), in this case 133.171 DOFs. Following this trend direct solvers soon will reach their limits. Thus, we pursued the idea to make use of iterative solvers.

As is seen in the plots 6.4, 6.5 and 6.6 there are serious flaws in the convergence process for obtaining acceptable current plots.

It turned out that the reason for these instabilities is the solutions accuracy of the linear system, as a part of the solving process of Newton-Raphson's method, were we switched over to iterative solvers.

Consequently, simply replacing the direct linear solver with an iterative one, e. g. a standard Generalized Minimal Residual Method (GMRES) turned out to cause more problems than it solves, i. e. it was difficult to get the GMRES converge to a desired accuracy not to say impossible. We hardly reached an residual of 10^{-5} in an appropriate amount of time and number of iterations.

The good news is that there are lots of techniques to improve the solving process

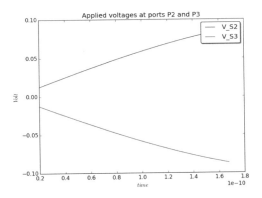

Fig. 6.4: Simulation results for the voltages at ports P2(V_{S2}) and P3(V_{S3}). They suit to the excitation given by the voltage sources.

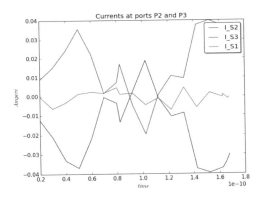

Fig. 6.5: Simulation results for the currents at ports P2(I_{S2}), P3(I_{S3}) and P1(I_{S1}). We observe numerical instabilities.

6.5.2 Complementary Implementations

In the following we introduce a collection of techniques that enhance the simulation process. On the one hand it is possible to exploit the system's structure, i. e. using Schur-decomposition to treat the network part different from the electromagnetic one. On the other hand we can also transform the linear systems matrix such that it obtains a lower condition number by various scaling techniques, for example variable scaling, time scaling and equation scaling.

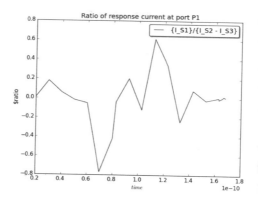

Fig. 6.6: Simulation results for the ratio of response current at ports P1($\frac{I_{S1}}{I_{S2}-I_{S3}}$). We observe numerical instabilities.

6.5.2.1 Schur Decomposition

In order to exploit the model based system structure one can make use of the so called Schur complement. For the block matrix

$$M = \begin{bmatrix} A & B \\ C & D \end{bmatrix}$$

we assume that the matrices A and D, belonging to the circuit system together with the field coupling and the field model respectively, are non singular. In that case we can decompose the equation system

$$M\begin{pmatrix} x \\ y \end{pmatrix} = \begin{pmatrix} r \\ s \end{pmatrix}$$

to

$$Dy = s - Cx$$
$$(A - BD^{-1}C)x = r - BD^{-1}s$$

The strategy then is solving $Dz = s$ in an iterative fashion, since D will be the biggest submatrix, just as $DZ = C$, followed by solving $(A - Z)x = r - Bz$ in a direct way and finally again solving $Dy = s - Cs$ iteratively.

This technique can be repeated a couple of times, focusing on special structured system blocks, e. g. symmetric positive definite ones, since iterative methods then are able to guarantee convergence.

A simple example of how we make use of this technique is demonstrated in Fig. 6.7. There is shown the fingerprint of a linear system's matrix that

arises during the solving process of the Newton-Raphson method within a
time integration step.

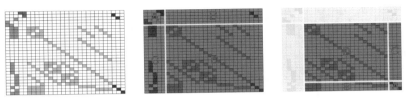

Fig. 6.7: Fingerprint of the Jacobian derived from a simple circuit-field coupled problem. The green, red, yellow and blue colums refer to circuit potentials, circuit internal, field internal and circuit coupling variables, respectively [5]. **Left**: Matrix without any modification. **Middle**: Schur decomposition applied on the matrix. **Right**: Another Schur decomposition applied
on block D.

6.5.2.2 Π-Elimination

Another form of exploiting the systems structure is the elimination of equations. In the field models equations we introduced the canonical momentum
$\Pi = \frac{\mathrm{d}}{\mathrm{d}t}A$. This leads to an equation block yielding two diagonal matrices in
the resulting linear equations matrix, e. g.

$$M = \begin{bmatrix} * & * & * & * \\ 0 & D_1 & D_2 & 0 \\ * & * & * & * \end{bmatrix},$$

with $D_1 = d_1 * I_m$ and $D_2 = d_2 * I_m$ where $d_1, d_2 \in \mathbb{R}\backslash\{0\}$ and $I_m \in \mathbb{R}^m$.
Algebraic transformation is now able to reduce this systems magnitude by
m, the number of discrete Π variables.

6.6 Scaling Issues

Since we deal with variables that have a direct link to SI units, their magnitude crucially varies. Take for example the frequency and capacitance difference; typical values in our field of application can be around 10^{10} and 10^{-16},
so do the entries in our equations. Additionally, they are scaled according to
the time integration's step width and method applied. An awful huge condition number of the resulting linear equations matrix is hardly surprising.
Thus, instead of solving $Mx = b$ we solve $MS_x\bar{x} = b$ such that MS_x has a

significant smaller condition number. Here S_x is a diagonal scaling matrix containing the usual magnitude of the corresponding variable. Finally, we recover the solution by $x = S_x \bar{x}$.

Another type of scaling is the time scaling. Recall the coupled system in its simplest form, a linear DAE,

$$A\frac{d}{dt}x(t) + Bx(t) - q(t) = 0.$$

During time integration we then have to solve $(\frac{\alpha_{n0}}{h}A + B)x = b$, where α_{n0} is the above mentioned BDF coefficient and h the step width, for some right-hand side b. With making use of time scaling by defining $\tau := s_t t$, we alternatively can solve

$$(s_t\frac{\alpha_{n0}}{h}A + B)\bar{x} = b$$

where the time scaling constant $s_t \in \mathbb{R}\backslash\{0\}$ is appropriately chosen. In the end we again recover $x(t) = \bar{x}(\tau)$.

6.7 Results

It turned out that the solution of the linear equations arising in the solving process of the discretized system (6.12) is a bottle neck in the whole solution process. Direct solvers require huge memory and iterative solvers often do not converge. So far, best results are obtained by the direct linear solver MUMPS (MUltifrontal Massively Parallel sparse direct Solver, c. f. [2]). The integration of this solver was rather cumbersome since a special framework needed to be provided on the machine. In order to make the MUMPS available in Python, it was necessary to install packages, such as BLACS, BLAS, MPI and ScaLAPACK. To access MUMPS via Python certain site-packages are required i. e. mpi4py and PyMUMPS. With using this solver, even in its single core version (sequential mode), the simulation got substantially faster and accurate.

Now it is possible to apply frequencies about 10GHz to the field model. For the coupled problem we took the netlist 6.8, a similar one as shown in Fig. 6.2. The results are shown in Fig. 6.9 -Fig. 6.11. In Fig. 6.12 - Fig. 6.14 is shown the magnetic inductance **B** in two different views around the Balun device.

```
Rs1 0 1 50
Vs2 0 2 sin(0 1 1e10 0 0)
Vs3 0 3 sin(0 0 -1e10 0 0)
$EM1 0 2 3 0 1 structure_balun.xml 0
```

Fig. 6.8: Netlist with a $-line that triggers a field solving approach for a netlist element. In this setup we apply one 10GHz voltage source to the field model.

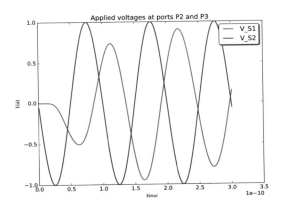

Fig. 6.9: Applied voltages at ports P2 and P3

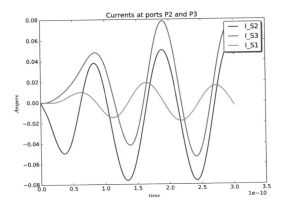

Fig. 6.10: Currents at ports P2 and P3 and P1

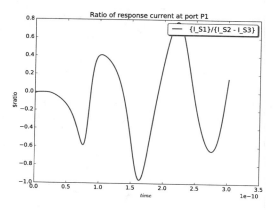

Fig. 6.11: Ratio of response currents at port P1

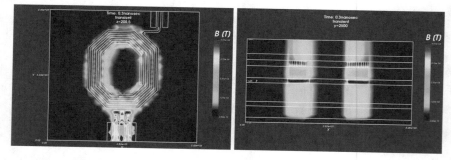

Fig. 6.12: **10GHz** benchmark of a Balun device at time point $t = 0.3$ nanoseconds. **Left**: Balun device top view. Shown is the magnetic inductance B. **Right**: Balun device cross-sectional area view. Shown is the magnetic inductance B.

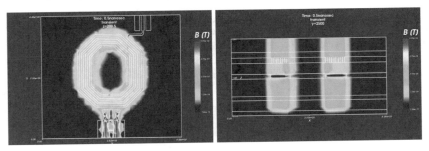

Fig. 6.13: **10GHz** benchmark of the Balun device at time point $t = 0.5$ nanoseconds. **Left**: Balun device top view. Shown is the magnetic inductance B. **Right**: Balun device cross-sectional area view. Shown is the magnetic inductance B.

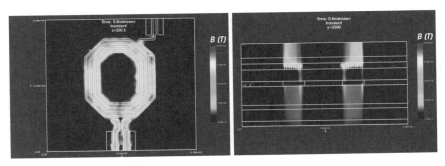

Fig. 6.14: **10GHz** benchmark of the Balun device at time point $t = 0.8$ nanoseconds. **Left**: Balun device top view. Shown is the magnetic inductance B. **Right**: Balun device cross-sectional area view. Shown is the magnetic inductance B,

References

1. GEAR, C.W.: *Numerical Initial Value Problems in Ordinary Differential Equations.* Prentice-Hall, Englewood Cliffs, N. J., 1971.
2. MUMPS: *A MUltifrontal Massively Parallel sparse direct Solver.* http://mumps.enseeiht.fr/, 2017.
3. SCHOENMAKER, W., BRACHTENDORF, H.-G., BITTNER, K., TISCHENDORF, C., AND STROHM, C.: *EM-Equations, Coupling to Heat and to Circuits.* Chapter in this Book, 2019.
4. SCHOENMAKER, W., DUPUIS, O., DE SMEDT, B., AND MEURIS, P.: *Fully-Coupled Electro-Thermal Power Device Fields.* In: RUSSO, G., CAPASSO, V., NICOSIA, G., AND ROMANO, V. (EDS.): *Progress in Industrial Mathematics at ECMI 2014*, Series

130   W. Schoenmaker, H.-G. Brachtendorf, K. Bittner, C. Tischendorf, C. Strohm

Mathematics in Industry Vol. 22, Springer International Publishing, pp. 829–834, 2016.

5. SCHOENMAKER, W., MEURIS, P., STROHM, C., AND TISCHENDORF, C.: *Holistic coupled field and circuit simulation.* In: *2016 Design, Automation Test in Europe Conference Exhibition (DATE),* pp. 307–312, March 2016.

Chapter 7
Non-Intrusive Methods for the Cosimulation of Coupled Problems

Sebastian Schöps, David J. Duque Guerra, Herbert De Gersem,
Andreas Bartel, Michael Günther, Roland Pulch

Abstract Many industrial applications require transient simulations of coupled problems. The coupling can involve multiples physical disciplines, multiple scales in space and time and different requirements on simulation accuracy and efficiency. In many cases, the coupling necessitates a decoupling of the time-stepping procedure, either because different software packages each have their own time stepper, or in order to improve the overall simulation efficiency, e.g. if multirate behaviour is present. In this chapter, two advanced cosimulation approaches are presented: dynamic iteration (waveform relaxation) and a fractional splitting equipped with error correction. These schemes provide an efficient numerical set-up despite large differences in time scales or the high nonlinearity of the coupling terms. The methods are illustrated for a field-circuit coupled magnetoquasistatic finite-element model of a power transformer and for an electrothermal finite-integration model of an electronic package.

7.1 Introduction

Recent trends in electrical engineering such as the proceeding miniaturisation and the continuing increase of operating frequencies have a consider-

Sebastian Schöps, David J. Duque Guerra, Herbert DeGersem
Technische Universität Darmstadt, Germany,
e-mail: {Schoeps,Duque}@gsc.tu-darmstadt.de;DeGersem@temf.tu-darmstadt.de

Andreas Bartel, Michael Günther
Bergische Universität Wuppertal, Germany,
e-mail: {Bartel,Guenther}@math.uni-wuppertal.de

Roland Pulch
Universität Greifswald, Germany,
e-mail: Roland.Pulch@uni-greifswald.de

© Springer Nature Switzerland AG 2019
E. J. W. ter Maten et al. (eds.), *Nanoelectronic Coupled Problems Solutions*,
Mathematics in Industry 29, https://doi.org/10.1007/978-3-030-30726-4_7

able impact on the requirements to design and simulation tools in electrical engineering. Nowadays, electromagnetic field simulation never comes alone. Thermal, structural-mechanic and fluid-mechanic effects necessitate according field simulation tools to accompany the electromagnetic field solver (*multiphysics* problems). Moreover, phenomena happen at different spatial and temporal scales (*multiscale* and *multirate* problems). Additionally, efficient computation schemes may require different formulations and discretisations for different model parts (*multidomain* models). The tendency towards *multi-X* problems and models comes with a price [14]. Despite the perception that the coupling of existing solution techniques was straightforward, the application of multi-X simulation approaches is hampered by failing guarantees on stability and numerical efficiency. In this chapter, this problem will be highlighted. Appropriate iterative coupling schemes for a few coupled formulations with a high relevance within the nanoCOPS project [35, 36] are developed and studied concerning their numerical properties.

Two challenging examples are presented:

- *Power transformers* in low-voltage equipment are more and more submitted to currents and voltages with a high harmonic distortion. Components at higher frequencies cause additional losses in windings and iron yoke parts which need to be cooled away or at least decrease the device's efficiency [58]. Moreover, voltage spikes cause insulation problems. A power transformer itself can only be modelled up to a sufficient accuracy by a 2D or 3D finite-element (FE) model. Because the surrounding circuitry heavily determines the transformer's performance, these parts need to be included in the model, preferably by a field-circuit coupling [17, 52]. Switching components (diodes, insulated gate bipolar transistors (IG-BTs) and metal oxide semiconductor field-effect transistors (MOSFETs)) cause phenomena in the electric circuitry at frequencies from 1 kHz up to 1 MHz, whereas the time constants in the magnetic circuit are typically in the range of 1 Hz to 1 kHz.
- *Nano-electronic devices* need to be designed according to their electric and thermal performance. Today, electrothermally coupled field simulation is indispensable for this task [46]. Thereby, the heat rise due to the Ohmic losses as well as the temperature dependence of the electric conductivity, the thermal conductivity and the volumetric heat capacity need to be accounted for.

For both problems, the relevant phenomena occur at different spatial and temporal scales. The overall simulation should nevertheless be embedded in an iterative design and optimisation process and thus needs to be computationally efficient.

The chapter is structured as follows. In Section 7.2, an attempt to a classification of coupled problems is presented. In Section 7.3, a coupled electroquasistatic-thermal formulation is set up and discretised by FEs. In Section 7.4, a field-circuit coupled magnetoquasistatic formulation is presented and again discretised by FEs. In Section 7.5, cosimulation with waveform relaxation iterating between two independent time steppers is worked out and discussed. In Section 7.6, a time-stepping algorithm using fractional splitting and error correction is developed and applied. Section 7.7 and Section 7.8 show the performance of waveform relaxation and fractional time stepping with error correction on the examples. The chapter ends with conclusions.

7.2 Classification and Framework for Coupled Problems

Single-physics stand-alone solvers are abandoned in favour of multiphysics, multiscale, multirate and multidomain field simulation approaches. Many coupling schemes have been described in international journal publications and are available in academic and commericial software packages. In this section, a (possibly incomplete) classification of coupled approaches involving electromagnetic field simulation is provided.

Distinctions are made according to five *categories*. It will become clear that a coupling according to one category may (e.g. for reasons of numerical efficiency) necessitate a further coupling according to another category. Examples will be given below.

7.2.1 Category 1: Multiphysics

When several physical phenomena strongly interact with each other, only multiphysics models are able to come up with a correct solution. In engineering practice, one often starts bottom up, i.e., existing solvers are combined. In particular, couplings of electromagnetics, structural dynamics, fluid dynamics and thermodynamics have been developed and are already indispensable for large classes of engineering problems [26, 36, 46]. Many commercial software packages offer multiphysics couplings. Recently, some simulation-software companies have merged in order to strengthen their position on the market concerning multiphysics simulations.

7.2.2 Category 2: Multiscale and Multirate Approaches

Many relevant problems involve spatial scales and time scales of different orders of magnitude [50]. In some situations, small-scale effects can be averaged and introduced in the coarse-scale models as, e.g. homogenised bulk materials or filtered excitation signals. For complicated phenomena, possibly including nonlinear effects, however, an a-priori determination of macroscopic surrogates correctly representing the fine-scale behaviour is impossible. Then, the overall model needs to be calculated on the fine scale or, whenever possible, an iteration between macroscopic and microscopic models should be set up [39]. Such models are called *multiscale* models when coupling different spatial scales and *multirate* models when coupling different time scales. A particular intrusive approach is given by MPDAEs as discussed in [41,42,44].

7.2.3 Category 3: Multifidelity Simulation

There is typically a correlation between the accuracy of obtained simulation results and the computational effort needed to obtain them [40]. Moreover, it does not make sense to simulate models up to higher accuracies than those that are given by the input data and those that are required in the actual stage of the design procedure. For different purposes and hence also in different stages of a design process, the requirements on accuracy are different. A further prospection of the design parameter space allows for fast and comparably inaccurate solution, whereas a final design check requires the maximally attainable accuracy. Also within one simulation run, the availability of several models for the same device but with different resolution can be advantageous. In algorithms that require repeated evaluations of large field models, the alternating use of a low- and high-fidelity model within a *multifidelity* simulation context can improve the computational efficiency [38]. Examples thereof are the use of surrogate models in optimisation procedures [9,20,25,27,30,32,60] and the use of a hierarchy of model with increasing accuracy in multilevel Monte Carlo models [7,24,56].

7.2.4 Category 4: Multidomain Modelling

Many models consist of several domains and model parts with significant differences in dimensions, materials, physical behaviour and numerical challenges. Therefore, it makes sense to adapt the applied modelling and simulation techniques to the particular properties of the subdomains, thereby necessitating their combination into a full model and an appropriate outer simulation procedure. Such *hybridisations* are many-sided. Subdomains with different physical quantities require a *hybrid formulation* [12,43]. Models with differences in geometric details or different resolution requirements are addressed by *hybrid discretisations* [16,31,55], a subclass thereof being field-circuit coupled models [17,52,57], which can also be considered as refined network modelling [5,44]. Multidomain models often come together with multiphysics, multiscale/multirate and multifidelity simulation. Multidomain approaches also offer possibilities for improving the overall simulation scheme, i.e., by reducing the overall number of DoFs, by improving the condition number of the underlying systems of equations and by providing possibilities for parallelisation. Even a decomposition of a homogeneous model, e.g. in overlapping or non-overlapping subdomains, may be advantageous as is shown in standard domain decomposition methods [3, 54]. Multidomain approach with largely different solver kernels motivate the application of cosimulation techniques [15, 50].

7.2.5 Challenges of Multi-X Simulation

Multi-X simulation approaches as classified in the previous sections seem to be intuitive. However, they come with considerable challenges at the numerical side. Only in a few cases, extended modelling features are achievable with small modifications and within an already existing numerical setup, e.g. the embedding of simple thin-sheet models in FE models requires a few matrix entries to be added as already existing positions and do not change the matrix properties nor the algorithms for solving the system of equations [45]. A further example is multiphysical behaviour which can be treated in a cascadic way, i.e. quantities from one field model are forwarded to a next field model, possibly after mesh interpolation (Fig. 7.1a) [34]. More mature multiphysics models, however, require the solution and time-stepping of mutually coupled transient processes, each of them described by an own set of PDEs, possibly on an own computational mesh [6, 33]. The interfaces between the model parts are then given along boundary conditions, sources, material coefficients and geometrical deformations.

7.2.6 Monolithic versus Iterative Coupling

An overall simulation model may be heavily heterogeneous, extremely large and badly conditioned. In general, considering the multi-X model as a whole, called a *monolithic* or *numerically strong* coupling (Fig. 7.1b) is impossible or at least cumbersome. This situation appears, e.g., when the electromagnetic and thermal field problems need to be simulated by two different simulation tools or when the simulation of one subproblem already touches the limit of the available computational capacity [10]. Instead, one is tempted to organise a straightforward *iterative* or *numerically weak* coupling (Fig. 7.1c), thereby motivated by the existence of already available software parts. Organising an iterative coupling can be done on the implementation level (coupling individual software packages), on the modelling level (coupling different models within a software package), on the simulation level (iterating between discre-

tised models) or even on the algebraic level (iterating between the blocks of a
coupled system of linear or nonlinear equations). Weak couplings have many
advantages. They allow to use different tools, different spatial and temporal
discretisation schemes, different resolutions and parallel computing facilities.
The main drawback of weak coupling schemes is the possibly slow or even
failing convergence of the iteration between the solvers. This drawback is
particularly pronounced when the overall problem is transient. In a standard
weak coupling, one is tempted to iterate until convergence at every time step.
A few iterations, however, already cause a raise of the overall simulation time
by more than an order of magnitude. Better approaches exist, which in some
sense tie the multiphysics iteration together with the time-stepping procedure. In this chapter, two such methods, i.e., dynamic iteration for cosimulation (Fig. 7.1d) and a time stepper based on fractional splitting and error
correction, are described and applied. In both cases, the convergence of such
methods is not guaranteed. A stagnation or a too slow convergence of the
iteration turns the weakly coupled multi-X model to become inapplicable in
practice. Only a rigorous a-priori analysis, mostly incorporating the physical
and numerical properties of the submodels, can alleviate the problem.

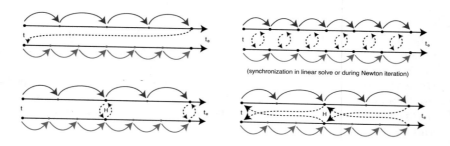

Fig. 7.1: Coupling schemes, schematics based on [14]: (a) Top-Left: *One-way*
coupling (Parameter extraction); (b) Top-Right: *Monolithic* coupling (Numerically strong coupling; (c) Bottom-Left: *Iterative* coupling (Numerically
weak coupling); (d) Bottom-Right: *Dynamic iteration* (Waveform relaxation).

7.2.7 Intrusive versus Non-Intrusive Coupling

The couplings underlying a multi-X simulation can be implemented in an *intrusive* or a *non-intrusive* way. An intrusive implementation necessitates a significant adaptation of the individual software parts, whereas a non-intrusive
implementation keeps the constituent parts more or less unchanged and sets
up a communicates between the parts. At first sight, a non-intrusive imple-

mentation strategy is preferred in order to set on existing, well-established tools. Then, a less performing algorithm and a slower convergence of the iteration is taken for granted. However, for physically strong couplings, an intrusive implementation offers to implement algorithms with better convergence properties, e.g., a monolithic coupling using a Newton scheme for linearising the nonlinear coupling terms. Then, the investment for adapting, merging and newly implementing major software parts is acceptable. In semiconductor industry, intrusive implementations of electrothermal field solvers are available and used within design platforms [46, 48].

7.2.8 Convergence of Coupled Simulations

A major concern in iteratively coupling two or more subproblems is the exchange of data between the subproblems. A loose coupling tends to lack information (large *splitting error*), whereas a tighter coupling guarantees a better exchange (smaller splitting error) and a monolithic coupling provides an optimal data exchange. In any case, there is a trade-off between the cost of the information exchange and the number of iterations needed to achieve convergence. A particular opportunity arises when the problems are only weakly coupled in time. Then, a small time lag is acceptable and the problems need only to be synchronised at a number of time instants which is substantially smaller than the number of time steps to be processed for the subproblems (Fig. 7.1c). This approach becomes unstable for problems where the subsystems are physically strong coupled. Increasing the frequency of data exchange does not cure the problem. A remedy for this is the *dynamic iteration* scheme discussed below (Fig. 7.1d).

A simple example how things can go seriously wrong is as follows. Consider the linear equations

$$y_1 + 2y_2 = 1; \tag{7.1}$$
$$y_1 + y_2 = 1, \tag{7.2}$$

which have as solution $[y_1, y_2] = [1, 0]$. Each equation can be resolved either towards y_1 or y_2. One can set up two straightforward cosimulation schemes of Gauss-Seidel type, i.e.,

$$\text{(i)} \quad y_2^{(0)} = 2 \ \rightarrow \ y_1^{(i)} = 1 - 2y_2^{(i-1)}, \qquad y_2^{(i)} = 1 - y_1^{(i)} \quad \rightarrow \ [\infty, \infty]; \tag{7.3}$$

$$\text{(ii)} \quad y_1^{(0)} = 2 \ \rightarrow \ y_2^{(i)} = \frac{1}{2}(1 - y_1^{(i-1)}), \quad y_1^{(i)} = 1 - y_2^{(i)} \quad \rightarrow \ [1, 0]. \tag{7.4}$$

Only the second scheme is convergent. This issue (already known to Gauss and Seidel [4]) demonstrates the necessity of an a-priori analysis for cosimulation schemes as will be presented in Section 7.5.

7.3 Electrothermal Finite-Element Model

This chapter describes the electroquasistatic formulation, its discretisation and suitable boundaries conditions.

7.3.1 Electrothermal Formulation

We intend to solve structures as shown in Fig. 7.2. Such structures consist of conducting vias, contacts and interconnects and of insulating dielectrics. We want to compute the currents, the electric fields and the temperature rise caused by them. The governing partial differential equations are

$$-\nabla \cdot (\sigma \nabla \phi) - \nabla \cdot \left(\varepsilon \nabla \frac{\partial \phi}{\partial t}\right) = 0; \tag{7.5}$$

$$-\nabla \cdot (\kappa(T)\nabla T) + \rho c_{\mathrm{e}} \frac{\partial T}{\partial t} = \dot{q}_{\mathrm{i}}. \tag{7.6}$$

Eq. (7.5) originates from the choice of the electric scalar potential ϕ to express the electric field strength $\boldsymbol{E} = -\nabla \phi$, the constitutive equation for the electric current density $\boldsymbol{J} = \sigma(T)\boldsymbol{E}$ with the conductivity $\sigma(T)$, the continuity equation for the electric flux density $\boldsymbol{D} = \varepsilon \boldsymbol{E}$ with the permittivity ε, the continuity equation $\nabla \cdot \boldsymbol{J} + \frac{\partial \varrho_{\mathrm{e}}}{\partial t} = 0$ in terms of the electric charge density ϱ_{e} and Gauss' law $\nabla \cdot \boldsymbol{D} = \varrho_{\mathrm{e}}$. Eq. (7.6) describes the heat conduction problem with ρc_{e} the volumetric heat capacity, $\kappa(T)$ the thermal conductivity, T the temperature and $\dot{q}_{\mathrm{i}}$ the heat source density. Both partial differential equations are coupled through the heat source $\dot{q}_{\mathrm{i}} = \boldsymbol{E} \cdot \boldsymbol{J}$ and through the temperature dependent electric conductivity $\sigma(T)$. Moreover, the heat equation is nonlinear because of the temperature dependent thermal conductivity $\kappa(T)$, the heat source density $\dot{q}_{\mathrm{i}}$ and possibly also because of the temperature-dependent volumetric heat capacity ρc_{e}.

7.3.2 Discretisation in Space

We discretise the electrothermal field problem using a tetrahedral mesh of the computational domain. The incidences between the mesh edges and the mesh nodes are stored in the discrete gradient matrix $\mathbf{G}$, i.e., when edge q is oriented from node i to node j, $\mathbf{G}_{qi} = 1$ and $\mathbf{G}_{qj} = -1$. The discrete gradient matrix allows to express electric voltages along the grid edges by $\widehat{\mathbf{e}} = -\mathbf{G}\boldsymbol{\varphi}$ in terms of the potentials $\boldsymbol{\varphi}$ at the grid nodes and allows to express temperature drops along the grid edges by $\widehat{\mathbf{t}} = -\mathbf{G}\mathbf{T}$ in terms of the temperatures $\mathbf{T}$ at the grid nodes. We can think about electric currents $\widehat{\widehat{\mathbf{j}}}$, electric fluxes $\widehat{\widehat{\mathbf{d}}}$ and

Fig. 7.2: Layout of a power transistor (stretched along the vertical direction) [35].

heat fluxes $\widehat{\widehat{\mathbf{q}}}$ along the same edges (or better said through the dual faces pairing with the primary grid edges). Charge and heat conservation is then expressed by the relations

$$\widetilde{\mathbf{S}}\widehat{\widehat{\mathbf{j}}} + \widetilde{\mathbf{S}}\frac{\mathrm{d}\widehat{\widehat{\mathbf{d}}}}{\mathrm{d}t} = \mathbf{0};\tag{7.7}$$

$$\widetilde{\mathbf{S}}\widehat{\widehat{\mathbf{q}}} = \mathbf{Q},\tag{7.8}$$

where $\mathbf{Q}$ denotes the heat production allocated at the grid nodes. Here, $\widetilde{\mathbf{S}}$ is the discrete divergence matrix for which is found that $\mathbf{G} = -\widetilde{\mathbf{S}}^{\top}$ [1,11,13,59].

We denote by $\mathbf{r}$ the spatial coordinate. We associate nodal shape functions $N_j(\mathbf{r})$ to the mesh nodes of the mesh and edge shape functions $\mathbf{w}_q(\mathbf{r})$ to the mesh edges. We assume that the edge shape functions are constructed from the nodal shape functions using the expression

$$\mathbf{w}_q(\mathbf{r}) = N_i(\mathbf{r})\nabla N_j(\mathbf{r}) - N_j(\mathbf{r})\nabla N_i(\mathbf{r}).\tag{7.9}$$

The constitutive equations are transferred in discrete material operators, i.e.,

$$\mathbf{M}_{\sigma;p,q} = \int_V \sigma \mathbf{w}_q \cdot \mathbf{w}_p \, \mathrm{d}\mathbf{r},\tag{7.10}$$

$$\mathbf{M}_{\varepsilon;p,q} = \int_V \varepsilon \mathbf{w}_q \cdot \mathbf{w}_p \, \mathrm{d}\mathbf{r},\tag{7.11}$$

$$\mathbf{M}_{\lambda;p,q} = \int_V \lambda \mathbf{w}_q \cdot \mathbf{w}_p \, \mathrm{d}\mathbf{r},\tag{7.12}$$

$$\mathbf{M}_{\rho c;i,j} = \int_V \rho c N_j N_i \, \mathrm{d}\mathbf{r},\tag{7.13}$$

where V is the computational domain. The discrete material operators connect the currents and fluxes with the voltages and temperature drops, i.e., $\widehat{\widehat{\mathbf{j}}} = \mathbf{M}_\sigma \widehat{\mathbf{e}}$, $\widehat{\widehat{\mathbf{d}}} = \mathbf{M}_\varepsilon \widehat{\mathbf{e}}$, $\widehat{\widehat{\mathbf{q}}} = \mathbf{M}_\lambda \widehat{\mathbf{t}}$ and $\mathbf{Q} = \mathbf{M}_{\rho c}\frac{\partial \mathbf{T}}{\partial t}$.

The space-discretised counterparts of (7.5) and (7.6) read as

$$\widetilde{\mathbf{S}}\mathbf{M}_\varepsilon \widetilde{\mathbf{S}}^\top \frac{d\boldsymbol{\varphi}}{dt} + \widetilde{\mathbf{S}}\mathbf{M}_\sigma(\mathbf{T})\widetilde{\mathbf{S}}^\top \boldsymbol{\varphi} = \mathbf{0}, \tag{7.14}$$

$$\mathbf{M}_{\rho c}\frac{d\mathbf{T}}{dt} + \widetilde{\mathbf{S}}\mathbf{M}_\lambda(\mathbf{T})\widetilde{\mathbf{S}}^\top \mathbf{T} = \mathbf{Q}_{el}(\boldsymbol{\varphi}). \tag{7.15}$$

The grid heat source is given by

$$\mathbf{Q}_{el,i} = \sum_p \sum_q \varphi_p \varphi_q \int_V \sigma \mathbf{w}_q \cdot \mathbf{w}_p N_i \, dr. \tag{7.16}$$

The temperature used for evaluating the temperature-dependent material data is

$$T(\mathbf{r}_m) = \sum_j T_j N_j(\mathbf{r}_m) \tag{7.17}$$

in the center points $\mathbf{r}_m$ of the grid cells.

7.3.3 Boundary conditions

We distinguish between Dirichlet boundary conditions (BCs) enforcing a prescribed voltage or temperature at particular surfaces in the model and homogeneous Neumann BCs where the total electric current density and the heat flux density are assumed to be zero. The implementation of these boundary conditions turns (7.14) and (7.15) into a system of equations of the form

$$\mathbf{M}_{eqs}\frac{d\mathbf{u}_{eqs}}{dt} + \mathbf{K}_{eqs}(\mathbf{u}_{th})\mathbf{u}_{eqs} = \mathbf{f}_{eqs}(\mathbf{u}_{th}), \tag{7.18a}$$

$$\mathbf{M}_{th}\frac{d\mathbf{u}_{th}}{dt} + \mathbf{K}_{th}(\mathbf{u}_{th})\mathbf{u}_{th} = \mathbf{f}_{th}(\mathbf{u}_{th}) + \mathbf{g}_{th}(\mathbf{u}_{eqs}). \tag{7.18b}$$

with the matrices $\mathbf{M}_{eqs} = \widetilde{\mathbf{S}}\mathbf{M}_\varepsilon\widetilde{\mathbf{S}}^\top$, $\mathbf{K}_{eqs}(\mathbf{u}_{th}) = \widetilde{\mathbf{S}}\mathbf{M}_\sigma(\mathbf{u}_{th})\widetilde{\mathbf{S}}^\top$, $\mathbf{M}_{th}$ and $\mathbf{K}_{th}(\mathbf{u}_{th})$ and the vectors $\mathbf{f}_{eqs}(\mathbf{u}_{th})$, $\mathbf{f}_{th}(\mathbf{u}_{th})$ and $\mathbf{g}_{th}(\mathbf{u}_{eqs})$. The vectors $\mathbf{u}_{eqs}$ and $\mathbf{u}_{th}$ contain the DoFs for the electric scalar potential and the DoFs for the temperature. The thermal subproblem is nonlinear as indicated by the dependencies on $\mathbf{u}_{th}$ in $\mathbf{K}_{th}(\mathbf{u}_{th})$ and $\mathbf{f}_{th}(\mathbf{u}_{th})$. The multiphysics coupling is represented by the dependencies on $\mathbf{u}_{th}$ and $\mathbf{u}_{eqs}$ in $\mathbf{K}_{eqs}(\mathbf{u}_{th})$, $\mathbf{f}_{eqs}(\mathbf{u}_{th})$ and $\mathbf{g}_{th}(\mathbf{u}_{eqs})$.

The coupled formulation (7.18a) and (7.18b) can, at least formally, be written as

$$\frac{d\mathbf{x}_1}{dt} = \mathbf{f}_1(\mathbf{x}_1, \mathbf{x}_2); \tag{7.19a}$$

$$\frac{d\mathbf{x}_2}{dt} = \mathbf{f}_2(\mathbf{x}_1, \mathbf{x}_2), \tag{7.19b}$$

where

$$\mathbf{f}_1(\mathbf{u}_{\mathrm{eqs}}, \mathbf{u}_{\mathrm{th}}) = \mathbf{M}_{\mathrm{eqs}}^{-1}\left(\mathbf{f}_{\mathrm{eqs}}(\mathbf{u}_{\mathrm{th}}) - \mathbf{K}_{\mathrm{eqs}}(\mathbf{u}_{\mathrm{th}})\mathbf{u}_{\mathrm{eqs}}\right); \tag{7.20a}$$

$$\mathbf{f}_2(\mathbf{u}_{\mathrm{eqs}}, \mathbf{u}_{\mathrm{th}}) = \mathbf{M}_{\mathrm{th}}^{-1}\left(\mathbf{f}_{\mathrm{th}}(\mathbf{u}_{\mathrm{th}}) + \mathbf{g}_{\mathrm{th}}(\mathbf{u}_{\mathrm{eqs}}) - \mathbf{K}_{\mathrm{th}}(\mathbf{u}_{\mathrm{th}})\mathbf{u}_{\mathrm{th}}\right). \tag{7.20b}$$

In practice, $\mathbf{f}_1(\mathbf{u}_{\mathrm{eqs}}, \mathbf{u}_{\mathrm{th}})$ and $\mathbf{f}_2(\mathbf{u}_{\mathrm{eqs}}, \mathbf{u}_{\mathrm{th}})$ are never evaluated as such. Instead, the time-discretised expression is reformulated into a system of equations to be solved for time instances of $\mathbf{u}_{\mathrm{eqs}}$ and $\mathbf{u}_{\mathrm{th}}$.

7.4 Magnetoquasistatic Field-Circuit Coupled Model

We consider electromagnetic problems at moderate frequencies including iron yokes for guiding the magnetic flux and coils as excitation thereof. Moreover, eddy current effects and ferromagnetic saturation effects are expected. The devices are connected to external circuitry in order to model parts outside the field part of the computational domain and for representing the sources (Fig. 7.3).

7.4.1 Formulation

A field-circuit coupled magnetoquasistatic (MQS) problem is expressed by

$$\nabla \times (\nu \nabla \times \mathbf{A}) + \sigma \frac{\partial \mathbf{A}}{\partial t} - \sum_q \sigma \chi_{\mathrm{sol},q} u_{\mathrm{sol},q} - \sum_p \chi_{\mathrm{str},p} i_{\mathrm{str},p} = 0; \tag{7.21}$$

$$-\int_V \sigma \chi_{\mathrm{sol},q} \cdot \frac{\partial \mathbf{A}}{\partial t}\, d\mathbf{r} + G_{\mathrm{sol},q} u_{\mathrm{sol},q} = i_{\mathrm{sol},q}; \tag{7.22}$$

$$\int_V \chi_{\mathrm{str},p} \cdot \frac{\partial \mathbf{A}}{\partial t}\, d\mathbf{r} + R_{\mathrm{str},p} i_{\mathrm{str},p} = u_{\mathrm{str},p}, \tag{7.23}$$

where $\mathbf{A}$ is the magnetic vector potential, ν the reluctivity, σ the conductivity, the subscripts "sol" and "str" stand for solid conductor and stranded conductor, respectively, $\chi_{\mathrm{sol},q}$ the connecting function for solid conductor q, $\chi_{\mathrm{str},p}$ the winding function for stranded conductor p [52], V the computational domain, $G_{\mathrm{sol},q}$, $u_{\mathrm{sol},q}$ and $i_{\mathrm{sol},q}$ the DC conductance, the voltage and the current of solid conductor q and $R_{\mathrm{str},p}$ $i_{\mathrm{str},p}$ and $u_{\mathrm{str},p}$ the DC resistance, the current and the voltage of stranded conductor p.

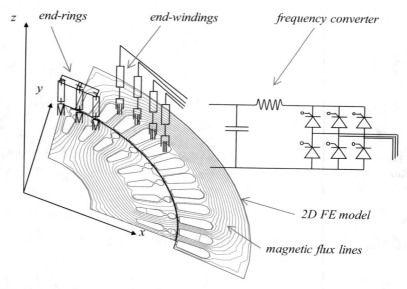

Fig. 7.3: 2D FE model of an asynchronous machine combined with external lumped parameters modelling the rotor end ring and the stator end windings and with an external circuit modelling the frequency converter exciting the system.

7.4.2 Discretisation in Space

The set of edge shape functions $\{\mathbf{w}_j\}$ is used to weigh the MQS equation (7.21) and to discretise the magnetic vector potential. The resulting algebraic equations are

$$\mathbf{K}_\nu \widehat{\mathbf{a}} + \mathbf{M}_\sigma \frac{\mathrm{d}\widehat{\mathbf{a}}}{\mathrm{d}t} - \mathbf{M}_\sigma \mathbf{X}_{\mathrm{sol}} \mathbf{u}_{\mathrm{sol}} - \mathbf{X}_{\mathrm{str}} \mathbf{i}_{\mathrm{str}} = \mathbf{0}; \tag{7.24a}$$

$$-\mathbf{X}_{\mathrm{sol}}^\top \mathbf{M}_\sigma \frac{\mathrm{d}\widehat{\mathbf{a}}}{\mathrm{d}t} + \mathbf{G}_{\mathrm{sol}} \mathbf{u}_{\mathrm{sol}} = \mathbf{i}_{\mathrm{sol}}; \tag{7.24b}$$

$$\mathbf{X}_{\mathrm{str}}^\top \frac{\mathrm{d}\widehat{\mathbf{a}}}{\mathrm{d}t} + \mathbf{R}_{\mathrm{str}} \mathbf{i}_{\mathrm{str}} = \mathbf{u}_{\mathrm{str}}, \tag{7.24c}$$

where

$$\mathbf{K}_{\nu,i,j} = \int_V \nu \nabla \times \mathbf{w}_j \cdot \nabla \times \mathbf{w}_i \, d\mathbf{r}; \tag{7.25}$$

$$\mathbf{M}_{\sigma,i,j} = \int_V \sigma \mathbf{w}_j \cdot \mathbf{w}_i \, d\mathbf{r}; \tag{7.26}$$

$$\mathbf{X}_{\text{sol},i,q} = \int_V \chi_{\text{sol},q} \cdot \mathbf{w}_i \, d\mathbf{r}; \tag{7.27}$$

$$\mathbf{X}_{\text{str},i,p} = \int_V \chi_{\text{str},p} \cdot \mathbf{w}_i \, d\mathbf{r}. \tag{7.28}$$

Using the discrete curl-curl matrix $\mathbf{K}_\nu$, it is possible to extract a lumped inductance matrix, i.e.,

$$\mathbf{L}_{\text{str}} = \mathbf{X}_{\text{str}}^\top \mathbf{K}_\nu^+ \mathbf{X}_{\text{str}}, \tag{7.29}$$

where $^+$ denotes a pseudo-inverse. In case of nonlinear materials in the field problem, $\mathbf{K}_\nu$ and hence also $\mathbf{L}_{\text{str}}$ depend on $\widehat{\mathbf{a}}$. A similar coupling scheme is obtained when discretising by the finite-integration technique [18].

The field-circuit coupled formulation (7.24) can be cast in the form of a DAE initial-value problem, i.e.,

$$\dot{\mathbf{x}}_1 = \mathbf{f}_1(\mathbf{x}_1, \mathbf{y}_1, \mathbf{x}_2, \mathbf{y}_2), \quad \text{with} \quad \mathbf{x}_1(t_0) = \mathbf{x}_{1,0}, \tag{7.30a}$$

$$0 = \mathbf{g}_1(\mathbf{x}_1, \mathbf{y}_1), \tag{7.30b}$$

$$0 = \mathbf{z}_1 - \mathbf{h}_1(\mathbf{x}_1, \mathbf{y}_1). \tag{7.30c}$$

Here, it is assumed that $\partial \mathbf{g}_1 / \partial \mathbf{y}_1$ is non-singular along the solution. Moreover, $\mathbf{x}_1 = \mathbf{P}\widehat{\mathbf{a}}$ are the differential variables related to the DoFs of the magnetic vector potential which are not in the kernel of $\mathbf{M}_\sigma$, i.e., $\mathbf{P} = \mathbf{I} - \mathbf{Q}$ where $\mathbf{Q}$ is a projector onto the kernel of $\mathbf{M}_\sigma$. $\mathbf{y}_1 = (\mathbf{Q}\widehat{\mathbf{a}}, \mathbf{i}_{\text{sol}}, \mathbf{i}_{\text{str}})^\top$ and $\mathbf{z}_1$ are algebraic variables, i.e., the algebraic DoFs of the magnetic vector potential and the variables expressing the field-circuit coupling. $\mathbf{x}_2(t)$ and $\mathbf{y}_2(t)$ are input functions from the circuit containing the voltages $\mathbf{u}_{\text{sol}}(t)$, $\mathbf{u}_{\text{str}}(t)$ [49,50].

7.4.3 Circuit Formulation

The electric circuit is formulated by the modified nodal analysis [23, 28], i.e.,

$$\mathbf{A}_C \frac{d\mathbf{q}}{dt} + \mathbf{A}_R \mathbf{G} \mathbf{A}_R^\top \mathbf{u} + \mathbf{A}_L \mathbf{I}_L + \mathbf{A}_V \mathbf{I}_V + \mathbf{A}_I \mathbf{I}(t) = \mathbf{0}; \qquad (7.31\text{a})$$

$$\frac{d\Phi}{dt} - \mathbf{A}_L^\top \mathbf{u} = \mathbf{0}; \qquad (7.31\text{b})$$

$$\mathbf{A}_V^\top \mathbf{u} - \mathbf{v}(t) = \mathbf{0}; \qquad (7.31\text{c})$$

$$\mathbf{q} - \mathbf{C} \mathbf{A}_C^\top \mathbf{u} = \mathbf{0}; \qquad (7.31\text{d})$$

$$\Phi - \mathbf{L} \mathbf{A}_L^\top \mathbf{I}_L = \mathbf{0}, \qquad (7.31\text{e})$$

where the circuit topology is given by incidence matrices $\mathbf{A}_\star$, the unknowns are the nodal potentials $\mathbf{u}$, the charges $\mathbf{q}$, the fluxes Φ and the currents $\mathbf{I}_L$ and $\mathbf{I}_V$. This model can be cast into a semi-explicit differential algebraic initial-value problem in the case no LI-cutsets and CV-loops are present [21], i.e.,

$$\dot{\mathbf{x}}_2 = \mathbf{f}_2(\mathbf{x}_2, \mathbf{y}_2, \mathbf{z}_1), \quad \text{with} \quad \mathbf{x}_2(t_0) = \mathbf{x}_{2,0} \qquad (7.32\text{a})$$

$$\mathbf{0} = \mathbf{g}_2(\mathbf{x}_2, \mathbf{y}_2, \mathbf{z}_1). \qquad (7.32\text{b})$$

Here, the differential variables are $\mathbf{x}_2 = (\mathbf{q}, \Phi)^\top$ and the algebraic variables are $\mathbf{y}_2 = (\mathbf{u}, \mathbf{I}_L, \mathbf{I}_V)^\top$. The input functions is $\mathbf{z}_1(t)$ which is coupled to the field system and may comprise any circuit parameter, e.g., the exciting currents $\mathbf{i}(t)$ or the inductances $\mathbf{L}_{\text{str}}(\mathbf{I}_L, t)$.

7.5 Cosimulation with Dynamic Iteration

To develop a cosimulation scheme with dynamic iteration, we consider semi-discrete coupled problems that are already discretised in space but not yet in time as are the systems of equations resulting from Section 7.3 and Section 7.4 (*method of lines* and *finite-element method*). For simplicity of notation, only semi-explicit problems are considered. The problem is expressed on a *time window* $[0, H]$ with time-window size H by

$$\frac{d\mathbf{y}}{dt} = \mathbf{f}(\mathbf{y}, \mathbf{z}); \qquad (7.33)$$

$$0 = \mathbf{g}(\mathbf{y}, \mathbf{z}), \qquad (7.34)$$

where the vectors of unknowns $\mathbf{y}$ and $\mathbf{z}$ have initial values $\mathbf{y}_0$ and $\mathbf{z}_0$, $\mathbf{f}(\mathbf{y}, \mathbf{z})$ and $\mathbf{g}(\mathbf{y}, \mathbf{z})$ are functions in both unknowns and the Jacobian of $\mathbf{g}(\mathbf{y}, \mathbf{z})$ is assumed to be regular, i.e., $\det \frac{\partial \mathbf{g}}{\partial \mathbf{z}} \neq 0$. The formulation is a first order differential algebraic equation of index 1. Many formulations can be cast in this framework, e.g., magnetoquasistatic, electroquasistatic, thermal field problems and combinations thereof. Also formulations with higher order time derivatives and formulations with time-dependent right-hand sides can be converted in (7.33) and (7.34). The generic problem is expressed as a function of the differential variables $\mathbf{y}$ and the algebraic variables $\mathbf{z}$, which exhibit a different numerical behaviour and therefore need a different treatment in the numerical analysis. In the implementation, however, this distinction does not always have to be made.

The vectors of unknowns $\mathbf{y}$ and $\mathbf{z}$ and the operators $\mathbf{f}$ and $\mathbf{g}$ are now divided into vectors of unknowns $\mathbf{y}_1,\ldots,\mathbf{y}_n$ and $\mathbf{z}_1,\ldots,\mathbf{z}_n$ and the operators $\mathbf{f}_1,\ldots,\mathbf{f}_n$ and $\mathbf{g}_1,\ldots,\mathbf{g}_n$ according to the n subproblems constituting the overall problem. A Gauss-Seidel type of iteration is expressed by

$$\frac{d\mathbf{y}_1^{(k+1)}}{dt} = \mathbf{f}_1\left(\mathbf{y}_1^{(k+1)},\mathbf{z}_1^{(k+1)},\mathbf{y}_2^{(k)},\mathbf{z}_2^{(k)},\ldots,\mathbf{y}_n^{(k)},\mathbf{z}_n^{(k)}\right); \tag{7.35}$$

$$0 = \mathbf{g}_1\left(\mathbf{y}_1^{(k+1)},\mathbf{z}_1^{(k+1)},\mathbf{y}_2^{(k)},\mathbf{z}_2^{(k)},\ldots,\mathbf{y}_n^{(k)},\mathbf{z}_n^{(k)}\right); \tag{7.36}$$

$$\vdots$$

$$\frac{d\mathbf{y}_n^{(k+1)}}{dt} = \mathbf{f}_n\left(\mathbf{y}_1^{(k+1)},\mathbf{z}_1^{(k+1)},\mathbf{y}_2^{(k+1)},\mathbf{z}_2^{(k+1)},\ldots,\mathbf{y}_n^{(k+1)},\mathbf{z}_n^{(k+1)}\right); \tag{7.37}$$

$$0 = \mathbf{g}_n\left(\mathbf{y}_1^{(k+1)},\mathbf{z}_1^{(k+1)},\mathbf{y}_2^{(k+1)},\mathbf{z}_2^{(k+1)},\ldots,\mathbf{y}_n^{(k+1)},\mathbf{z}_n^{(k+1)}\right), \tag{7.38}$$

where the superscripts (k) and $(k+1)$ denote the iteration steps. Hence, $\mathbf{y}_j^{(k)}$ and $\mathbf{z}_j^{(k)}$ denote the solution of the j-th subproblem $(j=1,\ldots,n)$ after k iterations $k=1,\ldots,m$ on time window $[0,H]$. The solutions according to operations $\mathbf{f}_i$ and $\mathbf{g}_i$ represent the time stepping of the i-th subproblem $(i=1,\ldots,n)$ on time window $[0,H]$. Notice that operation $(\mathbf{f}_i,\mathbf{g}_i)$ also needs the data $(\mathbf{y}_j,\mathbf{z}_j)$ for $j \neq i$ being the data exchanged between the subproblems. After m dynamic iterations necessary to reduce the splitting error below a user defined tolerance, the subsequent time window is addressed.

From the numerical analysis published in [2,4], it is known that the above formulation corresponds to a fixed point iteration where the splitting errors $\delta_y^{(k)}$ and $\delta_z^{(k)}$ for the differential variables $\mathbf{y}$ and the algebraic variables $\mathbf{z}$, respectively, converge for sufficiently small time windows H as given by

$$\begin{bmatrix} \delta_y^{(k+1)} \\ \delta_z^{(k+1)} \end{bmatrix} \leq \begin{bmatrix} \mathcal{O}(H) & \mathcal{O}(H) \\ \beta+\mathcal{O}(H) & \alpha+\mathcal{O}(H) \end{bmatrix} \begin{bmatrix} \delta_y^{(k)} \\ \delta_z^{(k)} \end{bmatrix}, \tag{7.39}$$

where α and β are two constants influencing the contraction. Convergence is obtained if α is sufficiently small which requires a weak coupling between the old and new algebraic variables, i.e. [4],

$$\left\|\left(\frac{\partial \mathbf{g}}{\partial \mathbf{z}}\left(\mathbf{y}^{(k+1)},\mathbf{z}^{(k+1)}\right)\right)^{-1}\frac{\partial \mathbf{g}}{\partial \mathbf{z}}\left(\mathbf{y}^{(k)},\mathbf{z}^{(k)}\right)\right\| < 1. \tag{7.40}$$

When in addition β is sufficiently small, the spectral radius of the fixed point iteration operator can be made arbitrarily small, which guarantees convergence. Further analysis in [2], [4] and [50] showed that the approach remains stable also in terms of the propagation of splitting errors on multiple time windows.

In many application, one can organise the systems $(\mathbf{f}_j, \mathbf{g}_j)$, $j = 1, \ldots, n$ such that there is no coupling between the old and the new algebraic variables. In that case, the contraction factor α vanishes completely. Then, as for the case of ordinary differential equations, the system of differential-algebraic equations does no longer suffer from a possibly divergent dynamic iteration. In that case as well, the dynamic iteration is proved to converge at a rate of $\mathcal{O}(H)$ which is often as good as the overall time stepping scheme itself, e.g., when using the implicit Euler method. Hence, weak coupling equipped with dynamic iteration does not slow down the time integration process, at least in its asymptotic order. Further analysis showed that the dynamic iteration attains a convergence rate up to order $\mathcal{O}(H^2)$ if the coupling interface is chosen adequately [4]. In [51], it is shown that by increasing the number of iterations per time window and by using on-the-fly adapted surrogate models, even higher convergence rates can be achieved.

7.6 Time Stepper with Fractional Splitting and Error Correction

In this section, we consider a more intrusive time-stepping approach, i.e., fractional splitting [61], here accomplished by an error correction procedure. The model problem reads

$$\frac{dx_1}{dt} = \mathbf{f}_1(\mathbf{x}_1, \mathbf{x}_2); \tag{7.41a}$$

$$\frac{dx_2}{dt} = \mathbf{f}_2(\mathbf{x}_1, \mathbf{x}_2), \tag{7.41b}$$

where $\mathbf{x}_1(t)$ and $\mathbf{x}_2(t)$ are differential variables and $\mathbf{f}_1(\mathbf{x}_1, \mathbf{x}_2)$ and $\mathbf{f}_2(\mathbf{x}_1, \mathbf{x}_2)$ are functions in both variables.

7.6.1 Fractional Splitting

The time integration is carried out on a staggered pair of grids in time. The primary grid is given by $\mathbf{t} = [t_1, t_2, \ldots, t_N]$ with time step h_n, whereas the dual grid is shifted in time by $h_n/2$. The main idea behind fractional splitting is to keep the time derivatives constant during a fraction of the time step. For the model problem (7.41), this amounts to

$$\frac{dx_1}{dt} \equiv 0 \quad \Rightarrow \quad \tilde{\mathbf{x}}_1^{(n+\frac{1}{2})} = \tilde{\mathbf{x}}_1^{(n)}, \qquad t_n \leq t < t_n + \frac{h_n}{2}, \tag{7.42a}$$

$$\frac{dx_2}{dt} \equiv 0 \quad \Rightarrow \quad \tilde{\mathbf{x}}_2^{(n+1)} = \tilde{\mathbf{x}}_2^{(n+\frac{1}{2})}, \quad t_n + \frac{h_n}{2} \leq t < t_n + h_n. \tag{7.42b}$$

Here, the DoFs discretised in time get a tilde and a superscript according to the time step. The procedure allows to decouple the thermal from the electroquasistatic equations, i.e.,

$$t_n \leq t < t_n + \frac{h_n}{2}; \quad \frac{dx_2}{dt} = \mathbf{f}_2(\tilde{\mathbf{x}}_1^{(n)}, x_2), \tag{7.43a}$$

$$t_n + \frac{h_n}{2} \leq t < t_n + h_n; \quad \frac{dx_1}{dt} = \mathbf{f}_1(x_1, \tilde{\mathbf{x}}_2^{(n+1)}). \tag{7.43b}$$

Now, the splitted system is discretised by the implicit Euler method, i.e.,

$$\frac{\tilde{\mathbf{x}}_2^{(n+1)} - \tilde{\mathbf{x}}_2^{(n)}}{\frac{h_n}{2}} = \mathbf{f}_2\left(\tilde{\mathbf{x}}_1^{(n)}, \tilde{\mathbf{x}}_2^{(n+1)}\right), \tag{7.44a}$$

$$\frac{\tilde{\mathbf{x}}_1^{(n+1)} - \tilde{\mathbf{x}}_1^{(n)}}{\frac{h_n}{2}} = \mathbf{f}_1\left(\tilde{\mathbf{x}}_1^{(n+1)}, \tilde{\mathbf{x}}_2^{(n+1)}\right). \tag{7.44b}$$

Together with (7.42), a sequence $\{\tilde{\mathbf{x}}_1^{(n)}, \tilde{\mathbf{x}}_2^{(n)}\}$ at the time instants t_n is defined. The time integration reads as

$$\frac{\tilde{\mathbf{x}}_2^{(n+1)} - \tilde{\mathbf{x}}_2^{(n)}}{\frac{h_n}{2}} = \mathbf{f}_2\left(\tilde{\mathbf{x}}_1^{(n)}, \tilde{\mathbf{x}}_2^{(n+1)}\right), \tag{7.45a}$$

$$\frac{\tilde{\mathbf{x}}_1^{(n+1)} - \tilde{\mathbf{x}}_1^{(n)}}{\frac{h_n}{2}} = \mathbf{f}_1\left(\tilde{\mathbf{x}}_1^{(n+1)}, \tilde{\mathbf{x}}_2^{(n+1)}\right). \tag{7.45b}$$

The procedure introduces two systematic errors, i.e., one coming from the fractional splitting and another one coming from the time discretisation. It has been shown that the error of such scheme is of order $\mathcal{O}(h_n)$ [61]. The procedure has, however, decisive advantages:

- The fractional splitting allows to solve for the individual systems successively, thereby avoiding the solution of the nonlinear coupled system of double size.
- The overall procedure still retains the order of accuracy in time giving by the implicit Euler method.
- The procedure allows to solve each of the subsystems by the most appropriate method. It is possible to accelerate the solution for one of the subsystems by, e.g., model order reduction.

7.6.2 Error Correction

This section additionally provides an error correction scheme to diminish the two errors mentioned above [29]. The Picard integral forms of (7.41a) and (7.41b) read as

$$\mathbf{x}_1(t) - \mathbf{x}_1(0) = \int_0^t \mathbf{f}_1(\mathbf{x}_1, \mathbf{x}_2) \, \mathrm{d}t, \tag{7.46a}$$

$$\mathbf{x}_2(t) - \mathbf{x}_2(0) = \int_0^t \mathbf{f}_2(\mathbf{x}_1, \mathbf{x}_2) \, \mathrm{d}t. \tag{7.46b}$$

When samples $\{\tilde{\mathbf{x}}_1^{(n)}, \tilde{\mathbf{x}}_2^{(n)}\}$ are available at the Chebyshev nodes $\mathbf{t} = [t_1, t_2, \ldots, t_N]$, approximate solutions for $\mathbf{x}_1(t)$ and $\mathbf{x}_2(t)$ can be constructed, i.e.,

$$\tilde{\mathbf{x}}_1(t) = \left[\tilde{\mathbf{x}}_1^{(1)} \ldots \tilde{\mathbf{x}}_1^{(N)} \right] \mathbf{L}, \tag{7.47a}$$

$$\tilde{\mathbf{x}}_2(t) = \left[\tilde{\mathbf{x}}_2^{(1)} \ldots \tilde{\mathbf{x}}_2^{(N)} \right] \mathbf{L}, \tag{7.47b}$$

using the Lagrange interpolation defined by

$$L_n(t) = \frac{1}{N} + \frac{2}{N} \sum_{m=1}^{N-1} T_m(t_n) T_m(t), \tag{7.48}$$

$$\mathbf{L} = \left[L_1(t) \ldots L_N(t) \right]^T, \tag{7.49}$$

where $T_m(t)$ denotes the Chebyshev polynomial of degree m. The errors on $\mathbf{x}_1(t)$ and $\mathbf{x}_2(t)$ are then given by

$$\mathbf{e}_1(t) = \mathbf{x}_1(t) - \tilde{\mathbf{x}}_1(t), \tag{7.50a}$$

$$\mathbf{e}_2(t) = \mathbf{x}_2(t) - \tilde{\mathbf{x}}_2(t). \tag{7.50b}$$

On the other hand, the approximate solutions $\tilde{\mathbf{x}}_1(t)$ and $\tilde{\mathbf{x}}_2(t)$ inserted in the Picard integral forms (7.46a) and (7.46b) lead to the residuals

$$\mathbf{r}_1(t) = \tilde{\mathbf{x}}_1(t) - \tilde{\mathbf{x}}_1^{(0)} - \int_0^t \mathbf{f}_1(\tilde{\mathbf{x}}_1, \tilde{\mathbf{x}}_2) \, \mathrm{d}t, \tag{7.51a}$$

$$\mathbf{r}_2(t) = \tilde{\mathbf{x}}_2(t) - \tilde{\mathbf{x}}_2^{(0)} - \int_0^t \mathbf{f}_2(\tilde{\mathbf{x}}_1, \tilde{\mathbf{x}}_2) \, \mathrm{d}t, \tag{7.51b}$$

where $\tilde{\mathbf{x}}_1^{(0)} = \tilde{\mathbf{x}}_1(0)$ and $\tilde{\mathbf{x}}_2^{(0)} = \tilde{\mathbf{x}}_2(0)$ are the initial conditions [19, 29]. Furthermore, two error-residual equations can be obtained by subtracting (7.51a) and (7.51b) from (7.46a) and (7.46b), i.e.,

$$-\mathbf{r}_1(t) \cong \mathbf{e}_1(t) - \int_0^t \left(\mathbf{f}_1\left(\mathbf{x}_1, \mathbf{x}_2\right) - \mathbf{f}_1\left(\tilde{\mathbf{x}}_1, \tilde{\mathbf{x}}_2\right)\right) \, \mathrm{d}t, \qquad (7.52a)$$

$$-\mathbf{r}_2(t) \cong \mathbf{e}_2(t) - \int_0^t \left(\mathbf{f}_2\left(\mathbf{x}_1, \mathbf{x}_2\right) - \mathbf{f}_2\left(\tilde{\mathbf{x}}_1, \tilde{\mathbf{x}}_2\right)\right) \, \mathrm{d}t. \qquad (7.52b)$$

If the expressions $\left(\mathbf{f}_1\left(\mathbf{x}_1, \mathbf{x}_2\right) - \mathbf{f}_1\left(\tilde{\mathbf{x}}_1, \tilde{\mathbf{x}}_2\right)\right)$ and $\left(\mathbf{f}_2\left(\mathbf{x}_1, \mathbf{x}_2\right) - \mathbf{f}_2\left(\tilde{\mathbf{x}}_1, \tilde{\mathbf{x}}_2\right)\right)$ would be linear in their arguments, they can be reformulated as functions of $\mathbf{e}_1(t)$ and $\mathbf{e}_2(t)$. In practice, only the linear parts can be reformulated as such. The remaining parts are resolved by fixed point iteration.

The error correction procedure is as follows. The errors $\mathbf{e}_1(t)$ and $\mathbf{e}_2(t)$ are expressed in terms of the Chebyshev polynomials as well, i.e.,

$$\mathbf{e}_1(t) = \left[\tilde{\mathbf{e}}_1^{(1)} \ldots \tilde{\mathbf{e}}_1^{(N)}\right] \boldsymbol{L}, \qquad (7.53a)$$

$$\mathbf{e}_2(t) = \left[\tilde{\mathbf{e}}_2^{(1)} \ldots \tilde{\mathbf{e}}_2^{(N)}\right] \boldsymbol{L}, \qquad (7.53b)$$

and inserted in the error-residual correction equations (7.52a) and (7.52b). Eqs. (7.52a) and (7.52b) are solved recursively. Inbetween the iteration steps, $\left[\tilde{\mathbf{x}}_1^{(1)} \ldots \tilde{\mathbf{x}}_1^{(N)}\right]$ and $\left[\tilde{\mathbf{x}}_2^{(1)} \ldots \tilde{\mathbf{x}}_2^{(N)}\right]$ are updated using the computed errors. The procedure is stopped when a convergence criterion is fulfilled.

7.7 Example with Dynamic Iteration: Field-Circuit Coupled Model of a Single-Phase Transformer

As an example for cosimulation with dynamic iteration, the start-up magnetisation of a single-phase transformer excited by a pulse-width-modulated (PWM) voltage is simulated. The transformer is surrounding by a small piece of circuitry (Fig. 7.4 (Left)). The transformer core is locally saturated, which necessitates its representation by a field model (Fig. 7.4 (Right)). The PWM voltage signal and the voltages at the transformer high-voltage terminal vary at largely different time scales (Fig. 7.5). The currents through the outer inductor and through the high-voltage winding only contain slowly varying components (Fig. 7.6). This motivates to use a multirate cosimulation scheme where the field model is time-stepped at a substantially slower rate.

The dynamic iteration does not deteriorate the convergence of the overall time stepping procedure, as is obvious from the convergence of the error calculated with respect to a reference solution obtained for a small time step $h = 10^{-6}$ s, shown in Fig. 7.7. For comparably small time windows, only one iteration over the time window is needed. For that reason, the algorithms with and without sweep control coincide. When a larger window $H = 2 \cdot 10^{-4}$ is used, the sweep control invokes additional waveform relaxation steps.

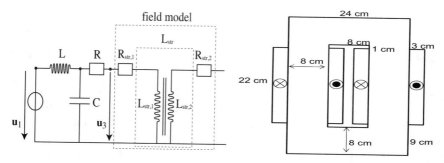

Fig. 7.4: Single-phase transformer example exhibiting different time constants in the voltages $\mathbf{u}_1$ and $\mathbf{u}_3$ due to a fast switching PWM voltage source and a RLC low-pass filter [50]: Left: Field model by circuit description; Right: Full 2D field model.

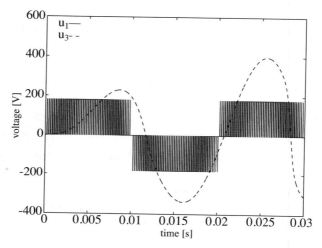

Fig. 7.5: Single-phase transformer: Voltages at nodes $\mathbf{u}_1$ and $\mathbf{u}_3$ [50].

The 2D transformer cross section is modelled in FEMM [37] and meshed by Triangle [53]. The simulation is carried out within the CoMSON demonstrator platform [22]. Both time steppers use the implicit Euler method, the one for the circuit problem with a fixed step size of 10^{-6} s, the one for the field problem with a time step equal to the time window size. The time window size is fixed and chosen between $H = 10^{-5}$ s and $H = 2 \cdot 10^{-4}$ s in order to study the convergence of the dynamic iteration scheme according to H. The computational effort is counted as the number of system solves required for the field problem part (Table 7.1). A distinction is made between the system solves for time stepping the field solution and the system solves for extracting

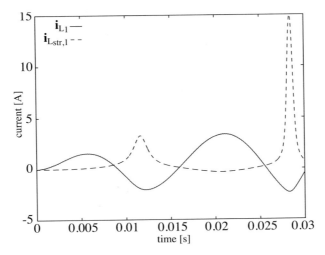

Fig. 7.6: Single-phase transformer: Currents through the inductors [50].

Table 7.1: Cosimulation with dynamic iteration: computational effort. "lin. solve" denotes the number of all linear system solved, this includes system due to implicit time integration plus systems for parameter extraction [50].

coupling method (parameters)	lin. solve	max. rel. error
strong (step size $h = 10^{-5}$s)	6574	5.8%
weak (sweep control, $H = 10^{-4}$s)	4282 + 560	5.7%
weak (no sweep, $H = 10^{-4}$s)	2196 + 300	6.0%
weak (sweep control, $H = 2 \cdot 10^{-4}$s)	2726 + 310	10.5%
weak (no sweep, $H = 2 \cdot 10^{-4}$s)	1274 + 150	12.7%

the inductances. The computational cost remains well below the cost of the monolithic time integration approach with a time step of $h = 10^{-5}$ s.

7.8 Example with Fractional Splitting: Chip Package

The time integration method with fractional splitting and error correction is applied to a chip package (Fig. 7.8). The package contains a brick-shaped compound with 8 metallic inclusions and a ground plane. The package model was generated with the PTM-ET software of MAGWEL [47]. Because the model only contains hexahedral pieces, the spatial discretisation was preferably carried out with a finite-difference method on a non-equidistant hexahedral grid. The obtained matrices share the properties of the matrices obtained by FE discretisation. The matrices were extracted from the PTM-ET software and loaded into Matlab^{TM}.

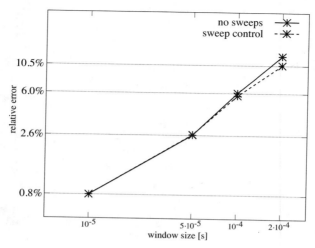

Fig. 7.7: Convergence of the time stepping error according to the time window size H [50].

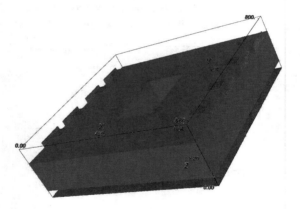

Fig. 7.8: Chip package serving as an example for the time integration method with fractional splitting and error correction.

The fractional time stepper follows the scheme expressed by (7.45a) and (7.45b). A reformulation into systems of equations gives

$$\left(\mathbf{K}_{\mathrm{th}}(\tilde{\mathbf{u}}_{\mathrm{th}}^{(n+1)}) + \frac{2}{h_n}\mathbf{M}_{\mathrm{th}}\right)\tilde{\mathbf{u}}_{\mathrm{th}}^{(n+1)} = \frac{2}{h_n}\mathbf{M}_{\mathrm{th}}\tilde{\mathbf{u}}_{\mathrm{th}}^{(n)} + \mathbf{f}_{\mathrm{th}}(\tilde{\mathbf{u}}_{\mathrm{th}}^{(n+1)}) + \mathbf{g}_{\mathrm{th}}(\tilde{\mathbf{u}}_{\mathrm{eqs}}^{(n)}),$$

$$\tag{7.54a}$$

$$\left(\mathbf{K}_{\mathrm{eqs}}(\tilde{\mathbf{u}}_{\mathrm{th}}^{(n+1)}) + \frac{2}{h_n}\mathbf{M}_{\mathrm{eqs}}\right)\tilde{\mathbf{u}}_{\mathrm{eqs}}^{(n+1)} = \frac{2}{h_n}\mathbf{M}_{\mathrm{eqs}}\tilde{\mathbf{u}}_{\mathrm{eqs}}^{(n)} + \mathbf{f}_{\mathrm{eqs}}(\tilde{\mathbf{u}}_{\mathrm{th}}^{(n+1)}), \tag{7.54b}$$

where the first system of equations is nonlinear because of the dependence of $\mathbf{K}_{th}$ and $\mathbf{f}_{th}$ on $\tilde{\mathbf{u}}_{th}^{(n+1)}$. The residuals are calculated according to (7.51a) and (7.51b), or in detail,

$$
\mathbf{r}_{eqs}(t) = \mathbf{M}_{eqs}\left(\tilde{\mathbf{u}}_{eqs}(t) - \tilde{\mathbf{u}}_{eqs}^{(0)}\right)
$$
$$
+ \int_0^t \mathbf{K}_{eqs}(\tilde{\mathbf{u}}_{th}(t))\tilde{\mathbf{u}}_{eqs}(t)\,\mathrm{d}t - \int_0^t \mathbf{f}_{eqs}(\tilde{\mathbf{u}}_{th}(t))\,\mathrm{d}t, \qquad (7.55a)
$$
$$
\mathbf{r}_{th}(t) = \mathbf{M}_{th}\left(\tilde{\mathbf{u}}_{th}(t) - \tilde{\mathbf{u}}_{th}^{(0)}\right)
$$
$$
+ \int_0^t \mathbf{K}_{th}(\tilde{\mathbf{u}}_{th}(t))\tilde{\mathbf{u}}_{th}(t)\,\mathrm{d}t - \int_0^t \mathbf{f}_{th}(\tilde{\mathbf{u}}_{th}(t))\,\mathrm{d}t - \int_0^t \mathbf{g}_{th}(\tilde{\mathbf{u}}_{eqs}(t))\,\mathrm{d}t.
$$
$$
(7.55b)
$$

The error correction procedure is carried out as shown in (7.52a) and (7.52b). Inserting the operators of the coupled electroquasistatic-thermal problem leads to the systems of equations

$$
-\mathbf{r}_{eqs} \cong \mathbf{M}_{eqs}\mathbf{e}_{eqs} + \int_0^t \left(\mathbf{K}_{eqs}(\mathbf{u}_{th})\mathbf{e}_{eqs} + \mathbf{K}_{eqs}^{(d)}\tilde{\mathbf{u}}_{eqs}\right)\,\mathrm{d}t - \int_0^t \mathbf{f}_{eqs}^{(d)}\,\mathrm{d}t, \quad (7.56a)
$$
$$
-\mathbf{r}_{th} \cong \mathbf{M}_{th}\mathbf{e}_{th} + \int_0^t \left(\mathbf{K}_{th}(\mathbf{u}_{th})\mathbf{e}_{th} + \mathbf{K}_{th}^{(d)}\tilde{\mathbf{u}}_{th}\right)\,\mathrm{d}t - \int_0^t \mathbf{f}_{th}^{(d)}\,\mathrm{d}t - \int_0^t \mathbf{g}_{th}^{(d)}\,\mathrm{d}t,
$$
$$
(7.56b)
$$

where

$$
\mathbf{K}_{eqs}^{(d)} = \mathbf{K}_{eqs}(\mathbf{u}_{th}) - \mathbf{K}_{eqs}(\tilde{\mathbf{u}}_{th}), \qquad (7.57a)
$$
$$
\mathbf{K}_{th}^{(d)} = \mathbf{K}_{th}(\mathbf{u}_{th}) - \mathbf{K}_{th}(\tilde{\mathbf{u}}_{th}), \qquad (7.57b)
$$
$$
\mathbf{f}_{eqs}^{(d)} = \mathbf{f}_{eqs}(\mathbf{u}_{th}) - \mathbf{f}_{eqs}(\tilde{\mathbf{u}}_{th}), \qquad (7.57c)
$$
$$
\mathbf{f}_{th}^{(d)} = \mathbf{f}_{th}(\mathbf{u}_{th}) - \mathbf{f}_{th}(\tilde{\mathbf{u}}_{th}), \qquad (7.57d)
$$
$$
\mathbf{g}_{th}^{(d)} = \mathbf{g}_{th}(\mathbf{u}_{eqs}) - \mathbf{g}_{th}(\tilde{\mathbf{u}}_{eqs}) \qquad (7.57e)
$$

are organised as functions of $\mathbf{e}_{eqs}$ and $\mathbf{e}_{th}$. An iteration is set up around Eqs. (7.56a) and (7.56b). New updates for $\left[\tilde{\mathbf{u}}_{eqs}^{(1)} \ldots \tilde{\mathbf{u}}_{eqs}^{(N)}\right]$ and $\left[\tilde{\mathbf{u}}_{th}^{(1)} \ldots \tilde{\mathbf{u}}_{th}^{(N)}\right]$ are obtained from the computed errors and then inserted to achieve new approximations for $\mathbf{K}_{eqs}^{(d)}$, $\mathbf{K}_{th}^{(d)}$, $\mathbf{f}_{eqs}^{(d)}$, $\mathbf{f}_{th}^{(d)}$ and $\mathbf{g}_{th}^{(d)}$ until the change of the updates drops below a prescribed tolerance.

The temperature dependence of the electric conductivity and the thermal conductivity has been modelled by polynomials. That simplifies the reconstruction of the matrices $\mathbf{K}_{eqs}$ and $\mathbf{K}_{th}$ and the vectors $\mathbf{f}_{eqs}$ and $\mathbf{f}_{th}$ for intermediate solutions $\tilde{\mathbf{u}}_{th}$ of the temperature. As boundary conditions, fixed potentials and fixed temperatures are applied to the metallic parts. The model contains 9193 degrees of freedom, i.e., 8071 for the temperature and 1122 for the electric scalar potential. The model is integrated in time between $t_0 = 0$ and $t_f = 10^{-3}$, which accommodates for the relevant time constants of the device which are in the range of $\tau_{th} = \|\mathbf{M}_{th}\|/\|\mathbf{K}_{th}\| \approx 10^{-5}$ s.

A reference solution for the electrothermal problem has been obtained with a very small time stepping of a high-order time integrator for differential-algebraic equations offered by Matlab^{TM}. Several own solutions have be calculated by the fractional-splitting approach with implicit Euler method for time steps with decreasing size. Each time, the solution is further treated by 1 up to 5 steps of the above described error-correction technique. The reference solution yields an error estimate of a high accuracy $\|\mathbf{e}_{th}\|$ (Fig. 7.9). The

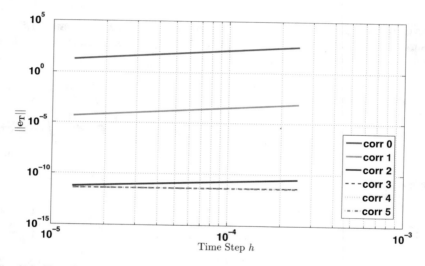

Fig. 7.9: Fractional time stepping: Correction error $\|\mathbf{e}_{th}\|$ as a function of the applied time step h for the implicit Euler method (the lines for corr 3, 4 and 5 coincide).

convergence of the time integrator is of the order $\mathcal{O}(h)$ with h the time step, as expected and indicated by the slopes of the lines in the figure. The first two error-correction steps diminish the error by approximately 10^{-5}, whereas decreasing the time step h by e.g. a factor 10 only brings a neglible improvement of the accuracy. After 4 error-correction steps, no further improvement is observed. As a conclusion, one can state that the error-correction procedure is capable of repairing the error introduced by the fractional splitting sufficiently.

7.9 Conclusion

Cosimulation equipped with dynamic iteration and fractional time stepping, the latter preferably accomplished by an error correction scheme, allows for a simulation of coupled problems by iterating between the constituting problem parts, thereby using separate implementations. Dedicated numerical analysis provides a guarantee on the convergence of the overall procedure. Moreover, both schemes can be organised such that the convergence order of the underlying single-problem time integrators is maintained. Both approaches are non-intrusive and thus available simulation packages can be applied straightforward to each subsystem. As a consequence, cosimulation with dynamic iteration and fractional time stepping methods are corner stones of multiphysical simulation procedures applied to electroquasistatic-thermal and magnetoquasistatic-thermal problems of industrial relevance.

Acknowledgements This work is a part of the project 'Nanoelectronic Coupled Problems Solutions' (nanoCOPS) funded by the European Union within FP7-ICT-2013 (grant no. 619166). The first and third authors are also supported by the 'Excellence Initiative' of the German Federal and State Governments and the Graduate School of Computational Engineering at Technische Universität Darmstadt.

References

1. ALOTTO, P., FRESCHI, F., AND REPETTO, M.: *Multiphysics problems via the cell method: The role of Tonti diagrams.* IEEE Trans. Magn. **46**(8), 2959–2962 (2010). doi:10.1109/TMAG.2010.2044487
2. ARNOLD, M., AND GÜNTHER, M.: *Preconditioned dynamic iteration for coupled differential-algebraic systems.* BIT **41**(1), 1–25 (2001). doi:10.1023/A:1021909032551
3. BARCHANSKI, A., CLEMENS, M., DE GERSEM, H., STEINER, T., AND WEILAND, T.: *Using domain decomposition techniques for the calculation of low-frequency electric current densities in high-resolution 3D human anatomy models.* COMPEL **24**(2), 458–467 (2005)
4. BARTEL, A., BRUNK, M., GÜNTHER, M., AND SCHÖPS, S.: *Dynamic iteration for coupled problems of electric circuits and distributed devices.* SIAM J. Sci. Comput. **35**(2), B315—B335 (2013). doi:10.1137/120867111
5. BARTEL, A., GÜNTHER, M.: *PDAEs in refined electric network modeling.* SIAM Rev. **60**(1), 56–91 (2018). doi:10.1137/17M1113643
6. BARTEL, A., AND PULCH, R.: *A concept for classification of partial differential algebraic equations in nanoelectronics.* In: BONILLA ET AL. [8]. doi:10.1007/978-3-540-71992-2_79
7. BARTH, A., SCHWAB, C., AND ZOLLINGER, N.: *Multi-level Monte Carlo finite element method for elliptic PDEs with stochastic coefficients.* Numer. Math. **119**, 123—161 (2011). doi:10.1007/s00211-011-0377-0
8. BONILLA, L.L., MOSCOSO, M., PLATERO, G., AND VEGA, J.M. (EDS.): *Progress in Industrial Mathematics at ECMI 2006*, Series Mathematics in Industry, Vol. 12. Springer, Berlin (2007). doi:10.1007/978-3-540-71992-2

9. BONTINCK, Z., DE GERSEM, H., AND SCHÖPS, S.: *Response surface models for the uncertainty quantification of eccentric permanent magnet synchronous machines.* IEEE Trans. Magn. **52**(3) (2016). doi:10.1109/TMAG.2015.2491607. Article #7203404

10. BORTOT, L., AUCHMANN, B., CORTES GARCIA, I., FERNANDO NAVARRO, A.M., MACIEJEWSKI, M., MENTINK, M., PRIOLI, M., RAVAIOLI, E., SCHÖPS, S., AND VERWEIJ, A.: *STEAM: A hierarchical co-simulation framework for superconducting accelerator magnet circuits.* IEEE Trans. Appl. Super. **28**(3) (2018). doi:10.1109/TASC.2017.2787665

11. BOSSAVIT, A., AND KETTUNEN, L.: *Yee-like schemes on staggered cellular grids: a synthesis between FIT and FEM approaches.* IEEE Trans. Magn. **36**(4), 861–867 (2000). doi:10.1109/20.877580

12. BÍRÓ, O., AND PREIS, K.: *Finite element analysis of 3-D eddy currents.* IEEE Trans. Magn. **26**(2), 418–423 (1990). doi:10.1109/20.106343

13. CLEMENS, M., GJONAJ, E., PINDER, P., AND WEILAND, T.: *Self-consistent simulations of transient heating effects in electrical devices using the finite integration technique.* IEEE Trans. Magn. **37**(5), 3375–3379 (2001). doi:10.1109/20.952617

14. CLEMENS, M., SCHÖPS, S., CIMALA, C., GÖDEL, N., RUNKE, S., AND SCHMIDTHÄUSLER, D.: *Aspects of coupled problems in computational electromagnetics formulations.* ICS Newsletter (International Compumag Society) **3** (2012)

15. CORTES GARCIA, I., SCHÖPS, S., BORTOT, L., MACIEJEWSKI, M., PRIOLI, M., FERNANDEZ NAVARRO, A.M., AUCHMANN, B., AND VERWEIJ, A.P.: *Optimized field/circuit coupling for the simulation of quenches in superconducting magnets.* IEEE J. Multiscale Multiphys. Comput. Tech. **2**(1), 97–104 (2017). doi:10.1109/JMMCT.2017.2710128

16. DE GERSEM, H., CLEMENS, M., AND WEILAND, T.: *Coupled finite-element, spectral-element discretisation for models with circular inclusions and far-field domains.* IET. Sci. Meas. Tech. **149**(5), 237–241 (2002). doi:10.1049/ip-smt:20020620

17. DE GERSEM, H., MERTENS, R., PAHNER, U., BELMANS, R.J.M., AND HAMEYER, K.: *A topological method used for field-circuit coupling.* IEEE Trans. Magn. **34**(5), 3190–3193 (1998). doi:10.1109/20.717748

18. DE GERSEM, H., AND WEILAND, T.: *Field-circuit coupling for time-harmonic models discretized by the finite integration technique.* IEEE Trans. Magn. **40**(2), 1334–1337 (2004). doi:10.1109/TMAG.2004.824536

19. DUTT, A., GREENGARD, L., AND ROKHLIN, V.: *Spectral deferred correction methods for ordinary differential equations.* BIT **40**(2), 241–266 (2000). doi:10.1023/A:1022338906936

20. ECHEVERRÍA, D., LAHAYE, D., ENCICA, L., LOMONOVA, E.A., HEMKER, P.W., AND VANDENPUT, A.J.A.: *Manifold-mapping optimization applied to linear actuator design.* IEEE Trans. Magn. **42**(4), 1183–1186 (2006). doi:10.1109/TMAG.2006.870969

21. ESTÉVEZ SCHWARZ, D., AND TISCHENDORF, C.: *Structural analysis of electric circuits and consequences for MNA.* Int. J. Circ. Theor. Appl. **28**(2), 131–162 (2000). doi:10.1002/(SICI)1097-007X(200003/04)28:2<131::AID-CTA100>3.0.CO;2-W

22. DE FALCO, C., DENK, G., AND SCHULTZ, R.: *A demonstrator platform for coupled multiscale simulation.* In: G. CIUPRINA, AND D. IOAN (EDS.): *Scientific Computing in Electrical Engineering SCEE 2006*, Series Mathematics in Industry, Vol. 11, pp. 63–71. Springer, Berlin (2007). doi:10.1007/978-3-540-71980-9_5

23. FELDMANN, U., AND GÜNTHER, M.: *CAD-based electric-circuit modeling in industry I: Mathematical structure and index of network equations.* Surv. Math. Ind. **8**(2), 97–129 (1999)

24. GILES, M.B.: *Multilevel Monte Carlo methods.* Acta. Num. **24**, 259–328 (2015). doi:10.1017/S09624929

25. GILLON, F., AND BROCHET, P.: *Screening and response surface method applied to the numerical optimization of electromagnetic devices.* IEEE Trans. Magn. **36**(4), 1163–1167 (2000). doi:10.1109/20.877647

26. HAMEYER, K., DRIESEN, J., DE GERSEM, H., AND BELMANS, R.J.M.: *The classification of coupled field problems.* IEEE Trans. Magn. **35**(3), 1618–1621 (1999). doi:10.1109/20.767304

27. HEMKER, T., VON STRYK, O., DE GERSEM, H., AND WEILAND, T.: *Mixed-integer nonlinear design optimization of a superconductive magnet with surrogate functions.* IEEE Trans. Magn. **44**(6), 1110–1113 (2008)

28. HO, C.W., RUEHLI, A.E., AND BRENNAN, P.A.: *The modified nodal approach to network analysis.* IEEE Trans. Circ. Syst. **22**(6), 504–509 (1975). doi:10.1109/TCS.1975.1084079

29. HUANG, J., JIA, J., AND MINION, M.: *Arbitrary order Krylov deferred correction methods for differential algebraic equations.* J. Comput. Phys. **221**(2), 739–760 (2007). doi:http://dx.doi.org/10.1016/j.jcp.2006.06.040

30. KOZIEL, S., AND LEIFSSON, L.: *Surrogate-Based Modeling and Optimization: Applications in Engineering.* Springer Science & Business Media (2013)

31. KURZ, S., FETZER, J., LEHNER, G., AND RUCKER, W.M.: *A novel formulation for 3D eddy current problems with moving bodies using a Lagrangian description and BEM-FEM coupling.* IEEE Trans. Magn. **34**(5), 3068–3073 (1998). doi:10.1109/20.717718

32. LI, J., LI, J., AND XIU, D.: *An efficient surrogate-based method for computing rare failure probability.* J. Comput. Phys. **230**(24), 8683–8697 (2011). doi:10.1016/j.jcp.2011.08.008

33. MACIEJEWSKI, M., BAYRASY, P., WOLF, K., WILCZEK, M., AUCHMANN, B., GRIESEMER, T., BORTOT, L., PRIOLI, M., FERNANDEZ NAVARRO, A.M., SCHÖPS, S., CORTES GARCIA, I., AND VERWEIJ, A.P.: *Coupling of magnetothermal and mechanical superconducting magnet models by means of mesh-based interpolation.* IEEE Trans. Appl. Super. **28**(3) (2018). doi:10.1109/TASC.2017.2786721

34. MASSCHAELE, B., DE GERSEM, H., ROGGEN, T., PODLECH, H., AND VANDEPLASSCHE, D.: *Simulation of the thermal deformation and the cooling of a four-rod radio frequency quadrupole.* In: 5th International Particle Accelerator Conference (IPAC14), pp. 376–378. www.jacow.org (2014)

35. TER MATEN, E.J.W., GÜNTHER, M., PUTEK, P., BENNER, P., FENG, L., SCHNEIDER, J., BRACHTENDORF, H.G., BITTNER, K., DELEU, F., WIEERS, A., JANSSEN, R., KRATOCHVÍL, T., GÖTTHANS, T., PULCH, R., LIU, Q., REYNIER, P., SCHOENMAKER, W., MEURIS, P., SCHÖPS, S., DE GERSEM, H., TISCHENDORF, C., AND STROHM, C.: *nanoCOPS: Nanoelectronic COupled Problem Solutions.* ECMI Newsletter Mathematics & Industry (56), 62–67 (2014)

36. TER MATEN, E.J.W., PUTEK, P.A., GÜNTHER, M., PULCH, R., TISCHENDORF, C., STROHM, C., SCHOENMAKER, W., MEURIS, P., DE SMEDT, B., BENNER, P., FENG, L., BANAGAAYA, N., YUE, Y., JANSSEN, R., DOHMEN, J.J., TASIĆ, B., DELEU, F., GILLON, R., WIEERS, R., BRACHTENDORF, H.G., BITTNER, K., KRATOCHVÍL, T., PETRZELA, J., SOTNER, R., GÖTHANS, T., DRÍNOVSKÝ, J., SCHÖPS, S., DUQUE, D.J., CASPER, T., DE GERSEM, H., RÖMER, U., REYNIER, P., BARROUL, P., MASLIAH, D., AND ROUSSEAU, B.: *Nanoelectronic COupled Problems Solutions - nanoCOPS: Modelling, multirate, model order reduction, uncertainty quantification, fast fault simulation.* Journal of Mathematics in Industry **7**(2) (2016). doi:10.1186/s13362-016-0025-5

37. MEEKER, D.: *Finite Element Method Magnetics User's Manual,* version 4.2 (09nov2010 build) edn. (2010). http://www.femm.info/

38. NARAYAN, A., GITTELSON, C., AND XIU, D.: *A stochastic collocation algorithm with multifidelity models.* SIAM J. Sci. Comput. **36**(2), A495–A521 (2014). doi:10.1137/130929461

39. NIYONZIMA, I., GEUZAINE, C., AND SCHÖPS, S.: *Waveform relaxation for the computational homogenization of multiscale magnetoquasistatic problems.* J. Comput. Phys. **327**, 416–433 (2016). doi:10.1016/j.jcp.2016.09.011

40. PEHERSTORFER, B., WILLCOX, K., AND GUNZBURGER, M.: *Survey of multifidelity methods in uncertainty propagation, inference, and optimization.* Tech. Rep. TR16-1, ACDL Technical Report (2016)

41. PELS, A., GYSELINCK, J., SABARIEGO, R.V., AND SCHÖPS, S.: *Solving nonlinear circuits with pulsed excitation by multirate partial differential equations.* IEEE Trans. Magn. **54**(3), 1–4 (2018). doi:10.1109/TMAG.2017.2759701

42. PULCH, R., GÜNTHER, M., AND KNORR, S.: *Multirate partial differential algebraic equations for simulating radio frequency signals.* Eur. J. Appl. Math. **18**, 709–743 (2007). doi:10.1017/s0956792507007188

43. REN, Z.: *On the complementary of dual formulations on dual meshes.* IEEE Trans. Magn. **45**(3), 1284–1287 (2009). doi:10.1109/TMAG.2009.2012596

44. ROYCHOWDHURY, J.: *Analyzing circuits with widely separated time scales using numerical PDE methods.* IEEE Trans. Circ. Syst. Fund. Theor. Appl. **48**(5), 578–594 (2001). doi:10.1109/81.922460

45. SCHMIDT, K., AND CHERNOV, A.: *A unified analysis of transmission conditions for thin conducting sheets in the time-harmonic eddy current model.* SIAM J. Appl. Math. **73**(6), 1980–2003 (2013). doi:10.1137/120901398

46. SCHOENMAKER, W., DUPUIS, O., DE SMEDT, B., MEURIS, P., OCENASEK, J., VERHAEGEN, W., DUMLUGÖL, D., AND PFOST, M.: *Fully-coupled 3D electro-thermal field simulator for chip-level analysis of power devices.* In: 19th International Workshop on Thermal Investigations of ICs and Systems (THERMINIC 2013), pp. 210–215. IEEE, Berlin, Germany (2013). doi:10.1109/THERMINIC.2013.6675199

47. SCHOENMAKER, W., MEURIS, P., JANSSENS, E., VERSCHAEVE, M., SEEBACHER, E., PFLANZL, W., STUCCHI, M., MANDEEP, B., MAEX, K., AND SCHILDERS, W.H.A.: *Simulation and measurement of interconnects and on-chip passives: Gauge fields and ghosts as numerical tools.* In: BONILLA ET AL. [8]. doi:10.1007/978-3-540-71992-2

48. SCHOENMAKER, W., SCHÖPS, S., FENG, L., BITTNER, K., AND TER MATEN, E.J.W.: *Software design for electro-thermal co-simulation.* Report D1.2, nanoCOPS FP7-ICT-2013-11/619166 (2014)

49. SCHÖPS, S.: *Field device simulator user's manual.* Manual, Bergische Universität Wuppertal (2009). http://cdn.bitbucket.org/schoeps/fides/downloads/fides_manual.pdf

50. SCHÖPS, S., DE GERSEM, H., AND BARTEL, A.: *A cosimulation framework for multirate time-integration of field/circuit coupled problems.* IEEE Trans. Magn. **46**(8), 3233–3236 (2010). doi:10.1109/TMAG.2010.2045156

51. SCHÖPS, S., DE GERSEM, H., BARTEL, A.: *Higher-order cosimulation of field/circuit coupled problems.* IEEE Trans. Magn. **48**(2), 535–538 (2012). doi:10.1109/TMAG.2011.2174039

52. SCHÖPS, S., DE GERSEM, H., AND WEILAND, T.: *Winding functions in transient magnetoquasistatic field-circuit coupled simulations.* COMPEL **32**(6), 2063–2083 (2013). doi:10.1108/COMPEL-01-2013-0004

53. SHEWCHUK, J.R.: *Triangle: Engineering a 2D quality mesh generator and Delaunay triangulator.* In: M.C. LIN, AND D. MANOCHA (EDS.): *Applied Computational Geometry: Towards Geometric Engineering*, Lecture Notes in Computer Science, Vol. 1148, pp. 203–222. Springer-Verlag (1996)

54. SMITH, B.F., BJØRSTAD, P.E., AND GROPP, W.D.: *Domain Decomposition: Parallel multilevel methods for elliptic partial differential equations.* Cambridge University Press (1996)

55. STEINMETZ, T., GÖDEL, N., WIMMER, G., CLEMENS, M., KURZ, S., AND BEBENDORF, M.: *Efficient symmetric FEM-BEM coupled simulations*

of electro-quasistatic fields. IEEE Trans. Magn. **44**(6), 1346–1349 (2008). doi:10.1109/TMAG.2008.915785

56. TECKENTRUP, A.L., SCHEICHL, R., GILES, M.B., AND ULLMANN, E.: *Further analysis of multilevel Monte Carlo methods for elliptic PDEs with random coefficients.* Numer. Math. **125**(3), 569–600 (2013). doi:10.1007/s00211-013-0546-4

57. TSUKERMAN, I.A., KONRAD, A., MEUNIER, G., AND SABONNADIÈRE, J.C.: *Coupled field-circuit problems: Trends and accomplishments.* IEEE Trans. Magn. **29**(2), 1701–1704 (1993). doi:10.1109/20.250733

58. VILLAR, I., VISCARRET, U., ETXEBERRIA-OTADUI, I., AND RUFER, A.: *Global loss evaluation methods for nonsinusoidally fed medium-frequency power transformers.* IEEE. Trans. Ind. Electron. **56**(10), 4132–4140 (2009). doi:10.1109/TIE.2009.2021174

59. WEILAND, T.: *Time domain electromagnetic field computation with finite difference methods.* Int. J. Numer. Model. Electron. Network. Dev. Field **9**(4), 295–319 (1996). doi:10.1002/(SICI)1099-1204(199607)9:4<295::AID-JNM240>3.0.CO;2-8

60. WROBEL, R., LUKANISZYN, M., JAGIELA, M., AND LATAWIEC, K.: *A new approach to reduction of the cogging torque in a brushless motor by skewing optimization of permanent magnets.* Electr. Eng. **85**, 59–69 (2003). doi:10.1007/s00202-002-0144-4

61. YANENKO, N.N.: *The method of fractional steps: The solution of problems of mathematical physics in several variables,* Lecture Notes in Mathematics, Vol. 91. Springer, Berlin (1971). doi:10.1007/978-3-642-65108-3

Chapter 8
Multirate Circuit - EM - Device Simulation

Kai Bittner, Hans-Georg Brachtendorf

Abstract Radio frequency (RF) integrated circuits are at the core of modern mobile communication. They basically comprise the analog front-end, the analog-to-digital and vice versa the digital-to-analog conversion, and the digital signal processing. These days, both the analog and digital parts, are integrated on the same die. The analog front-end mainly performs amplification, filtering and mixing to or vice versa from the RF regime to baseband. The waveforms of voltages and currents of such an IC are described by a system of ordinary Differential-Algebraic Equations (DAEs) resulting from the well-known Modified Nodal Analysis (MNA). Standard solvers for initial value problems, also referred to as transient analysis, are however prohibitively slow, since the time step or vice versa its inverse the sampling rate, are limited by Shannon's sampling theorem. The sampling theorem predicates that the sampling rate must be at least twice as high as the highest relevant frequency components of the spectra of all waveforms of the circuit. Since modern RF integrated circuits operate in the GHz range, solving the initial value problem of these DAEs is extremely slow. This chapter addresses the simulation problem of RF circuits by generalizing the method of the Equivalent Complex Baseband (ECB) for circuits and systems described by nonlinear DAEs.

8.1 Formulation of the Coupled Circuit/EM/Device Simulation Problem

Circuit simulators such as SPICE [45] formulate circuits as systems of Ordinary Differential Equations (ODEs) or ordinary Differential-Algebraic Equa-

Kai Bittner, Hans-Georg Brachtendorf
Fachhochschule Oberösterreich, Hagenberg im Mühlkreis, Austria, e-mail: {Kai.
Bittner,Hans-Georg.Brachtendorf}@fh-hagenberg.at

© Springer Nature Switzerland AG 2019   161
E. J. W. ter Maten et al. (eds.), *Nanoelectronic Coupled Problems Solutions*,
Mathematics in Industry 29, https://doi.org/10.1007/978-3-030-30726-4_8

tions (DAEs), resulting from Kirchhoff's voltage and current laws, the topology of the circuit, and the lumped or compact models of the constitutive devices. Lumped device models approximate the 3D device by their terminal currents, voltages, electric charges and fluxes through analytical expressions. Deriving analytical expressions for short channel transistors is however not a simple task [32, 50] and often done by curve fits only without any physical meaning. Such behavioral expressions often require more than hundred parameters which must be extracted in a cumbersome manner from measurements. Therefore, this design procedure needs standardized device designs which can only be slightly modified, e.g. the channel widths and lengths of transistors. If however a device shall be optimized within a circuit environment w.r.t. geometry variations in a wide range, physical properties such as the variation of the dielectric constant, doping profile etc., the modeling procedure outlined above is prohibitive. Therefore coupled circuit/EM/device simulation is used when devices in a circuitry are subject to optimization.

Furthermore, although valid at low frequencies, this compact model approximation looses its justification at more and more increasing frequencies, specifically in radio frequency (RF) circuitry. Hence, with the increased level of tight integration of Systems-on-Chip (SoC) and increased complexity of components, it becomes more and more cumbersome to represent the devices by a lumped/compact SPICE compatible model. This holds true specifically for passive structures such as on chip inductors, transmission lines, baluns, resonators etc.

Therefore, it has been advocated to avoid the compact modeling of the passive structures and to link the S-parameter data directly to the circuit simulator via so called S-parameter banks. The time domain characterization is then done by performing Fourier transformations on the frequency space data. Unfortunately, this route is error-prone due to a cascade of approximations related to the cut-off in frequency space and the fact that frequency domain techniques presuppose a linear device. Moreover, FFT based methods conflict with variable time stepping algorithms employed in transient analysis, which makes sophisticated interpolation techniques necessary.

In order to overcome these limitations, we propose to circumvent the frequency-based method by an incorporation of the passive components with their full 3D field equations directly into a circuit simulator. This technique is referred to as coupled or mixed level circuit/EM/device simulation. To this end, an electro-magnetic (EM) field TCAD (EM-TCAD) solver is coupled holistically with a circuit simulator, i.e., the EM solver provides all necessary information to the circuit simulator for a full Newton iteration. Whereas relaxation based methods are easy to implement, the convergence depends on the spectral radius near the correct solution [47]. Instead, Newton's method converges quadratic in almost all practical cases.

A technology CAD (TCAD) simulator employs Maxwell's equations for the 3D device model, together with some physical properties, resulting in Partial Differential Equations (PDEs) in time and space. The field equations are discretized in space by the TCAD solver and the resulting time dependent ODEs constitute a holistic part of the circuit equations. Several TCAD simulators have been developed in the past to optimize a single device w.r.t. geometries, doping profiles and material parameters such as the dielectric constant [41, 43, 49, 60, 61]. In [24] the TCAD simulator devEM has been presented, focused on passive RF distributed devices at very high frequencies, employing the scalar and vector potentials. The spatial discretization of the PDEs employs the Finite Integration Technique (FIT) [24, 43] by the EM-TCAD simulator devEM, resulting in huge systems of ordinary DAEs in time.

Since the advent of coupled circuit/TCAD simulators [29, 38, 39] several coupling strategies have been proposed [33, 35, 37, 57]. The method presented here differs from previous methods by time integration techniques dedicated to RF designs, namely the multirate technique, and the avoidance of relaxation based methods or hierarchical 2-level Newton techniques, but full (damped) Newton iterations instead [2–4, 27]. The latter guarantees in nearly all practical cases quadratic convergence for suitable initial conditions.

The circuit simulator LinzFrame serves as the master and provides AC, DC, transient, Periodic Steady-State (PSS) shooting and multirate simulation. In order to deal with RF circuitry at very high center frequencies, the multirate partial differential equation (PDE) technique is used [22, 46, 53, 55]. Whereas the former methods are industry standards, the latter method exploits the envelope modulated characteristic of radio frequency (RF) signals, whose spectra are centered around a carrier frequency f_c. The PDEs with mixed initial/boundary conditions are solved via Petrov-Galerkin techniques based on spline/wavelet methods [8].

For the passive RF components, we need a seamless integration of the EM field equations and the circuit equations. Therefore, the interface between the two domains needs to be based on voltages and currents. To this end, an EM field solver [30, 58] whose constitutive variables are potentials (both scalar and vector potentials) is coupled with a circuit simulator.

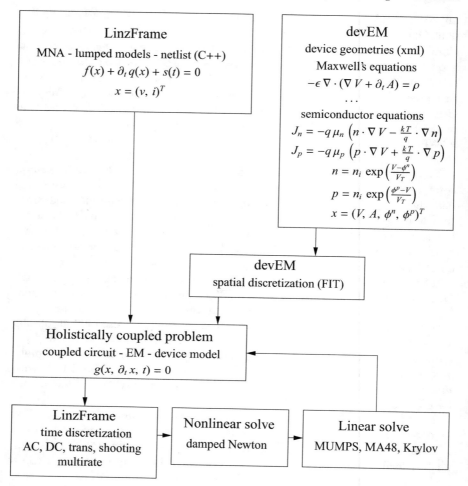

Fig. 8.1: Coupled simulation LinzFrame/devEM.

The concept of the coupled simulation is depicted in Fig. 8.1. On the one hand, the circuit simulator LinzFrame employs the Modified Nodal Analysis formulation, together with lumped device models to obtain a system of ordinary DAEs. The unknowns are the node potentials and some branch currents $(v, i)^T$. On the other hand, the devEM simulator performs a full 3D simulation of a single device employing Maxwell's equations and device equations such as the drift diffusion equations for semiconductors. Unknowns are the scalar and vector potentials and for semiconductors the quasi Fermi potentials too, i.e. $(V, \mathbf{A}, \Phi^n, \Phi^p)^T$. The spatial discretization is performed by devEM employing the Finite Integration Technique, providing to LinzFrame a system of ordinary DAEs in time. The numerical time integration is performed by LinzFrame. Beside standard techniques such as AC, DC and transient simulation, the circuit simulator provides special techniques dedicated to radio frequency systems, namely periodic steady state shooting and multirate simulation [11]. The latter technique is described in Sec. 8.2. Numerical time integration by Petrov-Galerkin techniques based on a spline/wavelet expansion is considered in Sec. 8.3. After numerical time integration, the generally nonlinear algebraic equations are solved by damped Newton techniques, resulting in huge but sparse linear systems. The linear systems are solved either by direct methods (MUMPS, MA48) [28, 44] or preconditioned iterative Krylov methods such as GMRES [56].

8.1.1 Formulation of the Circuit Equations

We consider the circuit DAE system of dimension N resulting from Kirchhoff's laws (KCL, KVL), employing the Modified Nodal Analysis (MNA) [34], the circuit topology and the lumped device models

$$\frac{d}{dt}q\big(x(t)\big) + \underbrace{i\big(x(t)\big) + b(t)}_{g(x(t),t)} = 0, \qquad (8.1)$$

where $x = (v, i)^T$ is the vector of unknown voltages and branch currents, $i : \mathbb{R}^N \to \mathbb{R}^N$ the vector sums of currents entering each node and $q : \mathbb{R}^N \to \mathbb{R}^N$ the vector sums of charges and fluxes. The derivation of (8.1) can be found in many textbooks, e.g. [25, 62]. Moreover x_0 is the vector of initial conditions and $b(t)$ the stimulus vector. If b is independent of time, the circuit is autonomous (e.g. an oscillator) and non-autonomous or driven otherwise.

8.1.2 Formulation of the 3D Field Equations

The electro-magnetic TCAD (EM-TCAD) simulator devEM employs the vector potential $\mathbf{A}$ such that the magnetic induction is $\mathbf{B} = \mathrm{rot}\,\mathbf{A}$ or $\mathbf{B} = \nabla \times \mathbf{A}$ and scalar potential V, such that $\mathbf{E} = -(\mathrm{grad}\,V + \partial_t \mathbf{A})$ or $\mathbf{E} = -(\nabla V + \partial_t \mathbf{A})$, where $\mathbf{E}$ is the electric field strength. The Maxwell Faraday law and the Gauss law for magnetic fields are then fulfilled. Furthermore we assume here linearity of the material, i.e., $\mathbf{D} = \epsilon \mathbf{E}$, where $\mathbf{D}$ is the dielectric displacement and $\mathbf{H} = \frac{1}{\mu}\mathbf{B}$, where $\mathbf{H}$ is the magnetic field strength. Moreover, we assume that the dielectric constant ϵ and the permeability μ are piecewise constant, and isotropic (scalars). To obtain systems of first-order PDEs in time, the quasi-canonical momentum $\mathbf{\Pi} = \partial_t \mathbf{A}$ is used as an additional variable. Further equations for incorporating different types of materials are required, which are considered next for metals, insulator and semiconductors.

8.1.2.1 Maxwell's Equations in Metals

From Ohm's law $\mathbf{J} = \sigma \mathbf{E}$ and $\mathbf{D} = \epsilon \mathbf{E}$ with current density $\mathbf{J}$, conductivity σ and the Maxwell-Ampère law $\nabla \times \mathbf{H} = \mathbf{J} + \partial_t \mathbf{D}$ and continuity $\nabla \cdot \mathbf{J} + \partial_t \varrho = 0$ laws one obtains

$$\nabla \times \left(\frac{1}{\mu} \nabla \times \mathbf{A} \right) = -\sigma \left(\nabla V + \mathbf{\Pi} \right) - \epsilon \frac{\partial}{\partial t} \left(\nabla V + \mathbf{\Pi} \right),$$

$$-\nabla \cdot \mathbf{J} - \dot{\varrho} = \nabla \cdot \left(\sigma \left(\nabla V + \mathbf{\Pi} \right) \right) + \frac{\partial}{\partial t} \left(\nabla \cdot \left(\epsilon \left(\nabla V + \mathbf{\Pi} \right) \right) \right) = 0,$$

with unknowns $(V, \mathbf{A}, \mathbf{\Pi})^T$.

8.1.2.2 Maxwell's Equations in Insulators

Since for the conductivity and charge density $\sigma, \varrho = 0$ is assumed, it follows that

$$\nabla \cdot \left(\epsilon \left(\nabla V + \mathbf{\Pi} \right) \right) = 0,$$

$$\nabla \times \left(\frac{1}{\mu} \nabla \times \mathbf{A} \right) = -\epsilon \frac{\partial}{\partial t} \left(\nabla V + \mathbf{\Pi} \right),$$

with unknowns $(V, \mathbf{A}, \mathbf{\Pi})^T$.

8.1.2.3 Maxwell's Equations in Semiconductors

For the densities n, p of free electrons/holes we use the drift-diffusion model (DD). For a given doping profile the donator/acceptor concentrations are N_D and N_A, respectively

$$-\nabla \cdot \left(\epsilon \left(\nabla V + \mathbf{\Pi} \right) \right) = \varrho, \quad \varrho = q \left(p - n + N_D - N_A \right),$$

$$\nabla \times \left(\frac{1}{\mu} \nabla \times \mathbf{A} \right) = \mathbf{J}_p + \mathbf{J}_n - \epsilon \frac{\partial}{\partial t} \left(\nabla V + \mathbf{\Pi} \right),$$

where $\mathbf{J}_n, \mathbf{J}_p$ are the current densities of electrons/holes, given by

$$\mathbf{J}_n = -q \mu_n \left(n \left(\nabla V + \mathbf{\Pi} \right) - V_T \cdot \nabla n \right),$$

$$\mathbf{J}_p = -q \mu_p \left(p \left(\nabla V + \mathbf{\Pi} \right) + V_T \cdot \nabla p \right),$$

with drift term proportional to $\sim \nabla V$ and diffusion terms $\sim \nabla n, \sim \nabla p$. Herein q is the elementary charge, μ_n, μ_p the mobilities of electrons and holes, $V_T = \frac{k_B T}{q}$ the thermal voltage, k_B the Boltzmann constant and T the absolute temperature. The mobilities are obtained from the effective masses and relaxation times of the electrons and holes, i.e.

$$\mu_n = \frac{q \tau_n}{m_n^*}, \quad \mu_p = \frac{q \tau_p}{m_p^*}.$$

The densities of electrons and holes read

$$n = n_i \exp\left(\frac{V - \phi^n}{V_T}\right), \quad p = n_i \exp\left(\frac{\phi^p - V}{V_T}\right),$$

where ϕ^n, ϕ^p are the respective quasi Fermi potentials for electrons/holes, respectively. The continuity equation holds for the electrons and holes separately, i.e.,

$$\nabla \cdot \boldsymbol{J}_n - q \frac{\partial n}{\partial t} = -q U(n,p), \quad \nabla \cdot \boldsymbol{J}_p + q \frac{\partial p}{\partial t} = q U(n,p),$$

with net generation rate $U(n,p) = G - R$. A familiar model for generation/recombination is, e.g., the Shockley, Read and Hall model. Unknowns are the scalar and vector potentials $(\boldsymbol{A}, \boldsymbol{\Pi} = \partial_t \boldsymbol{A})$ and moreover the quasi Fermi potentials, i.e. $(V, \boldsymbol{A}, \boldsymbol{\Pi}, \phi^n, \phi^p)^T$. Instead of the quasi Fermi potentials the electron/hole concentrations can be used as unknowns.

8.1.2.4 Gauge Equations

The solution of the field equations is still not unique. For uniqueness the system of equations is completed by the gauge condition

$$\frac{1}{\mu} \nabla (\nabla \cdot \boldsymbol{A}) + \xi \epsilon \nabla (\partial_t V) = 0. \tag{8.2}$$

For $\xi = 0$ one obtains the Coulomb and for $\xi = 1$ the Lorenz gauge as special cases.

8.1.3 Boundary/Interface Conditions within the 3D Device

As discussed above, we assume piecewise constant material parameters in the bulk. At boundaries/interfaces however the material parameters may change abruptly, requiring an individual treatment discussed next.

8.1.3.1 Metal-Semiconductor Contacts

At the metal-semiconductor contact charge neutrality is assumed, i.e., $p - n + N = 0$, $N = N_D - N_A$, with N_D, N_A the donator/acceptor concentrations, the abrupt voltage drop $\delta V = \Phi - V = V_{metal} - V$ reads

$$\delta V = V_T \ln \left(-\frac{N}{2n_i} \left(1 + \sqrt{1 + \frac{4n_i^2}{N^2}} \right) \right) \quad \text{p-dope,} \quad N < 0,$$

$$\delta V = -V_T \ln \left(\frac{N}{2n_i} \left(1 + \sqrt{1 + \frac{4n_i^2}{N^2}} \right) \right) \quad \text{n-dope,} \quad N > 0,$$

assuming that the Fermi potential coincides with the potential of the metal contact. The above expressions can be substituted by the arsinh function.

8.1.3.2 Insulator-Semiconductor Contacts

Continuity of the scalar potential V is assumed. The continuity equation for electrons and holes at the semiconductor side determines the potentials uniquely.

8.1.3.3 Metal-Insulator Contacts

The potentials $(V, \mathbf{A})^T$ are the sole unknowns. The surface charge at the contact is obtained from the gradient in normal direction $\nabla V \cdot n_\perp$.

8.1.3.4 Metal-Semiconductor-Insulator Triple Lines

From δV, given by the metal-semiconductor contact above, one obtains by averaging

$$V_{insul} = V_{metal} - \frac{1}{2}\delta V.$$

the contact voltage at the insulator side.

8.1.3.5 Boundary Conditions for the Vector Potentials

Since the magnetic induction is finite, the tangential component $\mathbf{A} \cdot n_\|$ must be continuous. For the normal component $\mathbf{A} \cdot n_\perp$ the argumentation is more intriguing. The continuity of the normal component $\mathbf{A} \cdot n_\perp$ follows directly for the Coulomb gauge from (8.2) for $\xi = 0$. For the Lorenz gauge with $\xi = 1$

one can show that the contribution of $\partial_t V$ vanishes in the limit, if V is continuously differentiable, then the continuity of $\mathbf{A} \cdot n_\perp$ follows immediately.

8.1.4 Boundary Conditions for Circuit/EM Coupling

A severe problem in TCAD simulation is to define the outer borders of the device where the "outer world" begins. It requires often some intuition. One assumes couplings to the outer world only through voltages and currents via a countable number of terminals and that offside terminal contacts there exists no electro-magnetic interaction to the outside world. Therefore Neumann boundary conditions are assumed offside from contact surfaces, which neither allow field nor displacement currents, i.e. $\nabla V \cdot n_\perp = 0$ and $(\nabla V + \boldsymbol{\Pi}) \cdot n_\perp$, where $n_\perp$ and $n_\parallel$ being the outer unit normal vectors transverse and parallel to the terminal contact boundary. Since the normal component $\frac{\partial}{\partial n} B = 0$ it follows that $\mathbf{A} \cdot n_\parallel = 0$.

8.1.4.1 Terminal Contacts

Let i_k be the k-th. current leaving the 3D device at the contact Γ_k. From KCL it follows that $\sum_k i_k = 0$. Moreover we assume that the normal component of the magnetic and the tangential component of the electric fields vanish, i.e. $(\nabla \times \mathbf{A}) \cdot n_\perp = 0, (\nabla V + \boldsymbol{\Pi}) \cdot n_\parallel = 0$. Under these assumptions the terminal currents are computed by

$$i_k = \int_{\Gamma_k} [\boldsymbol{J} - \partial_t(\epsilon(\nabla V + \boldsymbol{\Pi}))] \cdot n_\perp \, \mathrm{d}S.$$

For the terminal contacts, such as metal-semiconductor contacts, the same boundary conditions hold as for the internal boundaries/interfaces of the device given above.

8.1.5 The Coupled Circuit/Device/EM Simulator

The simulator LinzFrame [7, 8] follows a strictly modular concept as depicted in Fig. 8.2. The simulator kernel employs the Modified Nodal Analysis (MNA). Moreover it comprises an automatic differentiation suite which simplifies the implementation of new models significantly, because partial derivatives w.r.t. the state variables (required for Newton type methods) need not to be coded explicitly, but are computed automatically [42]. Furthermore, model libraries for linear devices, SPICE transistor models and a stimulus li-

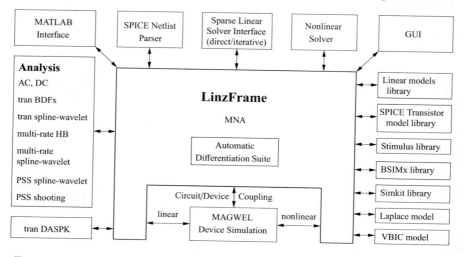

Fig. 8.2: Coupled simulation LinzFrame/devEM toolbox overview.

brary including modulated sources, libraries to industry relevant device models such as BSIMx, VBIC, and the Simkit library from NXP Semiconductors (MEXTRAM, MOS9, MOS11 etc.) are available. A Laplace transfer model interface, which allows the incorporation of compressed models from Model Order Reduction (MOR) complements the model libraries, represented by the ABCD matrices.

The analysis tool comprises standard methods such as in-house DC, AC and transient analysis with polynomial and trigonometric BDFx methods [20], an interface to the DASPK simulator [23] for solving higher index Differential-Algebraic Equations (DAEs) and the in-house tools of a spline-wavelet transient simulator. Several tools for multirate simulation [9, 15, 17], i.e., Harmonic Balance (HB), BDF and spline-wavelet techniques (both algebraic and trigonometric polynomials), are incorporated. The latter technique is superior when strong nonlinearities and/or sharp transients occur, which are efficiently resolved by an adaptive mesh. Periodic Steady-State (PSS) methods for driven and autonomous circuits such as oscillators complete the tool [6, 7, 10, 12, 20].

Moreover, interfaces to damped Newton solvers, homotopy methods [20], to several direct sparse linear solvers (e.g. MUMPS, MA48) [28, 44] as well as preconditioned Krylov subspace techniques (e.g. ILUPACK) are available. Finally, for a rapid prototyping and test of novel algorithms, a MATLAB interface is at hand. The simulator provides two interfaces to the devEM TCAD simulator [1]. Since for linear devices the space discretization must be performed only once by the TCAD simulator which delivers a matrix stamp to LinzFrame, a separate interface for linear devices has been developed. For a holistic simulation of nonlinear devices within a circuit, the TCAD simulator must evaluate the equations together with partial derivatives w.r.t. the unknowns at every Newton iteration which requires more communication between the circuit and device simulator.

8.2 Multirate Simulation Techniques

The multirate technique derived below reformulates the system of ordinary DAEs (8.1) by a system of hyperbolic PDEs. The solution of the underlying ordinary DAEs is obtained along a specific characteristic curve of the PDE. The method was originally developed for multitone steady state analysis [15]. In [46,54] the method has been generalized for initial value problems of driven circuits with fixed and a priori known fundamental frequencies. For further references see also [21,22]. For autonomous circuits such as oscillators the frequency is a priori unknown. Moreover, the instantaneous frequency changes during settling time. Frequency modulated (FM) signals occur in communication transceiver designs but also in pulse width modulation (PVM) circuitry, heavily used in control engineering. Therefore, in [16–18] a novel form of hyperbolic PDEs has been proposed for capturing frequency or phase modulated signals by introducing the instantaneous frequency (or phase) as additional unknown. Further variants of the method can be found in [40, 55] and [51,53]. Since the instantaneous frequency is an unknown, the resulting system is under determined. A good estimate of the instantaneous frequency is therefore crucial for a run time efficient solution of the hyperbolic DAE. The instantaneous frequency is defined via the Hilbert transformation of a signal. However numerical calculation of the Hilbert transform of signals with a large bandwidth is prohibitively slow. Therefore alternative techniques were developed which choose the instantaneous frequency such that the PDE solution is locally smooth. In [7] it could be proven that a slight modification of the method presented in [17, 18] is optimal w.r.t. smoothness of the solution using the Euclidean norm. The method is similar to the method presented in [52]. The derivation below follows [17].

8.2.1 Derivation of the Multirate Technique

We consider again (8.1) with initial condition x_0, namely

$$f(x(t),\dot{x}(t),t) = \frac{\mathrm{d}}{\mathrm{d}t}\,q(x(t)) + \underbrace{i(x(t)) + b(t)}_{g(x(t),t)} = 0, \quad x(0) = x_0, \qquad (8.3)$$

wherein $x \in \mathbb{R}^N$, $i, q \colon \mathbb{R}^N \to \mathbb{R}^N$ and $b \colon \mathbb{R} \to \mathbb{R}^N$. Moreover, it is assumed that (8.3) is uniquely solvable for all consistent initial conditions. We assume that oscillatory solutions with widely separated time scales occur. Examples include transceiver front ends in RF engineering such as mixers, oscillators etc. The multirate PDE technique essentially circumvents the bottle neck of the Nyquist's sampling theorem by introducing fast and slow time scales. It is presupposed that the ordinary DAE (8.3) has a unique solution for a family of initial conditions parametrized by a parameter vector Θ_0

$$f(x_\Theta(t), \dot{x}_\Theta(t), t) \equiv i(x_\Theta(t)) + \frac{\mathrm{d}}{\mathrm{d}t} q(x_\Theta(t)) + b_\Theta(t) = 0, \qquad (8.4)$$

$$x_\Theta(0) = x_0(\Theta_0), \ \Theta_0 \in \mathbb{R}^m.$$

In the equations above, $x_0(\Theta_0) \colon \mathbb{R}^m \to \mathbb{R}^N$ and $b_\Theta(t) \colon \mathbb{R} \to \mathbb{R}^N$ are periodic with normalized period $[2\pi]^m$ with respect to the vector $\Theta_0 = (\Theta_{1,0}, \ldots, \Theta_{m,0})$. Without loss of generality, let the initial condition for $\Theta_0 = 0$ of (8.4) be identical with the initial condition of (8.3), i.e. $x_0(0) = x_0$ and $b_0(t) \equiv b(t)$. The family of initial conditions $x_0(\Theta_0)$ and sources $b_\Theta(t)$ can be chosen arbitrarily for $\Theta_0 \neq 0$, if the periodicity constraint in $\Theta_{i,0}$, $i = 1, \ldots, m$ is fulfilled. This is considered in more detail in examples below. It can be shown that a special solution of the family of initial conditions coincides with the solution along a characteristic curve of a hyperbolic PDE or partial DAE. Let

$$\Omega := \{(\tau, t_1, \ldots, t_m) \mid \tau \in [0, \infty[, \ t_i \in [0, 2\pi], \ 1 \le i \le m\} \qquad (8.5)$$

be a domain.[1] A system of partial DAEs on $\Omega \subset \mathbb{R}^{m+1}$ of the form

$$\hat{f}(\hat{x}(\tau, t_1, \ldots, t_m), \nabla \hat{x}(\tau, t_1, \ldots, t_m), (\tau, t_1, \ldots, t_m))$$

$$= i(\hat{x}(\tau, t_1, \ldots, t_m)) + \nabla q(\hat{x}(\tau, t_1, \ldots, t_m)) + \hat{b}(\tau, t_1, \ldots, t_m) = 0, \qquad (8.6)$$

where $(\tau, t_1, \ldots, t_m) \in \Omega$, is defined with initial conditions $\hat{x}(0, t_1, \ldots, t_m)$. The initial conditions coincide with the initial conditions of the family of ordinary DAEs (8.4), i.e.,

$$\hat{x}(0, t_1, \ldots, t_m) = x_0(t_1, \ldots, t_m). \qquad (8.7)$$

Moreover, it is assumed that $\hat{x}(\tau, t_1, \ldots, t_m)$ satisfies 2π periodic boundary conditions in its arguments $t_1, \ldots, t_m$, i.e. $\hat{x}(\tau, \ldots, 0, \ldots) = \hat{x}(\tau, \ldots, 2\pi, \ldots)$. The differential operator ∇ is defined by

$$\nabla := \frac{\partial}{\partial \tau} + \sum_{i=1}^{m} \frac{\mathrm{d}(\tau \, \omega_i(\tau))}{\mathrm{d}\tau} \cdot \frac{\partial}{\partial t_i}, \quad \omega_i \in C^1(\mathbb{R}) \qquad (8.8)$$

and the stimulus $\hat{b}(\cdot)$ is given by

---

[1] It shall be noted that a mathematically more strict formulation defines the PDE on a hyper torus. However, for ease of presentation we use the formulation which is employed in the references.

$$\hat{b}(\tau,t_1,\ldots,t_m)\Big|_{\tau=t,\,t_1=\omega_1(t)\,t+\Theta_{1,0},\ldots,\,t_m=\omega_m(t)\,t+\Theta_{m,0}} := b_\Theta(t). \qquad (8.9)$$

The choice of m requires a priori knowledge of the physical background of the DAE as illustrated by subsequent examples. The parameters ω_i are estimates of the instantaneous frequencies. Their calculation is not part of the PDE formulation (8.7) and considered below.

The relation between the family of initial conditions (8.4) and the system of partial DAEs (8.6) is given by the following theorem.

Theorem 8.1 ([19]). *The partial DAE (8.6) with the ∇ operator defined by (8.8) has a solution $\hat{x}(\tau,t_1,\ldots,t_m)$ on the domain Ω (8.5) and initial conditions (8.7) if and only if the family of initial conditions (8.4) have a solution $x_\Theta(t)$ for consistent initial conditions $x_\Theta(0) = x_0(\Theta_0)$. The stimulus vector is given by (8.9). The relation*

$$x_\Theta(t) = \hat{x}(t,\Theta_1(t),\ldots,\Theta_m(t)), \qquad (8.10)$$

with

$$\Theta_i(t) = \omega_i(t)\,t + \Theta_{i,0}, \quad i = 1,\ldots,m, \qquad (8.11)$$

holds between the PDE solution and the solution of the ordinary DAE with the family of initial conditions $x_0(\Theta_0)$. Outside the domain Ω the solution $\hat{x}$ is periodically expanded with period 2π in the arguments $t_1,\ldots,t_m$.

Proof. Let $P: (t,\Theta_{1,0},\ldots,\Theta_{m,0}) \in \mathbb{R}^{m+1} \to (\tau,t_1,\ldots,t_m) \in \mathbb{R}^{m+1}$ be a bijective map of the form

$$\tau = t,$$
$$t_1 = \omega_1(t)\,t + \Theta_{1,0},$$
$$\vdots$$
$$t_m = \omega_m(t)\,t + \Theta_{m,0},$$

and $J_q(\hat{x}) := \left(\frac{\partial q_i(\hat{x})}{\partial \hat{x}_j}\right)$ the $N \times N$ Jacobian matrix of $q(\cdot)$. Introducing (8.10) into the DAE (8.4) leads to

$$i(x_\Theta(t)) + \frac{\mathrm{d}}{\mathrm{d}t}q(x_\Theta(t)) + b_\Theta(t)$$
$$= i(\hat{x}(P(t,\Theta_{1,0},\ldots,\Theta_{m,0}))) + \frac{\mathrm{d}}{\mathrm{d}t}q(\hat{x}(P(t,\Theta_{1,0},\ldots,\Theta_{m,0}))) +$$
$$\hat{b}(P(t,\Theta_{1,0},\ldots,\Theta_{m,0})),$$
$$= i(\hat{x}(P(t,\Theta_{1,0},\ldots,\Theta_{m,0}))) +$$
$$J_q(\hat{x}(P(t,\Theta_{1,0},\ldots,\Theta_{m,0}))) \cdot \frac{\mathrm{d}}{\mathrm{d}t}\hat{x}(P(t,\Theta_{1,0},\ldots,\Theta_{m,0})) +$$
$$\hat{b}(P(t,\Theta_{1,0},\ldots,\Theta_{m,0})),$$

$$= i(\hat{x}(P(t, \Theta_{1,0}, \dots, \Theta_{m,0}))) +$$

$$J_q(\hat{x}(P(t, \Theta_{1,0}, \dots, \Theta_{m,0}))) \cdot \nabla \hat{x}(\tau, t_1, \dots, t_m)\Big|_{(\tau, t_1, \dots, t_m) = P(t, \Theta_0)} +$$

$$\hat{b}(P(t, \Theta_{1,0}, \dots, \Theta_{m,0})),$$

$$= [i(\hat{x}(\tau, t_1, \dots, t_m)) + \nabla q(\hat{x}(\tau, t_1, \dots, t_m)) +$$

$$\hat{b}(\tau, t_1, \dots, t_m)]\Big|_{(\tau, t_1, \dots, t_m) = P(t, \Theta_0)},$$

which proofs the theorem.

Remark 8.1.
1. Since a unique solution of (8.4) has been assumed, the PDE (8.6) has a unique solution too.
2. The functions $\Theta_i(t) = \omega_i(t) t + \Theta_{i,0}$, $i = 1, \dots, m$ in (8.11) parametrize the characteristic curves of the PDE (8.6) as will be shown later.
3. The determination of m and $\omega_i(\tau)$, $1 \leq i \leq m$ requires a priori knowledge about the physical background of the DAE (8.3). This is not subject of the formulation of the partial DAE. Since the numerical effort for solving the PDE increases with m, in most practical cases $m = 1$ is chosen. This corresponds to waveforms, whose spectrum is centered around a carrier frequency, referred to as bandpass signals in the engineering literature.
4. The decoupling of the waveform into components exhibiting different frequencies or time scales is reflected by the anisotropy of the ∇ operator (8.8).

In what follows the theorem is illustrated by several examples.

Example 8.1 ([15, 21, 22]). The closed domain Ω is given by

$$\Omega := \{(t_1, \dots, t_m) \,|\, t_i \in [0, 2\pi], \, 1 \leq i \leq m\},$$

i.e. not dependency on the variable τ. This corresponds to a single ($m = 1$) or multitone steady state analysis. This example is strongly related to measurements. The standard method for solving the PDE is the Harmonic Balance technique. For non-autonomous systems the fundamental frequencies are known a priori, hence the equation

$$\omega_i - \omega_{i_0} = 0, \quad 1 \leq i \leq m,$$

with $\omega_{i_0} \equiv$ const. makes the system of partial DAEs unique. This example is referred to as the multitone steady state and no a periodic solution occurs, reflected by the fact that the PDE has no dependency on τ. The ω_{i_0} are a priori known fundamental frequencies. They are usually given by the signal sources. For the special case $m = 1$ one obtains the periodic steady state of the circuit, vice versa for $m > 1$ the quasiperiodic steady state for a multitone excitation. The solution is calculated on the hyper square $\Omega = [0, 2\pi]^m$ employing periodic boundary conditions along the coordinate axes. Along the characteristic curve one obtains the quasiperiodic steady state of the circuit under investigation. The corresponding partial DAE takes the form

$$\hat{f}(\hat{x}(t_1,\dots,t_m),\nabla\,\hat{x}(t_1,\dots,t_m),(t_1,\dots,t_m)) =$$

$$i(\hat{x}(t_1,\dots,t_m)) + \sum_{i=1}^{m}\omega_{i0}\frac{\partial}{\partial t_i}\,q(\hat{x}(t_1,\dots,t_m)) + \hat{b}(t_1,\dots,t_m) = 0. \quad (8.12)$$

For example the Harmonic Balance method for multitone signals employs a Ritz-Galerkin method based on trigonometric basis functions for solving (8.12). Alternatively, spline basis functions, wavelets as well as finite difference schemes are suitable for discretizing the PDE (8.12) as considered in Sec. 8.3.

For mixed autonomous/non autonomous circuits the PDE (8.12) is under determined, because only $l < m$ angular frequencies are known a priori. The remaining $m - l$ angular frequencies however can often well estimated when the physical background is known. Estimates of the instantaneous frequency are considered in Sec. 8.2.2.

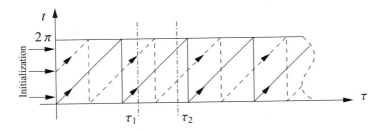

Fig. 8.3: Graphical illustration of the multirate PDE technique for $m = 1$ with two characteristic curves sketched.

Example 8.2 (non-autonomous envelope calculation [46,54]). Let the domain Ω be given by

$$\Omega := \{(\tau,t_1)\,|\,\tau\in[0,\infty[,\,t_1\in[0,2\pi]\},$$

with the frequency calibration $\omega_1 = \omega_0$, wherein ω_0 is a priori known from the stimulus (non-autonomous case). The partial DAE takes the form

$$\hat{f}(\hat{x}(\tau,t_1),\nabla\,\hat{x}(\tau,t_1),t_1) = i(\hat{x}(\tau,t_1)) + \left(\frac{\partial}{\partial\tau} + \omega_0\frac{\partial}{\partial t_1}\right)q(\hat{x}(\tau,t_1)) + \hat{b}(t_1) = 0.$$

$$(8.13)$$

Roychowdhury [54] and Ngoya et al [46] applied this PDE for simulating the transient response of non-autonomous systems with a priori known ω_0. Fig. 8.3 illustrates the method schematically. The PDE is solved along the strip Ω. The solution curves $x_\Theta(t)$ for the family of initial conditions are obtained along affine curves in $\mathbb{R}^2$. As stated in Theorem 8.1, these curves coincide with the characteristic curve of the hyperbolic PDE. Due to the presupposed periodicity along the axis t_1, the curves can be mapped onto the strip given by Ω. $x_\Theta(t) = \hat{x}(t, \omega_0 \cdot t + \Theta_0) = \hat{x}(t, \omega_0 \cdot t + \Theta_0 \mod 2\pi)$.

Example 8.3 ([16, 19]). Let Ω be again the domain

$$\Omega := \{(\tau, t_1) \,|\, \tau \in [0, \infty[, \; t_1 \in [0, 2\pi]\}.$$

In the general case with varying instantaneous frequency the partial DAE is given by

$$\hat{f}(\hat{x}(\tau, t_1), \nabla \hat{x}(\tau, t_1), (\tau, t_1))$$
$$= i(\hat{x}(\tau, t_1)) + \left(\frac{\partial}{\partial \tau} + \frac{\mathrm{d}(\tau \omega(\tau))}{\mathrm{d}\tau} \frac{\partial}{\partial t_1}\right) q(\hat{x}(\tau, t_1)) + \hat{b}(\tau, t_1) = 0. \quad (8.14)$$

For a constant stimulus $\hat{b}(\tau, t_1) = \text{const.}$ the underlying ordinary DAE is autonomous. The PDE (8.14) differs from (8.13) by the a priori unknown instantaneous frequency $\omega(\tau)$, or alternatively a phase modulation term $\Theta(\tau) = \omega(\tau)\tau + \Theta_0$ which parametrizes the characteristic curve of the PDE. The PDE (8.14) is employed for calculating, e.g., the initial transient of a free running oscillator or a driven circuit with phase/frequency modulated sources, such as in communication circuits. The transient response of, e.g., a high-quality quartz crystal oscillator is orders of magnitude longer than the oscillation period. The PDE (8.14) decouples the oscillation from its slowly varying envelope [16, 19]. For the special case $\omega(\tau) = \omega_0$ with parametrized characteristic curve $\Theta(\tau) = \omega_0 \tau + \Theta_0$ one obtains as a special case the non-autonomous envelope method (8.13).

The final examples deals with the simulation of injection locking, frequency entrainment or frequency locking. Entrainment is the basic principle of synchronization

Example 8.4. Let the domain Ω be given by

$$\Omega := \{(\tau, t_1, t_2) \,|\, \tau \in [0, \infty[, \; t_{1,2} \in [0, 2\pi]\}.$$

The calibration of one frequency is $\omega_1 - \omega_0 = 0$ with $\omega_0 = \text{const.}$ The other frequency is free running and a priori unknown. The source is assumed to be periodic in t_1. The PDE is given by

$$\hat{f}(\hat{x}(\tau, t_1, t_2), \nabla \hat{x}(\tau, t_1, t_2), (\tau, t_1, t_2))$$
$$= i(\hat{x}(\tau, t_1, t_2)) + \left(\frac{\partial}{\partial \tau} + \omega_0 \frac{\partial}{\partial t_1} + \frac{\mathrm{d}(\tau \omega_2(\tau))}{\mathrm{d}\tau} \cdot \frac{\partial}{\partial t_2}\right) q(\hat{x}(\tau, t_1, t_2)) +$$
$$\hat{b}(t_1) = 0.$$

If $\lim_{\tau \to \infty}(\omega_2(\tau) - \omega_0) = 0$, then the circuit is locked and the asymptotic waveform is periodic. Otherwise it is unlocked and the asymptotic waveform is quasiperiodic.

It can be shown that the waveforms $x_\Theta(t)$ corresponding to the family of initial conditions of the ordinary DAE coincide with the characteristics of the partial DAE.

Corollary 8.1 ([48]). *A solution* $x_\Theta(t), \Theta(0) = \Theta_0$ *of the parametrized ordinary DAE (8.4) is a solution of the partial DAE (8.6) along the characteristic curve*

$$\Gamma(\Theta_0) = \{(\tau(t), t_1(t), \ldots, t_m(t)) \,|\, t \in \mathbb{R}, \quad \tau = t, \quad t_i(t) = \Theta_i(t), \quad 1 \leq i \leq m\}.$$

Proof. The characteristic curves are evaluated by solving the differential equations $\tau'(t) = 1, t_i'(t) = (\omega_i(t)t)', 1 \leq i \leq m$, i.e. $\tau = t + \tau_0, t_i(t) = \Theta_i(t), 1 \leq i \leq m$. Solutions of the PDE along the characteristic curve $\Gamma(\Theta_0)$ are therefore solutions of (8.4) for all $\Theta_0 \in \mathbb{R}^m$. $\square$

8.2.2 Optimal Estimation of the Instantaneous Frequency

The instantaneous frequency is well defined via the analytical signal $x^+(t)$. However its calculation employs the Hilbert transformation of the signal which can be only hardly calculated within a broad frequency range. We resort therefore to a numerically easier method. The estimate of the instantaneous frequency follows the derivation in [7]. For ease of presentation, we consider only one single dominant oscillation, i.e., $m = 1$ with instantaneous frequency $\omega(\tau)$ and setting $t_1 = t$ for convenience. As mentioned above, the calculation of the frequency is not part of the PDE formulation. It can be easily verified that $\omega(\tau)$ can be any positive differentiable function, if the source term $\hat{b}(\tau, t)$ is chosen to satisfy (8.9).

Therefore, we will first investigate, what effect different selections of $\omega(\tau)$ have. Let $\omega_1(\tau)$ and $\omega_2(\tau)$ be two distinct choices. If $\hat{x}_1(\tau, t)$ and $\hat{x}_2(\tau, t)$ both satisfy (8.6), then it is obvious that

$$\hat{x}_2(\tau, t) = \hat{x}_1(\tau, t + S(\tau)), \tag{8.15}$$

where $S(\tau) = (\omega_1(\tau) - \omega_2(\tau)) \cdot \tau$. That is, for a fixed τ changing $\omega(t)$ results in a phase shift of $\hat{x}(\tau, t)$. A corresponding phase shift for the source term has to be performed in order to satisfy (8.9). The following Lemma describes how $\omega(\tau)$ affects the smoothness of $\hat{x}(\tau, t)$.

Lemma 8.1. *Let* $T(\tau)$ *be the unique solution of* $\Theta(\tau + T(\tau)) - \Theta(\tau) = 2\pi$ *for all* $\tau > 0$ *and* $L \geq \max_\tau T(\tau)$. *If*

$$\left\|\hat{x}(\tau+\delta,t)-\hat{x}(\tau,t)\right\|\leq\varepsilon,\tag{8.16}$$

for $\delta \in (0,L]$, $t \in [0,2\pi]$, then

$$\left\|x_\Theta\big(\tau+T(\tau)\big)-x_\Theta(\tau)\right\|\leq\varepsilon.\tag{8.17}$$

Proof. Following (8.10) we have

$$\begin{aligned} x_\Theta(\tau+T(\tau)) &= \hat{x}\big(\tau+T(\tau),\Theta(\tau+T(\tau))\big) = \hat{x}\big(\tau+T(\tau),\Theta(\tau)+2\pi\big),\\ &= \hat{x}\big(\tau+T(\tau),\Theta(\tau)\big). \end{aligned}$$

Since $x_\Theta(\tau) = \hat{x}\big(\tau,\Theta(\tau)\big)$ and $x_\Theta(\tau+T(\tau))$ correspond to function evaluations of the PDE solution $\hat{x}$ at points on the same horizontal curve in Fig. 8.3 parallel to the x axis. Since $T(\tau) < L$, (8.17) follows directly from (8.16). □

Lemma 8.1 states that we can only expect smoothness of $\hat{x}(\tau,t)$ in τ if x_Θ is nearly $T(\tau)$-periodic in a neighborhood of τ.

Since typical multirate signals behave locally like a periodic signal, i.e., $x_\Theta(t) \approx x_\Theta(t+T(t^*))$ as long as t is close to some t^*. That is, there should be a choice of $\omega(\tau)$ such that $\hat{x}(\tau_1,t)$ and $\hat{x}(\tau_2,t)$ do not differ much for sufficiently small $\tau_2 - \tau_1$. This leads to the additional condition

$$\int_0^P \left|\tfrac{\partial}{\partial\tau}\hat{x}(\tau,t)\right|^2 dt \;\to\; \min.\tag{8.18}$$

in order to determine $\omega(\tau)$.

It turns out that condition (8.18) yields the expected result for amplitude and frequency modulation of a periodic signal, i.e., for typical amplitude/phase modulated signals encountered in modern communication transceivers.

Lemma 8.2. *Assume $x_\Theta(t) = a(t)\,y(\Theta + \tilde{\Theta}(t))$ with $y(t) = y(t+2\pi)$ is a solution of (8.4) with non-trivial amplitude modulated waveform $a(t)$ and phase/frequency modulated signal $\tilde{\Theta}(t)$ and $y'(t) \neq 0$. The solution of the corresponding multirate problem (8.2) satisfies (8.18) if and only if $\Theta(\tau) = \tilde{\Theta}(\tau)$ and $\hat{x}(\tau,t) = a(\tau)y(t)$.*

Proof. The proof can be found in [7]. □

Applying the approach to more general problems permits to consider $\omega(t)$ as a generalization of the instantaneous angular frequency of the multirate problem if (8.18) is satisfied. For the numerical solution of the multirate problem we have to discretize (8.6) together with the smoothness condition (8.18).

8.3 A Spline Galerkin Method for Circuit Simulation

In [12, 13] an adaptive spline-wavelet method for the initial value problem (8.3) has been developed. For ease of presentation, the case of a dominant frequency $m = 1$ is considered here, hence we consider the mixed initial boundary value problem of (8.13) or (8.14), respectively. To this end, we will modify the approach in [12, 13] by expanding the solution in a periodic basis. Then the periodic boundary conditions are fulfilled automatically and has not to be enforced by additional equations. For a more comprehensive treatment we refer to [8].

8.3.1 Semi Discretization by Rothe's Method

Rothe's method applied to the mixed initial-boundary value problem (8.14) starts with discretizing the initial value problem first. In a second step the resulting boundary value problem is solved employing a spline method.

By discretizing (8.14) with respect to τ using, e.g., Gear's BDF method[2] of order s [31] one obtains

$$\sum_{i=0}^{s}\alpha_i^k q\big(X_{k-i}(t)\big) + \omega_k \frac{d}{dt}q\big(X_k(t)\big) + i\big(X_k(t)\big) + \hat{b}\big(\tau_k,t\big) = 0, \qquad (8.19)$$

$$X_k(t) = X_k(t+2\pi). \qquad (8.20)$$

Here we have to determine an approximation $X_k(t)$ of $\hat{x}(\tau_k,t)$ from known, approximate solutions $X_{k-i}(t)$ at previous time steps τ_{k-i}. $i = 1,\ldots,s$. The BDF coefficients α_i^k are chosen such that for any polynomial p of degree up to s the derivative is computed exactly, i.e.,

$$\sum_{i=0}^{s}\alpha_i^k p(\tau_{k-i}) = p'(\tau_k).$$

Since the solutions ω_{k-i} and $X_{k-i}(t)$ at previous time steps τ_{k-i} for $i > 0$ are already calculated, we define

$$f_k(x,t) := \alpha_0^k q(x) + i(x) + \hat{b}\big(\tau_k,t\big) + \sum_{i=1}^{s}\alpha_i^k q\big(X_{k-i}(t)\big). \qquad (8.21)$$

Then X_k is the solution of the periodic boundary value problem

---

[2] Other multistep integration methods such as the trapezoidal method can be used as well.

$$\omega_k \frac{d}{dt} q\big(x(t)\big) + f_k(x(t), t) = 0, \quad x(t) = x(t + 2\pi). \tag{8.22}$$

The new problem (8.22) is closely related to the original periodic steady state problem of the circuit, only modified by the additional 'source term' $\sum_{i=1}^{s} \alpha_i^k q(X_{k-i}(t))$.

If $\omega := \omega_k \approx \omega(\tau_k)$ is not fixed in advance, then an additional condition is needed. Following the treatment of Sec. 8.2 and [7], we use the discretized version

$$\int_0^{2\pi} \big| X_k(t) - X_{k-1}(t) \big|^2 \, dt \to \min \tag{8.23}$$

of (8.18).

8.3.2 A Petrov-Galerkin Method

We want to approximate the solution of (8.22) by a 2π periodic spline function of order m. For given grid points t_ℓ with

$$0 \le t_0 < t_1 < \ldots < t_{L-1} < 2\pi$$

we consider all $m - 2$-times differentiable, 2π periodic functions, which are polynomials[3] (both algebraic and trigonometric) of degree less than m at each sub-interval $(t_\ell, t_{\ell+1})$. The break points t_ℓ are called spline knots, which are periodically extended by $t_{iL+\ell} = t_\ell + 2\pi i$. Note, that the periodicity condition implies that the periodic spline is a piecewise polynomial also with respect to the extended grid.

A stable and computationally efficient basis for the linear space of spline functions is constituted by the B-splines $N_\ell^m(t)$, which are uniquely determined by their minimal support $[t_\ell, t_{\ell+m}]$ and the partition of unity $\sum_i N_i^m(t) = 1$ (for normalization). For more information on spline functions and B-splines as well as for efficient computational methods we refer the reader to the detailed description in [14, 59].

To expand the periodic solution of (8.22), we need periodic basis functions, which we obtain by the periodized B-splines

$$\varphi_\ell(t) = \sum_{i \in \mathbb{Z}} N_{\ell+iL}^m(t) = \sum_{i \in \mathbb{Z}} N_\ell^m(t - 2\pi i), \quad \ell = 1, \ldots, L,$$

which form a basis for the space of periodic spline functions.

Now we have to find a spline function

---

[3] In this Section m has the meaning of the polynomial order, whereas in Sec. 8.2 it has the connotation of the number of dominant frequencies.

$$\tilde{x}(t) = \sum_{\ell=1}^{L} c_\ell \varphi_\ell(t), \qquad (8.24)$$

which approximates the solution of (8.22). Since $c_\ell \in \mathbb{R}^N$ (N is the number of equations and unknowns in the circuit equations (8.3)), we have to determine $n \cdot L$ coefficients $c_{\ell,i}$. Thus, we have to derive from (8.22) $n \cdot L$ conditions, which ensure that $\tilde{x}$ indeed approximates the solution of (8.22). Taking the integral over L subintervals we obtain

$$0 = F_\ell(\boldsymbol{c}, \omega_k) := \int_{\hat{t}_{\ell-1}}^{\hat{t}_\ell} \omega_k \tfrac{d}{dt} q(\tilde{x}(t)) + f_k(\tilde{x}(t), t)\, dt \qquad (8.25)$$

$$= \omega_k \left(q(\tilde{x}(\hat{t}_\ell)) - q(\tilde{x}(\hat{t}_{\ell-1})) \right) + \int_{\hat{t}_{\ell-1}}^{\hat{t}_\ell} f_k(\tilde{x}(t), t)\, dt,$$

for $\ell = 1, \ldots, L$. This L vector valued equations determine the vector coefficients c_ℓ. Note, that the splitting points $\hat{t}_\ell$ do not coincide with the spline knots t_ℓ, but they have to be chosen in relation to the spline grid. In particular, we need due to periodicity that $\hat{t}_L - \hat{t}_0 = 2\pi$.

The nonlinear system (8.25) can be solved by Newton's method. If $\omega = \omega(\tau)$ is known in advance as in (8.13), we have to solve the linear system $\boldsymbol{A}\, \boldsymbol{d}_c = \boldsymbol{b}$ to determine the Newton correction $\boldsymbol{d}_c$ in

$$\boldsymbol{c}^{(k,j+1)} = \boldsymbol{c}^{(k,j)} - \boldsymbol{d}_c, \qquad (8.26)$$

where j is the Newton count. Starting from a sufficiently good initial guess, e.g. by extrapolation $\boldsymbol{c}^{(k,0)} = \boldsymbol{c}^{(k-1)}$, one obtains usually after sufficiently many steps a good approximation $\boldsymbol{c}^{(k)} = c^{(k,J)}$ of the solution of (8.25). In order to set up the linear system we have to compute the right hand side[4] $\boldsymbol{b} = F(\boldsymbol{c}, \omega) = (F_\ell(\boldsymbol{c}, \omega))_{\ell=1,\ldots,L}$ as well as the Jacobian (with respect to $\boldsymbol{c}$) $A = D_c F(\boldsymbol{c}, \omega)$ with the matrix block entries

$$\frac{\partial F_\ell(\boldsymbol{c}, \omega)}{\partial c_i} = \omega \left(C(\tilde{x}(\hat{t}_\ell))\, \varphi_i(\hat{t}_\ell) - C(\tilde{x}(\hat{t}_{\ell-1}))\, \varphi_i(\hat{t}_{\ell-1}) \right) +$$

$$\int_{\hat{t}_{\ell-1}}^{\hat{t}_\ell} G(\tilde{x}(t), t)\, \varphi_i((t))\, dt,$$

where the Jacobians $C := \partial_x q(x)$ and $G := \partial_x g(x, t)$ are available in any circuit simulator, which employs Newton's method. The integrals can be computed by a suitable quadrature rule. In practical computations where the spline

---

[4] For clarity of notation we use $\boldsymbol{c} = \boldsymbol{c}^{(k,j)}$ and $\omega = \omega_{k,j}$ in the sequel, if the relation is clear from the context.

order is chosen usually of order $m = 3$ or $m = 4$, one can use the Simpson rule or the two point Gauß quadrature formula.

If $\omega := \omega_k$ has to be determined during the computation, linearization of (8.25) results in the under determined system

$$\boldsymbol{A}\,\boldsymbol{d}_c + d_\omega\,\boldsymbol{z} = \boldsymbol{b}, \tag{8.27}$$

under condition (8.23). Here $\boldsymbol{z} = \big(\partial_\omega F_\ell(\boldsymbol{c},\omega)\big)_{\ell=1,\ldots,L}$, where

$$\partial_\omega F_\ell(\boldsymbol{c},\omega) = q\big(\tilde{x}(\hat{t}_\ell)\big) - q\big(\tilde{x}(\hat{t}_{\ell-1})\big) + \int_{\hat{t}_{\ell-1}}^{\hat{t}_\ell} \tfrac{d}{d\omega} f_k\big(\tilde{x}(t),t\big)\,dt.$$

For computational reasons we replace (8.23) by a similar condition on the spline coefficients, namely

$$\big\|\boldsymbol{c}^{(k)} - \boldsymbol{c}^{(k-1)}\big\|_2^2 \to \min. \tag{8.28}$$

Following [7], we obtain the solution of (8.27) which satisfies (8.28) by

$$\boldsymbol{d}_c = \tilde{\boldsymbol{b}} - d_\omega\,\tilde{\boldsymbol{z}},$$

where $\tilde{\boldsymbol{b}} = \boldsymbol{A}^{-1}\boldsymbol{b}$ and $\tilde{\boldsymbol{z}} = \boldsymbol{A}^{-1}\boldsymbol{z}$ are computed by solving the corresponding linear systems and

$$d_\omega = \frac{\tilde{\boldsymbol{z}}^T\big(\boldsymbol{c}^{(k,j-1)} - \boldsymbol{c}^{(k-1)} - \tilde{\boldsymbol{b}}\big)}{\tilde{\boldsymbol{z}}^T\tilde{\boldsymbol{z}}},$$

where in addition to (8.26) ω_k is determined by the iteration

$$\omega_{k,j+1} = \omega_{k,j} - d_\omega, \qquad \omega_{k,0} = \omega_{k-1}.$$

This choice of d_ω ensures the constraint (8.28), which is a slight modification of (8.23).

8.3.3 Adaptive Grid Generation

It is still open, how the spline knots t_i are to be chosen. In particular, for functions with sharp transients it is important that we place more grid points at the location of this refinement. Since those locations are often not known in advance, it is important that the grid is generated automatically by an adaptive refinement. For this refinement we use spline wavelets for non uniform grids introduced in [5].

Starting on some coarse initial grid we solve (8.25) by Newton's method and obtain a first approximation, which we denote as $X(t)$. Now we apply a fast wavelet transform introduced in [8] and obtain

$$X(t) = \tilde{X}(t) + \sum_{\ell} d_{\ell} \psi_{\ell}(t). \tag{8.29}$$

The spline $\tilde{X}(t)$ is an approximation of $X(t)$ using only the even spline knots $t_{2\ell}$. The ψ_{ℓ} are wavelets which carry detail information. In the setting of [5] the wavelets are compactly supported splines, with a chosen number of vanishing moments, which are localized near $t_{2\ell+1}$. Thus, the coefficients d_{ℓ} are a measure of the local approximation error, which describes also the smoothness of $X(t)$. Therefore we insert additional knots in the neighborhood of $t_{2\ell+1}$ if the coefficient d_{ℓ} exceeds a given threshold. The spline representation of $X(t)$ for the new grid can then be computed efficiently by the Oslo algorithm [26,36]. These refined spline expansion is used as a new initial guess for another Newton iteration.

This approach is repeated several times, with Newton tolerances and wavelet threshold reduced for each refinement. This leads to better and better approximations, which provide more and more information for the grid refinements, leading to an almost optimal grid. Furthermore, this approach is computationally efficient, since the first Newton iterations are computed on a relatively small grid. For the final refinements usually one Newton update is sufficient, since we have already an excellent initial guess. The whole process is stopped if the Newton iteration after the last refinement gets below a given error bound. This error bound is indeed a very good estimate of the achieved error. For details on the refinement algorithm we refer to [8].

Above we suggested to use the solution X_{k-1} of the previous time step τ_{k-1} as initial guess (predictor) for Newton's method. In fact any reasonable predictor should be based on solutions from previous time steps such that the following approach should be used for other predictors too. Due to the smoothness in τ, the difference between X_{k-1} and X_k should be small, and we can expect fast convergence of Newton's method. However, the refinements in the adaptive method described above will increase the number of spline knots and thus the size of the problem. Therefore, we will use an approximation of X_{k-1} on a coarser grid as initial guess. An efficient approximation can be achieved by wavelet thresholding as follows. From the wavelet expansion (8.29) of X_{k-1} we remove terms with small wavelet coefficients d_{ℓ}, i.e., an approximation is obtained by

$$Y(t) = \tilde{X}(t) + \sum_{|d_{\ell}| > \varepsilon} d_{\ell} \psi_{\ell}(t). \tag{8.30}$$

If the threshold ε is chosen in the order of magnitude of the error tolerance of the method, than we can still expect the initial guess $Y(t)$ to be as good as $X_k(t)$ itself, while the computational cost is reduced. Using the approach from [8] removing $\psi_{\ell}(t)$ is equivalent to remove the knot $t_{2\ell+1}$. Thus, the

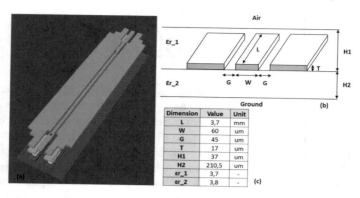

Dimension	Value	Unit
L	3,7	mm
W	60	um
G	45	um
T	17	um
H1	37	um
H2	210,5	um
er_1	3,7	-
er_2	3,8	-

Fig. 8.4: Transmission line 3D model.

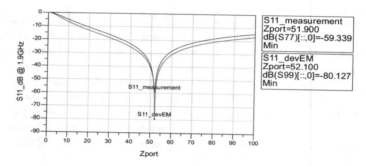

Fig. 8.5: TR line: Measurements vs. simulation results as function of the terminal values.

method will remove spline knots, which became dispensable due to a change in signal shape over several time steps τ_k, or which where not needed from the beginning due to the rough error estimate in the refinement described above. This leads to an almost optimal grid for any given error tolerance. For more details on this grid coarsening algorithm we refer to [8].

8.4 Numerical Examples

8.4.1 On-Chip Transmission Line

Simulation results of the LinzFrame/devEM coupled simulator has been compared with S-parameter measurements for the on-chip transmission line depicted in Fig. 8.4. The parameters of the device are also given in Fig. 8.4. The measurement frequency is $f = 1.9$ GHz. Fig. 8.5 shows the reflection

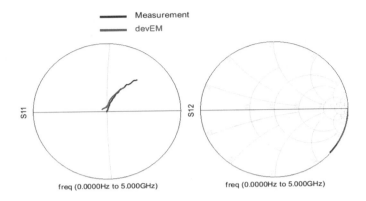

Fig. 8.6: Smith chart plots for the transmission line in the frequency range
0 : 5 GHz: Measurements vs. simulation results.

parameter S_{11} as function of the impedance of the port terminations. The characteristic impedance of the line is approx. 52 Ω. Fig. 8.6 depicts the S_{11}, S_{12} reflexion/transmission parameters from 0 : 5 GHz in the Smith chart. Simulations are in excellent agreement with measurements.

8.4.2 Differential Oscillator

Fig. 8.7 depicts the schematic of a differential oscillator, wherein the inductors are modeled as 3D devices. The holistic coupling is performed according to Fig. 8.1, i.e., the spatial discretization is realized by devEM and the multirate time integration by LinzFrame. Fig. 8.8 shows the multirate simulation result of the differential output waveform $V_2 - V_3$ of the PDE (8.14). One can see the settling of the envelope in the τ scale and the periodic oscillation in the other scale. The time domain solution is obtained along the characteristic curve $\Theta_0 = 0$, i.e., $\Theta(t) = \omega(t) t$, which solves (8.3) for initial condition $x(0) = x_0$.

8.4.3 Balun with Power Stage Circuit

Fig. 8.9 depicts a power stage circuit with its on-chip balun for a band I application (center frequency $f_c = 1 : 9$ GHz). The power stage operates in differential mode (DM). The DM has the advantages to be of wide bandwidth, to reduce the transformation ratio of the matching network, to be less sensitive to load, grounding and package variation and to reduce even order harmonics. Since the source and output is single ended, a first balun,

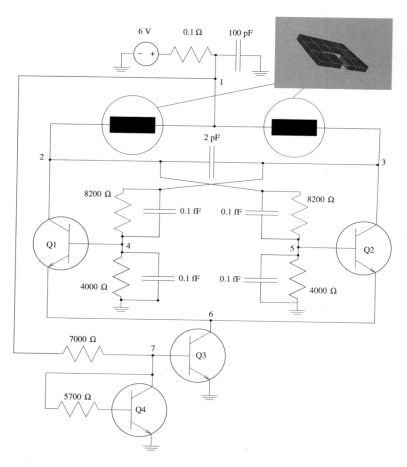

Fig. 8.7: Differential oscillator with a 3D inductor model.

operating at low power, together with a matching circuit is required. The critical device in the design is the balun at the output of the power stage and is modeled in full 3D employing Maxwell's equations above. The technology is bismaleimide-triazine (BT) with 4 layers as depicted in Fig. 8.10. Further technology relevant parameters are given in Fig. 8.11. The voltage waveforms at the in- and output and at the single ended side of the first balun Fig. 8.9 (left) are depicted in Fig. 8.12. Moreover, Fig. 8.13 shows the differential voltages at the output of the power amplifier (PA). All signals are marked in Fig. 8.9. The time domain waveforms correspond to the specific characteristic curve of the multirate PDE $x_\Theta(t)$ with $\Theta(0) = 0$.

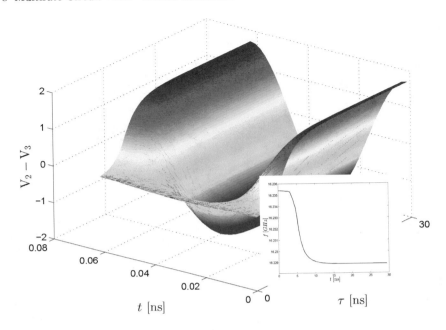

Fig. 8.8: Multirate PDE solution of Fig. 8.7 and instantaneous frequency
(window).

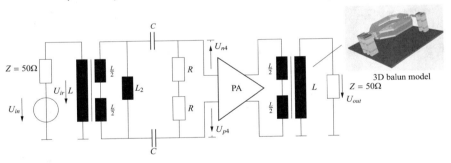

Fig. 8.9: On-chip baluns with power stage circuit and 3D model.

8.5 Conclusions

The multirate PDE method generalizes the method of the Equivalent Complex Baseband (ECB) to circuits and systems described by nonlinear differential-algebraic equations and thus circumvents the bottleneck of Shannon's
sampling theorem. The technique reformulates the system of ordinary DAEs
by partial DAEs through introduction of two or more time scales, one time
scale for capturing the radio frequency or fast oscillations and the other time

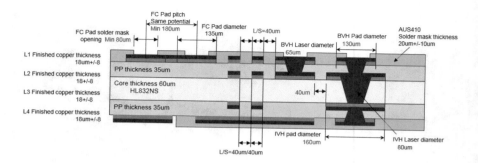

Fig. 8.10: Cross section of the balun in BT technology.

BT 4 layers (Measured stack)	Layer type	Thickness [μm]	εr	σ [S/m]	Material
Solder resist1 (B_SMT)	Oxide	20	3,7	-	PFR-800 AUS SR1
Metal: B_M1 (TOP)	Cu	20	-	5,90E+07	
PPreg1 (Via1)	Oxide	40	3,9	-	Mitsubishi Gas Chemical
Metal: B_M2	Cu	17	-	5,90E+07	
Core: (Via2)	Oxide	64	3,9	-	Mitsubishi Gas Chemical
Metal: B_M3	Cu	15	-	5,90E+07	
PPreg2: (Via3)	Oxide	38,5	3,9	-	Mitsubishi Gas Chemical
Metal: B_M4 (BOTTOM)	Cu	16	-	5,90E+07	
Solder resist2 (B_SMB)	Oxide	20	3,7	-	PFR-800 AUS SR1

Fig. 8.11: Parameters of the balun in BT technology.

scales for capturing the waveform envelopes. The solution of the ordinary DAE (the original problem) is obtained along a characteristic curve of the PDE. In most practical cases two time scales are sufficient.

The method, interpreted in the frequency domain, can be considered as a technique to exploit the sparsity of the spectrum. Unlike the Equivalent Complex Baseband (ECB) method, which also decouples the slowly varying envelope from the carrier frequency and hence also exploits the spectrum's sparsity, it is not limited to LTI systems, offering new capabilities in the simulation of digital transmission systems or at least parts thereof.

An efficient method for solving the partial DAEs is Rothe's method since it offers high flexibility w.r.t. numerical grid adaptation. Spline-wavelet methods are described to adaptively solve these systems by a Petrov-Galerkin method.

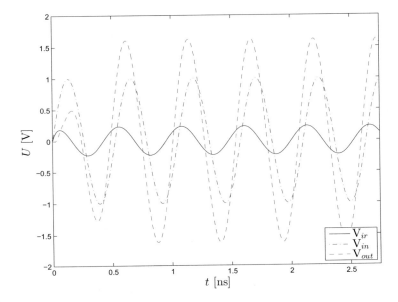

Fig. 8.12: Balun with power stage: simulation results.

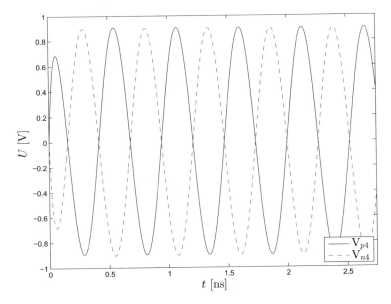

Fig. 8.13: Balun with power stage: simulation results.

Acknowledgements This project has been co-financed by the European Union using financial means of the European Regional Development Fund (INTERREG) for sustainable cross boarder cooperation. Further information on INTERREG Austria-Czech Republic is available at https://www.at-cz.eu/at.

References

1. *Nanoelectronic COupled Problems Solutions*, 2013-2016. FP7-ICT-2013.3.1, http://www.fp7-nanocops.eu.
2. BACH, K.H., DIRKS, H.K., MEINERZHAGEN, B., AND ENGL, W.L.: *A new nonlinear relaxation scheme for solving semiconductor device equations.* IEEE Transactions on Computer-Aided Design of Integrated Circuits and Systems, 10(9):1175–1186, Sep 1991.
3. BANK, R.E., AND ROSE, D.J.: *Parameter selection for Newton-like methods applicable to nonlinear partial differential equations.* SIAM Journal on Numerical Analysis, 17(6):806–822, 1980.
4. BANK, R.E., AND ROSE, D.J.: *Global approximate Newton methods.* Numerische Mathematik, 37(2):279–295, 1981.
5. BITTNER, K.: *Biorthogonal spline wavelets on the interval.* In: GUANRONG CHEN AND MING-JUN LAI (EDS.): *Wavelets and Splines: Athens 2005*, pages 93–104. Nashboro Press, Brentwood, TN, 2006.
6. BITTNER, K., AND BRACHTENDORF, H.-G.: *Adaptive multi-rate wavelet method for circuit simulation.* Radioengineering, 23(1):300–307, 2014 (open access).
7. BITTNER, K., AND BRACHTENDORF, H.-G.: *Optimal frequency sweep method in multi-rate circuit simulation.* COMPEL, 33(4):1189–1197, 2014.
8. BITTNER, K., AND BRACHTENDORF, H.-G.: *Fast algorithms for grid adaptation using non-uniform biorthogonal spline wavelets.* SIAM J. Scient. Computing, 37(2):283–304, 2015.
9. BITTNER, K., AND BRACHTENDORF, H.-G.: *Trigonometric integration methods in circuit simulation.* Presentation at *SIAM CSE-2015, Conference on Computational Science and Engineering*, Salt Lake City, March 14-16, 2015.
10. BITTNER, K., AND BRACHTENDORF, H.-G.: *Latency exploitation in wavelet-based multirate circuit simulation.* In: BARTEL, A., CLEMENS, M., GÜNTHER, M., AND TER MATEN, E.J.W. (EDS.): *Scientific Computing in Electrical Engineering – SCEE 2014*, Mathematics in Industry 23, pp. 13–20, Springer, Berlin, Heidelberg, 2016.
11. BITTNER, K., AND BRACHTENDORF, H.-G.: *Multirate Shooting Method with Frequency Sweep for Circuit Simulation.* In: LANGER, U., AMRHEIN, W., AND ZULEHNER, W. (EDS): *Scientific Computing in Electrical Engineering - SCEE 2016.* Springer International Publishing, Series Mathematics in Industry, Vol. 28, pp. 113–125, 2018. https://dx.doi.org/10.1007/978-3-319-75538-0_11.
12. BITTNER, K., AND DAUTBEGOVIC, E.: *Adaptive wavelet-based method for simulation of electronic circuits.* In: MICHIELSEN, B., AND POIRIER, J.-R. (EDS.): *Scientific Computing in Electrical Engineering – SCEE 2010*, Mathematics in Industry 16, pp. 321–328, Springer, Berlin, Heidelberg, 2012.
13. BITTNER, K., AND DAUTBEGOVIC, E.: *Wavelets algorithm for circuit simulation.* In: GÜNTHER, M., BARTEL, A., BRUNK, M., SCHÖPS, S., AND STRIEBEL, M. (EDS.): *Progress in Industrial Mathematics at ECMI 2010*, Mathematics in Industry 17, pp. 5–11, Springer, Berlin, Heidelberg, 2012.
14. DE BOOR, C.: *A Practical Guide to Splines.* Springer, New York, 1978.
15. BRACHTENDORF, H.-G.: *Simulation des eingeschwungenen Verhaltens elektronischer Schaltungen.* Shaker, Aachen, 1994.

16. BRACHTENDORF, H.-G.: *On the relation of certain classes of ordinary differential algebraic equations with partial differential algebraic equations.* Technical Report 1131G0-971114-19TM, Bell-Laboratories, 1997.

17. BRACHTENDORF, H.-G.: *Theorie und Analyse von autonomen und quasiperiodisch angeregten elektrischen Netzwerken. Eine algorithmisch orientierte Betrachtung.* Habilitationsschrift, Univ. Bremen, 2001.

18. BRACHTENDORF, H.-G., AND LAUR, R.: *A time-frequency algorithm for the simulation of the initial transient response of oscillators.* In: *Proc. IEEE Int. Symp. on Circuits and Systems,* Monterey, June 1998.

19. BRACHTENDORF, H.-G., AND LAUR, R.: *Transient simulation of oscillators.* Technical Report 1131G0-980410-09TM, Bell-Laboratories, 1998.

20. BRACHTENDORF, H.-G., MELVILLE, R., FELDMANN, P., LAMPE, S., AND LAUR, R.: *Homotopy method for finding the steady states of oscillators.* IEEE Transactions on Computer-Aided Design of Integrated Circuits and Systems, 33(6):867–878, June 2014.

21. BRACHTENDORF, H.-G., WELSCH, G., AND LAUR, R.: *A novel time-frequency method for the simulation of the steady state of circuits driven by multi-tone signals.* In: *Proc. IEEE Int. Symp. on Circuits and Systems,* pp. 1508–1511, Hongkong, June 1997.

22. BRACHTENDORF, H.-G., WELSCH, G., LAUR, R., AND BUNSE-GERSTNER, A.: *Numerical steady state analysis of electronic circuits driven by multi-tone signals.* Electronic Engineering, 79(2):103–112, April 1996.

23. BRENAN, K.E., CAMPBELL, S.L., AND PETZOLD, L.R.: *Numerical solution of initial-value problems in differential-algebraic equations.* SIAM, Philadelphia, 1996.

24. CHEN, Q., SCHOENMAKER, W., MEURIS, P., AND WONG, N.: *An effective formulation of coupled electromagnetic-tcad simulation for extremely high frequency onward.* IEEE Transactions on Computer-Aided Design of Integrated Circuits and Systems, 30(6):866–876, 2011.

25. CHUA, L.O., AND LIN, P.M.: *Computer-Aided Analysis of Electronic Circuits: Algorithms and Computational Techniques.* Prentice-Hall, Englewood Cliffs, N. J., 1975.

26. COHEN, E., LYCHE, T., AND RIESENFELD, R.: *Discrete B-splines and subdivision techniques in computer aided geometric design and computer graphics.* Comp. Graphics and Image Proc., 14:87–111, 1980.

27. DEUFLHARD, P., FIEDLER, B., AND KUNKEL, P.: *Efficient numerical path following beyond critical points.* SIAM J. Numer. Anal., 24(4):912–927, August 1987.

28. DUFF, I.S., AND REID, J.K.: *The design of MA48: A code for the direct solution of sparse unsymmetric linear systems of equations.* ACM Trans.Math. Softw., 22(2):187–226, 1996.

29. ENGL, W.L., LAUR, R., AND DIRKS, H.K.: *MEDUSA - A Simulator for Modular Circuits.* IEEE Trans. on Computer-Aided Design of Integrated Circuits and Systems, CAD-2, pages 85–93, April 1982.

30. GALY, P., AND SCHOENMAKER, W.: *In-depth electromagnetic analysis of ESD protection for advanced CMOS technology during fast transient and high-current surge.* IEEE Transactions on Electron Devices, 61(6):1900–1906, June 2014.

31. GEAR, C.W.: *Numerical Initial Value Problems in Ordinary Differential Equations.* Prentice-Hall, Englewood Cliffs, N. J., 1971.

32. DE GRAAFF, H.C., AND KLAASSEN, F.M.: *Compact transistor modelling for circuit design.* Computational Microelectronics. Springer, 1990.

33. GRASSER, T., AND SELBERHERR, S.: *Mixed-mode device simulation.* In: *2000 22nd International Conference on Microelectronics. Proceedings (Cat. No.00TH8400),* volume 1, pp. 35–42, 2000.

34. HO, C.W., RUEHLI, A.E., AND BRENNAN, P.A.: *The modified nodal approach to network analysis.* IEEE Trans. on Circuits and Systems, CAS-22:504–509, 1975.

35. KEITER, E.R., HUTCHINSON, S.A., HOEKSTRA, R.J., RANKIN, E.L., RUSSO, T.V., AND WATERS, L.J.: *Computational algorithms for device-circuit coupling*. Technical report, Sandia National Laboratories, 2003.

36. LYCHE, T., AND MOERKEN, K.: *Making the Oslo algorithm more efficient*. SIAM J. Numer. Anal., 23:663–675, 1986.

37. MAYARAM, K., AND PEDERSON, D.O.: *Coupling algorithms for mixed-level circuit and device simulation*. IEEE Transactions on Computer-Aided Design of Integrated Circuits and Systems, 11(8):1003–1012, Aug 1992.

38. MAYARAM, K., YANG, P., AND CHERN, J.-H.: *Transient three dimensional mixed-level circuit and device simulation: Algorithms and applications*. In: *IEEE Proceedings ISCAS*, pp. 112–115, 1991.

39. MCMACKEN, J.R.F., AND CHAMBERLAIN, S.G.: *Chord: a modular semiconductor device simulation development tool incorporating external network models*. IEEE Transactions on Computer-Aided Design of Integrated Circuits and Systems, 8(8):826–836, Aug 1989.

40. MEI, T., ROYCHOWDHURY, J., COFFEY, T.S., HUTCHINSON, S.A., AND DAY, D.M.: *Robust, stable time-domain methods for solving mpdes of fast/slow systems*. IEEE Transactions on Computer-Aided Design of Integrated Circuits and Systems,24(2):226–239, feb. 2005.

41. MEINERZHAGEN, B., AND ENGL, W.L.: *The influence of the thermal equilibrium approximation on the accuracy of classical two-dimensional numerical modeling of silicon submicrometer mos transistors*. IEEE Transactions on Electron Devices, 35(5):689–697, May 1988.

42. MELVILLE, R., MOINIAN, S., FELDMANN, P., AND WATSON, L.: *Sframe: An efficient system for detailed DC simulation of bipolar analog integrated circuits using continuation methods*. Analog Integrated Circuits and Signal Processing, 3(3):163–180, 1993.

43. MEURIS, P., SCHOENMAKER, W., AND MAGNUS, W.: *Strategy for electromagnetic interconnect modeling*. IEEE Transactions on Computer-Aided Design of Integrated Circuits and Systems, 20(6):753–762, Jun 2001.

44. MUMPS: *A MUltifrontal Massively Parallel sparse direct Solver*. http://mumps. enseeiht.fr/, 2017.

45. NAGEL, L.W.: *SPICE2: A simulation program with integrated circuit emphasis*. PhD-Thesis, Univ. of California, Berkeley, May 1975.

46. NGOYA, E., AND LARCHEVÈQUE, R.: *Envelope transient analysis: A new method for the transient and steady state analysis of microwave communication circuit and systems*. In: *Proc. IEEE MTT-S Int. Microwave Symp.*, pp. 1365–1368, San Francisco, 1996.

47. ORTEGA, J.M., AND RHEINBOLDT, W.C.: *Iterative solution of nonlinear equations in several variables*. Academic Press, San Diego, 1970.

48. OSWALD, P.: *Private communication*, 1998.

49. PINTO, M.R., RAFFERTY, C.S., AND DUTTON, R.W.: *PISCES-II: Poisson and continuity equation solver*. Technical report, Standford University, 1984.

50. POON, H.C.: *Modeling of bipolar transistor using integral charge-control model with application to third-order distortion studies*. IEEE Trans. on Electron Devices, 19(6):719–731, 1992.

51. PULCH, R.: *PDAE Methoden zur numerischen Simulation quasiperiodischer Grenzzyklen von Oszillatorschaltungen*. PhD-Thesis, Universität Wuppertal, 2003.

52. PULCH, R.: *Variational methods for solving warped multirate partial differential algebraic equations*. SIAM J. Scient. Computing, 31(2):1016–1034, 2008.

53. PULCH, R., AND GÜNTHER, M.: *A method of characteristics for solving multirate partial differential equations in radio frequency application*. Appl. Numer. Math., (42):399–409, 2002.

54. ROYCHOWDHURY, J.: *Efficient methods for simulating highly nonlinear multi-rate circuits*. In: *Proc. of the 34th IEEE Design Automation Conference*, pp. 269–274, 1997.

55. ROYCHOWDHURY, J.: *Analyzing circuits with widely separated time scales using numerical PDE methods*. IEEE Transactions on Circuits and Systems I: Fundamental Theory and Applications, 48(5):578–594, may 2001.

56. SAAD, Y., AND SCHULTZ, M.H.: *GMRES: A generalized minimal residual algorithm for solving nonsymmetric linear systems*. SIAM J. Sci. Stat. Comput., 7(3):856–869, 1986.

57. SCHOENMAKER, W., MEURIS, P., STROHM, C., AND TISCHENDORF, C.: *Holistic coupled field and circuit simulation*. In: *2016 Design, Automation Test in Europe Conference Exhibition (DATE)*, pp. 307–312, March 2016.

58. SCHOENMAKER, W., CHEN, Q., AND GALY, P.: *Computation of self-induced magnetic field effects including the Lorentz force for fast-transient phenomena in integrated-circuit devices*. IEEE Transactions on Computer-Aided Design of Integrated Circuits and Systems, 33(6):893–902, June 2014.

59. SCHUMAKER, L.L.: *Spline Functions: Basic Theory*. Wiley, New York, 1981.

60. SELBERHERR, S., SCHUTZ, A., AND POTZL, H.W.: *Minimos - a two-dimensional MOS transistor analyzer*. IEEE Journal of Solid-State Circuits, 15(4):605–615, Aug 1980.

61. THOMAS, V.A., JONES, M.E., AND MASON, R.J.: *Coupling of the Pisces device modeler to a 3-d Maxwell FDTD solver*. IEEE Transactions on Microwave Theory and Techniques, 43(9):2170–2172, Sep 1995.

62. VLACH,J., AND SINGHAL, K.: *Computer Methods for Circuit Analysis and Design*. Springer-Verlag, 1993.

Part III
Uncertainty Quantification

Part III covers the following topics.

- **Uncertainty Quantification: Introduction and Implementations**
  Authors:
  Roland Pulch (Universität Greifswald, Germany),
  Piotr Putek, E. Jan W. ter Maten (Bergische Universität Wuppertal, Germany),
  Wim Schoenmaker (MAGWEL, Leuven, Belgium).
- **Robust Shape Optimization under Uncertainties in Device Materials, Geometry and Boundary Conditions**
  Authors:
  Piotr Putek, E. Jan W. ter Maten, Michael Günther, Andreas Bartel (Bergische Universität Wuppertal, Germany),
  Roland Pulch (Universität Greifswald, Germany),
  Peter Meuris, Wim Schoenmaker (MAGWEL, Leuven, Belgium).
- **Going from Parameter Estimation to Density Estimation**
  Author:
  Alessandro Di Bucchianico (Eindhoven University of Technology, the Netherlands).
- **Inverse Modeling: Glue-Package-Die Problem**
  Authors:
  Roland Pulch (Universität Greifswald, Germany),
  Piotr Putek (Bergische Universität Wuppertal, Germany),
  Herbert De Gersem (Technische Universität Darmstadt, Germany),
  Renaud Gillon (ON Semiconductor, Oudenaarde, Belgium).

Chapter 9
Uncertainty Quantification: Introduction and Implementations

Roland Pulch, Piotr Putek, E. Jan W. ter Maten, Wim Schoenmaker

Abstract This chapter provides a short introduction to Uncertainty Quantification based on Monte Carlo simulations, on Generalized Polynomial Chaos expansions and on Worst Case Corner Analysis. Furthermore, it also covers remarks how the chosen implementation should fit within existing simulation environments.

9.1 Introduction

Modeling and simulation of coupled multiphysics problems often involves multiple scales and domains. Multiphysics models can be described by a system of Partial Differential Equations (PDEs) [48,49] or Differential-Algebraic Equations (DAEs) [2,5,24] that couples all relevant physical phenomena and describes the interaction between multidynamical time-varying behaviour. The coupling can be strong or weak. In the latter situation, co-simulation in time can be considered. In our applications the coupling can be between electromagnetics, semiconductor devices, electronic circuits and heat and can involve mechanical stress.

For example, the investigation of the thermal-electric coupling or analysis of thermal-electric-mechanic coupling in compound devices was conducted by [6,29], where, besides typical electric variables such as voltages and cur-

Roland Pulch, Piotr Putek
Universität Greifswald, Germany,
e-mail: {Roland.Pulch,Piotr.Putek}@uni-greifswald.de

Piotr Putek, E. Jan W. ter Maten
Bergische Universität Wuppertal, Germany,
e-mail: {Putek,terMaten}@math.uni-wuppertal.de;Piotr.Putek@gmail.com

Wim Schoenmaker
MAGWEL NV, Leuven, Belgium, e-mail: Wim.Schoenmaker@magwel.com

© Springer Nature Switzerland AG 2019

E. J. W. ter Maten et al. (eds.), *Nanoelectronic Coupled Problems Solutions*,
Mathematics in Industry 29, https://doi.org/10.1007/978-3-030-30726-4_9

rents, the purely non-electric quantities including temperature, stress, strain and displacement are taken into account. Apart from modeling errors, several parameters such as from geometry, or from coupling phenomena are subject to uncertainty. Uncertainties can be caused directly by measurements or a lack of knowledge. Furthermore, when dealing with heat, parameters show a nonlinear behaviour when varying temperature. In this way effects due to a priori known uncertainty can be taken into account. Apart from this an important source of uncertainty results from some imperfections in following steps like manufacturing processes including, for example, sub-wavelength lithography, lens aberration, and chemical-mechanical polishing [37]. A design must be robust to such additional variations. Thus, the Uncertainty Quantification (UQ) is a crucial key requirement in realistic and predictive simulation of coupled problems encountered in electronic devices.

9.1.1 Parameter Dependency

Let a dynamical system be given in the form of PDEs, including a time derivative for covering evolutionary effects and thus defining the dynamics of a solution $\mathbf{x}(t)$. The full Maxwell equations and the heat equation are an example of such a system. In these cases after semidiscretization in space a system of ODEs or DAEs arises. Circuit equations are a system of first-order DAEs. In general they are taken as a system on their own right, but they can also be considered as a lumped version of Maxwell equations. These separate systems can be coupled into one bigger system. Based on initial values and/or boundary values numerical procedures are able to solve for the solution $\mathbf{x}(t)$ at any time t and produce further output $\mathbf{y}(t)$ based on post-processing.

The output will vary at a given time t_0 when the initial/boundary values vary, or when parameters in the system vary. When the variation is described by a stochastic modeling of the uncertainties (f.i., due to measurements or estimates, or due to differences in manufacturing or construction) the output will also show a random variation. In general, the output $\mathbf{x}(t)$ will depend nonlinearly on initial/boundary values or on parameters in the system. A simple example is given by

$$x'(t) = p\,x(t), \quad x(0) = 1, \tag{9.1}$$

which has solution $x(t,p) = e^{pt}$ and depends nonlinearly on p. If the random parameter exhibits a normal probability distribution, then the randomness of $\mathbf{x}(t_0)$ will be a distorted distribution density. By (quasi) Monte-Carlo sampling [9, 39] one can approximate the expectation $\mathbb{E}_p[\mathbf{x}(t,p)]$ and variance $\mathrm{Var}_p[\mathbf{x}(t,p)]$ at a fixed time t. However, the convergence may be slow and may require advanced sampling methods.

We note that in (9.1) the solution $\mathbf{x}(t_0,p)$ depends smoothly on p. In such cases UQ by a generalized Polynomial Chaos (gPC) expansion, see Section 9.2.3 and [59], can efficiently deal with it.

In [58] a zero-crossing problem for the solution of the viscous Burgers' equation is discussed,

$$\begin{cases} x_t + x\,x_s = \nu\,x_{ss}, & s \in [-1,1], \\ x(-1) = 1 + p, & x(1) = -1. \end{cases} \tag{9.2}$$

Here $\nu = 0.05 > 0$ is the viscosity. The solution $x(t,s,p)$ will have a zero-crossing at time $t = z(p)$; if $p = 0$ then $z = 0$. We assume that $p = \mathcal{U}(0, 0.1)$, i.e., p is uniformly distributed in $(0, 0.1)$. The expectation reads as $\hat{z} = \mathbb{E}[z(p)] \approx 0.8$. A fourth order gPC approximation, that requires just five simulations, is able to determine a variance in z with an accuracy that coincides with a Monte-Carlo (MC) simulation using 10^4 samples [58]. Of course one may not need such an accuracy for the variance: one digit less accuracy may reduce the work by MC by a factor 100. But still the amount of only five deterministic simulations involved in gPC is remarkably small compared to those still needed by MC.

The zero-crossing problem reveals that one may not only be interested in the behaviour of the solution of the PDE, or DAE, or ODE, but also in post-processing results. We notice, for instance, that zero-crossings in circuit simulation are of interest for jitter quantification. Furthermore, determining zero-crossings are also important in the analysis of time-periodic steady state solutions. More general post-processing can be part of optimisation.

Monte Carlo sampling [9] is well established in industry: it is programmed in simple scripts around a simulator. By this, these scripts can be viewed to be relatively easy to maintain. It may be affected by dedicated scripting language to access special data from the simulator. With the increase of computer power MC runs can be done in parallel. In several cases, programs from Solido Design Automation[1], or from MunEDA[2] cover this communication and extend it to facilities for variation-aware design and characterization.

This chapter provides a short introduction to UQ. In particular we focus on approaches by gPC. The alternative methods used in industry are based on (Quasi) Monte Carlo and on (Worst Case) Corner Analysis.

9.1.2 Stochastic Coupled Problems Description

Let us consider a time-dependent coupled problem, in general described by two operators $\mathbf{F}_1$ and $\mathbf{F}_2$, which can represent ordinary differential equations (ODEs), differential-algebraic equations (DAEs), or partial differential equations (PDEs) after a semidiscretizations in space, in the form

---

[1] Now part of Mentor,
https://www.mentor.com/products/ic%5Fnanometer%5Fdesign/ solido-solutions/.

[2] https://www.muneda.com/.

$$\mathbf{F}_1(\mathbf{x}_1(t,\mathbf{p}),t,\mathbf{p}) = \mathbf{0}, \tag{9.3}$$
$$\mathbf{F}_2(\mathbf{x}_2(t,\mathbf{p}),t,\mathbf{p}) = \mathbf{0}, \tag{9.4}$$

where some parameters $\mathbf{p} \in \Pi \subseteq \mathbb{R}^d$ are involved. The time derivatives are included in each part of systems (9.3)-(9.4). Furthermore, we assume that each of the operators $\mathbf{F}_i$, for $i = 1, 2$, consists of n_i equations with given initial values for all $\mathbf{p} \in \Pi$. Then, the solutions of the system (9.3)-(9.4) are expressed by $\mathbf{x}_i : [t_0, t_{\mathrm{end}}] \times \Pi \to \mathbb{R}^{n_i}$. Here $\mathbf{x}_i = (\mathbf{y}_i^T, \mathbf{z}^T)^T$, $i = 1, 2$, where $\mathbf{z}$ are *the coupling variables*. Now, we assume, that some parameters exhibit a certain level of uncertainty in the above model. Therefore, for UQ, these parameters are replaced by independent random variables

$$\mathbf{p} : \Omega \to \Pi, \quad \mathbf{p}(\omega) = (p_1(\omega), \dots, p_d(\omega)) \tag{9.5}$$

on some probability space $(\Omega, \mathcal{A}, P)$ with a joint density $\rho : \Pi \to \mathbb{R}$. That is, we have d stochastic parameters with independent probability distributions such as Gaussian, uniform, or beta. Then, the solution of (9.3)-(9.4) becomes a time-dependent random process. The statistical information like the expectation value and the variance for a function $f : \Pi \to \mathbb{R}$ can be obtained by

$$\langle f(\mathbf{p}) \rangle := \mathbb{E}[f(\mathbf{p})] = \int_\Pi f(\mathbf{p})\rho(\mathbf{p})\, \mathrm{d}\mathbf{p}, \tag{9.6}$$

and

$$\mathrm{Var}[f(\mathbf{p})] := \mathbb{E}\left[(f(\mathbf{p}))^2\right] - (\mathbb{E}[f(\mathbf{p})])^2, \tag{9.7}$$

provided that the integrals (9.6) and $\mathbb{E}\left[(f(\mathbf{p}))^2\right]$ in (9.7) over the parameter space are finite. Consequently, for two functions $f, g : \Pi \to \mathbb{R}$ the expectation operator yields an inner product $\langle f, g \rangle := \mathbb{E}[f(\mathbf{p})g(\mathbf{p})]$ on $L^2(\Omega)$, see, e.g., [30, 59]. Numerical approximations of (9.6) and (9.7) are very often required in a practical computation.

9.2 Non-Intrusive Uncertainty Quantification

Non-intrusive methods belong to the sampling techniques, which require repetitive run of the deterministic solver in order to perform UQ. In each method, however, one of the aims is either to compute an approximation of the statistical moments like the mean and the standard deviation, for example, or, alternatively, to derive a response surface model [35], or a finite set of coefficient functions, when a generalized Polynomial Chaos expansion is looked for [28, 58] (in which case the expansion formula provides a response surface evaluation facility). In the latter cases, MC, applied to the response surface model, is much cheaper and faster than when applied to the original model. For a careful switching between evaluation of a response surface model and of the original model, see [32].

9.2.1 UQ by Monte Carlo Method

One of the most popular methods for uncertainty quantification is the Monte-Carlo (MC) simulation [26, 38]. The principles of MC methods are based on the Strong Law of Large Numbers and rely on repeated random sampling to obtain numerical results. Given K sample realizations $\boldsymbol{\xi}_1, \ldots, \boldsymbol{\xi}_K$ of the random variable $\mathbf{p} \in \mathbb{R}^d$, the mean value of a function f is estimated by

$$m_K := \frac{1}{K} \sum_{k=1}^{K} f(\boldsymbol{\xi}_k). \tag{9.8}$$

According to the Strong Law of Large Numbers, one can obtain $m_K \to \mathbb{E}[f(\mathbf{p})]$ in probability for $K \to \infty$, thus the expectation value is approximated by the sample mean $m_K \approx \mathbb{E}[f(\mathbf{p})]$ for sufficiently large K.

For Monte-Carlo sampling the convergence rate is $\mathcal{O}(1/\sqrt{K})$, with K being the sample size, and which rate is independent of the dimension d of the parameter space. To compare, for a rectangular kth order quadrature formula the convergence to approximate an integral is $\mathcal{O}(1/K^{k/d})$, which shows that Monte-Carlo sampling is asymptotically faster when $\frac{k}{d} < \frac{1}{2}$, see [9].

For quasi Monte-Carlo methods the convergence rate is $\mathcal{O}((\log(K))^k/K)$, in which $k = k(d)$ (one may take $k = d$) and which assumes a bit more smoothness of the solution with respect to the parameters than for Monte-Carlo sampling [9, 52].

In our computations, the MC method has been only applied for validation purposes.

9.2.2 UQ by Quadrature-Based Stochastic Collocation

Another approach, which relies also on repetitive run of the deterministic model, is the quadrature-based stochastic collocation method (SCM) [7, 40, 58, 59]. These numerical techniques can be directly applied to formula (9.6) or (9.7), when using appropriate quadrature rules for the integral approximation

$$\mathbb{E}[f(\mathbf{p})] = \int_{\Pi} f(\mathbf{p})\rho(\mathbf{p}) \, d\mathbf{p} \approx \hat{\mathbb{E}}[f(\mathbf{p})] := \sum_{k=1}^{K} f(\boldsymbol{\xi}_k) w_k, \tag{9.9}$$

where the quadrature grid points $\{\boldsymbol{\xi}_k\}_{k=1}^K$ and the weights $\{w_k\}_{k=1}^K$ correspond to the probability density function ρ. In this context, the MC or quasi MC techniques [9, 39] are a special case of sampling methods, where the weights are $w_k \equiv \frac{1}{K}$ for all k, compare (9.8). The effectiveness of the quadrature-based stochastic collocation method is strongly affected by the choice of the quadrature grid points.

In a similar way, the higher moments such as the variance can be approximated by

$$\mathrm{Var}[f(\mathbf{p})] \approx \sum_{k=1}^K (f(\boldsymbol{\xi}_k))^2 \, \omega_k - \left(\hat{\mathbb{E}}\,[f(\mathbf{p})] \right)^2. \qquad (9.10)$$

Both sampling methods (9.8) and (9.9) require only the repetitive run of the existing deterministic solver. However, they differ themselves in the choice of the quadrature grid points and the weights, which has an impact on their efficiency. In our computations, the Stroud-3 formula [54] and the so-called sparse grid method [53], which suffer less from the curse of the dimensionality [59], have been used to approximate (9.6) and (9.7) by (9.9) and (9.10), respectively.

9.2.3 UQ by gPC Series Expansions

For the gPC expansion we assume a complete orthonormal basis of polynomials $(\phi_i)_{i\in\mathbb{N}}$, $\phi_i : \mathbb{R}^d \to \mathbb{R}$, given with $<\phi_i,\phi_j> = \delta_{ij}$ $(i,j,\geq 0)$. When $d=1$, ϕ_i has degree i. To treat a uniform distribution (i.e., for studying effects caused by robust variations) one can use Legendre polynomials; for a Gaussian distribution one can use Hermite polynomials [30, 58, 59]. With $\mathbf{i} = (i_1,\ldots,i_d)^T$, a polynomial $\phi_{\mathbf{i}}$ on $\mathbb{R}^d$ can be defined from one-dimensional polynomials: $\phi_{\mathbf{i}}(\mathbf{p}) = \prod_{j=1}^d \phi_{i_j}(p_j)$. In [20, 21] one finds algorithms how to efficiently generate orthogonal polynomials for a given probability density function $\rho(p)$ that depends on a scalar p. This approach exploits a three-term recurrence for the determinants of a symmetric tridiagonal matrix. From the eigenvalue decomposition of this matrix the Gaussian quadrature points and the weights can be derived [22].

We will denote a dynamical system by

$$\mathbf{F}(\mathbf{x}(t,\mathbf{p}),t,\mathbf{p}) = \mathbf{0}, \quad \text{for } t \in [t_0,t_1]. \qquad (9.11)$$

Here $\mathbf{F}$ may contain differential operators. The solution $\mathbf{x} \in \mathbb{R}^n$ depends on t and on $\mathbf{p} \in \mathbb{R}^d$. In addition initial and/or boundary values are assumed. In general these may depend on $\mathbf{p}$ as well.

A solution $\mathbf{x}(t,\mathbf{p}) = (x_1(t,\mathbf{p}),\ldots,x_n(t,\mathbf{p}))^T$ of the dynamical system becomes a random process. We assume that second moments are finite: $\langle x_j^2(t,\mathbf{p}) \rangle < \infty$, for all $t \in [t_0,t_1]$ and $j = 1,\ldots,n$. We express $\mathbf{x}(t,\mathbf{p})$ in a gPC expansion

$$\mathbf{x}(t, \mathbf{p}) = \sum_{i=0}^{\infty} \mathbf{v}_i(t)\,\phi_i(\mathbf{p}), \tag{9.12}$$

where the coefficient functions $\mathbf{v}_i(t)$ are defined by

$$\mathbf{v}_i(t) = \langle \mathbf{x}(t, \mathbf{p}), \phi_i(\mathbf{p}) \rangle. \tag{9.13}$$

Continuity/smoothness follow from the solution $\mathbf{x}(t, \mathbf{p})$ and similarly the construction of expectation values and variances.
A finite approximation $\mathbf{x}^m(t, \mathbf{p})$ to $\mathbf{x}(t, \mathbf{p})$ is defined by

$$\mathbf{x}^m(t, \mathbf{p}) = \sum_{i=0}^{m} \mathbf{v}_i(t)\,\phi_i(\mathbf{p}). \tag{9.14}$$

For long time range integration m may have to be chosen larger than for short time ranges. The coefficient functions $\mathbf{v}_i(t)$ can be calculated or approximated in various ways [30, 58, 59].
Assuming exact coefficients $\mathbf{v}_i(t)$, for traditional probability distributions $\rho(.)$ convergence rates for $||\mathbf{x}(t, .) - \mathbf{x}^m(t, .)||$ including functions $\mathbf{x}(t, \mathbf{p})$, that smoothly depend on $\mathbf{p}$, are known. For instance, for single functions $x(p)$ that depend on a scalar parameter p such that $x^{(\ell)}$ is continuous (i.e., derivative w.r.t. p), one has

$$||x(\cdot) - x^m(\cdot)||_{L^2_\rho} \leq C \frac{1}{m^{\ell/2}}\,||x^{(\ell)}(\cdot)||_{L^2_\rho} \text{ (Hermite expansion)}, \tag{9.15}$$

$$||x(\cdot) - x^m(\cdot)||_{L^2_\rho} \leq C \frac{1}{m^{\ell}}\,\sqrt{\sum_{i=0}^{\ell} ||x^{(i)}(\cdot)||_{L^2_\rho}^2} \text{ (Legendre expansion)}, \tag{9.16}$$

see [3] and [59], respectively. Here the L^2_ρ-norms include the weighting/density function $\rho(.)$. Note that the upper bound in (9.16) actually involves a Sobolev-norm. In [10] one also finds upper bounds using seminorms (that involve only the highest derivatives). We did assume a complete orthonormal polynomial basis. Then they are dense by definition. For a lognormal variable each orthonormal polynomial basis is not dense and thus not complete then. Hence for more general distributions $\rho(.)$ convergence may not be true. For convergence one needs to require that the probability measure is uniqely determined by its moments [18], an issue that was also encountered by [31]. The statistics of the parameters often do not obey traditional probability distributions like Gaussian, uniform, beta or others. In such a case on may have to construct probability distributions or probability density functions, respectively, which approximate the true statistics at a sufficient accuracy. Thereby, one has to match corresponding data obtained from measurements and observations of electronic devices. The resulting probability distribution functions should be continuous and all moments of the random variables should be finite such that a broad class of methods like, e.g., Polynomial Chaos, is applicable. We approximate the $\mathbf{v}_i(t)$ in (9.13) by Stochastic Collocation, leading to

$$\tilde{\mathbf{v}}_i(t) := \sum_{k=1}^{K} w_k \mathbf{x}(t, \mathbf{p}_k) \phi_i(\mathbf{p}_k). \tag{9.17}$$

Then the mean, or expectation, and variance can be approximated by

$$\mathbb{E}[\mathbf{x}(t, \mathbf{p})] = \mathbf{v}_0(t) \approx \tilde{\mathbf{v}}_0(t), \tag{9.18}$$

$$\mathrm{Var}[\mathbf{x}(t, \mathbf{p})] = \sum_{i=1}^{\infty} |\mathbf{v}_i(t)|^2 \approx \sum_{i=1}^{m} |\mathbf{v}_i(t)|^2 \approx \sum_{i=1}^{m} |\tilde{\mathbf{v}}_i(t)|^2. \tag{9.19}$$

Here the equalities are based on the orthonormality of the polynomials. In (9.18), the equality also defines the unique constant polynomial ϕ_0. Note that mean and Var are vectors: we assumed the vector definition $|\boldsymbol{\delta}|^2 = (\dots, |\delta_i|^2, \dots)^T$. The approximations are due to the chosen quadrature and, in the case of the Variance, also due to truncation of the expansion. The approximations (9.18)-(9.19) are helpful in plotting realisations of $\mathbf{x}(t, \mathbf{p})$ and comparing it to $\kappa\sigma$ 'confidence' intervals (for σ being the standard deviation and κ a multiple, say $\kappa = 1, \dots, 6$) around the mean values. Note that a quadrature formula like (9.17) also allows for a substitution of the L^2 inner product directly by a discrete inner product involving chosen quadrature points and polynomials replaced by Lagrange interpolation polynomials [13, 56].

9.2.3.1 Sensitivity Analysis

Having a gPC expansion the sensitivity (matrix) w.r.t. $\mathbf{p}$ is easily obtained

$$\mathbf{S}_p(t, \mathbf{p}) = \left[\frac{\partial \mathbf{x}(t, \mathbf{p})}{\partial \mathbf{p}} \right] \approx \sum_{i=0}^{m} \mathbf{v}_i(t) \frac{\partial \phi_i(\mathbf{p})}{\partial \mathbf{p}} \tag{9.20}$$

(be aware that convergence for $m \to \infty$ is not given here in general). A relative sensitivity is defined by

$$\mathbf{S}_p^r(t, \mathbf{p}) = \left[\left(\frac{\partial x_i(t, \mathbf{p})}{\partial p_j} \cdot \frac{p_j}{x_i(t, \mathbf{p})} \right)_{ij} \right] = \mathbf{S}_p(t, \mathbf{p}) \circ \left[\left(\frac{p_j}{x_i(t, \mathbf{p})} \right)_{ij} \right]. \tag{9.21}$$

It describes the amplification of a relative error in p_j to the relative error in $x_i(t, \mathbf{p})$ (here $\circ$ denotes the Hadamard product of two matrices). One may restrict $\mathbf{S}_p(t, \mathbf{p})$ to $\mathbf{S}_p(t, \boldsymbol{\mu}_p)$, where $\boldsymbol{\mu}_p$ is constant and where $\frac{\partial \mathbf{x}(t, \mathbf{p})}{\partial \mathbf{p}}$ is the solution of the system that is differentiated w.r.t. $\mathbf{p}$ at $\mathbf{p} = \boldsymbol{\mu}_p$. For a scalar quantity x a 'stochastic influence' can be studied, based on

$$\max\left(\frac{\partial x}{\partial p_1}\sigma_{p_1},\ldots,\frac{\partial x}{\partial p_d}\sigma_{p_d}\right), \quad \sigma_{p_i}^2 = \mathrm{Var}[p_i]. \qquad (9.22)$$

The sensitivity matrix also is subject to stochastic variations. With a gPC expansion it is possible to determine a mean global sensitivity matrix by

$$\mathbf{S}_p = \mathbb{E}_p\left[\frac{\partial \mathbf{x}(t,\mathbf{p})}{\partial \mathbf{p}}\right] \approx \sum_{i=0}^{m} \mathbf{v}_i(t)\int_\Pi \frac{\partial \phi_i(\mathbf{p})}{\partial \mathbf{p}} \rho(\mathbf{p})\,\mathrm{d}\mathbf{p}. \qquad (9.23)$$

The integrals at the right-hand side can be determined in advance and stored in tables.

9.2.3.2 Dominant Contributions to the Variance

In order to quantify the impact of each uncertain parameter to the amount of uncertainty, variance-based sensitivity analysis has been developed. Actually, it is based on the Sobol indices, which allow for decomposing the total variance [55]. Let

$$\mathbb{I}_j = \{i \in \mathbb{N}_0 \mid \phi_i(.) \text{ is non-constant in } p_j\},$$
$$\tilde{\mathbb{I}}_j = \mathbb{I}_j \cap \{0,1,\ldots,m\}.$$

We approximate the variance by (9.19), i.e. by truncation of the series and, in practice, also by applying quadrature or another approximation. Next we define the relative contributions to the (time-varying) variance.

$$S_j = \frac{\sum_{k\in\tilde{\mathbb{I}}_j}|\tilde{\mathbf{v}}_k(t)|^2}{\sum_{k=1}^{m}|\tilde{\mathbf{v}}_k(t)|^2}, \quad j = 1,\ldots,d. \qquad (9.24)$$

Then $1 \leq S_j \leq 1$ and $1 \leq S_1 + S_2 + \ldots + S_d \leq d$. A *dominant* contribution to the variance is reflected by the largest S_j's. A relatively strong contribution to the uncertainty in $\mathbf{x}(t,\mathbf{p})$ is when $S_j \approx 1$. Note that in (9.24) the S_j are time-dependent. When considering problems in the frequency-domain, contributions in specific frequency bands may receive special attention. These sensitivities have shown to be helpful in better understanding influence of components [41] and in optimizing designs, while also reducing uncertainty [44–46].

9.2.3.3 Exploitation of the Dakota Toolbox

The flowchart of the implemented quadrature-based stochastic collocation algorithm for the UQ propagation using MAGWEL software (devEM or PTM-ET) [33] as black-box is shown on Fig. 9.1. It consists of two main part such

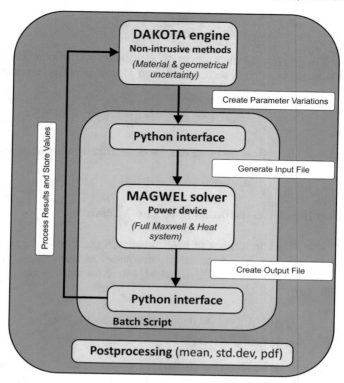

Fig. 9.1: Algorithm [43] for the Stochastic Collocation Method with the MAGWEL solver [33] serving as a 'black-box' simulation engine. In the flow indicated above we exploited Dakota [1], developed at Sandia National Laboratories.

as the Dakota Toolbox [1], developed at Sandia National Laboratories, and a MAGWEL simulator [33]. In the Python-C++ interface, the communication between both is established by using files. The Dakota Toolbox comprises three main sub-routines. In the first of them, so-called UQ settings, the input parameters such as the random parameter described by the mean and the standard deviation should be specified. Next, based on this information, in the UQ Preparation/Simulation main routines, first the quadrature points and the weights are generated. Then, after evaluation of the model parameters at these points, the solution of the deterministic problem using the MAGWEL electromagnetics solver [33] is computed at each quadrature grid point. In the post-processing stage, the statistical moments like the mean value and the variance are calculated using (9.9) and (9.10). Additionally, graphs of the probability density function and cumulative distribution function of the quantity of interest can be generated. The above approach is non-intrusive, i.e., the UQ set-up is made as a loop around the MAGWEL solver.

The efficiency of this algorithm can be improved by using a parallel technique, where the task related to deterministic calculation at grid points can be just sent to different cores or processors [23]. Another approach relies on utilizing the previous knowledge on the already computed solutions obtained for collocation points in order to efficiently approximate the remaining solutions by exploiting the smoothness of the solution with respect to the parameters [27, 34].

9.2.4 UQ by Worst-Case Corner Analysis

Let $\mathbf{X} = (X_1, \ldots, X_d)^T \in \mathbb{R}^d$ be process parameters that are normally distributed with covariance matrix $\mathbf{K_{XX}} \equiv \mathrm{Cov}(\mathbf{X}, \mathbf{X}) = \mathbb{E}[(\mathbf{X} - \boldsymbol{\mu})(\mathbf{X} - \boldsymbol{\mu})^T] = \mathbb{E}[\mathbf{X}\mathbf{X}^T] - \boldsymbol{\mu}\boldsymbol{\mu}^T$ in which $\boldsymbol{\mu} = (\mu_1, \ldots, \mu_d)^T \in \mathbb{R}^d$, $\mu_i = \mathbb{E}[X_i]$. On the diagonal the individual variances $\sigma_i^2(\mathbf{X}_i)$ are found. This leads to the correlation matrix $\mathbf{C_{XX}} \equiv [\mathrm{Diag}(\mathbf{K_{XX}})]^{-\frac{1}{2}} \mathbf{K_{XX}} [\mathrm{Diag}(\mathbf{K_{XX}})]^{-\frac{1}{2}}$. The matrix $\mathbf{R_{XX}} \equiv \mathbb{E}[\mathbf{X}\mathbf{X}^T]$ is the autocorrelation matrix of $\mathbf{X}$. For the process variations $\Delta\mathbf{X} = \mathbf{X} - \boldsymbol{\mu}$, the autocorrelation becomes $\mathbf{R}_{\Delta\mathbf{X}\Delta\mathbf{X}} \equiv \mathbb{E}[\Delta\mathbf{X}\Delta\mathbf{X}^T]$. The singular value decomposition gives $\mathbf{R}_{\Delta\mathbf{X}\Delta\mathbf{X}} = \mathbf{V}\boldsymbol{\Sigma}\mathbf{V}^T$ with orthonormal eigenvectors in $\mathbf{V}$ and in $\boldsymbol{\Sigma}$ the associated, non-negative, eigenvalues. Then $\Delta\mathbf{Y} = \boldsymbol{\Sigma}^{-\frac{1}{2}}\mathbf{V}^T\Delta\mathbf{X}$ provides ΔY_i that are independent, normally distributed, each with zero mean and unit variance.

In [61] Worst-Case Corner extraction for the $\Delta\mathbf{Y}$ is described starting from a quadratic response surface model for a scalar performance function

$$f(\Delta\mathbf{Y}) = \Delta\mathbf{Y}^T \mathbf{A} \, \Delta\mathbf{Y} + \mathbf{b}^T \Delta\mathbf{Y} + c, \qquad (9.25)$$

in which $\mathbf{A} \in \mathbb{R}^{d\times d}$, $\mathbf{b} \in \mathbb{R}^d$ and $c \in \mathbb{R}$ are calibrated from given sample pairs $\left((\Delta\mathbf{Y})^{(i)}, f^{(i)}\right)$. Without loss of generality, the matrix $\mathbf{A}$ is assumed to be symmetric.

A Worst-Case performance value f_{wc} is a thresh-hold value from below, or from above, that one, preferably, should not pass. A Process Corner is a $\Delta\mathbf{Y}^*$ for which $f(\Delta\mathbf{Y}^*) = f_{\mathrm{wc}}$. The most likely Process Corner is defined by

$$\max_{\Delta\mathbf{Y}\in\mathbb{R}^d \text{ s.t. } f(\Delta\mathbf{Y})=f_{\mathrm{wc}}} \left(\frac{1}{\sqrt{2\pi}}\right)^d \exp\left(-\frac{\|\Delta\mathbf{Y}\|_2^2}{2}\right). \qquad (9.26)$$

After relaxing the constraint to $f(\Delta\mathbf{Y}) \geq f_{\mathrm{wc}}$, or $f(\Delta\mathbf{Y}^*) \leq f_{\mathrm{wc}}$ one takes $\tilde{f} = \pm(f - f_{\mathrm{wc}})$ with constraint $\tilde{f}(\Delta\mathbf{Y}) \leq 0$, which leads to

$$\min_{\Delta\mathbf{Y}\in\mathbb{R}^d \text{ s.t. } \tilde{f}(\Delta\mathbf{Y})\leq 0} \|\Delta\mathbf{Y}\|_2^2. \qquad (9.27)$$

A Lagrangian $\mathcal{L}(\mathbf{z}, \lambda) = \|\Delta\mathbf{z}\|_2^2 + \lambda\tilde{f}(\mathbf{z})$ can be defined in which $\mathbf{z} = \Delta\mathbf{Y}$ and $\lambda \geq 0$ is a Lagrange multiplier. At the optimal value $\mathbf{z}^*, \lambda^*$ one has $\mathcal{L}_\mathbf{z}(\mathbf{z}^*, \lambda^*) = 0$, hence $\mathbf{z}^* = -\frac{\lambda^*}{2}(\mathbf{I} + \lambda^*\tilde{\mathbf{A}})^{-1}\tilde{\mathbf{b}}$, in which $\tilde{\mathbf{A}} = \pm\mathbf{A}$ and $\tilde{\mathbf{b}} = \pm\mathbf{b}$ (depending on the $\pm$ sign from the type of constraint). The optimum occurs at $\mathbf{z}^*$ for which the constraint is active, hence $\tilde{f}(\mathbf{z}) = 0$, or $f(\mathbf{z}) = f_{wc}$.

The problem (9.27) is not convex. Hence, for the actual optimisation, in [61] a convex, dual problem was considered, which allowed for applying the efficient semi-definite programming software from [8] and which led to a global minimum.

9.2.4.1 Asymptotic Probability Extraction

In [31] the probability density function of the values of f of (9.25) was approximated by the APEX (Asymptotic Probability Extraction) algorithm, that allowed in [61] to define a Worst-Case value f_{wc} as a quantile point of the associated cumulative distribution function. The chosen class of approximation for the probability density function of f in APEX was based on an observed analogy between a non-negative impulse response function $h(t)$ (and its Laplace transform $H(s)$) and the Gamma distribution $g_{\lambda,n}(f)$ (and its Laplace transform $G_{\lambda,n}(s)$) [11, 31]

$$
h(t) = \begin{cases} ae^{bt}, & t \geq 0 \\ 0, & t < 0 \end{cases} \qquad g_{\lambda,n}(t) = \begin{cases} \frac{\lambda^n t^{n-1} e^{-\lambda t}}{\Gamma(n)}, & t \geq 0, \ \Gamma(n) = (n-1)! \\ 0, & t < 0 \end{cases}.
$$
$$
H(s) = \int_0^\infty h(t)e^{-st}\mathrm{d}t = \frac{a}{s-b} \quad G_{\lambda,n}(s) = \left(\frac{\lambda}{\lambda+s}\right)^n \tag{9.28}
$$

Indeed, for $\lambda = 1$ and $n = 1$ one has $G_{1,1}(s) = 1/(1+s)$. Hence sums of functions of the form $h(t)$ may provide reasonable approximations for a probability density function

$$
h(t) = \begin{cases} \sum_{i=1}^M a_i e^{b_i t}, & t \geq 0 \\ 0, & t < 0 \end{cases}, \quad H(s) = \int_0^\infty h(t)e^{-st}\mathrm{d}t = \sum_{i=1}^M \frac{a_i}{s-b_i}. \tag{9.29}
$$

For the impulse function $h(t)$ in (9.29), the scalars a_i and b_i are called residue and pole, respectively. In [31] it was remarked that for the quadratic approximation in (9.25), in the case of uncorrelated and normally distributed parameters, the (stochastic) moments

$$
m_k \equiv \mathbb{E}[f^k] = \int_0^\infty f^k \phi(f)\mathrm{d}f, \tag{9.30}
$$

where $\phi(f)$ is the probability density function of f, can be calculated very efficiently. With $j = \sqrt{-1}$, the characteristic function $\Phi(\omega)$ is defined by the Fourier Transform of $\phi(f)$:

$$\Phi(\omega) = \int_{-\infty}^{\infty} \phi(f) e^{j\omega f} \mathrm{d}f = \int_{-\infty}^{\infty} \phi(f) \sum_{k \geq 0} \frac{(j\omega f)^k}{k!} \mathrm{d}f$$

$$= \sum_{k \geq 0} (-j\omega)^k \tilde{s}_k, \quad \tilde{s}_k = \frac{(-1)^k}{k!} m_k. \tag{9.31}$$

Also for the functions $h(t)$ and $H(s)$ in (9.29) moments and expansions can be made. We introduce

$$s_k \equiv \frac{(-1)^k}{k!} \int_0^{\infty} t^k h(t) \mathrm{d}t = -\sum_{i=1}^{M} \frac{a_i}{b_i^{k+1}} \tag{9.32}$$

and we obtain

$$H(s) = \sum_{i=1}^{M} \frac{a_i}{s - b_i} = \sum_{i=1}^{M} \frac{a_i}{b_i} \frac{1}{\frac{s}{b_i} - 1} = \sum_{i=1}^{M} -\frac{a_i}{b_i} \sum_{k=0}^{\infty} \left(\frac{s}{b_i}\right)^k$$

$$= \sum_{k=0}^{\infty} \left(-\sum_{i=1}^{M} \frac{a_i}{b_i^{k+1}}\right) s^k = \sum_{k=0}^{\infty} s_k s^k, \text{ for } s \text{ small enough.} \tag{9.33}$$

If $\tilde{s}_k = s_k$, for $k = 0, \ldots, K = 2M - 1$, the difference $\Phi(\omega) - H(-j\omega)$ has $K + 1$ cancelling moment order terms, thus making $H(-j\omega)$ a good approximation for $\Phi(\omega)$ for small ω, while $H(-j\omega)$ vanishes for large ω. Hence $h(f)$ (by inverse Fourier Transform) can be considered as a reasonable approximation for $\phi(f)$ (by inverse Laplace Transform). Note that $m_0 = \tilde{s}_0 = s_0 = -\sum_{i=1}^{M} \frac{a_i}{b_i}$. As a result the probability density function $\phi(f)$ and its cumulative distribution function of f are approximated by

$$\phi(f) = \mathrm{pdf}(f) = \begin{cases} \sum_{i=1}^{M} a_i e^{b_i f}, & f \geq 0 \\ 0, & f < 0 \end{cases}, \tag{9.34}$$

$$\int_{-\infty}^{f} \phi(f) \mathrm{d}f = \mathrm{cdf}(f) = \begin{cases} \sum_{i=1}^{M} \frac{a_i}{b_i} (e^{b_i f} - 1), & f \geq 0 \\ 0, & f < 0 \end{cases}. \tag{9.35}$$

We recognise that in (9.29) $h(t)$ directly follows analytically from $H(s)$. For the Laplace Transform one should have that $\mathrm{Re}(b - s) < 0$. For $a_i, b_i \in \mathbb{R}$ one immediately obtains $h(t) \in \mathbb{R}$. Then also $\overline{H(s)} = H(\bar{s})$ for $s \in j\mathbb{R}$. For complex poles b one encounters sums $H(s) = a_1/(s - b) + a_2/(s - \bar{b})$. For $a_2 = \overline{a_1} = \bar{a}$, with $\mathrm{Re}(a) > 0$, one preserves $\overline{H(s)} = H(\bar{s})$ for $s \in j\mathbb{R}$, as well as $h(t) = a \exp(bt) + \bar{a} \exp(\bar{b}t) \in \mathbb{R}^+$ (for $t \geq 0$).

9.2.4.2 Asymptotic Waveform Expansion and Vector Fitting

The APEX algorithm to determine the poles b_i and residues a_i such that $\tilde{s}_k = s_k$, for $k = 0, \ldots, K$, is based on the Asymptotic Waveform Expansion (AWE) algorithm [31]. AWE is well known in Model Order Reduction approaches. It involves some stability issues, because of involved Vandermonde matrices. In contrast, in Vector-Fitting (VF) techniques [57] in Model Order Reduction [15, 19], one also determines pole-residues (for a different class of approximations of $H(s)$ than above) and in a much better stable way. VF was developed for approximating the transfer function $H(s)$ for linear time-invariant systems, based on approximating known/measured transfer $H(z_k)$ at points $z_k \in j\mathbb{R}$. We indicate below how it may be used to also approximate a probability density function.

The series expansion of the function $\Phi(\omega)$ in (9.31) converges for a range of values of ω. From this one can define $H(z_k) = \Phi(-z_k/j)$. Alternatively, an approximation of $\phi(f)$ by a step function based on an Histogram and fit with an exponential tail can be made [51]. This allows for an analytic Laplace transform to derive values for $H(z_k)$. The 'VF-class', from which $H(s)$ is determined, is taken as

$$H(s) = \frac{\mathcal{N}[\mathbf{N}, \mathbf{b}](s)}{\mathcal{D}[\mathbf{D}, \mathbf{b}](s)}, \tag{9.36}$$

$$\mathcal{N}[\mathbf{N}, \mathbf{b}](s) = N_0 + \sum_{n=1}^{N} \frac{N_n}{s - b_n}, \quad \mathbf{N}^T = (N_0, N_1, \ldots, N_N), \quad \mathbf{b}^T = (b_1, \ldots, b_N),$$

$$\mathcal{D}[\mathbf{D}, \mathbf{b}](s) = 1 + \sum_{n=1}^{N} \frac{D_n}{s - b_n}, \quad \mathbf{D}^T = (D_1, \ldots, D_N).$$

If the coefficients of the denominator vanish ($\mathbf{D} = 0$), then the class of function reduces to same as assumed by the AWE algorithm. For VF [15, 19, 25] it is essential that $\mathbf{D} \neq 0$. An initial set of poles b_n has to be provided. The poles are iteratively updated by solving an eigenvalue problem for a diagonal matrix with a low-rank correction. Coefficients N_n, D_n are determined by an unconstrained Least-Squares approximation, based on given poles and including a particular weighting (so-called Sanathanian Koerner iteration). For our application, the data at the z_k either comes from an expansion, or by a step function approximation. The derived probability density function $\phi(.)$ has a broader range. In both cases, an extra check has to be made to assure that the approximation $\phi(.)$ is non-negative. In [51] this aspect is considered for the class of Exponential-Polynomial-Trigonometric functions of the form $h(t) = \mathbf{c}^T e^{\mathbf{A}t} \mathbf{b}$, where $\mathbf{b}, \mathbf{c} \in \mathbb{R}^n$, $\mathbf{A} \in \mathbb{R}^{n \times n}$, for $t \geq 0$ (we denote that the approximation provided by AWE fits this class). A necessary condition is that $\lambda_M \equiv \max_{\lambda \in \sigma(\mathbf{A})} \mathrm{Re}(\lambda) \in \sigma(\mathbf{A}) \subset \mathbb{C}^-$ (a real dominant eigenvalue of $\mathbf{A}$, or pole of the transfer function). But this is not sufficient. In general constraint optimization is needed and has to be included in the VF algorithm.

9.2.4.3 (Worst Case) Corner Analysis

Some industrial partners exploited a simplified (Worst Case) Corner Analysis (CA) for the quantification of uncertainties. For the CA, one considers for each single parameter p three predetermined values p_L, p_0, p_U (lower value, nominal value, upper value). It follows that 3^d (or 2^d when p_0 is skipped) combinations exist in the case of d varying parameters. Thus the computational work is the same as in a tensor product quadrature formula with three points per dimension. By an optimization procedure, one calculates only for those parameter values that lower or minimize some performance function. The result represents a biased look on the parameter space. Quadrature formulas in gPC found in literature are based on the underlying probability density function and usually reflect its symmetry. For worst case analyses, we may require biased schemes for the numerical computation of integrals. For comparing numerical effort we denote that the "Stroud-3" quadrature [54] in UQ or in Stochastic Collocation uses only $2d$ nodes, assuming a symmetric probability density function.

To the designer, the foundry provides (add-on Design Kit) for each electrical parameter (for example: resistance, capacitance, NMOS V_T, PMOS V_T) three values: p_L ("Slow": 'minimum' value over the wafer), p_0 ("Typical/Nominal": mean value) and p_U ("Fast": 'maximum' value over the wafer). From these values one builts Corners by mixing for each parameter "slow", "typ" or "fast" values, plus battery voltage and temperature (example: Corner$_1$ = R_{slow}, C_{fast}, NMOS$_{slow}$, PMOS$_{fast}$, VBAT3.5V, T65°). In this way, say up to 20 corners are defined to test a product in all conditions (in a later step in the design phase the amount of corners is increased). Simulation using corners yields the maximum and minimum performances one expects to be able to achieve (without mismatch and assembly effects). For a power amplifier (PA) mismatch has a low impact on power transistor performances (due to the large size). So for the RF part, corner simulation is sufficient.

Nevertheless, this is not the case for the analog part. For example: mismatch can have an impact on the voltage control or the reference current and so degrades after that the PA performances (by change of the PA biasing). Consequently, Monte-Carlo simulation is used for the analog part. The overall goal is to minimize the impact of process (corners) and mismatch (geometry) variations on the PA control functionalities and to preserve the PA performances. The analog functions should be well centered.

9.2.5 Graphical User Interface

As a kind of introduction to end-users and learning a Graphical User Interface (GUI) was designed, implemented and added to MAGWEL's software suite. It quantifies the impact of any design random parameter, described by assumed distributions on quantities of interest such as, e.g., S-parameters, drain/source currents and voltages, temperature of probe, heat flux, etc. For each input quantity the probability density could be defined. Next the method for statistics (MC, QMC, gPC, CA) could be selected. Finally results were drawn. For an impression, see Fig. 9.2 and Fig. 9.3.

In Section 9.2.4.3 already some remarks were made on efficiency: the number of simulations involved. For UQ this is related to the involved quadrature (nr points k in dimension $\mathbb{R}^d$). From [14] we have checked the following two cases: (i) The d-cube $[0,1]^d$ with weight function $w \equiv 1$ (uniform distribution); (ii) Euclidean space $\mathbb{R}^d$ with weight function $w = \exp(-r^2)$ (Gaussian distribution). In both cases, there exist the following quadrature formulas of some degree (multivariate polynomials up to this degree n are resolved exactly) with the proportionality of the number K of nodes to the space dimension d:

$$
\begin{aligned}
\text{degree } n{=}3: &\quad K = 2d \\
\text{degree } n{=}5: &\quad K \sim d^2 \\
\text{degree } n{=}7: &\quad K \sim d^3 \\
\text{degree } n{=}9: &\quad K \sim d^4 \\
\text{degree } n{=}11: &\quad K \sim d^5
\end{aligned}
\tag{9.37}
$$

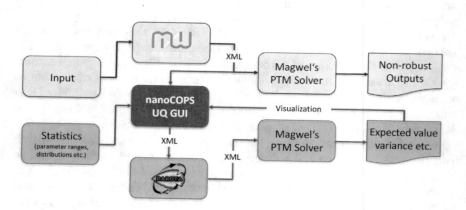

Fig. 9.2: Positioning of the Graphical User Interface to compare different statistical methods. The MAGWEL solver [33] was used as a 'black-box' simulation engine. Dakota [1], developed at Sandia National Laboratories, was used for UQ.

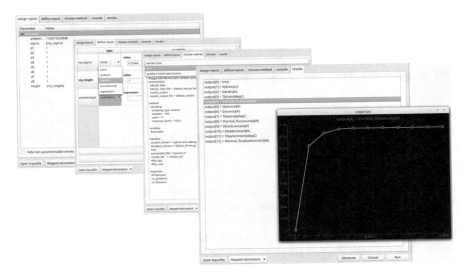

Fig. 9.3: Impression of the Graphical User Interface to compare different statistical methods. The results for electromagnetic, thermal and circuit quantities were provided by the MAGWEL solver [33], which served as a 'black-box' simulation engine. For UQ we exploited Dakota [1], developed at Sandia National Laboratories.

In gPC the accuracy of approximated results comes from limiting the expansion, as well as from quadrature. When one is interested only in the mean and in the variance, in practise, one does not need degree 9 or higher. The weights are positive for degree 3. For all higher degrees, negative weights appear as a price for the efficiency of the quadrature. An exception is the d-cube with degree 5, where also a formula exists with positive weights. Note that sparse grids typically involve also negative weights. We mostly exploited the Stroud-3 and Stroud-5 formulae for the d-cube, which have the minimum numbers of nodes required for a polynomial exactness of degree 3 and 5, respectively. The efficiency of these formulas (relatively low number of nodes for a given degree) follows from the symmetry of the weight function. Hence the symmetry of the weight function is necessary for these properties. Formulas of the above type may exist for symmetric beta-distributions, but surely not for unsymmetric beta-distributions.

We conclude that for an amount of about 250–1000 simulations, UQ can treat (if all proportionality factors are taken equal)

- 10 parameters and exploit an accuracy of degree 7.
- 30 parameters and exploit accuracy of degree 5.
- 500 parameters and exploit accuracy of degree 3.

Thus we may say that one can use 20 parameters (relaxing the assumption of the proportionality a bit) in an amount of work that is comparable to a typical standard amount of (a first series of) Monte-Carlo simulations. We remark, however, that the proportionality has some impact: for $d = 22$, Stroud-5 uses 969 nodes; for $d = 23$, one requires 1059 nodes in Stroud-5.

9.3 UQ for a Transient Nonlinear Field/Circuit Coupled Problem

As an introductory impression for UQ outcomes, we solved a stochastic field/circuit problem [4, 48], shown in Fig. 9.4, exploiting FEMM [36] as (publicly available) prototype field simulator, for a monolithic approach. The stochastic eddy-current field problem, defined in a spatial domain $D \in \mathbb{R}^2$ and a time domain $[t_0, t_{\text{end}}]$, is governed by the random quasi-linear equation

$$\sigma \frac{\partial}{\partial t} \mathbf{a} + \nabla \times \left(\left(\nu(|\nabla \times \mathbf{a}|) \, \kappa(x, \omega) \right) \nabla \times \mathbf{a} \right) = \boldsymbol{\chi} \bullet \mathbf{j}, \qquad (9.38)$$

where $\omega \in \Omega$ denotes the random inputs of the model, $\mathbf{a} := \mathbf{a}(\mathbf{x}, t, \omega)$ is the magnetic vector potential (with homogeneous Dirichlet conditions), σ and ν are conductivity and reluctivity, respectively. Additionally, the function $\kappa(\mathbf{x}, \omega)$ has been introduced in order to model the local degradation of material (reluctivity). It has been parameterized by a truncated Karhunen-Loève Expansion (KLE) [59] with N terms, i.e.,

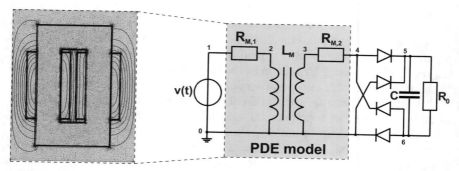

(a) 2D transformer FEMM   **(b) rectifier circuit with embeded FEMM model**

Fig. 9.4: Nonlinear Field/Circuit configuration [4, 48] using FEMM [36] as (publicly available) prototype field simulator.

$$\kappa(\mathbf{x},\omega) \approx \overline{\kappa}(\mathbf{x}) + \sum_{i=1}^{N} \hat{\kappa}_i(\mathbf{x}) Z_i(\omega), \tag{9.39}$$

where $\overline{\kappa}(\mathbf{x})$ denotes the mean of the random field and where summation describes the variations. The functions $\hat{\kappa}_i(\mathbf{x})$ are determined by the eigenvalues and eigenfunctions of the assumed covariance function, e.g., the exponential covariance function. The $Z_i(\omega)$ are uncorrelated random variables.

In (9.38), the winding operator $\boldsymbol{\chi} = [\boldsymbol{\chi}_1,\dots,\boldsymbol{\chi}_l,\dots,\boldsymbol{\chi}_L]^\top$ collects local, directional, winding functions $\boldsymbol{\chi}_l$, which are functions of space that distribute the lumped currents j_l, collected in $\mathbf{j} = [j_1,\dots,j_l,\dots,j_L]^\top$, in the 2D domain, see [4, 48] and [50, Sections 4.3 and 4.4 on stranded conductors]. The right-hand side in (9.38) is defined by $\boldsymbol{\chi} \bullet \mathbf{j} = \sum_l \boldsymbol{\chi}_l j_l$. To establish the circuit coupling we additionally have

$$\frac{\partial}{\partial t} \int_D \boldsymbol{\chi}_l \cdot \mathbf{a} \, d\mathbf{x} + R_l j_l = v_l \qquad l = 1,\dots,L, \tag{9.40}$$

in which $R_l j_l$ is some voltage loss. Each conductor has its own conductance and R_l is given by $\int_D \frac{1}{\sigma_l} \boldsymbol{\chi}_l \cdot \boldsymbol{\chi}_l d\mathbf{x}$, see [48, Eq. (17)]. The v_l in (9.40) are the couplings to the circuit system of stochastic DAEs.

$$\mathbf{A}_C \frac{d}{dt} \mathbf{q}_C(\mathbf{A}_C^\top \mathbf{u}, t) + \mathbf{A}_R \mathbf{g}_R(\mathbf{A}_R^\top \mathbf{u}, t) + \mathbf{A}_L \mathbf{i}_L + \mathbf{A}_M \mathbf{j} + \mathbf{A}_V \mathbf{i}_V + \mathbf{A}_I \mathbf{i}_s(t) = \mathbf{0},$$

$$\frac{d}{dt} \boldsymbol{\Phi}_L(\mathbf{i}_L, t) - \mathbf{A}_L^\top \mathbf{u} = \mathbf{0},$$

$$\mathbf{A}_V^\top \mathbf{u} - \mathbf{v}_s(t) = \mathbf{0},$$

with incidence matrices $\mathbf{A}_*$ such that $v_l = \mathbf{A}_l^\top \mathbf{u}$, or $\mathbf{v} = (v_1, \dots, v_L)^\top = \mathbf{A}^\top \mathbf{u}$, for suitable $\mathbf{A}_l$ (or $\mathbf{A}$) and constitutive laws for (nonlinear) conductances, inductances and capacitances (random functions with subscripts R, L and C, determined by, e.g., the statistical moments), independent sources $\mathbf{i}_s$ and $\mathbf{v}_s$, unknowns are the potentials $\mathbf{u} := \mathbf{u}(t, \omega)$ and currents $\mathbf{i}_L := \mathbf{i}_L(t, \omega)$ and $\mathbf{i}_V := \mathbf{i}_V(t, \omega)$.

To perform the UQ analysis, the nominal parameters of all the four diodes as well as the resistance and the capacitance of the low pass filter have been replaced by

$$\tilde{I}_S(p_1) := I_S(1 + 0.05p_1),$$
$$\tilde{V}_{th}(p_2) := V_{th}(1 + 0.05p_2),$$
$$\tilde{R}(p_2) := R(1 + 0.1p_3),$$
$$\tilde{C}(p_4) := C(1 + 0.1p_4)$$

with independent, uniformly distributed, random variables $p_j \in [-1, 1]$, for $j = 1, \dots, 4$. Hence, a relatively high uncertainty of 10% or 5% is considered for each parameter. The same procedure has been conducted for other stochastic

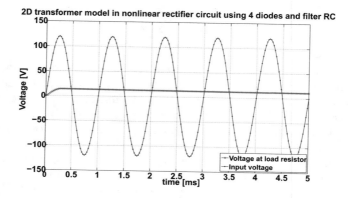

Fig. 9.5: Deterministic output/input relation in a bridge rectifier with an RC filter.

parameters (with the variance of 10% for each component) related to the first seven components of the KLE that capture 95% of the random field energy. For the purpose of modeling of the transformer core, we consider a random field of reluctivity with mean one and an exponential two-point covariance function

$$C(\mathbf{x}, \mathbf{y}) = \delta^2 \exp(-\|\mathbf{x} - \mathbf{y}\|_2 / L), \qquad (9.41)$$

with $\delta = 0.1$ and $L=10$ being the correlation length. The total number of parameters is 11. Consequently, Stroud-3 quadrature involves 22 nodes, resulting in 22 deterministic transient simulations of a rectifier circuit model. In each simulation, the degradation of an iron core of the transformer has been taken into account using a KLE dimension-reduction technique. Fig. 9.5 gives an impression of the output/input relation in the bridge rectifier with an RC filter. Fig. 9.6 presents the approximation of the mean and standard deviation obtained by the numerical simulation using the non-intrusive method (SCM).

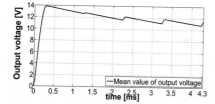 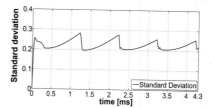

Fig. 9.6: Expectation value (left) and standard deviation (right) of the solution of a bridge rectifier with random parameters.

The result of the variance-based sensitivity analysis, see [41], is depicted in Fig. 9.7. It shows that the output voltage is more or less equally sensitive to all considered parameters.

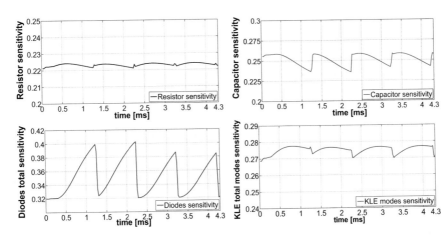

Fig. 9.7: The variance-based sensitivities calculated for the resistance, capacitor and diodes parameters and for the KLE components.

9.4 Conclusion

This chapter provided an introduction to Uncertainty Quantification based on generalized Polynomial Expansions. It has shown itself as a robust technique amidst competing approaches like (Quasi) Monte-Carlo and Corner Analysis. In this book UQ was further shown for electro-heat coupling in bondwire modeling [17], and in a transient analysis for a Power Transistor Model [47]. In [47] this was used for robust topology optimization of the Power Transistor Model.

In the current chapter the response surface models considered were either quadratic (based on fitting parameters) or implied by the truncated generalized polynomial expansion. Alternatively the book chapter [60] studies UQ based on a reduced-order model obtained by parametric Model Order Reduction.

Interest arose to directly estimate probability density functions. Designers do not always have reliable descriptions of uncertainties (like geometrical ones) in electrical components in ICs, as actual production is done externally. This has led to the chapters [16] (for other approaches than considered in the current chapter) and [42] (for estimating probability density functions by inverse modeling).

References

1. ADAMS, B.M., EBEIDA, M.S., ELDRED, M.S., GERACI, G., JAKEMAN, J.D., MAUPIN, K.A., MONSCHKE, J.A., STEPHENS, J.A., SWILER, L.P., VIGIL, D.M., WILDEY, T.M., BOHNHOFF, W.J., DALBEY, K.R., EDDY, J.P., FRYE, J.R., HOOPER, R.W., HU, K.T., HOUGH, P.D., KHALIL, M., RIDGWAY, E.M., WINOKUR, J.G., AND RUSHDI, A.: *Dakota, A Multilevel Parallel Object-Oriented Framework for Design Optimization, Parameter Estimation, Uncertainty Quantification, and Sensitivity Analysis: Version 6.9 User's Manual.* Sandia Technical Report SAND2014-4633, July 2014. Updated November 13, 2018. Sandia National Laboratories, Albuquerque, New Mexico, USA. https://dakota.sandia.gov/.
2. ARNOLD, M., AND GÜNTHER, M.: *Preconditioned Dynamic Iteration for Coupled Differential-Algebraic Systems.* Numerical Mathematics 41, 1–25, 2001.
3. AUGUSTIN, F., GILG, A., PAFFRATH, M., RENTROP, P., AND WEVER, U.: POLYNOMIAL CHAOS FOR THE APPROXIMATION OF UNCERTAINTIES: CHANCES AND LIMITS. Euro. J. of Appl. Maths, Vol. 19, pp. 149–190, 2008.
4. BARTEL, A., BRUNK, M., GÜNTHER, M., AND SCHÖPS, S.: *Dynamic iteration for coupled problems of electric circuits and distributed devices.* SIAM J. Sci. Comput., 35:2, B315–B335, 2013.
5. BARTEL, A., AND GÜNTHER, M.: *PDAEs in Refined Electrical Network Modeling.* SIAM Review 60-1, pp. 56–91, 2018.
6. BARTEL, A., GÜNTHER, M., AND SCHULZ, M.: *Modeling and discretization of a thermal-electric test circuit.* In: K. ANTREICH, R. BULIRSCH, A. GILG, AND P. RENTROP (Eds.): *Modeling, Simulation and Optimization of Integrated Circuits.* Book Series ISNM Vol. 146, pp. 187–201. Birkhäuser, 2003.
7. BENNER, P., AND SCHNEIDER, J.: *Uncertainty quantification for Maxwell's equations using stochastic collocation and model order reduction.* International Journal for Uncertainty Quantification, 5-3, pp. 195–208, 2015.
8. BOYD, S., AND VANDENBERGHE, L.: *Convex Optimization.* Cambridge Univ. Press, 6th edition, 2008. See also http://web.stanford.edu/%7Eboyd/cvxbook/. For the cvx-code, see http://web.stanford.edu/%7Eboyd/software.html.
9. CAFLISCH, R.E.: *Monte Carlo and quasi-Monte Carlo methods.* Acta Numerica 7, pp. 1–49, 1998.
10. CANUTO, C., HUSSAINI, M.Y., QUARTERONI, A., AND ZANG, T.A.: *Spectral methods – Fundamentals in single domains.* Springer, Berlin, 2010.
11. CELIK, M., PILEGGI, L., AND ODABASIOGLU, A.: *IC Interconnect Analysis.* Kluwer Academic Publishers, New York, 2002.
12. CHAUVIÉRE, C., HESTHAVEN, J.S., AND LURATI, L.: *Computational modeling of uncertainty in time-domain electromagnetics.* SIAM J. Sci. Comput., 28-2, pp. 751–775, 2006.
13. CONSTANTINE, P.: *A primer on stochastic Galerkin methods.* Project Document and MATLAB code, http://www.stanford.edu/%7Epaulcon/projects/SGS%5FPrimer/, Stanford Univ., 2007.
14. COOLS, R.: *An Encyclopaedia of Cubature Formulas.* J. Complexity, 19: 445–453, 2003. http://dx.doi.org/10.1016/S0885-064X(03)00011-6. See also: http://nines.cs.kuleuven.be/ecf/.

15. DESCHRIJVER, D., AND DHAENE, T.: *Data-driven model order reduction using orthonormal vector fitting.* In: SCHILDERS, W.H.A., VAN DER VORST, H.A., AND ROMMES, J. (Eds.): *Model Order Reduction: Theory, Research Aspects and Applications.* Series Mathematics in Industry, Vol. 13, Springer, pp. 341–359, 2008.

16. DI BUCCHIANICO, A.: *Going from Parameter Estimation to Density Estimation.* Chapter in this Book, 2019.

17. DUQUE GUERRA, D.J., CASPER, T., SCHÖPS, S., DE GERSEM, H., RÖMER, U., GILLON, R., WIEERS, A., KRATOCHVIL, T., GOTTHANS, T., AND MEURIS, P.: *Bond Wire Models.* Chapter in this Book, 2019.

18. ERNST, O.G., MUGLER, A., STARKLOF, H.-J., AND ULLMANN, E.: *On the convergence of generalized polynomial chaos expansions.* ESAIM: Mathematical Modelling and Numerical Analysis, 46, pp. 317–339, 2012.

19. FERRANTI, F., DESCHRIJVER, D., KNOCKAERT, L., AND DHAENE, T.: *Data-driven parameterized model order reduction using z-domain multivariate orthonormal vector fitting technique.* In: BENNER, P., HINZE, M., AND TER MATEN, E.J.W. (Eds.): *Model reduction for circuit simulation.* Series Lecture Notes in Electrical Engineering, Vol. 74,Springer-Verlag, pp. 147–155, 2011.

20. GAUTSCHI, W.: *OPQ: A MATLAB suite of programs for generating orthogonal polynomials and related quadrature rules.* http://www.cs.purdue.edu/archives/2002/wxg/codes, 2002.

21. GAUTSCHI, W.: *Orthogonal polynomials (in Matlab).* J. of Comput. and Applied Maths., 178, pp. 215–234, 2005.

22. GOLUB, G.H., AND VAN LOAN, C.F.: MATRIX COMPUTATIONS, FOURTH EDITION. The Johns Hopkins University Press, Baltimore, MD, 2013.

23. GRAMA, A., GUPTA, A., KARYPIS, G., AND KUMAR, V.: *Introduction to Parallel Computing.* Addison-Wesley, 2003.

24. GÜNTHER, M., FELDMANN, U., AND TER MATEN, J.: *Modelling and discretization of circuit problems.* In: SCHILDERS, W.H.A., TER MATEN, E.J.W. (Eds.): *Handbook of Numerical Analysis*, Vol. 13, pp. 523–659, Elsevier, 2005.

25. GUSTAVSEN, B., AND HEITZ, C.: *Fast Realization of the Modal Vector Fitting Method for Rational Modeling With Accurate Representation of Small Eigenvalues.* IEEE Trans. on Power Delivery, 24-3, pp. 1396–1405, 2009.

26. HAMMERSLEY, J.M., AND HANDSCOMB, D.C.: *Monte Carlo Methods*, Methuen & Co., London, and John Wiley & Sons, New York, 1964. Also available in Series Monographs on Statistics and Applied Probability, Springer Netherlands. http://www.cs.fsu.edu/%7Emascagni/Hammersley-Handscomb.pdf.

27. JANSEN, L., AND TISCHENDORF, C.: *Effective numerical computation of parameter dependent problems.* In: MICHIELSEN, B.L., AND POIRIER, J.-R. (Eds.): *Scientific Computing in Electrical Engineering SCEE 2010.* Mathematics in Industry, Vol. 16, Springer, pp. 49–57, 2012.

28. KAINTURA, A., DHAENE, T., AND SPINA, D.: *Review of Polynomial Chaos-Based Methods for Uncertainty Quantification in Modern Integrated Circuits.* Electronics, 7, 30, 2018. Online: https://www.mdpi.com/2079-9292/7/3/30/pdf.

29. KAUFMANN, C., GÜNTHER, M., KLAGGES, D., KNORRENSCHILD, M., RICHWIN, M., SCHÖPS, S., AND TER MATEN, E.J.W.: *Efficient Frequency-Transient Co-simulation of Coupled Heat-Electromagnetic Problems.* Journal of Mathematics in Industry, 4:1, 2014. Online: http://www.mathematicsinindustry.com/content/4/1/1.

30. LE MAÎTRE, O.P., AND KNIO, O.M.: *Spectral methods for uncertainty quantification, with applications to computational fluid dynamics.* Springer, Science+Business Media B.V., Dordrecht, 2010.

31. LI, X., LE, J., GOPALAKRISHNAN, P, AND, PILEGGI, L.T.: *Asymptotic probability extraction for nonnormal performance distributions.* IEEE Trans. on Comput.-Aided Design of Intergr. Circuits and Systems (TCAD) 26-1, pp. 16–37, 2007.

32. LI, J., AND XIU, D.: *Evaluation of failure probability via surrogate models.* J. Comput. Phys. 229:23, pp. 8966–8980, 2010.

33. MAGWEL NV: *Device-Electro-Magnetic Modeler (devEM) and Power Transistor Modeler: PTM.* MAGWEL NV, Leuven, Belgium, 2018, http://www.magwel.com/
34. TER MATEN, E.J.W., PULCH, R., SCHILDERS, W.H.A., AND JANSSEN, H.H.J.M.: *Efficient Calculation of Uncertainty Quantification.* In: FONTES, M., GÜNTHER, M., AND MARHEINEKE, N. (Eds): *Progress in Industrial Mathematics at ECMI 2012.* Series Mathematics in Industry Vol. 19, Springer, pp. 361–370, 2014.
35. MCCONAGHY, T., AND GIELEN, G.G.E.: GLOBALLY RELIABLE VARIATION-AWARE SIZING OF ANALOG INTEGRATED CIRCUITS VIA RESPONSE SURFACES AND STRUC-TURAL HOMOTOPY. IEEE Trans. on Comput.-Aided Design of Integr. Circuits and Systems (TCAD), 28-11, pp. 1627–1640, 2009.
36. MEEKER, D.: *FEMM – Finite Element Method Magnetics.* Reference and User's Manual, Version 4.2, 2018. http://www.femm.info/wiki/HomePage
37. MOHANTY, S.P., AND KOUGIANOS, E.: *Incorporating manufacturing process variation awareness in fast design optimization of nanoscale CMOS VCOs.* IEEE Trans. on Semiconductor Manuf. 27, pp. 22–31, 2014.
38. MOONEY, C.Z.: *Monte Carlo Simulation.* Series Quantitative Application in the Social Sciences, Vol. 116, SAGE Publications, Inc., Thousand Oaks, CA, USA, 1997.
39. MOROKOFF, W.J., AND CAFLISCH, R.E.: *Quasi-Monte Carlo integration.* J. Comput. Physics 122, pp. 218–230, 1995.
40. PULCH, R.: *Stochastic collocation and stochastic Galerkin methods for linear differential algebraic equations.* Journal of Computational and Applied Mathematics, 262, pp. 281–291, 2014.
41. PULCH, R., TER MATEN, E.J.W., AND AUGUSTIN, F.: *Sensitivity analysis and model order reduction for random linear dynamical systems.* Mathematics and Computers in Simulation 111, pp. 80–95, 2015. DOI: http://doi.org/10.1016/j.matcom.2015.01.003.
42. PULCH, R., PUTEK, P., DE GERSEM, H., AND GILLON, R.: *Inverse Modeling: Glue-Package-Die Problem.* Chapter in this Book, 2019.
43. PULCH, R., PUTEK, P., TER MATEN, J., DUQUE, D., SCHÖPS, S., RÖMER, U., CASPER, T., YUE, Y., FENG, L., BENNER, P., AND SCHOENMAKER, W.: *D2.8 – Public Report (b) on Uncertainty Quantification,* Deliverable Project nanoCOPS: nanoelectronic COupled Problems Solutions, FP7-ICT-2013-11/619166, 2016. http://fp7-nanocops.eu/.
44. PUTEK, P., MEURIS, P., PULCH, R., TER MATEN, E.J.W., SCHOENMAKER, W., AND GÜNTHER, M.: *Uncertainty quantification for robust topology optimization of power transistor devices.* IEEE Transactions on Magnetics 52-3, article# 1700104, (2015/2016). DOI: http://doi.org/10.1109/TMAG.2015.2479361.
45. PUTEK, P.A., TER MATEN, E.J.W., GÜNTHER, M., AND SYKULSKI, J.K.: *Variance-based Robust Optimization of Permanent Magnet Synchronous Machine.* IEEE Transactions on Magnetics, Volume 54, Issue 3, 2018 (article 8102504, 2017). DOI: http://dx.doi.org/10.1109/TMAG.2017.2750485.
46. PUTEK, P., JANSSEN, R., NIEHOF, J., TER MATEN, E.J.W., PULCH, R., TASIĆ, B., AND GÜNTHER, M.: *Nanoelectronic COupled Problems Solutions: Uncertainty Quantification for Analysis and Optimization of an RFIC Interference Problem.* Journal Mathematics in Industry 8:12, 2018. Online. DOI: https://doi.org/10.1186/s13362-018-0054-3.
47. PUTEK, P., TER MATEN, E.J.W., GÜNTHER, M., BARTEL, A., PULCH, R., MEURIS, P., AND SCHOENMAKER, W.: *Robust Shape Optimization under Uncertainties in Device Materials, Geometry and Boundary Conditions.* Chapter in this Book, 2019.
48. SCHÖPS, S., DE GERSEM, H., AND BARTEL, A.: *A Cosimulation Framework for Multirate Time Integration of Field/Circuit Coupled Problems.* IEEE Transactions on Magnetics, 46, pp. 3233–3236, 2010.
49. SCHÖPS, S., DE GERSEM, H., AND BARTEL, A.: *Higher-Order Cosimulation of Field/Circuit Coupled Problems.* IEEE Transactions on Magnetics, 48, pp. 535–538, 2010.

50. SCHÖPS, S., DE GERSEM, H., AND WEILAND, T.: *Winding functions in transient magnetostatic field-circuit coupled simulations.* COMPEL: The Int. Journal for Computation and Mathematics in Electrical and Electronic Engineering, 32-6, pp. 2063–2083, 2013.

51. SEXTON, C., OLIVI, M., AND HANZON, B.: *Rational Approximation of Transfer Functions for Non-Negative EPT Densities.* 16th IFAC Symposium on System Identification, Bruxelles, Belgium. IFAC Proceedings Volumes, 45-16, pp. 716–721, 2012. See also: https://hal.inria.fr/hal-00763205/document.

52. SINGHEE, A., SINGHAL, S., AND RUTENBAR, R.A.: *Practical, Fast Monte Carlo Statistical Static Timing Analysis: Why and How.* IEEE/ACM Int. Conf. on Computer-Aided Design (ICCAD), San Jose, CA, USA, 2008. https://doi.org/10.1109/ICCAD.2008.4681573

53. SMOLYAK, S.A.: *Quadrature and interpolation formula for tensor products of certain classes of functions.* Soviet. Math. Dokl. 4:240–243, 1963.

54. STROUD, A.H.: *Remarks on the Disposition of Points in Numerical Integration Formulas.* Mathematical Tables and Other Aids to Computation, 11(60):257–261, 1957. For a Library, see also at: https://people.sc.fsu.edu/%7Ejburkardt/m%5Fsrc/stroud/stroud.html.

55. SUDRET, B.: *Global sensitivity analysis using polynomial chaos expansion.* Reliab. Eng. Syst. Saf. 93(7), pp. 964–979, 2008.

56. XIU, D.: *Efficient Collocational Approach for Parametric Uncertainty Analysis.* Commun. Comput. Phys., Vol. 2, No. 2, pp. 293–309, 2007.

57. *The Vector Fitting Web Site.* https://www.sintef.no/projectweb/vectfit/, 2019.

58. XIU, D.: *Fast numerical methods for stochastic computations: a review.* Commun. Comput. Phys., Vol. 5, No. 2-4, pp. 242–272, 2009.

59. XIU, D.: *Numerical methods for stochastic computations – A spectral method approach.* Princeton Univ. Press, Princeton, NJ, USA, 2010.

60. YUE, Y., FENG, L., BENNER, P., PULCH, R., AND SCHÖPS, S.: *Reduced Models and Uncertainty Quantification.* Chapter in this Book, 2019.

61. ZHANG, H., CHEN, T.-H., TING, M.-Y., AND LI, X.: *Efficient Design-Specific Worst-Case Corner Extraction for Integrated Circuits.* Proceedings of the 46th Design Automation Conference DAC'09, San Francisco, CA, USA, 2009. http://dx.doi.org/10.1145/1629911.1630013.

62. ZHANG, Z., YANG, X., OSELEDETS, I.V., KARNIADAKIS, G.E., AND DANIEL, L.: *Enabling High-Dimensional Hierarchical Uncertainty Quantification by ANOVA and Tensor-Train Decomposition.* IEEE Trans. on Comp.-Aided Design of Integr. Circuits and Systems (TCAD), 34-1, pp. 63–76, 2015. See also https://arxiv.org/abs/1407.3023.

Chapter 10
Robust Shape Optimization under Uncertainties in Device Materials, Geometry and Boundary Conditions

Piotr Putek, E. Jan W. ter Maten, Michael Günther, Andreas Bartel,
Roland Pulch, Peter Meuris, Wim Schoenmaker

Abstract We address the shape optimization problem of electronic and electric devices under geometrical and material uncertainties. Thereby, we aim at reducing undesirable phenomena of these devices such as hot-spots or torque fluctuations. The underlying minimization is based on the computation of a direct problem with random input data. To investigate the propagation of uncertainties through two- and three-dimensional, spatial models the stochastic collocation method (SCM) has been used in our work. In particular, uncertainties, which result from imperfections of an industrial production, are modelled by random variables with known probability distributions. Then, the polynomial chaos expansion (PCE) is used to construct a suitable response surface model, which can be effectively incorporated into the robust optimization framework. Correspondingly, the gradient directions of a cost functional, comprised of the expectation and the variance value, are calculated using the continuum design shape sensitivity and the PCE in conjunction with the SCM. Finally, the optimization results for the relevant electronics/electrical engineering problems demonstrate that the proposed method is robust and efficient. Overall, this work demonstrates, how recent techniques from shape and topology optimization can be combined with uncertainty quantification to solve complicated real-life problems.

Piotr Putek, E. Jan W. ter Maten, Michael Günther, Andreas Bartel
Bergische Universität Wuppertal, Germany,
e-mail: {Putek,terMaten,Guenther,Bartel}@math.uni-wuppertal.de;Piotr.Putek@gmail.com

Roland Pulch
Universität Greifswald, Germany,
e-mail: Roland.Pulch@uni-greifswald.de

Peter Meuris, Wim Schoenmaker
MAGWEL NV, Leuven, Belgium,
e-mail: {Peter.Meuris,Wim.Schoenmaker}@magwel.com

© Springer Nature Switzerland AG 2019
E. J. W. ter Maten et al. (eds.), *Nanoelectronic Coupled Problems Solutions*,
Mathematics in Industry 29, https://doi.org/10.1007/978-3-030-30726-4_10

10.1 Introduction

In the nanoCOPS project, we dealt with robust shape and topology optimization for finding an optimal design of electronics and distributed devices under geometrical and material uncertainties. Within this project, we were inspired by real-life applications from automotive industry such as power devices, an integrated RFCMOS transceiver [43–45] or transducer devices [42, 46, 50]. Particularly, in the shape optimization we have taken into account uncertainties that inherently exist in the design variables, geometric and material parameters, excitation terms, or the initial/boundary conditions, etc. This allowed us to avoid those optimization results that were very sensitive to variations of the deterministically treated nominal parameters. That is, a robust optimization result is obtained.

To be more precise, we considered a power MOS transistor device as a case study. It plays a key role in efficiently exploiting resources and energy in power electronics with respect to both an energy harvesting and distribution, as well as in applications for automotive industry. Apparently, the physical domain of power devices consists of several thousands of parallel channel devices. It cannot be determined precisely due to manufacturing processes such as the sub-wavelength lithography, the lens aberration, and the chemical-mechanical polishing. In particular, these industrial imperfections directly influence both the yield and the performance [3] and, as a consequence, they have also impact on the acceptability, reliability and profitability of power electronic systems [34]. In fact, the statistical variations in input parameters such as the localized imperfections inside the die can result in the formation of a current density overshoots and may lead to a thermal destruction of the device due to thermal runaway [5, 53]. Moreover, many reliability failure mechanisms strongly accelerate at high temperature, including, for example, the voltage breakdown [14]. In this respect, there was a need for a robust shape optimization method, which aimed at reducing the heat dissipation such that the hot spot phenomena in the sense of the current density overshoots was eliminated, taking into account an electro-thermal behavior of power devices [48, 49].

In addition, we report on a shape optimization of an electric machine under geometrical and material uncertainties, which is treated as the extension of shape optimization problem to a nonlinear case [42, 46, 50].

The outline of this chapter is as follows. We start with the definition of the shape optimization problem in deterministic settings. in our work, this problem is solved using the material derivative method, presented in Section 10.2.1. Then, we show the link between the shape and the topological derivative in Section 10.2.3. This allows us to explore two types of algorithms for the shape/topology optimization, that is, the topological derivative method and the level set method (LSM), demonstrated in Section 10.2.4. In the latter, we show how to incorporate the geometrical uncertainties, represented by the material variations, into the LSM framework. Section 10.3

is devoted to the robust shape optimization problem, which is constrained by an according system of partial differential equations with random input data. Furthermore, we exploit the Stochastic Collocation (SC) technique with polynomial chaos and the material derivative method to find the solution to the shape optimization problem under uncertainties. This algorithm is presented in Section 10.3.3. Finally, the results for simulation under uncertainty and robust shape optimization are shown in Sections 10.4–10.6.

10.2 Shape and sensitivity-based topology optimization

We consider a partial differential equation (PDE), which we represent by some operator $e(\Omega, u) = 0$ in terms of the corresponding solution $u : D \to \mathbb{R}$ and some spatial subdomain $\Omega \subset D$. Moreover, the overall domain $D \subset \mathbb{R}^d$ shall be bounded with Lipschitz boundary. On Ω, the PDE model may be different w.r.t. $D\backslash\Omega$. Now, a non-parametric shape optimization problem seeks an optimal domain Ω^* within a set of admissible shapes $\mathcal{A}$, where for any $\Omega \in \mathcal{A}$ holds $\Omega \subset D$. Optimality is reached by minimizing a domain-dependent objective functional $\mathcal{J}(\Omega, u)$, while fulfilling the PDE constraint, i.e.,

$$\inf_{\Omega \in \mathcal{A}} \mathcal{J}(\Omega, u), \tag{10.1a}$$

$$\text{s.t.} \quad e(\Omega, u) = 0. \tag{10.1b}$$

Next, we roughly explain the shape differentiability of the functional $\mathcal{J}$ (10.1a), see [57]. To this end, let $\boldsymbol{V} : \mathbb{R}^d \to \mathbb{R}^d$ denote a sufficiently often differentiable vector field/velocity field with $\text{supp}(\boldsymbol{V}) \subset D$. Furthermore, let $\boldsymbol{F}_t(\boldsymbol{V}) : D \to D$ be a corresponding family of transformations of Ω for a pseudo time $t \in [0, \epsilon)$, such that we have a perturbed domain $\Omega_t = F_t(\boldsymbol{V})(\Omega)$ for any $t \in [0, \epsilon)$ with $\Omega = \Omega_0$. For some visualization, see Fig. 10.1.

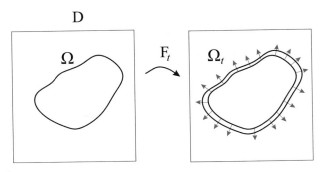

Fig. 10.1: Transformation of a generic domain Ω with $\boldsymbol{F}_t(\boldsymbol{V})(\boldsymbol{X})$ into Ω_t for a given $t > 0$. Notice that for any $\boldsymbol{X} \in \partial D$ holds $\boldsymbol{F}_t(\boldsymbol{V})(\boldsymbol{X}) = \boldsymbol{X}$.

The functional $\mathcal{J}$ is said to be shape differentiable if the limit

$$dJ(\Omega)[V] = \lim_{t \to 0+} \frac{\mathcal{J}(\Omega_t) - \mathcal{J}(\Omega)}{t} \tag{10.2}$$

exists and the map $V \mapsto dJ(\Omega)[V]$ is linear and continuous.

There are at least six main approaches, which allow for solving the shape optimization problem with PDE constraint, (10.1), and prove the existence of the shape derivative (10.2):

   (i) Hadamard's normal variation method [29],
   (ii) the perturbation of the identity method by Simon [54],
   (iii) the material derivative (velocity, speed) method [7, 19, 55],
   (iv) the minimax formulation [16, 57],
   (v) the Lagrange method invented by Céa [8],
   (vi) the rearrangement method introduced in [23].

In our work, we applied the material derivative method (iii) for the solution of the shape optimization problem. Furthermore, using the link between the shape and the topological derivative and the spectral representation of the Polynomial-Chaos Expansion (PCE) of u, we are enabled to explore thesensitivity-based algorithm for shape optimization of electronics and electrical devices under material uncertainties.

10.2.1 Material and shape derivative method

The speed method, called also the material derivative method in the engineering terminology, was originally developed around 1979 at the University of Nice by the French scientist J. Céa in application to the structural and civil engineering [28]. Since then, a lot of scientists have investigated this topic, see, e.g., [19, 55] and references therein for overview.

Here, we consider the domain Ω as the design variable. Therefore, we can use the material derivative concept of continuum mechanics to represent the deformation process of the domain $\Omega(\subset D)$ into the perturbed domain $\Omega_t(\subset D)$. Let $\Omega \in \mathcal{A}$ be fixed. We consider the mapping $F_t(V) : \mathbb{R}^d \to \mathbb{R}^d$ in the form

$$F_t(V)(X) := X + tV(X) \qquad X \mapsto F_t(V)(X) =: x \tag{10.3}$$

for $t \geq 0$ and with $V(X) = (V_1(X), \dots, V_d(X))^\top \in (C^{1,1}(\mathbb{R}^2))^d$ as well as $V = 0$ on ∂D. To distinguish between the functions with fixed domain, depending on X, and those having domain in the moving frame, depending on x, we introduce a superscript $(\Box)^t$ and a subscripts $(\Box)_t$, respectively. E.g. for the composition, we have

$$u^t(\boldsymbol{X}) = u_t(\boldsymbol{x}) = u_t(\boldsymbol{F}_t(\boldsymbol{X})).$$

The pointwise material derivative of the state variable u at $\boldsymbol{X} \in \Omega$ is defined by [19]

$$\dot{u}(\boldsymbol{X}) = \dot{u}(\boldsymbol{X}; \Omega, \mathbf{V}) \equiv \frac{d}{dt} u^t(\boldsymbol{X} + t\boldsymbol{V}(\boldsymbol{X}))\Big|_{t=0} = \lim_{t \to 0+} \left[\frac{u^t(\boldsymbol{X} + t\boldsymbol{V}(\boldsymbol{X})) - u(\boldsymbol{X})}{t} \right]$$
(10.4)

if the limit exists. Moreover, under assumption that $u^t(\mathbf{x})$ has a regular extension to a neighborhood of $\overline{\Omega}_t$, the material derivative of (10.4) can be split into two contributions as follows [19]

$$\dot{u}(\boldsymbol{X}) = \lim_{t \to 0+} \left[\frac{u^t(\boldsymbol{X}) - u(\boldsymbol{X})}{t} \right] + \lim_{t \to 0+} \left[\frac{u^t(\boldsymbol{X} + t\boldsymbol{V}(\boldsymbol{X})) - u^t(\boldsymbol{X})}{t} \right]$$
$$= u'(\boldsymbol{X}) + \boldsymbol{V}(\boldsymbol{X}) \cdot \nabla u(\boldsymbol{X}),$$

where u' is the partial derivative of u for fixed $\boldsymbol{X}$, while $\boldsymbol{V} \cdot \nabla u$ denotes the convective term. The pointwise material derivative $\dot{u}(\boldsymbol{X})$ in (10.4) can be defined only for the solution to the strong formulation [19]. In the case of solutions for the variational problem, the material derivative $\dot{u}$ needs to be defined in the Sobolev norm sense as follows [55]

$$\lim_{t \to 0+} \left\| \frac{u^t(\boldsymbol{X} + t\boldsymbol{V}(\boldsymbol{X})) - u(\boldsymbol{X})}{t} - \dot{u}(\boldsymbol{X}) \right\|_{H^1(\Omega)}.$$

Let the symbol D before a vector-valued function $\boldsymbol{f} : \mathbb{R}^d \to \mathbb{R}^d$ denote the Jacobian, i.e.,

$$D\boldsymbol{f} = \begin{pmatrix} \partial_{x_1} f_1 & \partial_{x_2} f_1 & \cdots & \partial_{x_d} f_1 \\ \partial_{x_1} f_2 & \partial_{x_2} f_2 & \cdots & \partial_{x_d} f_2 \\ \vdots & \vdots & & \vdots \\ \partial_{x_1} f_d & \partial_{x_2} f_d & \cdots & \partial_{x_d} f_d \end{pmatrix} = \begin{pmatrix} \nabla f_1 \\ \nabla f_2 \\ \vdots \\ \nabla f_d \end{pmatrix}.$$

For $\boldsymbol{F}_t$ (10.3), we have $D\boldsymbol{F}_t = Id + t\nabla\boldsymbol{V}(\boldsymbol{X})$, that is, only derivatives with respect to $\boldsymbol{X}$ are meant. Next, we introduce the following basic material derivative formulas [22]:

$$I_t := \det(D\boldsymbol{F}_t), \quad M_t := (D\boldsymbol{F}_t)^{-1},$$
$$A_t := M_t M_t^\top I_t, \quad A := \nabla \cdot \boldsymbol{V} Id - (D\boldsymbol{V}^\top + D\boldsymbol{V}),$$
(10.5)
$$\frac{I_t - 1}{t}\Big|_{t=0} = \nabla \cdot \boldsymbol{h}, \quad \frac{A_t - Id}{t}\Big|_{t=0} = A.$$
(10.6)

Moreover, we have [22]

$$\nabla u_t = M_t \nabla u^t.$$

Let the domain-dependent functional $\mathcal{J} = \mathcal{J}(\Omega_t)$ be defined as an integral over Ω_t

$$\mathcal{J} = \int_{\Omega_t} g_t(\boldsymbol{x})d\boldsymbol{x},$$

with g_t a regular function defined in Ω_t. Based on the definitions (10.4) and (10.5) one can derive the material derivative of $\mathcal{J}$. Assuming that $\Omega \in C^k$, we have [19]

$$\mathcal{J}' = \frac{d}{dt}\int_{\Omega_t} g_t(\boldsymbol{x})d\boldsymbol{x}\Big|_{t=0} = \frac{d}{dt}\int_{\Omega} g^t(\boldsymbol{X}+t\boldsymbol{V}(\boldsymbol{X}))I_t d\boldsymbol{X}\Big|_{t=0} \qquad (10.7)$$

$$= \int_{\Omega} \dot{g}(\boldsymbol{X})+g(\boldsymbol{X})\nabla\cdot\boldsymbol{V}(\boldsymbol{X})d\boldsymbol{X}$$

$$= \int_{\Omega} g'(\boldsymbol{X})+\nabla g(\boldsymbol{X})^\top\boldsymbol{V}(\boldsymbol{X})+g(\boldsymbol{X})\nabla\cdot\boldsymbol{V}(\boldsymbol{X})d\boldsymbol{X}$$

$$= \int_{\Omega} g'(\boldsymbol{X})+\nabla\cdot(g(\boldsymbol{X})\boldsymbol{V}(\boldsymbol{X}))d\boldsymbol{X} = \int_{\Omega} g'(\boldsymbol{X})d\boldsymbol{X}+\int_{\partial\Omega} g(\boldsymbol{X})V_n d\boldsymbol{X},$$

using the transformation of the integral domain Ω_t to Ω and for the last step the divergence theorem with $V_n = \boldsymbol{V}\cdot\boldsymbol{n}$ denoting the normal component of the velocity field on the boundary $\partial\Omega$ (and outer unit normal $\boldsymbol{n}$).

10.2.2 Material and shape derivative applied to electronics and to electrical devices

Based on [12, 22], we give a brief sketch of the material and shape derivative for a bilinear and a linear form. In our project, the mathematical models of electronics and electrical devices were based on a system of linear/quasilinear elliptic equations. Thus, we consider a mathematical model of the device described by the linear Poisson equation in terms of the spatial variable $\boldsymbol{x} \in \mathrm{D} \subset \mathbb{R}^d$ for $d=2$ (with Lipschitz boundary). Of course, the case $d=3$ is similar. The weak formulation of this boundary value problem is given by:

find $u \in H_0^1(D)$ such that $(\nu\nabla u, \nabla\varphi) = (f,\varphi)$ for all $\varphi \in H_0^1(D)$ (10.8)

with the scalar product $(u_1, u_2) = \int_D u_1 u_2\, d\boldsymbol{x}$ (for real valued functions u_1 and u_2). We assume that the coefficient function $\nu(\boldsymbol{x}) \in L^\infty(\mathrm{D})$ is defined as

$$\nu(\boldsymbol{x}) = \begin{cases} \nu_1(\boldsymbol{x}), & \boldsymbol{x} \in \Omega, \\ \nu_2(\boldsymbol{x}), & \boldsymbol{x} \in \mathrm{D}\backslash\Omega. \end{cases} \qquad (10.9)$$

The coefficient ν can either represent the permittivity ϵ, the electric conductivity σ, the thermal conductivity κ, or the reluctivity $\upsilon(B^2)$. In general, the latter will be nonlinearly depending on the magnetic flux density $B = |\boldsymbol{B}|$. However, for the needs of this derivation, we assume that ν is a real-valued bounded function:

$$0 < \nu_{\min} \leq \nu_j \leq \nu_{\max} < \infty, \qquad j = 1, 2.$$

The right-hand side function f of (10.8) can either denote the electrical charge ρ, the power density Q, or the excitation current density of an electric machine J_{s}. In case of a semiconductor device, the excitation of the device would be given via a suitable boundary condition, i.e., a non-homogeneous boundary condition. For simplicity, we disregard this case, here. Finally, the state variable $u \in H_0^1(\mathrm{D})$ is a scalar function, which describes either the scalar electric potential, the temperature or the third component of the magnetic vector potential.

For the purpose of optimization, we define a domain-dependent cost functional

$$\mathcal{J}(\Omega, u) = \int_D g(u) d\boldsymbol{x}, \qquad (10.10)$$

via a sufficiently smooth function g. The function g can be either the power dissipation Q due to Joule's law, which result in the self-heating of semiconductor devices, or the magnetic energy W. Then, our specific shape optimization problem reads as

$$\inf_{\Omega \in \mathcal{A}} \mathcal{J}(\Omega, u) \qquad (10.11a)$$

$$\text{s.t.} \quad (\nu \nabla u, \nabla \varphi) = (f, \varphi) \qquad \text{for all } \varphi \in H_0^1. \qquad (10.11b)$$

To solve this problem via shape derivatives, we employ deformations Ω_t of the initial domain Ω (in terms of the pseudo time t). This affects the weak formulation of the PDE constraint (10.11b) as follows:

$$(\nu_t \nabla u_t, \nabla \varphi_t) = (f_t, \varphi_t). \qquad (10.12)$$

For simplicity of the notation, we suppress the spatial dependence of ν. Now, we can "pull back" our model (10.12) by performing the change of variables $\boldsymbol{x} = \boldsymbol{F}_t(\boldsymbol{X})$ with $\Omega_0 = \Omega_{t=0} = \Omega$, $\nu_{t=0} = \nu$ and $f_{t=0} = f$. Then, using (10.5), the weak problem formulation (10.12) can be equivalently written as

$$\left(\nu A_t \nabla u^t, \nabla \varphi^t\right) = \left(I_t f^t, \varphi^t\right). \qquad (10.13)$$

Next, we subtract the direct problem (10.12) for the time instance $t = 0$ from the last equation, and then we divide the result by t:

$$\underbrace{\left(\nu A_t \frac{\nabla u^t - \nabla u}{t}, \nabla\varphi\right)}_{=:a^t(\varpi,\varphi)} + \underbrace{\left(\nu \frac{A_t - I_t}{t}\nabla u, \nabla\varphi\right)}_{=:b_1^t(\varphi)} = \underbrace{\left(\frac{f^t - f}{t}I_t, \varphi\right)}_{=:b_2^t(\varphi)} + \underbrace{\left(\frac{I_t - 1}{t}f^t, \varphi\right)}_{=:b_3^t(\varphi)},$$

where the test functions are simply denoted by φ. Using the shorthand $\varpi := (u^t - u)/t$, the latter equation can be written in a more compact way as

$$a^t\left(\varpi^t, \varphi\right) = b^t(\varphi), \tag{10.14}$$

with $b^t(\varphi) = b_1^t(\varphi) + b_2^t(\varphi) + b_3^t(\varphi)$ (notice that the dependence of b_1^t on u is suppressed).

Solvability of the weak problem. The uniqueness of solutions for (10.14) for every positive t can be proven by using the Lax-Milgram theorem under the condition of the continuity and coercivity of $a^t\left(\varpi^t, \varphi\right)$ together with the continuity of $b^t(\varphi)$. Now, from $\|\nabla u\|_{L^\infty(D)} \leq C$, we have the continuity of $a^t\left(\varpi^t, \varphi\right)$ such that $|(a^t\left(\varpi^t, \varphi\right))| \leq \nu_{\max}\|\nabla\varpi^t\|_2\|\nabla\varphi\|_2$, which implies (coercivity)

$$|(a^t\left(\varpi^t, \varpi^t\right))| \geq \nu_{\min}\|\nabla\varpi^t\|_2^2.$$

Next, we address the first term b_1^t of b. From (10.6), we have

$$\|t^{-1}(A_t - I)\|_{L^\infty(D)} \leq C.$$

Together with L^∞ boundedness of ∇u, we obtain the boundedness and thus continuity of b_1. Likewise, we obtain that b_3 is bounded and continuous. And finally, the smoothness assumption of f_t guarantees the boundedness and the continuity of b_2.

Material derivative. To derive an equation for $\dot{u}$, we follow [12, 22]. Thus, let (t_n) be any sequence with $t_n \in [0,\epsilon)$, which converges to zero. The boundedness of $(\varpi^n) = (\varpi^{t_n})$ gives a weakly convergent subsequence, also denote by (ϖ^n), with limit $\dot{u}$ in $H^1(D)$. Then it holds:

$$\lim_{n\to\infty} \left|a_1^{t_n}(\varpi^n, \varphi) - (\nu\nabla\dot{u}, \nabla\varphi)\right| = 0, \qquad \lim_{n\to\infty} \left|b_1^{t_n}(\varphi) - (\nu A\nabla u, \nabla\varphi)\right| = 0,$$

$$\lim_{n\to\infty} \left|b_2^{t_n}(\varphi) - (-\nabla f \cdot \mathbf{V}, \varphi)\right| = 0, \qquad \lim_{n\to\infty} \left|b_3^{t_n}(\varphi) - (-f\nabla \cdot \mathbf{V}, \varphi)\right| = 0.$$

Hence, there exists the weak material derivative $\dot{u}$, which satisfies

$$a(\dot{u}, \varphi) + b(\varphi) = 0, \tag{10.15}$$

where

$$a(\dot{u}, \varphi) = (\nu\nabla\dot{u}, \nabla\varphi), \qquad b(\varphi) = (\nu A\nabla u, \nabla\varphi) + (-\nabla \cdot (f\mathbf{V}), \varphi).$$

The continuity and coercivity of a and b can be shown in a similar way as in the case of a^t and b^t, see [12, 22] for details.

Shape derivative. Next, to derive an equation for the shape derivative in the form $u' = \dot{u} - \mathbf{V} \cdot \nabla u$, we follow [12,22] and investigate

$$R := a(u',\varphi) + (\nabla \cdot (\nu \nabla u), \mathbf{V} \cdot \nabla \varphi),$$

where assume that u and φ are sufficiently regular. After some manipulation [12,22], we obtain (using $D \subset \mathbb{R}^2$ and (10.15))

$$R = -(\nabla \times [\nu(V_2\,u_x - V_1 u_y)], \nabla \varphi) + (-\nabla \cdot (f\mathbf{V}), \varphi). \tag{10.16}$$

Then, we use Green's theorem (in $\mathbb{R}^2$) for the first term of (10.16), split the domain into the subdomains Ω and $D \backslash \Omega$, and employ $\nabla \cdot (\nu \nabla u) = -f$. This way, we obtain the final formula of the shape derivative

$$a(u',\varphi) = \left([\nu^+ - \nu^-] \nabla u \mathbf{V} \cdot \mathbf{n}, \nabla \varphi \right)_{\partial \Omega}, \tag{10.17}$$

where superscripts $^+$ and $^-$ are used to indicate the approaching of the boundary $\partial \Omega$ from outside and inside of the considered domain Ω.

Applying (10.7) to the domain-dependent cost functional (10.10) for the magnetic energy density $g(\nabla u) = w(\nabla u) = v|\nabla u|^2$, we obtain

$$d\mathcal{J}(\Omega, u) = \lim_{t \to 0+} \frac{\mathcal{J}(\Omega_t) - \mathcal{J}(\Omega)}{t} = \int_{D_1} v \nabla u \nabla u' d\mathbf{x}. \tag{10.18}$$

Here, $D_1 \subset D$ denotes the subdomain, where the magnetic energy density is calculated. Now, we can define an adjoint problem

$$a(\lambda, \psi) = \int_{D_1} v \nabla u \cdot \nabla \psi d\mathbf{x} = (v \nabla u, \nabla \psi)_{D_1}, \tag{10.19}$$

where $\lambda, \psi \in H_0^1$. Then, after taking the following test functions $\varphi = \lambda$ in (10.17) and $\psi = u'$ in (10.19) yields

$$d\mathcal{J}(\Omega, u) = \left([\nu^+ - \nu^-] \nabla u \mathbf{V} \cdot \mathbf{n}, \nabla \lambda \right)_{\partial \Omega}. \tag{10.20}$$

Moreover, under the assumption that also the optimization of the source term f is included, an integral over the respective boundary Γ (where f is applied) is included into the cost functional of our optimization problem [48,49]. Thus, the derivative of the domains dependent cost functional becomes

$$d\mathcal{J}(\Omega, \Gamma, u) = \left([\nu^+ - \nu^-] \nabla u \mathbf{V} \cdot \mathbf{n}, \nabla \lambda \right)_{\partial \Omega} - \left([f_s^+ - f_s^-] u \mathbf{V} \cdot \mathbf{n}_\Gamma, \lambda \right)_\Gamma. \tag{10.21}$$

Apparently, we could also benefit from the fact that magnetic energy operator W is self-adjoint [24]

$$W(u) = W_m + W'_m = \int_{D_1} \boldsymbol{B} \cdot \mathbf{H} d\boldsymbol{x} = \int_{D_1} \left[\int \mathbf{H} d\mathbf{B} + \int \mathbf{B} d\mathbf{H} \right] d\boldsymbol{x} = \int_{\Gamma} f \cdot u \, d\boldsymbol{x},$$

where W_m and W'_m are the stored magnetic energy and co-energy, and $\mathbf{H}$ and $\mathbf{B}$ denote the magnetic field intensity and the magnetic flux density, respectively. We recall that f is the current density, while u denotes the third component of the magnetic vector potential.

Finally, adopting the last result into our problem leads to the conclusion that both systems (primary and adjoint) are the same, due to the form of the right-hand side of (10.19), which is defined by $(f, u)_\Gamma$. Hence, there is no need to formulate and solve an adjoint problem. Likewise, the electric power density P or power Q at least in the steady state regime as energy-like quantities belongs to self-adjoint operators. This fact, for example, was used in optimizing a Power MOS device and an electric machine [40, 46, 49].

10.2.3 Topological derivative method and sensitivity-based topology optimization

The topological derivati (TD) method was originally invented by the authors of [7, 9, 17]. It provides the sensitivity information of a design-dependent functional with respect to a topological change in the domain. In particular, the paper by Eschenauer et al. [17] is widely recognized and introduces a method, which allows for an iterative positioning of new holes inside some structure or material by solving some variational problem. For example, in magnetics, one could be interested in inserting and modifying an air hole inside some ferromagnetic component in the domain of interest. Thus, the TD method is more general than shape optimization.

A further development is the so-called asymptotic expansion method [56], which yields the TD asymptotic expansion for a wide range of 2D/3D linear and nonlinear problems, see, e.g., [1, 18, 32]. Thus, recently, the TD methodology is been widely applied in many areas of industry such as electrical and electronic engineering applications to solve both deterministic and stochastic optimization problems, e.g., [5, 26, 30, 51].

To roughly present the idea of the asymptotic expansion of the TD, we follow [32] and we consider an underlying PDE problem with corresponding solution $u : \Omega \subset \mathbb{R}^d \to \mathbb{R}$, where the bounded Ω needs to be optimized. The problem shall be well-posed for any admissible Ω. Let $j(\Omega) := \mathcal{J}(\Omega, u)$ denote the respective cost functional, which needs to be minimized w.r.t. Ω. Furthermore, let $B(\boldsymbol{x}_0, r)$ denote the d-dimensional ball of radius $r > 0$ and with center $\boldsymbol{x}_0 \in \Omega$, such that $B(\boldsymbol{x}_0, r)$ is completely contained inside Ω, i.e.,

$$B(\boldsymbol{x}_0, r) \subset \Omega \quad \text{and} \quad \bar{B}(\boldsymbol{x}_0, r) \cap \partial \Omega = \{\}.$$

We remark that we consider small values of $r > 0$. As a shorthand, we use $\Omega_{\boldsymbol{x}_0,r} := \Omega \backslash B(\boldsymbol{x}_0, r)$ where the ball is removed from Ω. Then, we need to assign also a boundary condition at the hole: $\Gamma_{\boldsymbol{x}_0,r} := \partial B(\boldsymbol{x}_0, r)$. For simplicity, we assume a homogeneous Neumann condition on $\Gamma_{\boldsymbol{x}_0,r}$ be given, such that the problem is still well posed. Then, the asymptotic expansion of the TD is defined as [32]

$$j(\Omega_{\boldsymbol{x}_0,r}) - j(\Omega) = h(r)g(\boldsymbol{x}_0) + o(h(r)), \tag{10.22}$$

with some positive function $h(r)$, such that $\lim\limits_{r \to 0+} h(r) = 0$, some function g and the (little) Landau symbol o.

The derivation of the TD for the linear elliptic problem based on the asymptotic expansion method can be found in, e.g., [56]. Here, we explore a link between the shape and topological derivative [9], which allows us to provide a brief derivation of the TD for the linear optimization problem (10.11), analysed in Section 10.2.1.

In a large class of the shape optimization problems, when the boundary is sufficiently smooth, the shape derivative can be written in the form of Hadamard's formula

$$d\mathcal{J}(\Omega, \boldsymbol{V}) = \int_{\partial\Omega} L(s)V_n ds, \tag{10.23}$$

where $L \in L^1(\partial\Omega)$ is a an integrable function defined on the boundary and V_n is the normal component of the velocity field $\boldsymbol{V}$. Now, let the cost functional depend only on the radius r, i.e., $J(r) = \mathcal{J}(\Omega_{\boldsymbol{x}_0,r})$, and let a Neumann condition be imposed on $\Gamma_{\boldsymbol{x}_0,r}$ (to ensure the convergence for $r \to 0$). We consider small perturbations, which keep the outer boundary ∂D invariant, i.e, $\boldsymbol{V} = 0$ on ∂D, but it increases the radius r of the ball, that is $\boldsymbol{V} = \boldsymbol{n}$ on $\Gamma_{\boldsymbol{x}_0,r}$ as schematically shown in Fig. 10.2. Under these assumptions and using the shape derivative (10.20) as the scalar function in the Hadamard formula

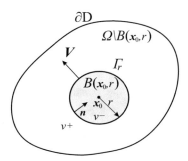

Fig. 10.2: The concept of topological derivative including a domain with one circular hole $B(\boldsymbol{x}_0, r)$.

(10.23)

$$L(\nabla u_r, \nabla \lambda_r)(s) = [\nu^+ - \nu^-] \nabla u_r(s)^\top \cdot \nabla \lambda_r(s)$$

with $u_r = u_{\Omega \setminus B(\boldsymbol{x}_0, r)}$ and an adjoint variable $\lambda_r = \lambda_{\Omega \setminus B(\boldsymbol{x}_0, r)}$, the derivative of functional $dJ(r)$ takes the form

$$J'(r) = -\int_{\Gamma_{\boldsymbol{x}_0, r}} L(\nabla u_r, \nabla \lambda_r)(s) ds.$$

Furthermore, the application of a local expansion of u_r and λ_r as in [9, 21], yields [26]

$$J'(r) = -4\pi r [\nu^+ - \nu^-] \nabla u(\boldsymbol{x}_0)^\top \cdot \nabla \lambda(\boldsymbol{x}_0) + \mathcal{O}(r).$$

Consequently, the left-hand side of (10.22) can be expressed by

$$J(r) - J(0) = \int_0^r J'(\zeta) d\zeta = -2\pi r^2 [\nu^+ - \nu^-] \nabla u(\boldsymbol{x}_0)^\top \cdot \nabla \lambda(\boldsymbol{x}_0) + \mathcal{O}(r^2).$$

Finally, the topological derivative with $h(r) = -\pi r^2$ [9], due to the imposed Neumann boundary condition on $\Gamma_{\boldsymbol{x}_0, r}$ takes the form

$$g(\boldsymbol{x}_0) = 2[\nu^+ - \nu^-] \nabla u(\boldsymbol{x}_0)^\top \mathcal{P}\left(\omega, \nu^+/\nu^-\right) \nabla \lambda(\boldsymbol{x}_0), \qquad (10.24)$$

where additionally the concept of the polarization matrix $\mathcal{P}\left(\omega, \nu^+/\nu^-\right)$ in the linear setting for the unit disc is involved [11]

$$\mathcal{P}\left(\omega, \nu^+/\nu^-\right) = \frac{2|\omega|}{1 + \nu^+/\nu^-} Id = \frac{2\pi \nu^-}{\nu^+ + \nu^-} Id$$

with identity matrix Id and being the identity matrix and the area of the unit disc $|\omega|$.

There are mainly two distinguished methods, which allows for including the topological derivative into the optimization task (10.11): (i) the sensitivity-based topology optimization method proposed by Céa [7, 9] and recently reinvented by the group of the Japanese scientist Takahashi, the so-called ON/OFF method [35] and (ii) the topological derivative-based level set methods [2, 62].

We conclude this section by the sensitivity-based Alg. 10.1, where the topological derivative is used for solving the shape optimization problem. Following the work [9], where the analysed algorithm has been proposed, we consider here a decreasing sequence of the area constraint $|\Omega_{k \geq 0}|$ with $|\Omega_0|$. The process of optimization stops, when a constraint expressed by $\tau |\Omega_0|$ is achieved.

The next section is devoted to the multi-level set method, which is based on the shape derivative.

Algorithm 10.1 Sensitivity based algorithm for topology optimization.

1: Formulate the optimization problem with the definition of the objective functional
$\mathcal{J}(\Omega, u)$ (10.11a) and areas constraints $|\Omega_{end}| = \tau|\Omega_0|$, $\tau \in (0, 1)$.
2: Determine the design domains and divide the initial domains into voxels using, e.g.,
a finite element discretisation.
3: BoolDecrease = True. BoolAreaConstraintsSatisfied = False. $k = 0$.
4: **while** BoolDecrease and (not BoolAreaConstraintsSatisfied) **do**
5:     $k = k + 1$.
6:     Solve the primary and adjoint systems (10.11b) and (10.18) for u and λ.
7:     Calculate the TD using (10.24) for each voxel in the design area.
8:     Sort the TD values and remove material from the design area with respect to the
assumed threshold ratio.
9:     Calculate the Area Constraint $|\Omega_k|$.
10:    BoolDecrease = $|\Omega_k| < |\Omega_{k-1}|$.
11:    BoolAreaConstraintsSatisfied = $|\Omega_k| < \tau|\Omega_0|$.
12: **end while**

10.2.4 Multilevel Set Approaches

Here, we aim at finding the solution of (10.11) in the two-dimensional setting
$D = \Omega^+ \cup \Omega^- \subset \mathbb{R}^2$, when the level set method is used for the description
of the geometry. The domains Ω^+, Ω^- represent different materials and the
interface between the two domains Γ is described by the zero level function as
shown on Fig. 10.3. For an overview of this method we refer, e.g., to [22, 36].

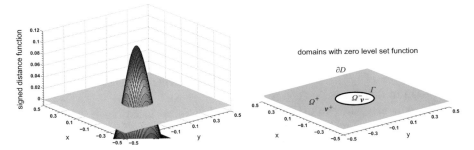

Fig. 10.3: Distribution of the level set function with the zero-level set representing interface Γ.

Let $\phi = \phi(t, \boldsymbol{x})$, $t \geq 0$, $\boldsymbol{x} \in \mathbb{R}^2$ be the level set function, which can be defined,
e.g., as a signed distance function [10]. Again, t denotes a pseudo time. It
generates a family of interfaces

$$\Gamma_t = \{\boldsymbol{x} \in \mathbb{R}^2 : \phi(t, \boldsymbol{x}) = 0\}$$

and a family of domains

$$\Omega_t^+ = \{\boldsymbol{x} \in D : \phi(t, \boldsymbol{x}) < 0\}, \quad \Omega_t^- = \{\boldsymbol{x} \in D : \phi(t, \boldsymbol{x}) > 0\}.$$

Then, the domain dependent parameter ν is represented by:

$$\nu(\phi(t, \boldsymbol{x})) = \begin{cases} \nu^+ & \text{if } \boldsymbol{x} \in \Omega_t^+, \\ \nu^- & \text{if } \boldsymbol{x} \in \Omega_t^-. \end{cases}$$

This can be written in terms of the Heaviside function H:

$$\nu(\phi(t, \boldsymbol{x})) = H(\phi(t, \boldsymbol{x})) \nu^+ + [1 - H(\phi(t, \boldsymbol{x}))] \nu^-. \tag{10.25}$$

For a smoother transition between domains Ω^+ and Ω^-, it is recommended to employ the smeared out version of the Heaviside function [36]:

$$H_k(\phi) = \frac{1}{2} + \frac{1}{\pi} \arctan(k\phi)$$

with prescribed k, which controls the smoothness [13].

Now, we can either put the level set representation into the PDE constraint (10.11b) and repeat the procedure related to the material derivative-based derivation or we can use a chain rule in order to incorporate shape derivative (10.20) into the level set framework. Since we consider here only linear materials represented by the coefficients ν^+ and ν^-, we decided to use the second option. Thus, the domain-dependent functional with the level-set approach has the structure $\mathcal{J}(\Omega(\phi))$. Using (10.20) and the chain rule, we find

$$\begin{aligned} \partial_\phi \mathcal{J}(\Omega(\phi)) &= \partial_\nu \mathcal{J}(\Omega(\phi)) \partial_\phi \nu \\ &= (\nu^+ - \nu^-) \partial_\nu \mathcal{J}(\Omega(\phi)) \delta(\phi) \\ &= (\nu^+ - \nu^-) \nabla u \nabla \lambda \delta(\phi), \end{aligned} \tag{10.26}$$

where $\delta(\phi) = H'(\phi)$ is the derivative of the Heaviside function H' (the so-called Dirac delta function).

Consequently, the evolution of the signed distance functions ϕ is controlled by the Hamilton-Jacobi-type equation [36]

$$\partial_t \phi = -\nabla \phi(t, \boldsymbol{x}) d_t \boldsymbol{x} = V_n |\nabla \phi| \tag{10.27}$$

with pseudo-time t, normal component of the zero-level set velocity V_n and such that the domain-dependent functional $\mathcal{J}(\Omega(\phi))$ is minimized.

Multilevel set method. Obviously, this concept can be generalized to a shape optimization problem comprised of many domains by the so-called multilevel set method [10]. For instance, the application of two level set function ϕ_1 and ϕ_2 allows for splitting the whole region into four domains Ω_1, Ω_2, Ω_3 and Ω_4, Fig. 10.4, via

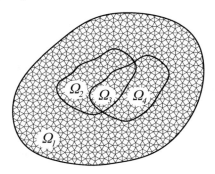

Fig. 10.4: The partition of domains into 4 regions.

$$\Omega_1 = \{\mathbf{x} \in \Omega | \phi_1 > 0 \text{ and } \phi_2 > 0\}, \quad \Omega_2 = \{\mathbf{x} \in \Omega | \phi_1 > 0 \text{ and } \phi_2 < 0\},$$
$$\Omega_3 = \{\mathbf{x} \in \Omega | \phi_1 < 0 \text{ and } \phi_2 > 0\}, \quad \Omega_4 = \{\mathbf{x} \in \Omega | \phi_1 < 0 \text{ and } \phi_2 < 0\}.$$
$$(10.28)$$

As an example, we consider the shape optimization problem (10.11), where also the source term f, is taken into account in the optimization process. Using (10.28), the model functions ν and f in (10.11) have the multilevel representation:

$$\nu(\boldsymbol{\phi}) = \nu_1 H(\phi_1) H(\phi_2) + \nu_2 H(\phi_1)(1 - H(\phi_2)) + \tag{10.29}$$
$$+ \nu_3 (1 - H(\phi_1)) H(\phi_2) + \nu_4 (1 - H(\phi_1))(1 - H(\phi_2)),$$
$$f(\boldsymbol{\phi}) = f_1 H(\phi_1) H(\phi_2) + f_2 H(\phi_1)(1 - H(\phi_2)) + \tag{10.30}$$
$$+ f_3 (1 - H(\phi_1)) H(\phi_2) + f_4 (1 - H(\phi_1))(1 - H(\phi_2)).$$

In this situation, the shape derivative (10.21) can be extended to four regions, which leads to [39]

$$\partial_{\phi_1} J(\Omega(\boldsymbol{\phi})) = \nabla u \nabla \lambda \delta(\phi_1) \left[(\nu_1 - \nu_3) H(\phi_2) + (\nu_2 - \nu_4)(1 - H(\phi_2)) \right]$$
$$- u \lambda \delta(\phi_1) \left[(f_1 - f_3) H(\phi_2) + (f_2 - f_4)(1 - H(\phi_2)) \right]$$
$$\partial_{\phi_2} J(\Omega(\boldsymbol{\phi})) = \nabla u \nabla \lambda \delta(\phi_2) \left[(\nu_1 - \nu_2) H(\phi_1) + (\nu_3 - \nu_4)(1 - H(\phi_1)) \right]$$
$$- u \lambda \delta(\phi_2) \left[(f_1 - f_2) H(\phi_1) + (f_3 - f_4)(1 - H(\phi_1)) \right].$$

Here, the evolution of the level set functions is governed by $\partial_t \phi_i = -V_{n,i} |\nabla \phi_i|$ for $i = 1, 2$.

We finally remark that this framework can be also extended to the stochastic shape optimization problem, where, for example, geometric uncertainties are represented by a random-dependent version of the level set [42, 50]. Summarizing, the following steps are needed to construct the standard multilevel-set-based Algorithm 10.2 [10, 36].

Algorithm 10.2 Multilevel-set-based algorithm for shape optimization.

1: Formulate the optimization problem with the definition of the objective functional $\mathcal{J}(\Omega(\phi), u)$ (10.11a).
2: Determine the design domains and initialize the level set distributions.
3: BoolAreaConstraintsSatisfied = False.
4: **while** not BoolAreaConstraintsSatisfied **do**
5:     BoolOptimized = False.
6:     **while** not BoolOptimized **do**
7:         Solve the primary and adjoint systems (10.11b) and (10.18) for u and λ.
8:         Determine $\partial_\phi \mathcal{J}$ based on the velocity field (10.26).
9:         Calculate the correction of the level sets and introduce it to the model.
10:        BoolOptimized = Objective function reaches the assumed level.
11:    **end while**
12:    Update BoolAreaConstraintsSatisfied.
13: **end while**

10.3 Robust Optimization Algorithm

In the stochastic setting, the mathematical models of devices are described by a system of the partial differentials with random input data. This influences both the direct problem and shape optimization problem due to the propagation of the uncertainties.

10.3.1 Direct and shape optimization problem with random input data

Next, our deterministic shape optimization problem (10.11) is extended to include input uncertainties. These uncertainties are modelled by material and geometrical variations, which result from manufacturing imperfections. They are represented by some random fields $\boldsymbol{p}(\boldsymbol{\xi})$, where the random vector $\boldsymbol{\xi}$ is defined on a suitable probability space $(\mathbb{A}, \mathbb{F}, \mathbb{P})$ with event space $\mathbb{A}$, σ-algebra $\mathbb{F}$ and probability measure $\mathbb{P}$. Therein, the random variables $\boldsymbol{\xi} = (\xi_1, \ldots, \xi_q)^\top$ are assumed to be q independent and identically uniformly distributed on the interval $[-1, 1]$. As usual [61], we denote by $\Gamma_n := p_n(\mathbb{A})$ the image of p_n and $\Gamma := \prod_{n=1}^q \Gamma_n \subset \mathbb{R}^q$. Moreover, we assume that the random variable $\boldsymbol{p}(\boldsymbol{\xi})$ has the joint probability density function $\rho : \Gamma \to \mathbb{R}^+$. Hence, the probability space $(\mathbb{A}, \mathbb{F}, \mathbb{P})$ can be replaced by $(\Gamma, \mathcal{B}^q, \rho d\mathbb{P})$ with the q-dimensional Borel space.

The expected value and the variance of a random function $y : \Gamma \to \mathbb{R}$ with $y \in L^2_\rho(\Gamma)$ is given by

$$\mathbb{E}[y(\mathbf{p}(\boldsymbol{\xi}))] := \int_{\Gamma} y(\mathbf{p})\rho(\mathbf{p})\,\mathrm{d}\mathbf{p}, \tag{10.31}$$

$$\mathrm{Var}[y(\cdot)] := \mathbb{E}[y(\cdot)^2] - (\mathbb{E}[y(\cdot)])^2 \tag{10.32}$$

with corresponding inner product

$$\langle y(\cdot), z(\cdot) \rangle := \mathbb{E}\left(y(\cdot)z(\cdot)\right), \qquad y,\, z \in L^2_\rho(\Gamma). \tag{10.33}$$

This allows us to consider the weak formulation of the direct problem with random input data in a finite-dimensional space

$$V_\rho = L^2_\rho(\Gamma) \otimes L^2\left(H^1_0(\mathrm{D})\right) := \{u : \Gamma \times \mathrm{D} \to \mathbb{R} : \int_{\Gamma} \|u\|^2_{H^1_0(\mathrm{D})}\rho(\boldsymbol{p})d\boldsymbol{p} < \infty\}$$

as follows:

$$\text{find } u \in V_\rho \text{ such that } \langle \nu\nabla u, \nabla\varphi \rangle = \langle f,\varphi \rangle \qquad \text{for all } \varphi \in V_\rho, \tag{10.34}$$

with corresponding inner product $\langle u,v \rangle = \int_{\Gamma}\int_{D} u \cdot v\,dx\,\rho dp$. For the solution of this direct problem (10.34), we used the Stochastic Collocation Method (SCM) combined with the Polynomial-Chaos Expansion (PCE) technique, the so-called pseudo-spectral approach [61].

Consequently, the robust shape optimization problem is given by

$$\inf_{\Omega \in \mathcal{A}} \mathbb{E}\left[\mathcal{J}(\Omega, u(\cdot))\right] + \iota\sqrt{\mathrm{Var}\left[\mathcal{J}(\Omega, u(\cdot))\right]} \tag{10.35a}$$

$$\text{s.t. } \langle \nu\nabla u, \nabla\varphi \rangle = \langle f,\varphi \rangle \qquad \text{for all } \varphi \in V_\rho, \tag{10.35b}$$

with, e.g., $\iota = 3$ (allowing for $\iota \cdot \sigma$−variation [27]). The above problem (10.35) is solved using either the topological sensitivity-based method or the multi-level set method, which were the subject of the previous two sections, compound with the pseudo-spectral approach. The latter is the content of the next section.

10.3.2 Exploiting Stochastic Collocation

Recently, the pseudo-spectral approach [60] has gained the particularly large attention not only to quantify the uncertainty in electrical engineering applications but also to assess the reliability and robustness in the design of electric devices with respect to uncertain parameters, see, e.g., [6, 37, 46, 47, 49, 52].

We follow [61]. Under the assumption that a joint density function ρ exists, we again denote the selected parameters with uncertainty by $\boldsymbol{p}(\boldsymbol{\xi})$ and random variable $\boldsymbol{\xi} = (\xi_1, \ldots, \xi_q)^\top$ with respective distribution. According to the distribution, one selects an orthonormal basis for the PCE, which is optimal. For a uniform distribution, one has the Legendre polynomials.

Given a function $y : \Gamma \to \mathbb{R}$ with $y \in L^2(\Gamma)$, we define a response surface model of y in the form of the truncated series of the PCE [60]

$$y(\boldsymbol{p}) \doteq \sum_{i=0}^{N} \alpha_i \Phi_i(\boldsymbol{p}), \qquad (10.36)$$

where α_i are a priori unknown coefficients, while Φ_i denote the predetermined basis polynomials with the orthonormality property $\mathbb{E}[\Phi_i \Phi_j] = \mathbb{E}[\Phi_i^2] \delta_{ij}$. Additionally, to speed up the calculation of the unknown coefficients α_i, we combine a pseudo-spectral approach with the Stroud quadrature formula of order 3, see, e.g., [4, 61]. Even though, the use of Stroud rules leads to a very small number of quadrature node, it yields as well a fixed accuracy [58]. Here, the basic concept is to run repetitively the deterministic problem, defined by (10.35), in order to compute a solution at each quadrature nodes $\boldsymbol{p}^{(k)} \in \Gamma$, $k = 1, \dots, K$. Then, the multi-dimensional quadrature rule with associated weights w_k can be applied to calculate an approximation of the exact projection of y onto the basis polynomials

$$\alpha_i \doteq \sum_{k=1}^{K} w_k \, y\left(\boldsymbol{p}^{(k)}\right) \Phi_i\left(\boldsymbol{p}^{(k)}\right). \qquad (10.37)$$

Finally, the statistical moments such as the mean and variance are approximated by

$$\mathbb{E}[y(\cdot)] \doteq \alpha_0, \qquad \mathrm{Var}[y(\cdot)] \doteq \sum_{i=1}^{N} |\alpha_i|^2, \qquad (10.38)$$

see [60]. Furthermore, if the PCE coefficients are known, we can deduce global and local sensitivities, which quantify the variations of the output y with respect to the random variable $\boldsymbol{p}$. The local sensitivity is provided by the partial derivatives of the smooth function y evaluated at the mean value of the random value by

$$\left.\frac{\partial \tilde{y}}{\partial p_j}\right|_{\boldsymbol{p}=\overline{\boldsymbol{p}}} = \sum_{i=0}^{N} \alpha_i \frac{\partial \Phi_i}{\partial p_j} \left.\frac{\partial \mathbf{p}}{\partial \xi_j}\right|_{\boldsymbol{p}=\overline{\boldsymbol{p}}}, \quad j = 1, \dots, q, \qquad (10.39)$$

see, e.g., [60], again. Alternatively, the variance-based sensitivity analysis can be performed. Thereby, the Sobol decomposition yields the normalized variance-based sensitivity coefficients [59] in the form

$$S_j := \frac{\mathrm{V}_j^d}{\mathrm{Var}(y)} \quad \text{with} \quad \mathrm{V}_j^d := \sum_{i \in I_j^d} |\alpha_i|^2, \ j = 1, \dots, q, \qquad (10.40)$$

with sets

$$I_j^d := \{j \in \mathbb{N} : \Phi_j(p_1, \ldots, p_q) \text{ is not constant in } p_j \text{ and } \text{degree}(\Phi_i) \leq d_{\text{PC}}\},$$

where d_{PC} is the maximum degree of the polynomials in the truncated PCE (10.36). Notice that $0 \leq S_j \leq 1$. A value S_j close to 1 means a large contribution to the variance.

We can use the same procedure together with a shape derivative, a topological derivative, or derivative of level set function to provide the derivative of the expected value and the standard deviation [15, 42, 48–50].

Algorithm 10.3 Robust algorithm for shape optimization.

1: Formulate the optimization problem with the definition of the objective functional $\mathcal{E}(\Omega(\phi), u)$ (10.11a) and/or the areas constraints $|\Omega_{\text{end}}| = \tau |\Omega_0|$, $\tau \in (0, 1)$.
2: **if** The topological derivative used **then**
3:     Determine the design domain & divide it into voxels.
4: **end if**
5: **if** The multilevel set method used **then**
6:     Determine the design domain & initialize the LSM distributions.
7: **end if**
8: Fix the finite set of random parameters $Q_q = (p_1, \ldots, p_q)$. ▷ For UQ (Uncertainty Quantification)
9: Describe the inputs variations by suitable probability distributions. ▷ For UQ
10: BoolAreaConstraintsSatisfied = False.
11: **while** not BoolAreaConstraintsSatisfied **do**
12:     BoolOptimized = False.
13:     **while** not BoolOptimized **do**
14:         **for** $k = 1, \ldots, K$ **do** ▷ Quadrature points loop in (10.37)
15:             Solve primary and adjoint systems (10.11b) and (10.18) for u and λ.
16:             Calculate statistical moments and their derivatives.
17:             **if** The topological derivative used **then**
18:                 Calculate the TD using (10.24) for each voxel in the design area.
19:                 Sort the TD values and remove material from the design area with respect to the assumed threshold ratio.
20:             **end if**
21:             **if** The multilevel set method used **then**
22:                 Determine $\partial_\phi \mathcal{E}$ based on the velocity field (10.26).
23:                 Calculate the correction of ϕ_i and introduce it to model (10.25).
24:             **end if**
25:         **end for**
26:         BoolOptimized = Objective function reaches the assumed level.
27:     **end while**
28:     Update BoolAreaConstraintsSatisfied
29: **end while**

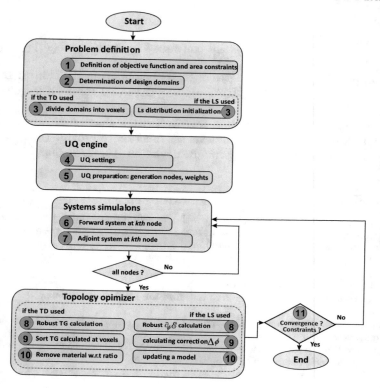

Fig. 10.5: Robust algorithm for shape optimization.

10.3.3 Robust sensitivity-based algorithm for shape and topology optimization

We are now able to formulate our robust sensitivity-based algorithm for shape optimization in Alg. 10.3. To provide a better explanation, the optimization flow is also schematically shown on Fig. 10.5. We remark, that the numbers in Alg. 10.3 refer to Fig. 10.5.

10.4 Results of the UQ Transient Analysis for the Power Transistor Model

In our research as a case study, we used a multi-finger MOSFET power transistor with a stripe cell structure. It consists of several thousands of parallel channels, depicted in Fig. 10.6. The device is stretched in the vertical direction layout with three metal layers as shown on Fig. 10.7. In this structure,

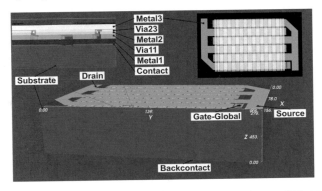

Fig. 10.6: Topology of a power transistor device [47–49].

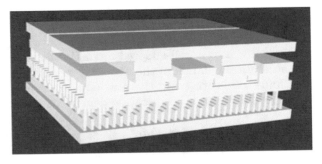

Fig. 10.7: Typical layout of a power transistor with its complex geometry (vertically stretched) [47–49].

the source and drain contacts are located on the top metal finger. A series of metal stripes transports the current from drain to source in each individual channel. This is modelled in terms of the charge density ρ_c and the electric scalar potential V by the Poisson equation and charge transport

$$-\nabla \cdot [\epsilon \nabla V] = \rho_c, \quad \partial_t \rho_c(\theta) - \nabla \cdot \sigma \nabla V = 0, \tag{10.41}$$

and using the material parameters of the electric conductivity σ and the permittivity ϵ on the bounded domain $D \subset \mathbb{R}^3$ and $t \in [0, t_{\text{end}}]$ (with respective boundary conditions and initial conditions).

Moreover, these structures are thermally sensitive. Thus, we need to model the heat evolution. To this end, let T denote the temperature and T_0 the constant environment temperature. The heat conduction reads

$$\partial_t (C_v T) = \nabla \cdot (\kappa \nabla T) + Q_e, \tag{10.42}$$

using thermal capacity C_v, heat conductivity κ and the source term for self-heating Q_e. We remark that the thermal flux is given by $Q = -\kappa \nabla T$.

Now, the above problems (10.41) and (10.42) are coupled via the source term Q_e, which is given by Joule's law $Q_e = \sigma |\nabla V|^2$ and via temperature dependent material parameters.

Finally, we include uncertainties (see Section 10.3.2) and we represent selected parameters via suitable random variables $\boldsymbol{\xi} = (\xi_1, \ldots, \xi_q)^\top$ with corresponding event space Γ (from a suitable probability space), where we assume the existence of a joint probability density. Thus, the respective coupled stochastic thermal-electric problem has the form: for $\theta := (\boldsymbol{x}, t, \boldsymbol{\xi}) \in D \times [0, t_{\text{end}}] \times \Gamma$

$$
\left\{
\begin{aligned}
-\nabla \cdot [\epsilon(\theta) \nabla V(\theta)] &= \rho_c(\theta), \\
\partial_t \rho_c(\theta) - \nabla \cdot (\sigma(\theta) \nabla V(\theta)) &= 0, \\
\partial_t (C_v(\theta) T(\theta)) - \nabla \cdot (\kappa(\theta) \nabla T(\theta)) &= \sigma(\theta) |\nabla V(\theta)|^2 = Q_e(\theta),
\end{aligned}
\right.
\tag{10.43}
$$

for stochastic processes $\rho_c(\theta)$, $V(\theta)$ and $T(\theta)$.

For the solution of (10.43), we used the MAGWEL software [31]. On the one hand, this simulator uses a well-adopted mesh for the substrate, which is crucial for an accurate simulation of the temperature distribution. On the other hand, the Joule self-heating and the heat flow in a metal is modelled together with the following linear temperature-dependent electrical conductivity

$$
\sigma = W_l \sigma_l,
$$

with layer size W_l and layer number l. Analogously, for the respective thermal conductivity and thermal capacitance.

As a result, within the used self-consistent approach, every electric transport inside the MOS channels is dealt with a compact model, which means that the drain to source current flows are described by $I_{\text{DS}} = f(V_{\text{DS}}, V_{\text{GS}})$ using drain-source voltage V_{DS} and gate-source voltage V_{GS}. Moreover, the dissipated power in the channel is computed based on the channel resistance:

$$
Q_e = V_{\text{DS}} \cdot I_{\text{DS}}(V_{\text{DS}}, V_{\text{GS}}).
$$

Consequently, it is possible to solve such a complex system in an efficient way.

Now addressing uncertainties, we assumed that the random parameters are as follows [47]: (cf. Fig. 10.7)

- the electrical conductivity of Metal1, Metal2 and Metal3, denoted by $\sigma_1(\xi_1)$, $\sigma_2(\xi_2), \sigma_3(\xi_3)$,
- the thermal capacitance and the thermal conductivity of the Metal3, represented by $C_{v0}(\xi_4), \kappa_0(\xi_5)$,
- the thickness of Metal1, Via12, Metal2, Via23, Metal3 denoted by $W_1(\xi_6)$, $W_2(\xi_7)$, $W_3(\xi_8)$, $W_4(\xi_9)$, $W_5(\xi_{10})$.

That is, we consider $q = 10$ random variables. In our simulations input variations are model as follows: we used uniform distributions on $[-1, 1]$

with 10% variation around given mean values for $\sigma_1, \sigma_2, \sigma_3$, 5% variation around mean values for C_{v0}, κ_0 and 3% variation around mean values for W_1, W_2, W_3, W_4, W_5.

As a quadrature rule we applied the Stroud-3 formula, which results in $2q = 20$ transient deterministic simulations. To find the statistical moments of relevant output quantities such as I(drain), thermal flux(drain), V(probe) and T(drain), we first provide the deterministic solutions in every generated quadrature grid point $k = 1, \ldots, K$. Next, we apply the multi-dimensional quadrature scheme (10.37) to find the coefficients of the used truncated Legendre PCE. Finally, the expectation and standard deviation are computed by (10.38). The results are shown in Fig. 10.8. Additionally, we calculate the variance, based on the sensitivity analysis (10.40). The results are depicted in Fig. 10.9 and quantify the influences of the input variation. For comparison purpose, we also carried out the mean gradient sensitivity analysis (10.39), which is presented in Fig. 10.10.

This way, it was possible to design an efficient algorithm for solving the coupled electro-thermal problem with random input data (10.43). For this purpose, we used the deterministic PTM-ET field solver by MAGWEL [31] and the DAKOTA package by Sandia National Laboratories [15]. Based on the provided local and global sensitivity analysis, we could identify that the Metal3 layer, has a predominant impact on the output functions, in our example. Furthermore, this information was used for the deterministic and robust optimization of a Power-MOS transistor in order to eliminate a 'hot-spot' phenomenon. It is the topic of the next Section.

10.5 Results for the Robust Topology Optimization of a Steady-State Power Transistor Model

In [33, 48, 49], we dealt with the topology optimization of a power transistor device including geometrical and material uncertainties. On the one hand, we could use the advanced PTM-ET solver (from MAGWEL) [31] for the simulation of the complicated structure of a power MOS device to predict the behavior of the device under critical conditions. On the other hand, our methodology allowed for solving a real engineering problem such as an anomalous failure due to a thermal runway. Specifically, we reduced the thermal instability by optimizing the geometry within the device layout, while taking both the conductive power losses and the shape variations of source/drain into account. In [49] and [48], we mainly focused on a shape/topology optimization problem of a power MOS device with three metal layers under geometrical and material uncertainties to reduce the current density overshoot. This problem occurs in the automotive industry. It yields a stochastic electro-thermal problem, which is described in Section 10.4.

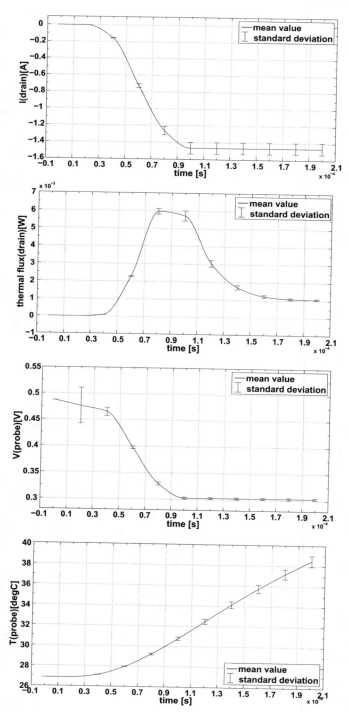

Fig. 10.8: The mean value and standard deviation of chosen output functions due to the input variations $(\sigma_1, \sigma_2, \sigma_3, C_{v0}, K_0, W_1, \ldots, W_5)$.

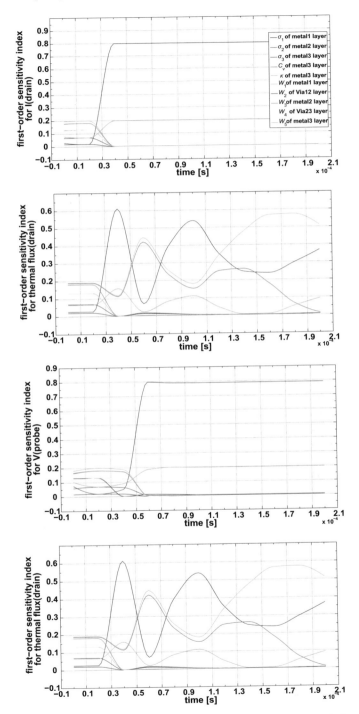

Fig. 10.9: The variance-based sensitivity analysis performed for a power MOS transistor. Due to the normalization, a value close to 1 means the large contribution to the variance.

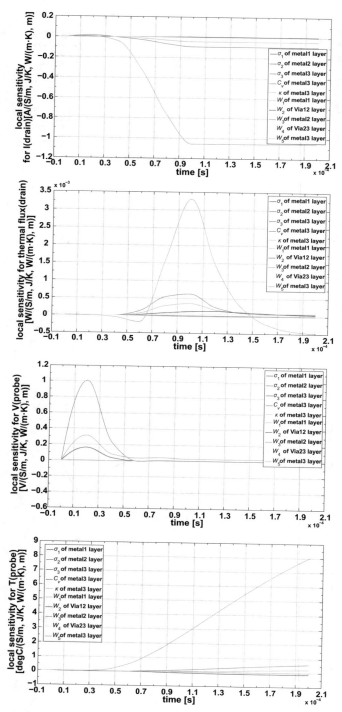

Fig. 10.10: The mean gradient sensitivity analysis for a power MOS transistor.

In particular, the provided sensitivity analysis, presented in the previous section, gave us the opportunity to find those parameters of our power MOS transistor, which are most responsible for variations of the performance functions. This, in turn, allowed us to conduct the optimization of the power MOS transistor in such a way that we did reduce the current density overshoot. Hence, a thermal destruction of the device was avoided.

Here, we consider a random-dependent steady-state counterpart of the system (10.43), which serves as a stochastic constraint in the formulation of the respective robust optimization problem. To this end, we use $\chi := (\boldsymbol{x}, \boldsymbol{\xi}) \in D \times \Gamma$, composed of the space variable $\boldsymbol{x}$ and random vector $\boldsymbol{\xi}$. In terms of ρ_c, V and T, we have

$$\begin{cases} -\nabla \cdot [\epsilon(\chi) \nabla V(\chi)] = \rho_c, \\ -\nabla \cdot [\sigma(\chi) \nabla V(\chi)] = 0, \\ -\nabla \cdot [\kappa(\chi) \nabla T(\chi)] = Q_e(\chi) = \sigma(\chi) |\nabla V(\chi)|^2, \end{cases} \qquad (10.44)$$

equipped with random Dirichlet boundary conditions:

$$V_D(\chi) = V_{D0}(\boldsymbol{x}, \boldsymbol{\xi}) \quad \text{on } \Gamma_D, \qquad V_S(\chi) = V_{S0}(\boldsymbol{x}, \boldsymbol{\xi}) \quad \text{on } \Gamma_S,$$

which describes the potentials of the drain Γ_D and source pads Γ_S, respectively. Alternatively, one can consider Neumann boundary conditions

$$\partial_n J_D(\chi) = J_{D0}(\boldsymbol{x}, \boldsymbol{\xi}) \quad \text{on } \Gamma_D, \qquad \partial_n J_S(\chi) = J_{S0}(\boldsymbol{x}, \boldsymbol{\xi}) \quad \text{on } \Gamma_S.$$

Either way, its solution enables to investigate the propagation of uncertainties through a 3D model, which affect the yield and the performance of, e.g., a power transistor. In particular, for our power device in Fig. 10.7, we have as stochastic parameters in (10.44): the conductivity of the Metal3 layer, σ_3, the thickness of the Metal2, W_2, the thermal conductivity of the Via12, κ, the drain and source contacts are considered, see also Fig. 10.11. That is,

$$\boldsymbol{p}(\boldsymbol{\xi}) = (\sigma_3(\xi_1), W_2(\xi_2), \kappa(\xi_3), V_D(\xi_4), V_S(\xi_5))^\top$$

with respective probability spaces.

We aim at reducing the current density overshoots in the area of the contact layer of the power device. The corresponding design variable is defined as $\Omega = \Omega_{\text{FD}} \cup \Omega_{\text{FS}} \cup \Gamma_D \cup \Gamma_S$ and it consists of the Metal3 finger structures for drain and source, as well as the drain and source contacts, respectively (cf., Fig. 10.12). Thus, the basic random-dependent cost functional reads

$$\mathcal{P}(\Omega, V(\chi)) = w_1 \int_{D_1} Q_e[V(\chi)] \, d\boldsymbol{x} + w_2 \int_{\Gamma_1} P_h[V(\chi)] \, d\boldsymbol{s} \qquad (10.45)$$

with the dissipated power Q_e analysed in the area of the Metal3 layer $D_1 \subset \mathbb{R}^3$, and P_h denotes the power density on the boundary Γ_1 of the source and drain pads. Last, the weights w_1 and w_2 are used to incorporate a priori information about the objectives.

Finally, the robust optimization problem constrained by the stochastic PDEs is given by

$$\inf_{\Omega \in \mathcal{A}} \; \mathcal{E}(\Omega, V(\chi)) = \mathbb{E}\left[\mathcal{P}(\Omega, V(\chi))\right] + \iota\sqrt{\mathrm{Var}\left[\mathcal{P}(\Omega, V(\chi))\right]}$$

$$\text{s.t.} \quad \begin{cases} -\nabla \cdot \left[\epsilon(\chi)\nabla V(\chi)\right] = \rho_c, \\ -\nabla \cdot \left[\sigma(\chi)\nabla V(\chi)\right] = 0, \\ -\nabla \cdot \left[\kappa(\chi)\nabla T(\chi)\right] = Q_e(\chi), \\ u_D(\chi) = u_{D0}(\chi), \\ u_S(\chi) = u_{S0}(\chi), \end{cases} \qquad (10.46)$$

where $\iota = 3$ was taken.

The SCM, based on the PCE [60], provides a response surface model to estimate the expectation and standard deviation

$$\mathbb{E}\left[\mathcal{P}(\Omega, V(\chi))\right], \qquad \sqrt{\mathrm{Var}\left[\mathcal{P}(\Omega, V(\chi))\right]}.$$

To find the solution of the robust formulation (10.46), the sensitivity-based topology optimization method was applied. In this way, we could reduce the hot spot phenomena in an efficient way, see Fig. 10.11, Fig. 10.12 and Fig. 10.13 (see also [48,49]). Additionally, we present also the iterative change of the total conductance, the total current and the total power during the optimization in Figs. 10.14–10.16, respectively.

Summarizing, we have implemented an algorithm for the robust and deterministic shape optimization of a Power-MOS transistor. For this purpose, the PCE-based Stochastic Collocation (SC) Method has been incorporated in the topological derivative method. This allows for finding the optimized configuration of the Power-MOS transistor under geometrical and material uncertainties. As a result of the optimization, the current density overshoot has been eliminated in contact layer. Additionally, we obtained a decrease of temperature in the metal3 layer by 32 °C, while for the contact layer the temperature reduction of 8 °C was achieved.

10.6 An extension to Shape and Topology Optimization of a Permanent-Magnet Machine under Uncertainties

Next, we show our results on a shape optimization problem, which occurs in the quasilinear magnetostatic. This kind of the optimization problems are treated here as the natural extension of the proposed methodology in

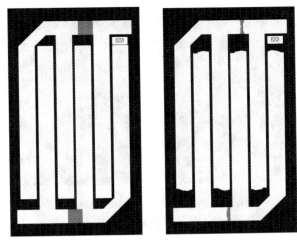

Fig. 10.11: Shapes of the metal3 layer as well as the drain and source pads (red color) for the initial configuration (left), and for the optimized model in the 18th iteration (right) [49].

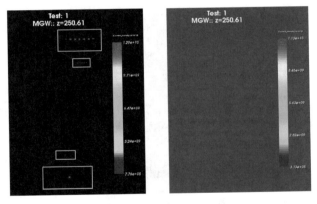

Fig. 10.12: The current density for the initial model in the contact layer - hot spots, represented, by 8 red dots in the enlarged blue box (left) and the current density in the contact layer after optimization (right) [48, 49].

Section 10.5 to the case, where a nonlinear interface problem is considered. We intend to underline the main differences between a linear and nonlinear shape optimization. Results have been presented in [42, 46, 50].

Our ultimate goal was to conduct a shape optimization for a permanent-magnet machine, while including material nonlinearity and its uncertainties to minimize noise and vibration caused by the cogging torque. The mathematical model was described by a quasilinear curl-curl equation in a two

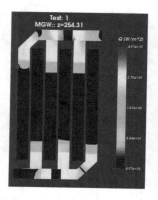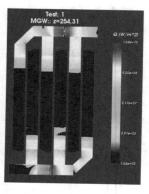

Fig. 10.13: The heat flux density (HFD) in the Metal3 layer for initial model (left) and the HFD in the optimized topology (right) [49].

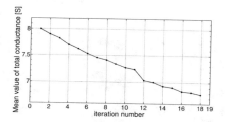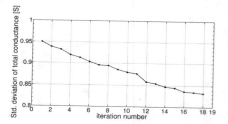

Fig. 10.14: The total conductance as a function of the iteration steps: mean value (left) and standard deviation (right) [49].

dimensional setting $\boldsymbol{x} \in D \subset \mathbb{R}^2$ with sufficiently regular boundaries $\partial D \in C^2$. It becomes the nonlinear Poisson equation if considered for the third component of the magnetic vector potential u. Its weak formulation is similar to (10.8) and reads as:

find $u \in H_0^1$ such that $\left(\nu(\boldsymbol{x}, |\nabla u|^2)\nabla u, \nabla \varphi\right) = (f, \varphi)$ for all $\varphi \in H_0^1$.
$$(10.47)$$

Here the function f is given by $f = J + \nu_1 \nabla \cdot \mathbf{M}$, where J, $\mathbf{M}$ and ν_1 denote the current density, the magnetization and the reluctivity of a permanent magnet, respectively. In general, for soft iron material, the reluctivity $\nu : D \times \mathbb{R}_0^+ \to \mathbb{R}^+$ can be modelled as

$$\nu\left(\boldsymbol{x}, |\nabla u|^2\right) = \begin{cases} \nu_0\left(\boldsymbol{x}\right), & \boldsymbol{x} \in D_{\mathrm{air}}, \\ \nu_1\left(\boldsymbol{x}\right), & \boldsymbol{x} \in D_{\mathrm{PM}}, \\ \nu_2\left(\boldsymbol{x}, |\nabla u|^2\right) & \boldsymbol{x} \in D_{\mathrm{FE}}. \end{cases} \qquad (10.48)$$

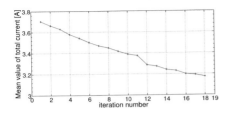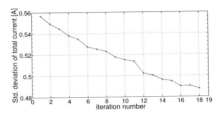

Fig. 10.15: The total current as a function of the iteration steps: mean value (left) and standard deviation(right) [49].

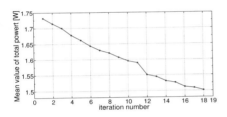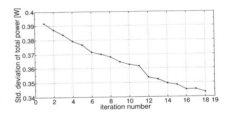

Fig. 10.16: The total power as a function of the iteration steps: mean value (left) and standard deviation (right) [49].

with vacuum reluctivity ν_0 and computational domain $D = D_{air} \cup D_{PM} \cup D_{FE}$ composed of the area of air, the region of the permanent magnet and the iron domain, respectively.

Now, we focus on modifying the material distribution in the area of the iron made rotor pole, denoted by $\Omega_{FE} \subset D_{FE}$ as well as the shape of permanent magnet pole, denoted by $\Omega_{PM} \subset D_{PM}$, represented by the nonlinear coefficient $\nu_2\left(\boldsymbol{x}, |\nabla u|^2\right)$. In fact, we seek optimal shapes of Ω_{FE} and Ω_{PM}, which mitigate the cogging torque (in the overall region D).

For this purpose, we construct the random- and domain-dependent cost functional with $\Omega = \Omega_{FE} \cup \Omega_{PM}$ in terms of the magnetic energy,

$$\mathcal{W}(\Omega, u(\chi)) = \frac{1}{2}\int_D \nu |\nabla u(\chi)|^2 d\boldsymbol{x}, \qquad (10.49)$$

which is superior for the minimization of the cogging torque, because it is a self-adjoint quantity. Thus, the robust shape optimization problem for

$$V_\rho = L_\rho^2(\Gamma) \otimes L^2\left(H_0^1(\mathrm{D})\right) := \{u : \Gamma \times \mathrm{D} \to \mathbb{R} : \int_\Gamma \|u\|_{H_0^1(\mathrm{D})}^2 \rho(\boldsymbol{p}) d\boldsymbol{p} < \infty\}$$

is given by

$$\inf_{\Omega \in \mathcal{A}} \mathcal{E}(\Omega, u(\chi)) = \mathbb{E}[\mathcal{W}(\Omega, u(\chi))] + \iota\sqrt{\mathrm{Var}[\mathcal{W}(\Omega, u(\chi))]} \qquad (10.50a)$$

s.t. u satisfies $\left(\nu(|\nabla u(\chi)|^2)\nabla u(\chi), \nabla\varphi(\chi)\right) = (f, \varphi(\chi))$ for all $\varphi \in V_\rho$
$$(10.50b)$$

with $\iota = 3$. For the solution of the above-mentioned problem the multilevel set method, described in Section 10.2.4, was used. The shape derivative is given by

$$d\mathcal{E}(\Omega) = d\mathbb{E}[\mathcal{W}(\Omega, u(\chi))] + \iota d\sqrt{\mathrm{Var}[\mathcal{W}(\Omega, u(\chi))]}$$

$$= \left([v^+(|\nabla u(\chi)|^2) - v^-(|\nabla u(\chi)|^2)]\nabla u(\chi)\, \mathbf{V} \cdot \mathbf{n}_{\Omega_1}, \nabla\lambda(\chi)\right)_{\partial\Omega_{\mathrm{FE}}}$$
$$(10.51a)$$

$$- \left(v(\chi)_{\mathrm{PM}}[\mathbf{M}^+ - \mathbf{M}^-]\mathbf{V} \cdot \mathbf{n}_{\Omega_2}, \nabla\lambda(\chi)\right)_{\partial\Omega_{\mathrm{PM}}}, \qquad (10.51b)$$

where an adjoint variable $\lambda \in V_\rho$ is the solution of the random-dependent dual problem for given test function $\phi \in V_\rho$, defined as

$$a(\lambda, \phi) = (d\mathbb{E}[\mathcal{W}(\Omega, u(\chi))] + \iota d\sqrt{\mathrm{Var}[\mathcal{W}(\Omega, u(\chi))]}, \phi). \qquad (10.52)$$

Here,

$$a(\lambda, \phi) = \left(\nu\left(|\nabla u|^2\right)\nabla\lambda, \nabla\phi\right) + 2\left(\nu'\left(|\nabla u|^2\right)\nabla\lambda \cdot \nabla u\nabla u, \nabla\phi\right),$$

was derived using shape and material derivative methods in [13].

Finally, to solve the optimization problem defined by (10.50), the multilevel set method was used [41, 42, 50]. Here the boundaries $\partial\Omega_{\mathrm{FE}}$ and $\partial\Omega_{\mathrm{PM}}$ are represented by a zero level set of functions ϕ_1 and ϕ_2. Thus, to minimize the shape functional

$$\mathcal{E}(\Omega, u(\chi)) = \mathbb{E}[\mathcal{W}(\Omega, u(\chi))] + \iota\sqrt{\mathrm{Var}[\mathcal{W}(\Omega, u(\chi))]},$$

the interfaces $\partial\Omega_{\mathrm{FE}}$ and $\partial\Omega_{\mathrm{PM}}$ need to be moved in the direction of V_n, defined by (10.51). This is achieved by solving the Hamilton-Jacobi equation

$$\phi_{t,i} + V_n|\nabla\phi_i| = 0.$$

The result for the robust shape optimization is depicted in Fig. 10.17 (see also [42, 50], where we can see the Electrically Controlled Permanent Magnet Excited Synchronous Machine (ECPSM) structure before and after shape optimization. The chosen output quantities such as the magnetic flux density under the magnet and iron poles, the cogging torque, the back electromotive force, the electromagnetic torque were shown on Fig. 10.18 and Fig. 10.19.

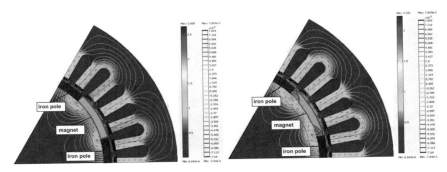

Fig. 10.17: The configuration of the ECPSM before (left) and after (right) optimization [42, 50].

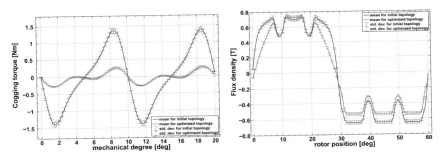

Fig. 10.18: Mean and standard deviation for the initial and for the optimized topology of the ECPSM [50]. Left: The magnetic flux density in the air-gap. Right: The cogging torque.

10.7 Conclusion

We have demonstrated how to combine efficiently the stochastic collocation method with the multi-level set method in order to accomplish the robust topology optimization of a permanent magnet PM synchronous machine. For this purpose, we have taken variations with respect to manufacturing tolerances/imperfections into account by assuming a random field for the reluc-

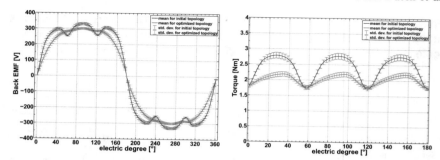

Fig. 10.19: Mean and standard deviation for the initial and for the optimized topology of the ECPSM [50]. Left: The back EMF. Right: The electromagnetic torque.

tivities. The provided optimized configuration of a PM machine yields the significant reductions of both the rms of the cogging torque CT (82%) and the mean value of the standard deviation (43%). Moreover, the waveform of the back electromagnetic force has been considerably improved by 32%, while the torque ripple has been reduced by 52%. As a result, it allows for suppressing the level of noise and vibrations in the considered electric machine. We studied Robust Optimization based on minimizing $\mathbb{E} + \kappa\sqrt{\mathrm{Var}}$, for $\kappa \in [1,6]$. By iteratively shifting the center of the assumed probability density function for the best setting, the approach was also successfully applied to minimize unwanted couplings from digital parts to analogue parts on an RFIC chip [43–45].

References

1. AMSTUTZ, S.: *Topological sensitivity analysis for some nonlinear PDE systems*. J. de Mathématiques Pures et Appliquées, **85**, 540–557, 2006.
2. AMSTUTZ S., AND ANDRÄ, H.: *A new algorithm for topology optimization using a level-set method*. Journal of Computational Physics, **216**, 573–588, 2006.
3. BAN, Y., AND PAN, D.Z.: *Modeling of Layout Aware Line-Edge Roughness and Poly Optimization for Leakage Minimization*. IEEE Journal on Emerg. and Select. Topics in Circuits and Systems, **1**, 150–159, 2011.
4. BENNER, P., AND SCHNEIDER, J.: *Uncertainty Quantification for Maxwell's equations using Stochastic Collocation and Model Order Reduction*. J. Uncertainty Quantification **5** 195–208, 2015.
5. BIONDI, T., GRECO, G., BAZZA, G., AND RINAUDO, S.: *Effect of layout parasitics on the current distribution of power MOSFETs operated at high switching frequency*. J. Comput. Electron., **5** 149–153, 2006.
6. BARTEL, A., DE GERSEM, H., HÜLSMANN, T., RÖMER, U., SCHÖPS, S., AND WEILAND, T.: *Quantification of uncertainty in the field quality of magnets originating from material measurements*. IEEE Trans.on Magn., **49**, 2367–2370, 2013.

7. CÉA, J., GIOAN, A., AND MICHEL, J.: *Adaptation de la méthode du gradient à un probléme d'identification de domaine*. In: Lecture Notes in Computer Science, **11**, Springer, Berlin, 371–402, 1974. Open access: https://link.springer.com/content/pdf/10.1007%2F3-540-06769-8_19.pdf.

8. CÉA, J.: *Conception optimale ou identification de formes, calcul rapide de la dérivée directionelle de la fonction coût*. ESAIM M2AN Math. Model. Numer. Anal., **20**, 371–402, 1986. Open access:https://doi.org/10.1051/m2an/1986200303711

9. CÉA, J., GARREAU, S., GUILLAUME, P., AND MASMOUDI, M.: *The shape and topological optimizations connection*. Comput. Methods in Appl. Mech. and Engrg., **118**, 713–726, 2000.

10. CHAN, T.F., AND TAI, X.-C.: *Level set and total variation regularization for elliptic inverse problems with discontinuous coefficients*. Journal of Computational Physics, **193**, 40–66, 2004.

11. CHAABANE, S., MASMOUDI, M., AND MEFTAHI, H.: *Topological and shape gradient strategy for solving geometrical inverse problems*. J. Math. Anal. Appl., **400**, 724–742, 2013.

12. CIMRÁK, I.: *Inverse thermal imaging in materials with nonlinear conductivity by material and shape derivative method*. Mathematical Method in the Applied Science, **34**, 2303–2317, 2011.

13. CIMRÁK, I.: *Material and shape derivative method for quasi-linear elliptic systems with applications in inverse electromagnetic interface problems*. SIAM Journal of Numerical Analysis, **50**, 1086–1110, 2012.

14. CONSOLI, A., GENNARO, F., TESTA, A., CONSENTINO, G., FRISINA, F., LETOR, R., AND MAGRI A.: *Thermal Instability of Low Voltage Power-MOSFET's*. IEEE Trans. on Pow. Electron., **15**, 575–581, 2000.

15. DAKOTA 6.3, https://dakota.sandia.gov/.

16. DELFOUR, M.C., AND ZOLESIO, J.-P.: *Shapes and Geometries*. Adv. Des. Control 22, 2nd ed., SIAM, Philadelphia, 2011.

17. ESCHENAUER, H.A., KOBELEV, V.V., AND SCHUMACHER, A.: *Bubble method for topology and shape optimization of structures*. Structural optimization, **8**, 42–51, 1994.

18. GANGL, P., AMSTUTZ, S., AND LANGER, U.: *Topology optimization of electric motor using topological derivative for nonlinear magnetostatic*. IEEE Trans. on Magn. **52** 7201104, 2016.

19. HAUG, E.J., CHOI, K.K., AND KOMKOV, V.: *Design Sensitivity Analysis of Structural System*. Academic Press, New York, 1986.

20. HETTLICH, F., AND RUNDEL, W.: *Identification of a discontinuous source in the heat equation*. Inverse Problems **17** 1465–1482, 2001.

21. IL'IN, A.M.: *Matching of asymptotic expansions of solutions of boundary value problems*. Translations of Mathematical Monographs 102, AMS, Providence, RI, 1992.

22. ITO, K., KUNISCH, K., AND LI, Z.: *Level-set function approach to an inverse interface problem*. Inverse Problem, **17**, 1225–1242, 2001.

23. ITO, K., KUNISCH, K., AND PEICHL, G.H.: *Variational approach to shape derivatives*. ESAIM Control Optim. Calc. Var., **14**, 517–539, 2008.

24. KIM, D.H., LOWTHER D.A., AND SYKULSKI, J.K.: *Efficient force calculations based on continuum sensitivity analysis*. IEEE Trans.and Magn., **41**, 1404–1407, 2005.

25. KIM, D.H., PARK, I.H., SHIN, M.CH., AND SYKULSKI, J.K.: *Generalized Continuum Sensitivity Formula for Optimum Design of Electrode and Dielectric Contours*. IEEE Trans. on Magn., **39**, 1281–1284, 2003.

26. KIM, D., SYKULSKI, J., AND LOWTHER, D.: *The implications of the use of composite materials in electromagnetic device topology and shape optimization*. IEEE Trans. on Magn. **45**, 1154–1156, 2009.

27. KOCH, P.N., YANG, R.J., AND GU, L.: *Design for six sigma through robust optimization*. Structural and Multidisciplinary Optimization, **26**, 235–248, 2004.

28. KOMKOV, V.: *Sensitivity of Functionals with Applications to Engineering Sciences.* Springer-Verlag, Berlin Heidelberg New York Tokyo, 1984.

29. KOZONO, H., AND USHIKOSHI, E.: *Hadamard Variational Formula for the Green's Function of the Boundary Value Problem on the Stokes Equations.* Archive for Rational Mechanics and Analysis, **208**, 1005–1055, 2013.

30. LI, M., AND LOWTHER D.A.: *Topological Sensitivity Analysis for Steady State Eddy Current Problems with an Application to Nondestructive Testing.* IEEE Trans. on Magn. **47**, 1294–1297, 2011.

31. MAGWEL NV, Leuven, Belgium, 2018. http://www.magwel.com/.

32. MASMOUDI, M., POMMIER J., AND SAMET, B.: *The topological asymptotic expansion for the Maxwell equations and some applications.* Inverse Problems, **21**, 547–564, 2005.

33. TER MATEN, E.J.W., PUTEK, P., GÜNTHER, M., PULCH, R., TISCHENDORF, C., STROHM, C., SCHOENMAKER, W., MEURIS, P., DE SMEDT, B., BENNER, P., FENG, L., BANAGAAYA, N., YUE, Y., JANSSEN, R., DOHMEN, J.J., TASIĆ, B., DELEU, F., GILLON, R., WIEERS, A., BRACHTENDORF, H.-G., BITTNER, K., KRATOCHVÍL, T., PETŘZELA, J., ŠOTNER, R., GÖTTHANS, T., DŘÍNOVSKÝ, J., SCHÖPS, S., DUQUE GUERRA, D.J., CASPER, T., DE GERSEM, H., RÖMER, U., REYNIER, P., BARROUL, P., MASLIAH, D., AND ROUSSEAU, B.: *Nanoelectronic COupled Problems Solutions – nanoCOPS: Modelling, Multirate, Model Order Reduction, Uncertainty Quantification, Fast Fault Simulation.* Journal Mathematics in Industry 7:2, 2016. Open access: http://dx.doi.org/10.1186/s13362-016-0025-5.

34. MOHANTY, S.P., KOUGIANOS, E.: *Incorporating Manufacturing Process Variation Awareness in Fast Design Optimization of Nanoscale CMOS VCOs.* IEEE Trans. on Semicond. Manufact., **27**, 22–31, 2014.

35. OKAMOTO, Y., AND TAKAHASHI, N.: *A novel topology optimization of nonlinear magnetic circuit using ON/OFF method.* Trans. on Fund. and Mat., **125**, 549–553, 2005.

36. OSHER, S.J., AND SETHIAN, J.A.: *Fronts propagating with curvature dependent speed: algorithms based on Hamilton-Jacobi formulations.* J. Comput. Phys, **79**, 12–49, 1988.

37. PULCH, R.: *Stochastic collocation and stochastic Galerkin methods for linear differential algebraic equations.* J. of Comput. and Appl. Math., **262**, 281–291, 2014.

38. PULCH, R., PUTEK, P., DE GERSEM, H., AND GILLON, R.: *Identification of probabilistic input data for a glue-die-package problem.* In: QUINTELA, P., BARRAL, P., GÓMEZ, D., PENA, F.J., RODRÍGUEZ, J., SALGADO, P., AND VÁZQUEZ-MENDÉZ, M.E. (EDS.): *Progress in Industrial Mathematics at ECMI 2016.* Series Mathematics in Industry Vol. 26, Springer International Publishing AG, pp. 255–262, 2017. http://dx.doi.org/10.1007/978-3-319-63082-3_40.

39. PUTEK, P.: *Mitigation of the cogging torque and loss minimization in a permanent magnet machine using shape and topology optimization.* Engineering Computation **33**, 831–854, 2016.

40. PUTEK, P.: *Nonlinear magnetoquasistatic interface problem in a permanent-magnet machine with stochastic partial differential equation constraints.* Engineering Optimization, pp. 1-24, 2019. https://doi.org/10.1080/0305215X.2019.1577403.

41. PUTEK, P., CREVECOEUR, G., SLODIČKA, M., VAN KEER, R., VAN DE WIELE, B., AND DUPRÉ, L.,: *Space mapping methodology for defect recognition in eddy current testing – type NDT.* COMPEL – The international journal for computation and mathematics in electrical and electronic engineering, 31-3, pp. 881–894, 2012. http://dx.doi.org/10.1108/03321641211209771.

42. PUTEK, P., GAUSLING, K., BARTEL, A., GAWRYLCZYK, K.M., TER MATEN, E.J.W., PULCH, R., AND GÜNTHER, M.: *Robust topology optimization of a Permanent Magnet synchronous machine using level set and stochastic collocation methods.* In: BARTEL, A., CLEMENS, M., GÜNTHER, M., AND TER MATEN, E.J.W. (EDS.):

Scientific Computing in Electrical Engineering – SCEE 2014. Series Mathematics in Industry, Vol. 23, Springer, Berlin, 233–242, 2016. http://dx.doi.org/10.1007/978-3-319-30399-4_23.

43. PUTEK, P., JANSSEN, R., NIEHOF, J., TER MATEN, E.J.W., PULCH, R., GÜNTHER, M., AND TASIĆ, B.: *Robust optimization of an RFIC isolation problem under uncertainties.* In: LANGER, U., AMRHEIN, W., AND ZULEHNER, W. (EDS.): *Scientific Computing in Electrical Engineering - SCEE 2016.* Series Mathematics in Industry, Vol. 28, Springer International Publishing AG, 177–186, 2018. http://dx.doi.org/10.1007/978-3-319-75538-0_17.

44. PUTEK, P., JANSSEN, R., NIEHOF, J., TER MATEN, E.J.W., PULCH, R., TASIĆ, B., AND GÜNTHER, M.: *Nanoelectronic coupled problem solutions: Uncertainty Quantification of RFIC Interference.* In: QUINTELA, P., BARRAL, P., GÓMEZ, D., PENA, F.J., RODRÍGUEZ, J., SALGADO, P., AND VÁZQUEZ-MÉNDEZ, M.E. (EDS.): *Progress in Industrial Mathematics at ECMI 2016.* Series Mathematics in Industry, Vol. 26, Springer International Publishing AG, 271–279, 2018. http://dx.doi.org/10.1007/978-3-319-63082-3_42.

45. PUTEK, P., JANSSEN, R., NIEHOF, J., TER MATEN, E.J.W., PULCH, R., TASIĆ, B., AND GÜNTHER, M.: *Nanoelectronic COupled Problems Solutions: Uncertainty Quantification for Analysis and Optimization of an RFIC Interference Problem.* Journal Mathematics in Industry 8:12, 2018. Open access: http://dx.doi.org/10.1186/s13362-018-0054-3.

46. PUTEK, P., TER MATEN, E.J.W., GÜNTHER M., AND SYKULSKI, J.K.: *Variance-Based Robust Optimization of a Permanent Magnet Synchronous Machine.* IEEE Transactions on Magnetics, **54:3**, 2018 (article 8102504, 2017). http://dx.doi.org/10.1109/TMAG.2017.2750485.

47. PUTEK, P., MEURIS, P., GÜNTHER, M., TER MATEN, E.J.W., PULCH, R., WIEERS, A., AND SCHOENMAKER, W.: *Uncertainty quantification in electro-thermal coupled problems based on a power transistor device.* IFAC-PapersOnLine, **48-1**, 938–939, 2015. Open access: http://dx.doi.org/10.1016/j.ifacol.2015.05.206.

48. PUTEK, P., MEURIS, P., PULCH, R., TER MATEN, E.J.W., GÜNTHER, M., SCHOENMAKER, W., DELEU, F., AND WIEERS, A.: *Shape optimization of a power MOS device under uncertainties.* In: Proceedings of 2016 Design, Automation & Test in Europe Conference & Exhibition (DATE), Dresden, 319–324, 2016. http://dx.doi.org/10.3850/9783981537079_0998.

49. PUTEK, P., MEURIS, P., PULCH, R., TER MATEN, E.J.W., SCHOENMAKER, W., AND GÜNTHER, M.: *Uncertainty Quantification for Robust Topology Optimization of Power Transistor Devices.* IEEE Transactions on Magnetics, **52** 1–4, 2016.

50. PUTEK, P., PULCH, R., BARTEL, A., TER MATEN, E.J.W., AND GÜNTHER, M., AND GAWRYLCZYK, K.M.: *Shape and topology optimization of a permanent-magnet machine under uncertainties.* Journal of Mathematics in Industry, **6:11**, 2016. Open access: http://dx.doi.org/10.1186/s13362-016-0032-6.

51. PUTEK, P., SLODIĆKA, M., PAPLICKI, P., AND PAŁKA, R.: *Minimization of Cogging Torque in Permanent Magnet Machines Using the Topological Gradient and Adjoint Sensitivity in Multi-objective Design.* International Journal of Applied Electromagnetics and Mechanics, **38**, 933–940, 2012.

52. RÖMER, U. SCHÖPS, S., AND WEILAND, T.: *Approximation of moments for the nonlinear magnetoquasistatic problem with material uncertainties.* IEEE Trans. on Magn. **50**, 417–420, 2014.

53. SHUE, J.L., AND LEIDECKER, H.W.: *Power MOSFET Thermal Instability operation characterization support.* Report NASA/TM-2010-216684, 2010. Open access. https://ntrs.nasa.gov/archive/nasa/casi.ntrs.nasa.gov/20100014777.pdf.

54. SIMON, J.: *Differentiation with respect to the domain in boundary value problems.* Numerical Functional Analysis and Optimzation, **2** 649–687, 1980.

55. SOKOLOWSKI, J., AND ZALESIO, P.: *Introduction to Shape Optimization: Shape Sensitivity Analysis.* Springer, New York, 1992.

56. SOKOŁOWSKI, J., AND ŻOCHOWSKI, A.: *Topological derivatives for elliptic problems.* Inverse Problems, **15**, 123-124, 1999.

57. STURM, K.: *Minimax Lagrangian approach to the differentiability of nonlinear PDE constrained shape functions without saddle point assumption.* SIAM Journal on Control and Optimization, **53**:2017–2039, 2015.

58. STROUD, A.H.: *Approximate Calculation of Multiple Integrals.* Prentice-Hall Inc., Englewood Cliffs, NJ, USA, 1971.

59. SUDRET, B.: *Global sensitivity analysis using polynomial chaos expansions.* Rel. Eng. Syst. Safety, **93** 964–979, 2008.

60. XIU, D.: *Numerical Methods for Stochastic Computations: a Spectral Method Approach.* Princeton University Press, 2010.

61. XIU, D.: *Efficient Collocational Approach for Parametric Uncertainty Analysis.* Commun. in Comput. Phys. **2**, 293–309, 2007.

62. YAMADA, T., IZUI, K., NISHIWAKI, S., AND TAKEZAWA, A.: *A topology optimization method based on the level set method incorporating a fictitious interface energy.* Computer Methods in Applied Mechanics and Engineering, **199**, 2876–2891, 2010.

Chapter 11
Going from Parameter Estimation to Density Estimation

Alessandro Di Bucchianico

Abstract A common approach to model variability in integrated circuits is to select a normal distribution for input variables and express variability as a non-linear function of the input variables. Even for simple non-linear functional forms as quadratic polynomials this causes the variability to no longer be normally distributed. It is thus important to be able to estimate the probability distribution of the output. In this chapter we give a brief introduction to the statistical theory of density estimation, which allows to estimate the density of a probability distribution without assuming a specific form of this density. In order to give a self-contained story accessible for non-statisticians, we first present the basic definitions and results of parameter estimation. Then we go beyond parametric estimation by discussing kernel density estimators in detail. We will indicate links with common notions from mathematical analysis such as convolution, Fourier analysis and approximation theory.

11.1 Introduction

A common approach to model variability is to select a parametrized family of probability distributions for parameters of interest and then use data to determine the most appropriate member of that family of probability distributions. This is the approach taken in Uncertainty Quantification. An explicit example of this can be found in Chapter 12 of this book [9], where this occurs in the context of a nanoelectronics problem. This approach is called parameter estimation in statistics. An example is to assume normal distributions and then select the normal distribution with mean equal to the sample mean and variance equal to the sample variance. This approach can be justified by

Alessandro Di Bucchianico
Eindhoven University of Technology, the Netherlands, e-mail: A.D.Bucchianico@tue.nl

© Springer Nature Switzerland AG 2019   261
E. J. W. ter Maten et al. (eds.), *Nanoelectronic Coupled Problems Solutions*,
Mathematics in Industry 29, https://doi.org/10.1007/978-3-030-30726-4_11

the maximum likelihood method, which can be proven to be optimal in an asymptotic sense for general probability distributions under mild regularity conditions (see e.g., [1, Section 9.4]).

11.2 Parameter estimation

In this section we introduce the simple case of statistical estimation when we directly observe variables without a transformation from input variables to an output variable. In practice one often uses classes of probability distributions like the normal distributions. Such classes depend on one or more parameters. It is the task of statistics to choose and validate choices of classes of probability models and given such a choice, to extract as well as possible information on these parameters from data. Assume n independent identically distributed random variables[1] X_k, $k = 1, \ldots, n$ (such a set of random variables is called a random sample in statistics). We denote their common finite mean and finite variance by μ and σ^2, respectively. The sample mean $\widehat{\mu}_n$[2] is defined by

$$\widehat{\mu}_n = \frac{1}{n} \sum_{k=1}^{n} X_k. \tag{11.1}$$

It is sometimes useful to compute the sample mean sequentially according to the recursive formula

$$\widehat{\mu}_n = \frac{1}{n} \left((n-1)\widehat{\mu}_{n-1} + X_n \right). \tag{11.2}$$

Since $\widehat{\mu}_n$ depends on the random sample $X_1, \ldots, X_n$, it is a random variable too. A random variable (or random vector in a multidimensional setting) like $\widehat{\mu}_n$ that is constructed to get an idea of a theoretical, unknown parameter (here μ) is called an estimator. The observed value of an estimator is called estimate. This distinction is similar to the common distinction in analysis between a function f and a function value $f(x)$. An estimate is thus a number or in a multidimensional setting a vector. Note the difference between the daily use of the verb estimate and the statistical use here. For any sample from a distribution with a finite mean, the estimator (11.1) is always (i.e., not depending on the actual probability distribution of the X_i's as long as all expectations are finite) *unbiased*, meaning,

---

[1] The terminology random variable is a historical misnomer. The term random function should be more appropriate, but random functions have obtained the meaning of an indexed collection of random variables.

[2] In statistics, this estimator is usually denoted as $\overline{X}_n$. The common usage is to use Greek letters only for theoretical, true quantities like the mean and variance.

$$E\left(\widehat{\mu}_n\right) = E\left(\frac{1}{n}\sum_{k=1}^{n}X_k\right) = \frac{1}{n}\sum_{k=1}^{n}E\left(X_k\right) = \frac{1}{n}n\mu = \mu, \qquad (11.3)$$

where E is the expectation operator. The expectation expresses whether there is a systematic deviation from the true, unknown mean. In order to assess the accuracy (fluctuations) of an estimator, we need to consider the variance too:

$$\text{Var}\left(\widehat{\mu}_n\right) = \text{Var}\left(\frac{1}{n}\sum_{k=1}^{n}X_k\right) = \frac{1}{n^2}\sum_{k=1}^{n}\text{Var}\left(X_k\right) = \frac{1}{n^2}n\sigma^2 = \frac{\sigma^2}{n}. \qquad (11.4)$$

The ideal estimator is unbiased (expectation equal to the target parameter) with minimal variance. However, there are cases known in which a slightly unbiased estimator has a much lower variance than the optimal unbiased estimator and is thus a preferred estimator (one often refers to this as the bias-variance trade-off). Therefore the MSE (Mean Squared Error), which can be shown to be the sum of the variance and the square of the bias, is often used to measure the quality of an estimator.

If Y is a random variable, then expansion of brackets in the definition $\text{Var}(Y) = E(Y - E(Y))^2$ yields that $E(Y^2) = \text{Var}(Y) + (E(Y))^2 = \sigma^2 + \mu^2$. Hence,

$$E\left(\sum_{k=1}^{n}(X_k - \widehat{\mu}_n)^2\right) = E\left(\sum_{k=1}^{n}(X_k^2 - 2\widehat{\mu}_n X_k + \widehat{\mu}_n^2)\right)$$

$$= E\left(\sum_{k=1}^{n}X_k^2 - n\widehat{\mu}_n^2\right)$$

$$= \sum_{k=1}^{n}(\mu^2 + \sigma^2) - n\left(\mu^2 + \frac{\sigma^2}{n}\right) = (n-1)\sigma^2. \qquad (11.5)$$

We now introduce the *sample variance* $\widehat{\sigma_n^2}$ as estimator for the variance σ^2 (in the statistical literature the sample variance is usually denoted by S^2)

$$\widehat{\sigma_n^2} = \frac{1}{n-1}\sum_{k=1}^{n}(X_k - \widehat{\mu}_n)^2. \qquad (11.6)$$

The use of $n-1$ instead of n is explained by the following consequence of (11.5)

$$E\left(\widehat{\sigma_n^2}\right) = \frac{1}{n-1}E\left(\sum_{k=1}^{n}(X_k - \widehat{\mu}_n)^2\right) = \frac{1}{n-1}(n-1)\sigma^2 = \sigma^2. \qquad (11.7)$$

Clearly $\widehat{\sigma_n^2}$ is unbiased. Note that unbiasedness of the sample variance does not hold for its square root, the sample standard deviation. In general $E(\widehat{\sigma_n}) \neq \sigma$. From the recursion (11.2) we observe that

$$(n-1)(\widehat{\mu}_n - \widehat{\mu}_{n-1}) = X_n - \widehat{\mu}_n, \tag{11.8}$$

$$n(\widehat{\mu}_n - \widehat{\mu}_{n-1}) = X_n - \widehat{\mu}_{n-1}. \tag{11.9}$$

With this we obtain a practical recursive formula for $\widehat{\sigma_n^2}$, which can be viewed as a parallel to the recursion for mean values (11.2)

$$\widehat{\sigma_n^2} = \frac{n-2}{n-1}\widehat{\sigma_{n-1}^2} + \frac{(X_n - \widehat{\mu}_n)(X_n - \widehat{\mu}_{n-1})}{n-1}. \tag{11.10}$$

All formulas presented so far are valid for arbitrary distributions as long as all means and variances are finite. In case the distribution of the sample $X_1, \ldots, X_n$ is known, then one may obtain more specific results. For example, if the sample is from a normal distribution with mean μ and variance σ^2, then the sample mean is again normally distributed with mean μ and variance σ^2/n (cf. (11.3) and (11.4)), while the sample variance $(n-1)\widehat{\sigma_n^2}/\sigma^2$ has a χ^2-square distribution with $n-1$ degrees of freedom. This yields the extra information

$$\text{Var}\left(\widehat{\sigma_n^2}\right) = \frac{\sigma^4}{(n-1)^2}\text{Var}\left(\frac{(n-1)\widehat{\sigma_n^2}}{\sigma^2}\right) = \frac{\sigma^4}{(n-1)^2}2(n-1) = \frac{2\sigma^4}{n-1}.$$

One may also prove under normality that $E(\widehat{\sigma_n})$ is a constant times σ, where the constant depends on the sample size n but not on the mean μ.

There are several ways to check whether a sample follows a given class of probability distributions. There are graphical checks like quantile-quantile plots (for normal distributions, this is often called the normal probability plot), but also so-called goodness-of-fit tests. For the normal distributions, there are dedicated tests like the Shapiro-Wilks test.

We now present an example of an estimator for a parameter which is not related to means and variances. Here we sample from a uniform distribution on an interval $[0,\theta]$, where the right-end of the interval is unknown and must be estimated from observations $X_1, \ldots, X_n$. An obvious estimator here is $\widehat{\Theta} := \max(X_1, \ldots, X_n)$. The distribution function[3] of a random variable X is defined as $F_X(x) = P(X \leq x)$. If X has an a density f (or stated in analysis terminology, if the probability distribution of X is absolutely continuous with Radon-Nikodym derivative f), then $F' = f$ and $F_X(x) = \int_{-\infty}^{x} f(u)\,du$. Recall that an estimator is also a random variable. It thus follows from independence of the observations X_i that

---

[3] Strictly speaking, one should speak about the cumulative distribution function, but the adjective cumulative is often suppressed.

$$F_{\widehat{\Theta}}(x) = P(\max(X_1,\ldots,X_n) \leq x)$$
$$= P(X_1 \leq x)\ldots P(X_n \leq x) = (x/\theta)^n, \quad \text{for } 0 \leq x \leq \theta.$$

Hence, $f_{\widehat{\Theta}}(x) = F'_{\widehat{\Theta}}(x) = n\left(\frac{x}{\theta}\right)^{n-1}\frac{1}{\theta}$ for $0 \leq x \leq \theta$ and thus

$$\mathrm{E}\left(\widehat{\Theta}\right) = \int_0^\theta x\, f_{\widehat{\Theta}}(x)\, dx = \int_0^\theta x\, n\left(\frac{x}{\theta}\right)^{n-1}\frac{1}{\theta}\, dx = \int_0^\theta n\left(\frac{x}{\theta}\right)^n dx = \frac{n}{n+1}\theta.$$

It now immediately follows that $\widetilde{\Theta} := \frac{n+1}{n}\widehat{\Theta}$ is an unbiased estimator for θ

$$\mathrm{E}\left(\widetilde{\Theta}\right) = \frac{n+1}{n}\mathrm{E}\left(\widehat{\Theta}\right) = \frac{n+1}{n}\frac{n}{n+1}\theta = \theta.$$

It is not surprising to see that $\max(X_1,\ldots,X_n)$ is systematically underestimating θ, but it is surprising that there is a factor depending on the sample size n only that may be used to compensate for this.

There is a huge literature on estimation theory. We only briefly mention that there are systematic approaches for developing estimators (maximum likelihood, moment methods, entropy methods). Maximum Likelihood is popular because it is asymptotically optimal in the sense that in an asymptotic sense, maximum likelihood estimators are unbiased and have minimum variance. However, maximum likelihood estimators may be intractable in some case. Therefore moment estimators are also used. There are also methods to investigate for finite sample sizes whether an unbiased estimator has minimal variance (Cramér-Rao lower bound for variances, see [1, Chapter 9], Lehmann-Scheffé theorem, see [1, Chapter 10]).

11.3 Nonparametric estimation of probabilities

In the previous section we discussed estimators when we could assume a functional form for the underlying probability distribution. In many applications, one cannot make such assumptions and thus require to have estimation procedures that make no such assumptions. Therefore we show in this section examples of estimators that do not require a specific assumed form of probability distributions. Such methods are often referred to as nonparametric methods, also technically speaking distribution free would be a more appropriate adjective. There is of course a trade-off: nonparametric methods are more flexible but require more data than parametric models. Let $Y_1,\ldots,Y_n$ be a sample and A be a (measurable) set on the real line. We define new random variables X_i by

$$X_i = 1_{\{Y_i \in A\}} = \begin{cases} 1 & \text{if } Y_i \in A \\ 0 & \text{if } Y_i \notin A \end{cases}.$$

Random variables like X_i are called indicator random variables in statistics (we will see later that statisticians use the word characteristic function in another context). The mean of such indicator random variables is the probability $P(Y \in A)$. In the special case of indicator random variables, we may use the following approach. We write $p = P(X \in A)$ and define

$$\widehat{p} = \frac{\sum_{i=1}^n X_i}{n} = \frac{\#\{i \mid Y_i \in A\}}{n}. \tag{11.11}$$

Note that $n\widehat{p} \sim \text{Bin}(n,p)$, i.e., $n\widehat{p}$ is binomially distributed with n trials and success probability p. Hence,

$$E(\widehat{p}) = \frac{1}{n} np = p, \tag{11.12}$$

$$\text{Stand. dev}(\widehat{p}) = \frac{1}{n}\sqrt{np(1-p)} = \sqrt{\frac{p(1-p)}{n}}. \tag{11.13}$$

Since a binomial distribution is a sum of independent Bernoulli random variables, the Central Limit Theorem yields

$$P(|\widehat{p}-p| > \varepsilon) = P\left(\frac{|\widehat{p}-p|}{\sqrt{\frac{p(1-p)}{n}}} > y\right) \approx 2\Phi(-y), \tag{11.14}$$

Such formulas can be used to determine beforehand how many Monte Carlo simulation runs one needs, to obtain a required accuracy.

We now show examples of of estimators for a function rather than a parameter. The empirical distribution function of a sample $X_1, \ldots, X_n$ is the estimator

$$\widehat{F}_n(x) = \frac{1}{n}\{\#i \mid X_i \le x\} = \frac{1}{n}\sum_{i=1}^n I_{\{X_i \le x\}}. \tag{11.15}$$

The empirical distribution function is an estimator for the distribution function. The empirical distribution function can also be interpreted of the distribution function of the discrete probability distribution with masses $1/n$ at the points $x_1, \ldots, x_n$. It is easy to see that $n\widehat{F}_n(x) \sim \text{Bin}(n, F(x))$ (a binomial distribution with n trials and success probability $F(x)$), from which it directly follows that $\widehat{F}_n(x)$ is an unbiased estimator for $F(x)$ for any distribution function F. The famous Glivenko-Cantelli Theorem from the 1930's shows that the empirical distribution function uniformly converges to the true distribution function F of the $X_1, \ldots, X_n$:

$$\lim_{n\to\infty} \sup_{x\in\mathbb{R}} |\widehat{F}_n(x) - F(x)| = 0.$$

The empirical distribution function is implicitly playing an important role in Monte Carlo simulations. MATLAB offers the procedures `cdfplot` and `ecdf` to plot and compute the empirical distribution function, respectively.

Related to the empirical distribution function is the empirical characteristic function. Please note the different terminology between analysis and statistics. A characteristic function in statistics is a Fourier-Stieltjes transform (statisticians use the word indicator function for what is called a characteristic function in analysis). The empirical characteristic function of a sample $X_1, \ldots, X_n$ is the estimator

$$\widehat{C}_n(r) = \frac{1}{n} \sum_{i=1}^{n} e^{ir X_i}. \tag{11.16}$$

The empirical characteristic function was implicitly introduced in [8]. Convergence properties like almost sure uniform convergence to the characteristic function was proved in [4]. An overview of various applications like symmetry testing and goodness-of-fit testing can be found in [14]. A different application of the empirical characteristic function was proposed in [6], where an advanced moment matching procedure is applied to the empirical characteristic function of a random response surface in order to estimate the probability distribution of the performance of a circuit.

11.4 Density estimation

The distribution function fully characterizes a probability distribution. However, by definition the distribution function is a global description. It is thus not a convenient way to study local properties like the shape of a distribution (peaks, asymmetries etc.). For such properties it is more convenient to turn to the density if the probability distribution is absolutely continuous. Let $X_1, \ldots, X_n$ be a random sample with a common distribution function F with continuous derivative $F' = f$. As before, we denote the empirical distribution function by $\widehat{F}_n$. Since $\widehat{F}_n$ is piecewise constant, we cannot simply obtain a density estimator by differentiating $\widehat{F}_n$ directly. It is also intuitively clear that density estimation being a local procedure requires higher sample sizes than estimation of the distribution function.

A widely used density estimator (although it is not always recognized as such) is the histogram. We denote the empirical distribution function by P_n, i.e. $P_n(A) := \frac{1}{n} \{\#i \mid X_i \in A\}$. Let I be a compact interval on $\mathbb{R}$ and suppose that the intervals $I_1, \ldots, I_k$ form a partition of I. The histogram of $X_1, \ldots, X_n$ with respect to the partition $I_1, \ldots, I_k$ is defined as

$$H_n(x) := \sum_{j=1}^{k} \frac{P_n(I_j)\, 1_{I_j}(x)}{|I_j|},$$

where $|I_j|$ denotes the length of the interval I_j. It is clear that the histogram is a stepwise constant function. Two major disadvantages of the histogram are

- the stepwise constant nature of the histogram
- the fact that the histogram heavily depends on the choice of the partition

In order to illustrate the last point, consider Figure 11.1 where the two histograms are made from the same data set.

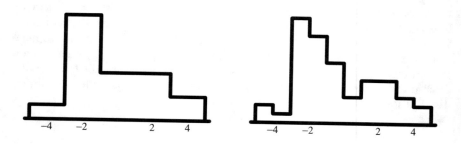

Fig. 11.1: Two histograms of the same sample of size 50 from a mixture of 2 normal distributions.

It is because of this phenomenon that histograms are not to be recommended. A natural way to improve on histograms is to get rid of the fixed partition by putting an interval around each point. If $h > 0$ is fixed, then

$$\widehat{N}_n(x) := \frac{P_n((x-h, x+h))}{2h} \qquad (11.17)$$

is called the *naive density estimator* which was introduced in 1951 by Fix and Hodges in an unpublished report (reprinted in [3]) dealing with discriminant analysis. The motivation for the naive estimator is that for small h we approximately have

$$P(x-h < X < x+h) = \int_{x-h}^{x+h} f(t)\, dt \approx 2\,h\,f(x). \qquad (11.18)$$

It can also be interpreted as a symmetric differential quotient of the empirical distribution function. Note that the naive estimator is a local procedure; it uses only the observations close to the point at which one wants to estimate the unknown density. Compare this with the empirical distribution function, which uses all observations to the right of the point at which one is estimating.

It is intuitively clear from (11.18) that the bias of $\widehat{N}_n$ decreases as h tends to 0. However, if h tends to 0, then one is using less and less observations, and hence the variance of $\widehat{N}_n$ increases. This phenomenon occurs often in density estimation. The optimal value of h is a compromise between the bias and the variance. We will return to this topic of great practical importance when we discuss the MSE.

The naive estimator is a special case of the following class of density estimators. Let K be a *kernel function*, that is a nonnegative function such that

$$\int_{-\infty}^{\infty} K(x)\, dx = 1. \tag{11.19}$$

The *kernel estimator* with kernel K and bandwidth h is defined by

$$\widehat{f}_n(x) := \frac{1}{n} \sum_{i=1}^{n} \frac{1}{h} K\left(\frac{x - X_i}{h}\right). \tag{11.20}$$

Thus, the kernel indicates the weight that each observation receives in estimating the unknown density. It is easy to verify that kernel estimators are densities and that the naive estimator is a kernel estimator with kernel

$$K(x) = \begin{cases} \frac{1}{2} & \text{if } |x| < 1 \\ 0 & \text{otherwise.} \end{cases}$$

The kernel estimator can also be written in terms of the empirical distribution function $\widehat{F}_n$:

$$\widehat{f}_n(x) = \int_{-\infty}^{\infty} \frac{1}{h} K\left(\frac{x - y}{h}\right) d\widehat{F}_n(y),$$

where the integral is a Stieltjes integral.

Examples of other kernels are given in Table 11.1. Kernel density estimators are available in MATLAB through the command `ksdensity`, including an automatic choice of the bandwidth h (we will see more about bandwidth choices in the next section. An example of an explicit formula is (11.34).

11.5 Mathematical background on density estimators

In this section we give an introduction to the mathematical background of the kernel density estimators defined in the previous section. We will see that the underlying theory is closely linked to the early work of Bochner on approximation theory. A good impression of kernel estimation is given by the

name	function
Gaussian	$\dfrac{1}{\sqrt{2\pi}}e^{-\frac{1}{2}x^2}$
naive/rectangular	$\dfrac{1}{2}1_{(-1,1)}(x)$
triangular	$(1-|x|)1_{(-1,1)}(x)$
biweight	$\dfrac{15}{16}(1-x^2)^2\,1_{(-1,1)}(x)$
Epanechnikov	$\dfrac{3}{4}(1-x^2)\,1_{(-1,1)}(x)$

Table 11.1: Well-known kernels for density estimators.

books [11] and [13]. For other types of density estimators, we refer to [11] and [12].

11.5.1 Finite sample behaviour of density estimators

In order to assess point estimators, we look at properties like unbiasedness and efficiency (as discussed in Section 11.2). In kernel density estimation, it is very important to know the influence of the bandwidth h (cf. our discussion of the naive estimator). To combine the assessment of these properties, the Mean Squared Error (MSE) is used. We now discuss the analogues of these properties for density estimators. The difference is that the estimate is not a single number, but a function. However, we start with pointwise properties.

Theorem 11.1. Let $\widehat{f}_n$ be a kernel estimator with kernel K. Then

$$\mathrm{E}\left(\widehat{f}_n(x)\right) = \frac{1}{h}\int_{-\infty}^{\infty} K\left(\frac{x-y}{h}\right)f(y)\,dy = \frac{1}{h}\int_{-\infty}^{\infty} K\left(\frac{y}{h}\right)f(x-y)\,dy. \tag{11.21}$$

Proof: This follows from the fact that for a random variable X with density f, we have $\mathrm{E}\left(g(X)\right) = \int_{-\infty}^{\infty} g(x)f(x)\,dx$. $\qquad\square$.

Theorem 11.2. Let $\widehat{f}_n$ be a kernel estimator with kernel K. Then

$$\mathrm{Var}\left(\widehat{f}_n(x)\right) = \frac{1}{nh^2}\int_{-\infty}^{\infty} K^2\left(\frac{x-y}{h}\right)f(y)\,dy$$

$$- \frac{1}{nh^2}\left\{\int_{-\infty}^{\infty} K\left(\frac{x-y}{h}\right)f(y)\,dy\right\}^2. \tag{11.22}$$

Proof: It is easy to see that

$$\left(\widehat{f}_n(x)\right)^2 = \frac{1}{n^2}\sum_{i=1}^{n}\frac{1}{h^2}K^2\left(\frac{x-X_i}{h}\right) + \frac{1}{n^2}\sum_{\substack{i,j=1\\i\neq j}}^{n}\frac{1}{h^2}K\left(\frac{x-X_i}{h}\right)K\left(\frac{x-X_j}{h}\right).$$

Then

$$\mathrm{E}\left(\widehat{f}_n(x)\right)^2 = \frac{1}{nh^2}\int_{-\infty}^{\infty}K^2\left(\frac{x-y}{h}\right)f(y)dy$$
$$+ \frac{n-1}{nh^2}\left(\int_{-\infty}^{\infty}K\left(\frac{x-y}{h}\right)f(y)dy\right)^2.$$

The results now follows from using (11.21) and the well-known fact that $\mathrm{Var}\,(X) = \mathrm{E}\left(X^2\right) - (\mathrm{E}\,(X))^2$. □

The following general result due to Rosenblatt (see [10] for a slightly more general result) shows that we cannot have unbiasedness for all x.

Theorem 11.3 (Rosenblatt). *A kernel estimator cannot be unbiased for all $x \in \mathbb{R}$.*

Proof: We argue by contradiction. Assume that $\mathrm{E}\left(\widehat{f}_n(x)\right) = f(x)$ for all $x \in \mathbb{R}$. Then $\int_a^b \widehat{f}_n(x)\,dx$ is an unbiased estimator for $F(b) - F(a)$, since

$$\mathrm{E}\left(\int_a^b \widehat{f}_n(x)\,dx\right) = \int_a^b \mathrm{E}\left(\widehat{f}_n(x)\right)\,dx = \int_a^b f(x)\,dx = F(b) - F(a),$$

where the interchange of integrals is allowed since the integrand is positive. Now it can be shown that the only unbiased estimator of $F(b) - F(a)$ symmetric in $X_1, \ldots, X_n$ is $\widehat{F}_n(b) - \widehat{F}_n(a)$. This leads to a contradiction, since it would imply that the empirical distribution function is differentiable. □

For point estimators, the Mean Squared Error is a useful concept as it combines bias and variance as noted in Section 11.2. We now generalize this concept to density estimators.

Definition 11.1. The Mean Squared Error at x of a density estimator $\widehat{f}$ is defined as

$$\mathrm{MSE}_x(\widehat{f}) := \mathrm{E}\left(\widehat{f}(x) - f(x)\right)^2. \tag{11.23}$$

For further use we note that the MSE can be rewritten as

$$\mathrm{MSE}_x(\widehat{f}) = \mathrm{Var}\left(\widehat{f}(x)\right) + \left(\mathrm{E}\widehat{f}(x) - f(x)\right)^2 \tag{11.24}$$

The Mean Integrated Squared Error of a density estimator $\widehat{f}$ is defined as

$$\mathrm{MISE}(\widehat{f}) := \mathrm{E}\left(\int_{-\infty}^{\infty}\left(\widehat{f}(x) - f(x)\right)^2 dx\right). \tag{11.25}$$

For further use, we state the following expressions for the MSE and MISE.

Theorem 11.4. *For a kernel density estimator $\widehat{f}_n$ with kernel K the* MSE *and* MISE *can be expressed as:*

$$\mathrm{MSE}_x(\widehat{f}_n) = \frac{1}{nh^2} \int_{-\infty}^{\infty} K^2\left(\frac{x-y}{h}\right) f(y)\, dy$$

$$- \frac{1}{nh^2} \left\{ \int_{-\infty}^{\infty} K\left(\frac{x-y}{h}\right) f(y)\, dy \right\}^2$$

$$+ \left(\frac{1}{h} \int_{-\infty}^{\infty} K\left(\frac{x-y}{h}\right) f(y)\, dy - f(x) \right)^2. \quad (11.26)$$

$$\mathrm{MISE}(\widehat{f}_n) = \frac{1}{nh^2} \int_{-\infty}^{\infty} \left(\int_{-\infty}^{\infty} K^2\left(\frac{x-y}{h}\right) f(y)\, dy \right.$$

$$\left. - \left\{ \int_{-\infty}^{\infty} K\left(\frac{x-y}{h}\right) f(y)\, dy \right\}^2 \right) dx$$

$$+ \int_{-\infty}^{\infty} \left(\frac{1}{h} \int_{-\infty}^{\infty} K\left(\frac{x-y}{h}\right) f(y)\, dy - f(x) \right)^2 dx. (11.27)$$

Proof: Combination of (11.24) with (11.21) and (11.22) yields the formula for the MSE. Integrating this formula with respect to x, we obtain the formula for the MISE. □

In general, the above formulas can not be evaluated explicitly. When both the kernel and the unknown density are Gaussian, then straightforward but tedious computations yield explicit formulas as shown in [5]. These formulas were extended in [7] to the case of mixtures of normal distributions. It is claimed in [7] that the class of mixture of normal distributions is very rich and that it is thus possible to perform exact calculations for many distributions. These calculations can be used to choose an optimal bandwidth h (see [7] for details). For other examples of explicit MSE calculations, we refer to [2].

We conclude this section with a note on the use of Fourier analysis. Formulas (11.21) and (11.22) show that $\mathrm{E}\left(\widehat{f}_n(x)\right)$ and $\mathrm{Var}\left(\widehat{f}_n(x)\right)$ can be expressed in terms of convolutions of the kernel with the unknown density. Another (even more important) use of Fourier transforms is the computation of the kernel estimate itself. Computing density estimates directly from the definition is very time consuming. However, the Fourier transform of the kernel estimator is a convolution of the empirical characteristic function defined at the end of Section 11.2 with the Fourier transform of the kernel itself:

$$\int_{-\infty}^{\infty} \widehat{f_n}(x)\, e^{ikx}\, dx = \frac{1}{h} \int_{-\infty}^{\infty} \int_{-\infty}^{\infty} K\left(\frac{x-y}{h}\right) e^{ikx}\, d\widehat{F_n}(y)\, dx$$

$$= \frac{1}{h} \int_{-\infty}^{\infty} \int_{-\infty}^{\infty} K\left(\frac{x}{h}\right) e^{ik(x+y)}\, d\widehat{F_n}(y)\, dx$$

$$= \int_{-\infty}^{\infty} e^{ikhz} K(z)\, dz \int_{-\infty}^{\infty} e^{iky}\, d\widehat{F_n}(y).$$

Using the Fast Fourier Transform (FFT), one can in this way efficiently compute good approximations to the kernel estimates. For details we refer to [11, pp. 61-66] and [13, Appendix D].

11.5.2 Asymptotic behaviour of kernel density estimators

We have seen in the previous section that it is possible to evaluate exactly the important properties of kernel density estimators. However, the unknown density f appears in a complicated way in exact calculations, which limits the applicability. Such calculations are very important for choosing the optimal bandwidth h. Therefore, much effort has been put in obtaining asymptotic results in which the unknown density f appears in a less complicated way. In this section we give an introduction to these results. Many of these results can be found in [8] and [10]. For an overview of more recent results, we refer to the monographs [11] and [13].

Theorem 11.5 (Bochner). *Let K be a bounded kernel function such that $\lim_{|y| \to \infty} y\, K(y) = 0$. Define for any absolutely integrable function g the functions*

$$g_n(x) := \frac{1}{h_n} \int_{-\infty}^{\infty} K\left(\frac{y}{h_n}\right) g(x-y)\, dy,$$

where $(h_n)_{n \in \mathbb{N}}$ is a sequence of positive numbers such that $\lim_{n \to \infty} h_n = 0$. If g is continuous at x, then we have

$$\lim_{n \to \infty} g_n(x) = g(x). \tag{11.28}$$

Proof: Since $\int_{-\infty}^{\infty} \frac{1}{h} K\left(\frac{y}{h}\right) dy = \int_{-\infty}^{\infty} K(y)\, dy = 1$, we may write

$$|g_n(x) - g(x)| = \left| g_n(x) - g(x) \int_{-\infty}^{\infty} \frac{1}{h_n} K\left(\frac{y}{h_n}\right) dy \right|$$

$$\leq \int_{-\infty}^{\infty} \left| \{g(x-y) - g(x)\} \frac{1}{h_n} K\left(\frac{y}{h_n}\right) \right| dy.$$

Let $\delta > 0$ be arbitrary. We now split the integration interval into 2 parts: $\{y : |y| \geq \delta\}$ and $\{y : |y| < \delta\}$. The first integral can be bounded from above by

$$\int_{|y| \geq \delta} \frac{|g(x-y)|}{y} \frac{y}{h_n} K\left(\frac{y}{h_n}\right) dy + |g(x)| \int_{|y| \geq \delta} \frac{1}{h_n} K\left(\frac{y}{h_n}\right) dy$$

$$\leq \frac{\sup_{|v| \geq \delta/h_n} |v K(v)|}{\delta} \int_{|y| \geq \delta} |g(x-y)| dy + |g(x)| \int_{|t| \geq \delta/h_n} K(t) dt$$

$$\leq \frac{\sup_{|v| \geq \delta/h_n} |v K(v)|}{\delta} \int_{-\infty}^{\infty} |g(u)| du + |g(x)| \int_{|t| \geq \delta/h_n} K(t) dt.$$

Letting $n \to \infty$ and using that K is absolutely integrable, we see that these terms can be made arbitrarily small. The integral over the second region can be bounded from above by

$$\sup_{|y| < \delta} |g(x-y) - g(x)| \int_{|y| < \delta} K(y) dy \leq \sup_{|y| < \delta} |g(x-y) - g(x)|.$$

Since this holds for all $\delta > 0$ and g is continuous at x, the above expression can be made arbitrarily small.   □.

As a corollary, we obtain the following asymptotic results (taken from [8]) for the mean and variance of the kernel estimator at a point x.

Corollary 11.1 (Parzen). *Let $\widehat{f}_n$ be a kernel estimator such that its kernel K is bounded and satisfies $\lim_{|y| \to \infty} y K(y) = 0$. Then $\widehat{f}_n$ is an asymptotically unbiased estimator for f at all continuity points x if $\lim_{n \to \infty} h_n = 0$.*

Proof: Apply Theorem 11.5 to (11.21) and (11.22).   □.

In the above corollary, there is no restriction on the rate at which $(h_n)_{n \in \mathbb{N}}$ converges to 0. The next corollaries show that if $(h_n)_{n \in \mathbb{N}}$ converges to 0 slower than n^{-1}, then the MSE of $\widehat{f}_n(x)$ converges to 0.

Corollary 11.2 (Parzen). *Let $\widehat{f}_n$ be a kernel estimator such that its kernel K is bounded and satisfies $\lim_{|y| \to \infty} y K(y) = 0$. If $\lim_{n \to \infty} h_n = 0$ and x is a continuity point of the unknown density f, then*

$$\lim_{n \to \infty} n h_n \operatorname{Var}\left(\widehat{f}_n(x)\right) = f(x) \int_{-\infty}^{\infty} K^2(y) dy.$$

Proof: First note that since K is bounded, the kernel $K^2 / \int_{-\infty}^{\infty} K^2(u) du$ also satisfies the conditions of Theorem 11.5. Hence, the result follows from applying Theorem 11.5 to (11.22).   □.

Corollary 11.3 (Parzen). Let $\widehat{f}_n$ be a kernel estimator such that its kernel K is bounded and satisfies $\lim_{|y| \to \infty} y K(y) = 0$. If $\lim_{n \to \infty} h_n = 0$, $\lim_{n \to \infty} n h_n = \infty$ and x is a continuity point of the unknown density f, then

$$\lim_{n \to \infty} \mathrm{MSE}_x(\widehat{f}_n) = 0.$$

Proof: It follows from Corollary 11.2 that $\lim_{n \to \infty} \mathrm{Var}\, \widehat{f}_n(x) = 0$. The result now follows by combining Corollary 11.1 and (11.24). $\qquad\square$.

Although the above theorems give insight in the asymptotic behaviour of density estimators, they are not sufficient for practical purposes. Therefore, we now refine them by using Taylor expansions and adding a third order moment condition.

Theorem 11.6. Let $\widehat{f}_n$ be a kernel estimator such that its kernel K is bounded and symmetric and such that $\int_{-\infty}^{\infty} |t^3| K(t)\, dt$ exists and is finite. If the unknown density f has a bounded third derivative, then we have that

$$\mathrm{E}\left(\widehat{f}_n(x)\right) = f(x) + \frac{1}{2} h^2 f''(x) \int_{-\infty}^{\infty} t^2 K(t)\, dt + o(h^2), \ h \downarrow 0, \quad (11.29)$$

$$\mathrm{Var}\left(\widehat{f}_n(x)\right) = \frac{1}{nh} f(x) \int_{-\infty}^{\infty} K^2(t)\, dt + o\left(\frac{1}{nh}\right),$$
$$\text{for } h \downarrow 0 \text{ and } nh \to \infty, \quad (11.30)$$

$$\mathrm{MSE}_x(\widehat{f}_n) = \frac{f(x)}{nh} \int_{-\infty}^{\infty} K^2(t)\, dt + \frac{1}{4} h^4 \left(f''(x) \int_{-\infty}^{\infty} t^2 K(t)\, dt \right)^2$$
$$+ o\left(\frac{1}{nh}\right) + o\left(h^4\right), \text{ for } h \downarrow 0 \text{ and } nh \to \infty. \quad (11.31)$$

Proof: By (11.21) and a change of variables, we may write

$$\mathrm{E}\left(\widehat{f}_n(x)\right) - f(x) = \int_{-\infty}^{\infty} K(t) \left\{ f(x - th) - f(x) \right\} dt.$$

Now Taylor's Theorem with the Lagrange form of the remainder says that

$$f(x - th) = f(x) - th f'(x) + \frac{(th)^2}{2} f''(x) - \frac{(th)^3}{3!} f'''(\xi),$$

where ξ depends on x, t, and h and is such that $|x - \xi| < |th|$. Since $\int_{-\infty}^{\infty} K(t)\, dt = 1$, it follows that

$$\mathrm{E}\left(\widehat{f}_n(x)\right) - f(x) = \int_{-\infty}^{\infty} K(t) \left(-th f'(x) + \frac{(th)^2}{2} f''(x) - \frac{(th)^3}{3!} f'''(\xi) \right) dt,$$

Because of the symmetry of K, this expression simplifies to

$$\mathrm{E}\left(\widehat{f}_n(x)\right) - f(x) = \int_{-\infty}^{\infty} K(t)\left(\frac{(th)^2}{2}f''(x) - \frac{(th)^3}{3!}f'''(\xi)\right)dt.$$

If M denotes an upper bound for f''', then the first result follows from

$$\left|\mathrm{E}\left(\widehat{f}_n(x)\right) - f(x) - \frac{1}{2}h^2 f''(x)\int_{-\infty}^{\infty} t^2 K(t)\,dt\right| \leq \frac{h^3}{3!}\int_{-\infty}^{\infty}\left|t^3 K(t) f'''(\xi)\right|dt$$

$$\leq M\frac{h^3}{3!}\int_{-\infty}^{\infty}\left|t^3 K(t)\right|dt,$$

where the last term obviously is $o(h^2)$. The asymptotic expansion of the variance follows immediately from Corollary 11.2. In order to obtain the asymptotic expansion for the MSE, it suffices to combine (11.24) with (11.29) and (11.30). □

These expressions show that the asymptotic expressions are much easier to interpret than the exact expression of the previous section. For example, we can now clearly see that the bias decreases if h is small and that the variance decreases if h is large (cf. our discussion of the naive density estimator).

Theorem 11.6 is essential for obtaining optimal choices of the bandwidth. If we assume that f'' is square integrable, then it follows from (11.31) that

$$\mathrm{MISE}(\widehat{f}_n) = \frac{1}{nh}\int_{-\infty}^{\infty} K^2(t)\,dt + \frac{1}{4}h^4\int_{-\infty}^{\infty}(f'')^2(x)\,dx\left(\int_{-\infty}^{\infty} t^2 K(t)\,dt\right)^2$$
$$+ o\left(\frac{1}{nh}\right) + o(h^4), \quad h \downarrow 0 \text{ and } nh \to \infty.$$

The expression

$$\frac{1}{nh}\int_{-\infty}^{\infty} K^2(t)\,dt + \frac{1}{4}h^4\int_{-\infty}^{\infty}(f'')^2(x)\,dx\left(\int_{-\infty}^{\infty} t^2 K(t)\,dt\right)^2 \qquad (11.32)$$

is called the *asymptotic MISE*, often abbreviated as AMISE. Note that (11.32) is much easier to use than (11.27) because it allows us to balance the squared bias and variance in order to obtain a choice of h that minimizes the MISE. It boils down to minimizing an expression of the form $\frac{a}{nh} + bh^4$ with respect to h. This leads to

$$h_{\mathrm{AMISE}} = \left(\frac{\int_{-\infty}^{\infty} K^2(t)\,dt}{4n\left(\int_{-\infty}^{\infty} t^2 K(t)\,dt\right)^2 \int_{-\infty}^{\infty}(f'')^2(x)\,dx}\right)^{1/5}. \qquad (11.33)$$

Substituting this value for h in the asymptotic expression for the MISE, one sees that the rate of convergence for density estimators is $n^{-4/5}$.

An important drawback of (11.33) is that it depends on $\int_{-\infty}^{\infty} (f'')^2 (x)\, dx$, which is unknown. However, there are good methods for estimating this quantity. For details, we refer to the literature (see e.g., [11] and [13]). An example of a simple method is the following. If f is a normal density with parameters μ and σ^2, then one can prove that

$$h_{\mathrm{AMISE}} = \left(\frac{8\sqrt{\pi} \int_{-\infty}^{\infty} K^2(t)\, dt}{3n \left(\int_{-\infty}^{\infty} t^2 K(t)\, dt \right)^2} \right)^{1/5} \sigma. \qquad (11.34)$$

Given an optimal choice of the bandwidth h, we may wonder which kernel gives the smallest MISE. It turns out that the Epanechnikov kernel is the optimal kernel. However, the other kernels perform nearly as well, so that the optimality property of the Epanechnikov kernel is not very important in practice. For details, we refer to [11] and [13].

11.6 Conclusion

In this chapter we gave an overview of the basic statistical estimation techniques when normality cannot be assumed and one has to go beyond estimation of means and variances. After an introductory section on performance measures for estimation, we briefly introduced nonparametric estimation of probabilities. This is usually a global procedure. In many cases it is important to estimate local properties of probability distributions through the density function. We showed how to go from unsatisfactory histograms to kernel density estimators. Both finite sample and asymptotic results for kernel density estimators were derived. The asymptotic results yielded ways to determine suitable bandwidth choices. Wherever possible, references to the underlying techniques from analysis like approximation theory and Fourier analysis were provided in order to put the statistical approach in a wider context.

References

1. BAIN, L.J., AND ENGELHARDT, M.: *Introduction to Probability and Mathematical Statistics*. Duxbury, Belmont, California, 2nd edition, 1992.
2. DEHEUVELS, P.: *Estimation nonparametrique de la densité par histogrammes generalisés*. Revue de Statistique Appliquée, **35**, pp. 5–42, 1977.
3. FIX, E., AND HODGES, J.L.: *Discriminatory analysis – nonparametric discrimination: consistency properties*. Int. Stat. Rev., **57**, pp. 238–247, 1989.
4. FEUERVERGER, A., AND MUREIKA, R.A.: *The empirical characteristic function and its applications*. Ann. Statistics, **5 (1)**, pp. 88–97, 1997.
5. FRYER, M.J.: *Some errors associated with the non-parametric estimation of density functions*. J. Inst. Math. Appl., **18**, pp. 371–380, 1976.

6. LI, X., LE, J. , GOPALAKRISHNAN, P., AND PILLEGE, L.T.: *Asymptotic Probability Extraction for Nonnormal Performance Distributions*. IEEE Trans. Computer-Aided Design Integrated Circuits Systems, **26 (1)**, pp. 16–37, 2007.
7. MARRON, J.S., AND WAND, M.P.: *Exact mean integrated squared error*. Ann. Stat., **20**, 712–736, 1992.
8. PARZEN, E.: *On estimation of a probability density function and mode*. Ann. Math. Stat., **33**, pp. 1065–1076, 1962.
9. PULCH, R., PUTEK, P., DE GERSEM, H., AND GILLON, R.: *Inverse Modeling: Glue-Package-Die Problem*. Chapter in this Book, 2019.
10. ROSENBLATT, M.: *Remarks on some nonparametric estimates of a density function*. Ann. Math. Stat., **27**, pp. 827–837, 1956.
11. SILVERMAN, B.W.: *Density Estimation for Statistics and Data Analysis*. Chapman & Hall, London, 1986.
12. TAPIA, R.A., AND THOMPSON, J.R.: *Nonparametric Probability Density Estimation*. The John Hopkins University Press, Baltimore, 1978.
13. WAND, M.P., AND JONES, M.C.: *Kernel Smoothing*. Chapman & Hall, London (1995)
14. YU, J.: *Empirical characteristic function estimation and its applications*. Econ. Rev., **23 (2)**, pp. 93–123, 2004.

Chapter 12
Inverse Modeling: Glue-Package-Die Problem

Roland Pulch, Piotr Putek, Herbert De Gersem, Renaud Gillon

Abstract In mathematical modeling, the physical or geometrical parameters are often affected by uncertainties. For example, imperfections of an industrial manufacturing generate undesired variations in the produced devices. We consider an uncertainty quantification, where parameters are replaced by random variables. Consequently, the probability distributions of the parameters have to be predetermined as an input to the stochastic model. However, the variability of input parameters is often not directly accessible by measurements, whereas the output quantities are available. We investigate a problem from nanoelectronics: a piece of glue connecting a die and a package. A randomness in both the formation and the quality of the piece of glue cause uncertain geometrical parameters and material parameters. The task consists in a fitting of input probability distributions for the random parameters to measurements of the output. This problem can be seen as a form of a stochastic inverse problem. The cumulative distribution function is approximated by a piecewise linear function. We apply a minimization, which yields a nonlinear least squares problem. Numerical results are illustrated for this test problem.

Roland Pulch
Universität Greifswald, Germany, e-mail: Roland.Pulch@uni-greifswald.de

Piotr Putek
Bergische Universität Wuppertal, Germany, e-mail: Putek@math.uni-wuppertal.de; Piotr.Putek@gmail.com

Herbert De Gersem
Technische Universität Darmstadt, Germany, e-mail: DeGersem@temf.tu-darmstadt.de

Renaud Gillon
ON Semiconductor Belgium BVBA, Oudenaarde, Belgium, e-mail: Renaud.Gillon@onsemi.com

© Springer Nature Switzerland AG 2019
E. J. W. ter Maten et al. (eds.), *Nanoelectronic Coupled Problems Solutions*, Mathematics in Industry 29, https://doi.org/10.1007/978-3-030-30726-4_12

12.1 Introduction

In mathematical models of physical or technical applications, the involved parameters often exhibit uncertainties. Concerning nanoelectronics, miniaturization causes variations of parameters in an industrial production, see [4, 6, 13]. On the one hand, variations can be examined by a worst-case-corner-analysis , see [15]. On the other hand, Uncertainty Quantification (UQ) is often based on stochastic models, where random variables or random processes replace varying parameters, see [14].

For a numerical simulation of a stochastic model, the probability distributions of the parameters have to be predetermined. Measurements of the input parameters are often not feasible. Alternatively, measurements of output quantities are accessible. Therefore, we will fit probabilistic input data to measurements of outputs. This strategy represents a kind of an inverse problem or a backward problem instead of the forward problem, where inputs are propagated to outputs.

We discuss a test problem relevant for the production of electronic circuits. A simplified model of a piece of glue between a die and a package is considered, which was also examined in [10]. The material parameters and the geometrical parameters of the glue are uncertain. The aim is to identify the probability distributions of five random parameters. Therein, independent probability distributions are assumed. If the parameters exhibit dependencies, then one can try to derive independent model parameters by a Rosenblatt transformation [11].

Each probability distribution typically includes on some parameters, for example, the expected value and the variance in the case of Gaussian distributions. These unknown parameters can be determined by the method of moments, the maximum likelihood principle or Bayesian estimations, see [7]. Polynomials of Gaussian-distributed variables are fitted by Fleishman's approach, cf. [1, 5], for example. However, these methods require that the type of the input probability distribution (uniform, Gaussian, beta, etc.) is known a priori.

Alternatively, we construct a numerical technique for the fitting of arbitrary cumulative distribution functions, where just the continuity of the function is assumed. The method employs a piecewise linear approximation of the cumulative distribution function. The approach yields a minimization problem, more precisely a nonlinear least squares problem. We apply pseudo random numbers, see [9], for both the evaluation of the objective function by sampling and the determination of grid points in the domain of the unknowns.

Finally, we apply this numerical method to a glue-die-package problem. Therein, measurements are still generated artificially. The results demonstrate the feasibility of the technique. In contrast to [10], we included new numerical simulations using beta distributions instead of uniform distributions in this chapter.

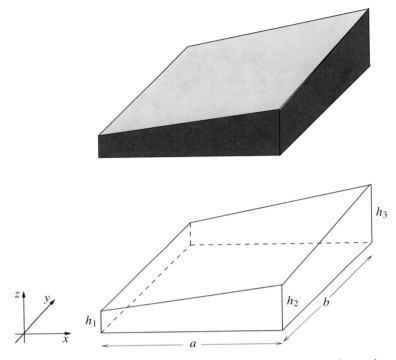

Fig. 12.1: Simplified geometry of a piece of glue connecting a die and a package. (Figure represents a modification from [10].)

12.2 Modeling of Glue-Package-Die Configuration

In the production of electronic devices, glue is applied to hold a die and a package together. On the one hand, metallic parts and semiconducting parts of devices can often be produced with a high precision. On the other hand, a piece of glue represents a cheap substance, where both the material parameters and the formation undergo significant variations. The material properties as well as the geometry are important for the heat transfer between the layers.

We consider a simplified geometry for a piece of glue, as shown in Fig. 12.1. The bottom layer is a fixed rectangle with the lengths a, b. The top layer is not parallel due to the variations. Five parameters specify the setting: the thermal conductivity λ and the volumetric heat capacity ρc (actually a product of two parameters, but we treat it as one parameter) as material parameters, the average height h_0 and the two angles α, β between the bottom and the top layer in x- and y-direction, respectively, as geometrical parameters. Table 12.1 illustrates the parameters.

Table 12.1: Parameters in glue-die-package problem.

parameter	symbol	unit	minimum value	maximum value
thermal conductivity	λ	[W/mK]	1.6	2.4
volumetric heat capacity	ρc	[J/m^{3}K]	$2.5 \cdot 10^6$	$3.75 \cdot 10^6$
average height	h_0	[m]	$5.0 \cdot 10^{-4}$	$7.5 \cdot 10^{-4}$
first angle	α	[rad]	0.	0.175
second angle	β	[rad]	0.	0.175

The output data consists of measurements in L different patches between the bottom layer and the top layer. Figure 12.2 illustrates the partition of the bottom layer into the patches. The thermal conductance G_ℓ and the heat capacitance C_ℓ for each patch read as

$$G_\ell = \frac{\lambda \, a_\ell \, b_\ell}{h_\ell}$$

$$C_\ell = \frac{\rho c \, h_\ell \, a_\ell \, b_\ell}{3} \qquad (12.1)$$

for $\ell = 1, \ldots, L$. The constants a_ℓ, b_ℓ denote the lengths of the rectangle for a single patch and h_ℓ is the average height in the patch. For $h_\ell \to 0$ the thermal conductance becomes unbounded, because the thermal resistance converges to zero. It holds that

$$h_\ell = h_0 + \tan \alpha \, x_\ell + \tan \beta \, y_\ell$$

$$\approx h_0 + \alpha \, x_\ell + \beta \, y_\ell \qquad (12.2)$$

with known coordinates x_ℓ, y_ℓ for $\ell = 1, \ldots, L$. The approximation (12.2), which is appropriate for small angles, is inserted into the model functions (12.1).

For example, independent beta distributions with fixed intervals for each input parameter can be predetermined. The minimum and maximum values of the intervals are given in Table 12.1. The model functions (12.1) allow for the simulation of measurements, where measurement errors can be added to test the techniques. Now our aim is to recover the original probability distributions from the measurements only.

12.3 Fitting of Cumulative Distribution Functions

We design a method for the approximation of any continuous cumulative distribution function. Consequently, the strategy can also be applied for prob-

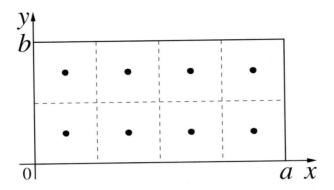

Fig. 12.2: Partition into patches on bottom side of glue geometry. The points specify the locations of the measurements.

lems, where the type of the input probability distributions is unknown a priori.

We make the following assumptions. On the one hand, k output quantities are given by a model function $f : \Pi \to \mathbb{R}^k$, i.e.,

$$y = f(p), \qquad y_i = f_i(p_1,\dots,p_n) \qquad \text{for } i = 1,\dots,k \qquad (12.3)$$

with the input parameters $p \in \Pi \subset \mathbb{R}^n$. The parameter domain is a cuboid

$$\Pi = \prod_{i=1}^{n} [p_{i,\min}, p_{i,\max}] \subset \mathbb{R}^n. \qquad (12.4)$$

In practice, the range of a parameter is finite in most of the cases. If the range of a parameter is infinite (for example, assuming a Gaussian distribution), then a truncation to a compact interval is required, which introduces an additional approximation error. Nevertheless, this error will be small provided that the interval is chosen sufficiently large. Given a realization of the input parameters, the outputs can be evaluated using (12.3). On the other hand, the output quantities can be measured simultaneously, which yields the measurements $\{\tilde{y}^{(1)},\dots,\tilde{y}^{(M)}\} \subset \mathbb{R}^k$ for different unknown realizations of the input parameters. These values are typically affected by measurement errors. Kernel smoothing can be used to approximate the probability density functions of the output quantities, see [3]. However, we want to identify the probability distributions of the input parameters, where no samples are available.

The random variables p are assumed to be independent. To derive our approach, we consider a single parameter $p \in [p_{\min}, p_{\max}]$. Let $F : [p_{\min}, p_{\max}] \to [0,1]$ be its cumulative distribution function. It holds that

$$F(p_{\min}) = 0 \qquad \text{and} \qquad F(p_{\max}) = 1.$$

Furthermore, the function is monotone. Thus its inverse distribution function can be defined by the quantile function

$$F^{-1}(x) = \inf \{ p \in [p_{\min}, p_{\max}] \; : \; x \le F(p) \} \qquad \text{for } x \in [0,1]. \tag{12.5}$$

We construct a piecewise linear approximation of the cumulative distribution function by

$$\tilde{F}(p) := \begin{cases} 0 & \text{for } p < q_1 \\ \dfrac{j-1}{m-1} \cdot \dfrac{q_{j+1}-p}{q_{j+1}-q_j} + \dfrac{j}{m-1} \cdot \dfrac{p-q_j}{q_{j+1}-q_j} & \text{for } p \in [q_j, q_{j+1}), \; j \in \{1, \ldots, m-1\} \\ 1 & \text{for } p \ge q_m \end{cases}$$

$$\tag{12.6}$$

using discrete points

$$p_{\min} \le q_1 < q_2 < \cdots < q_m \le p_{\max} \tag{12.7}$$

for some integer m. The function increases by $\frac{1}{m}$ in each subinterval. This construction yields a strictly monotone function $\tilde{F}$. For any continuous cumulative distribution function F and $\varepsilon > 0$, there is an integer m and an associated choice of points $q_1, \ldots, q_m$ such that

$$\max_{p \in [p_{\min}, p_{\max}]} \left| F(p) - \tilde{F}(p) \right| < \varepsilon.$$

For example, this property follows from the choices $q_j = F^{-1}(\frac{j-1}{m-1})$ for $j = 1, \ldots, m$ using the inverse function (12.5) and $\frac{1}{m} < \varepsilon$. However, just relatively small integers m can be applied in practice.

The grid points $q_1, \ldots, q_m$ represent the unknowns now. For given starting values, an arbitrary number N of samples for p can be generated by the approximation (12.6). Therefore the inverse function $\tilde{F}^{-1}$ is evaluated at uniformly distributed (pseudo) random numbers in $[0,1]$.

The above construction is used for each random variable separately. For simplicity, we arrange the same number m for all n input parameters, which is not necessary in general. Let $\mathcal{K} := \Pi^m \subset \mathbb{R}^{m \times n}$ with the parameter domain (12.4). The unknown degrees of freedom are written as a matrix $Q \in \mathcal{K}$, whose ith column is associated to the ith parameter. Let $\mathcal{K}' \subset \mathcal{K}$ be the subset of all matrices satisfying the ascending order (12.7) in each column. The ordering can be ensured by a transformation of the problem to variables

$$\Delta q_j := q_{j+1} - q_j \qquad \text{for } j = 1, \ldots, m-1, \tag{12.8}$$

which are required to be positive.

On the one hand, the parameter-dependent output quantities (12.3) are measured simultaneously. The range $y_i \in [y_{i,\min}, y_{i,\max}]$ is partitioned into r_i equidistant intervals for $i = 1, \ldots, k$. Let M be the total number of measurements and s_{ij} for $j = 1, \ldots, r_i$ be the number of measurements in the jth interval for the ith output quantity.

On the other hand, we consider N realizations of the tuple of random parameters, which are computed by inverse functions $\tilde{F}_1^{-1}, \ldots, \tilde{F}_n^{-1}$ assuming some starting values in $\mathcal{K}'$. The realizations $\{p^{(1)}, \ldots, p^{(N)}\}$ yield associated output quantities $\{y^{(1)}, \ldots, y^{(N)}\}$ using the model function (12.3). Let t_{ij} for $j = 1, \ldots, r_i$ be the number of values in the jth interval for the ith output quantity.

Comparing the ratio from sampling and the ratio from measurements for each interval and each output quantity, a nonlinear least squares problem arises for the unknown grid points in the approximations (12.6). The problem reads as

$$\min_{Q \in \mathcal{K}'} \sum_{i=1}^{k} \sum_{j=1}^{r_i} \left(\frac{s_{ij}}{M} - \frac{t_{ij}(Q)}{N} \right)^2. \tag{12.9}$$

The dimensionality of the unknowns is nm. The number M of measurements is often restricted in practice. In contrast, the number N of realizations can be chosen arbitrarily large limited by the computational effort for the evaluations (12.3) only.

12.4 Numerical Solution of the Minimization Problem

We address the numerical solution of the nonlinear least squares problem (12.9) now. In the case of general minimization problems, there are two types of iteration schemes:

- *derivative-based algorithms*,
  for example, steepest-descend method and Levenberg-Marquardt method, see [2],
- *derivative-free algorithms*,
  for example, Nelder-Mead method, see [8].

Unfortunately, typical numerical methods, where the values $\Delta q_j > 0$ from (12.8) are considered as unknowns, yield poor approximations of this minimization problem. This property occurs both in the Levenberg-Marquardt method and the Nelder-Mead technique. An explanation of this critical behavior is still missing.

Hence we construct a simple alternative, which can be applied only if the evaluations of the model function (12.3) are relatively cheap. The idea is to evaluate the objective function in (12.9) on a finite set $\mathcal{G} \subset \mathcal{K}'$ and to determine the minimum value in the set. Thus the approximation becomes

$$\hat{Q} = \arg\min_{Q \in \mathcal{G}} \sum_{i=1}^{k} \sum_{j=1}^{r_i} \left(\frac{s_{ij}}{M} - \frac{t_{ij}(Q)}{N} \right)^2 \qquad (12.10)$$

An advantage of this approach is that starting values for the unknown solution are not required.

Although nm unknowns appear, the dimensions are not independent of each other. The problem consists of n groups of unknowns. For the ith group, we generate a trial set $\{q_1, \ldots, q_m\}$ by (pseudo) random numbers with respect to a uniform distribution in $[p_{i,\min}, p_{i,\max}]$. Afterwards the values are sorted in ascending order. An arbitrary number $N_{\mathrm{grid}} = |\mathcal{G}|$ of grid points in $\mathcal{K}'$ can be produced by this approach.

Furthermore, numerical tests showed that the approximation becomes worse if sequences of low discrepancy in an n-dimensional space, cf. [9], are used as grid points instead of pseudo random numbers. The reason is that the tuples $(q_1, \ldots, q_m) \subset \mathcal{K}'$ often become close to an equal spacing of the points in each group. We require pseudo random numbers to obtain also clusters of points at steeper gradients of the cumulative distribution function.

12.5 Numerical Results

We apply the technique designed in Sect. 12.3 to the test problem from Sect. 12.2 now. The required computations were done in the software package MATLAB [12]. In particular, we utilized the built-in routine **rand** for the generation of all pseudo random numbers.

In the model problem, we arrange a rectangle of the (standardized) size $a = 4$, $b = 2$ and $L = 8$ patches shown in Figure 12.2. Hence the number of output quantities (12.1) becomes $k = 2L = 16$. We arrange independent beta distributions for the five input parameters, where the probability density functions read as

$$\rho_i(p_i) = C_i(p_i - p_{i,\min})^{\mu_i} (p_{i,\max} - p_i)^{\nu_i} \qquad \text{for } p_i \in [p_{i,\min}, p_{i,\max}] \quad (12.11)$$

and $i = 1, \ldots, 5$ with exponents $\mu_i, \nu_i \geq 0$ and constants $C_i > 0$ for standardization. The used minimun and maximum parameter values are depicted in Table 12.1. We produce artificial measurements of the outputs by samples from the probability distributions of the input parameters using pseudo random numbers. Moreover, measurement errors up to 0.1% are included by adding pseudo random numbers of this magnitude. We employ $M = 100$ measurements in each patch. The ranges $G \in [0, 0.015]$ and $C \in [0, 0.001]$ are supposed for each patch. Furthermore, we select $r_i = 40$ for all $i = 1, \ldots, 5$. Hence the nonlinear least squares problem (12.9) includes a sum with 640 terms.

The minimum and maximum values of the random parameters, see Table 12.1, are considered as unknowns. Thus slightly larger intervals $[p_{i,\min}, p_{i,\max}]$ are arranged in the numerical method. The number $m = 10$ is used within the discretization of each cumulative distribution function (12.6) for all parameters. We produce $N = 1000$ samples of the output quantities for each evaluation of the objective function.

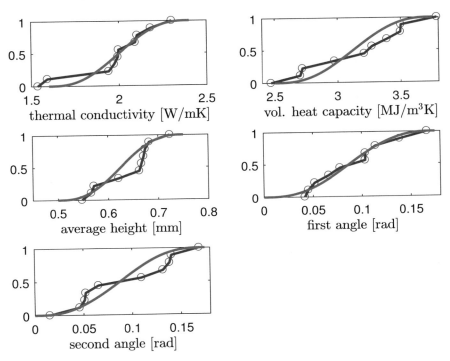

Fig. 12.3: Exact cumulative distribution functions (red lines) and piecewise linear approximations (blue lines) for the five input parameters using symmetric beta distributions.

The simple variant (12.10) of our numerical method for the nonlinear least squares problem (12.9) is applied using $N_{\text{grid}} = 10\,000$ grid points in $\mathcal{K}'$. We apply independent beta distributions (12.11) in two different cases of exponents:

i) *symmetric choice*: $\mu_i = \nu_i = 2$ for all i, and
ii) *asymmetric choice*: $\mu_i = 1$, $\nu_i = 3$ for all i.

Figure 12.3 and Figure 12.4 illustrate the true cumulative distribution functions of the input parameters and the computed approximations. We observe a moderate approximation of the exact distributions. However, our approach does not require a priori assumptions on the type of the distribution, except for the minimum and maximum values. The Levenberg-Marquardt method as well as the Nelder-Mead method yield worse approximations, even though good starting values were chosen in view of the known probability distributions in our test.

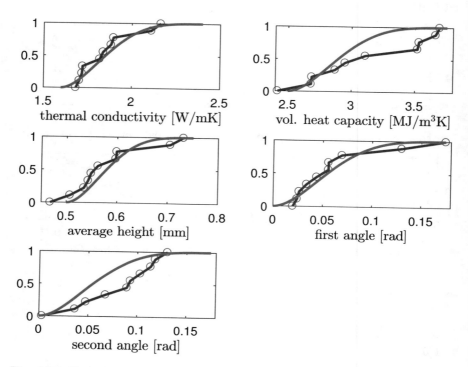

Fig. 12.4: Exact cumulative distribution functions (red lines) and piecewise linear approximations (blue lines) for the five input parameters using asymmetric beta distributions.

12.6 Conclusions

We designed a method to identify an approximation of cumulative distribution functions for random input parameters, which implies a nonlinear least squares problem. This minimization problem causes divergence in traditional optimization algorithms. Yet a simplified approach yields moderate approximations.

The package-die-glue problem can be considered in a more realistic form by two modifications. Firstly, the artificially computed measurements will be replaced by real measured data. Secondly, the simple model functions (12.1) will be substituted by parabolic partial differential equations describing the heat transfer. In this case, the evaluation of the model functions becomes costly.

References

1. FLEISHMAN, A.I.: *A method for simulating non-normal distributions*, Psychometrika, 43, pp. 521–532, 1978.
2. GAVIN, H.: *The Levenberg-Marquardt method for nonlinear least squares curve-fitting problems.* http://www.people.duke.edu/~hpgavin/ce281/lm.pdf, 2016. Online, accessed 20-December-2016.
3. HOROVA, I., KOLACEK, J., AND ZELINKA, J.: *Kernel Smoothing in MATLAB: Theory and Practice of Kernel Smoothing*, World Scientific, 2012.
4. LI, P., LIU, F., LI, X., PILEGGI, L., AND NASSIF, S.: *Modeling interconnect variability using efficient parametric model order reduction*. In: Proc. Design, Automation and Test Conference in Europe (DATE), pp. 958–963, 2005.
5. LUO, H.: *Generation of non-normal data – a study of Fleishman's power method*, research report, Department of Statistics, Uppsala University, Sweden, 2011.
6. MOHANTY, S.P., AND KOUGIANOS, E.: *Incorporating manufacturing process variation awareness in fast design optimization of nanoscale CMOS VCOs*, IEEE Transactions on Semiconductor Manufacturing, 27, pp. 22–31, 2014.
7. MONTGOMERY, D., AND RUNGER, G.: *Applied Statistics and Probability for Engineers*, John Wilex & Sons, Inc., 3rd ed., 2003.
8. NELDER, J., AND MEAD, R.: *A simplex method for function minimization*, Comput. J., 7, pp. 308–313, 1965.
9. NIEDERREITER, H.: *Random Number Generation and Quasi-Monte Carlo Methods*, SIAM Philadelphia, 1992.
10. PULCH, R., PUTEK, P., DE GERSEM, H., AND GILLON, R.: *Identification of probabilistic input data for a glue-die-package problem*. In: QUINTELA, P., GÓMEZ, D., PENA, F., RODRÍGUEZ, J., SALGADO, P., AND VÁZQUEZ-MENDÉZ, M. (EDS): *Progress in Industrial Mathematics at ECMI 2016*, Series Mathematics in Industry, Vol. 26, Springer International Publishing AG, pp. 255–262, 2017. http://dx.doi.org/10.1007/978-3-319-63082-3_40.
11. ROSENBLATT, M.: *Remarks on multirate transformation*, Ann. Math. Statist., 23, pp. 470–472, 1952.
12. THE MATHWORKS INC.: *MATLAB, version 9.1.0 (R2016b)*, Natick, Massachusetts, 2016.
13. WANG, Z., LAI, X., AND ROYCHOWDHURY, J.: *PV-PPV: Parameter variability aware, automatically extracted, nonlinear time-shifted oscillators macromodels.* In Proc. IEEE Design Automation Conference DAC 2007, pp. 142–147, 2007.
14. XIU, D.: *Numerical Methods for Stochastic Computations: A Spectral Method Approach*, Princeton University Press, 2010.
15. ZHANG, H., CHEN, T., TING, M., AND LI, X.: *Efficient design-specific worst-case corner extraction for integrated circuits*. In Proc. IEEE Design Automation Conference DAC 2009, pp. 386–389, 2009.

Part IV
Model Order Reduction

Part IV covers the following topics.

- **Parametric Model Order Reduction for Electro-Thermal Coupled Problems**
  Authors:
  Lihong Feng, Peter Benner (Max Planck Institute, Magdeburg, Germany).
- **Sparse (P)MOR for Electro-Thermal Coupled Problems with Many Inputs**
  Authors:
  Nicodemus Banagaaya, Lihong Feng, Peter Benner (Max Planck Institute, Magdeburg, Germany).
- **Reduced Models and Uncertainty Quantification**
  Authors:
  Yao Yue, Lihong Feng, Peter Benner (Max Planck Institute, Magdeburg, Germany),
  Roland Pulch (Universität Greifswald, Germany),
  Sebastian Schöps (Technische Universität Darmstadt, Germany).

Chapter 13
Parametric Model Order Reduction for Electro-Thermal Coupled Problems

Lihong Feng, Peter Benner

Abstract We consider automatic parametric model order reduction for electro-thermal (ET) coupled problems arising from (nano-)microelectronic simulations. We show that the PMOR method based on multi-moment matching proposed in [2] can be applied to the discretized ET models. The error bound in [6] is used to check the accuracy of the reduced models. Based on the error bound, an adaptive algorithm for computing the reduced ET models is proposed in [7]. As a result, the reduced ET models are constructed automatically and reliably. This chapter reviews the PMOR method, the error bound, the adaptive algorithm, and their applications to ET models.

13.1 Introduction

With the scaling down of integrated circuits, thermal issues have attracted increasingly more attention and have become a major consideration in design of integrated circuits. Electro-thermal (ET) simulation is an integrated approach for the analysis of the interaction between the electrical and thermal dynamics of the system.

Parameter variations have become essential in many analyses such as optimization and uncertainty quantification, where ET simulation at many values of the parameters are unavoidable. Due to the large-scale of the ET model after discretization, using, e.g. finite element discretization, finite-integration, etc., simulation of ET models in the above multi-query tasks is time and memory demanding [4, 8, 10].

Parametric Model Order Reduction (PMOR) [3] obtains Reduced-Order Models (ROMs) from the large-scale systems, which can be simulated much

Lihong Feng, Peter Benner
Max Planck Institut für Dynamik komplexer technischer Systeme, Magdeburg, Germany,
e-mail: {Feng,Benner}@mpi-magdeburg.mpg.de

© Springer Nature Switzerland AG 2019
E. J. W. ter Maten et al. (eds.), *Nanoelectronic Coupled Problems Solutions*,
Mathematics in Industry 29, https://doi.org/10.1007/978-3-030-30726-4_13

293

faster than the original models. Given a sound error estimator, the outputs (state vectors) computed from the ROMs are guaranteed to be reliable for a specified error tolerance.

This chapter reviews PMOR applied to the ET models. Especially, we consider PMOR for the electrical and the thermal subsystems separately. It is discussed in [7] that the error bound [5, 6] for the transfer function of the reduced model can be applied to measure the error of the reduced ET models. The ROMs are automatically constructed from a greedy algorithm using the error bound. The key steps of automatic PMOR are reviewed in this chapter, and some details for application to the ET-models are further explored.

In the next section, we describe mathematical models of the ET-coupled problems and the discretized systems. Section 13.3 discusses the PMOR method based on multi-moment matching for linear parametric systems. It shows that the method can be applied to the electrical and the thermal subsystems separately. We present the error bound for the reduced ET models, as well as the adaptive greedy algorithm in Section 13.4. Numerical results are given thereafter. Section 13.6 concludes the chapter.

13.2 ET Coupled Problems and the Discretized Systems

We use two applications arising from electro-thermal simulations to show the ET-coupled problems we are considering. One is a package shown in Fig. 13.1(Left), whose purpose is to allow easy handling and assembly onto printed circuit boards and to protect the devices from damage. The other is a Power-MOS device shown in Fig. 13.1(Right), which is commonly used in energy harvesting, where energy from external sources like light and environmental heat is collected in order to power small devices such as implanted biosensors [7].

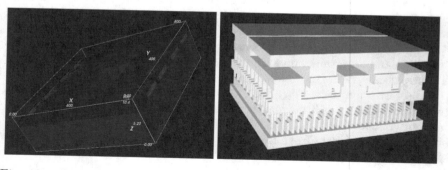

Fig. 13.1: Physical models considered in numerical tests. **Left**: A package. **Right**: A Power-MOS device (stretched in the vertical direction).

The dynamics of both applications can be described by the same governing equations. The electrical subsystem is

$$\nabla \cdot J + \frac{\partial \rho}{\partial t} = 0,$$

$$J = \sigma \cdot E, \quad E = -\nabla U,$$

$$\rho = -\nabla \cdot (\epsilon \nabla U),$$

where J is the current density, E is the electrical field, U is the electrical potential, σ is the electrical conductivity, ϵ is the permittivity, and ρ is the charge density. Currently, we ignore local charging, i.e., $\rho = 0$, and the dependence of the conductivity on temperature, i.e., the electrical subsystem is independent of the thermal subsystem, and obtain the following time-independent partial differential equation:

$$\nabla \cdot (\sigma \cdot \nabla U) = 0.$$

The thermal subsystem is governed by similar governing equations:

$$\nabla \cdot \phi_q + \frac{\partial w(T)}{\partial t} = Q,$$

$$\phi_q = -\kappa \nabla T,$$

$$w(T) = C_T(T - T_{\text{ref}}),$$

where ϕ_q is the heat flux, w is the local energy storage, C_T is the thermal capacitance. For the thermal subsystem, we also ignore the dependence of the thermal capacitance on the temperature.

Q represents heat sources or sinks, it is the coupling term from the electrical subsystem: the Joule self-heating that is of great importance in power-aware design of integrated circuits:

$$Q = Q_{\text{SH}} = E \cdot J.$$

In this case, the whole system is one-way coupled: the thermal subsystem depends on the electrical subsystem, while the electrical subsystem is independent of the thermal system. After spatial discretization, the state space representation of the electro-thermal model is:

$$
\begin{aligned}
A_E(p)x_E(p,t) &= -B_E(p)u(t) \\
E_T(p)\dot{x}_T(p,t) &= A_T(p)x_T(p,t) + B_T(p)u(t) \\
&\quad + F(p) \times_2 x_E(p) \times_3 x_E(p), \\
y(p,t) &= C_E(p)x_E(p,t) + C_T(p)x_T(p,t) + D(p)u(t), \\
x_T(p,0) &= x_T^0, \\
x_E(p,0) &= x_E^0.
\end{aligned}
\tag{13.1}
$$

where $A_E(p) \in \mathbb{R}^{n_E \times n_E}$, $B_E(p) \in \mathbb{R}^{n_E \times l_I}$, $E_T(p) \in \mathbb{R}^{n_T \times n_T}$, $A_T(p) \in \mathbb{R}^{n_T \times n_T}$, $B_T(p) \in \mathbb{R}^{n_T \times l_I}$, $C_E(p) \in \mathbb{R}^{l_O \times n_E}$, $C_T(p) \in \mathbb{R}^{l_O \times n_T}$, $D_T(p) \in \mathbb{R}^{l_O \times l_I}$, and the tensor $F(p) \in \mathbb{R}^{n_T \times n_E \times n_E}$, which can be considered as n_T slices of n_E by n_E matrices $F_i(p) \in \mathbb{R}^{n_E \times n_E}$, $i = 1, \ldots, n_T$, represents the nonlinear coupling of the electrical part with the thermal part. The product $F(p) \times_2 x_E(p) \times_3 x_E(p)$ is a vector of length n_T, whose i-th component is the standard vector-matrix-vector product $x_E(p)^T F_i(p) x_E(p)$. Here, the i-mode tensor-matrix product is denoted by $\times_i$ [9]. In this formulation, the algebraic equation defining x_E describes the electrical part, the ordinary differential equation for x_T describes the thermal part, in which the tensor $F(p)$ describes Joule self-heating. The third equation computes the output obtained from the electrical and thermal state vectors. The last two specify the initial conditions. Theoretically, Joule self-heating should be modeled by two tensor products: $F(p) \times_2 x_E(p) \times_3 x_E(p)$ and $G(p) \times_2 x_E(p) \times_3 u(t)$. However, the influence of the second part is rather limited, and is therefore ignored. Instead of a single coupled system, we write out the electrical and thermal subsystems explicitly to show the one-way coupling. Furthermore, if we use a single set of differential algebraic equations, our numerical results proved that PMOR becomes much less computationally efficient than dealing with the subsystem of algebraic equations and the subsystem of ordinary differential equations separately.

For the package model shown in Fig. 13.1(Left), the parameter is chosen to be the top layer thickness of the package h. A finite-integration technique (FIT) for modeling of the package leads to thermal fluxes that are proportional to the dual areas of the mesh cells and inversely proportional to the lengths of the edges in the mesh cells. Therefore, when considering meshes that are topologically equivalent for different package thicknesses, the parametric dependence of the matrices will take the form

$$M(h) = M_0 + hM_1 + \frac{1}{h}M_2, \quad (M = A_E, B_E, E_T, A_T, B_T, F, C_E, C_T, D).$$

The second term originates from the linear dependence of dual areas corresponding to the cell edges perpendicular to the thickness, whereas the third term originates from dual areas associated to cell edges tangential to the thickness orientation [7].

For the Power-MOS circuit model shown in Fig. 13.1(Right), the conductivity σ of the third metal layer is chosen to be the parameter. A finite-integration technique (FIT) assembles fluxes that are proportional to the the conductivity of each mesh cell material, and therefore, the parametric dependence of the matrices will take the form

$$M(\sigma) = M_0 + \sigma M_1, \quad (M = A_E, B_E, E_T, A_T, B_T, F, C_E, C_T, D).$$

13.3 PMOR for the ET models

This section first briefly presents the idea of the PMOR method based on multi-moment matching for general linear parametric systems. Afterwards, application of the method to the ET coupled models is discussed in details.

13.3.1 Basic idea

The PMOR method based on multi-moment matching [2] computes reduced-order models of linear parametric systems. Regardless of first-order systems,

$$E(p)\frac{dx(p,t)}{dt} = A(p)x(p,t) + B(p)u(t),$$
$$y(p,t) = C(p)x(p,t), \tag{13.2}$$

or second order systems,

$$M(p)\frac{d^2x(p,t)}{dt^2} + D(p)\frac{dx(p,t)}{dt} + T(p)x(p,t) = B(p)u(t),$$
$$y(p,t) = C(p)x(p,t), \tag{13.3}$$

where $x(p,t) \in \mathbb{R}^n$, $M(p), D(p), T(p), E(p), A(p) \in \mathbb{R}^{n \times n}$, $B(p) \in \mathbb{R}^{n \times l_I}$, $C(p) \in \mathbb{R}^{n \times l_O}$, a projection matrix $V \in \mathbb{R}^{n \times r}$ is computed using the Laplace transformed systems in the following form (assume zero initial conditions)

$$(E_0 + \mu_1 E_1 + \mu_2 E_2 + \ldots + \mu_l E_l)x(p,s) = B(p)u(s),$$
$$y(p,s) = C(p)x(p,s). \tag{13.4}$$

Here, $p = (p_1, \ldots, p_q)$ is the vector of the original physical or geometrical parameters. $u(s)$ is the Laplace transform of the input signal $u(t)$. $\mu = (\mu_1, \ldots, \mu_l)$ includes parameters $\mu_i, i = 1, \ldots, l$, which might be some functions (rational, polynomial) of the original parameters p_i and the Laplace variable s. The $E_i \in \mathbb{R}^{n \times n}$ are constant matrices.

The reduced-order models of either (13.2) or (13.3) can then be obtained from Galerkin projection as

$$V^T E(p)V\frac{dz}{dt} = V^T A(p)Vz + V^T B(p)u(t),$$
$$\hat{y} = C(p)Vz, \tag{13.5}$$

or

$$V^T M(p)V\frac{d^2z}{dt^2} + V^T D(p)V\frac{dz}{dt} + V^T T(p)Vz = V^T B(p)u(t),$$
$$\hat{y} = C(p)Vz. \tag{13.6}$$

Here $z \in \mathbb{R}^r$. $r(\ll n)$ is also called the order of the ROM. The state vector x of the original system can be approximately computed as $x \approx Vz$.

The first step of the method is to expand the state $x(p,s)$ in (13.4) into a power series w.r.t the parameters μ_i. Given an expansion point $\mu_0 = (\mu_1^0, \ldots, \mu_l^0)$,

$$
\begin{aligned}
x(p,s) &= [I - (\sigma_1 M_1 + \ldots + \sigma_l M_l)]^{-1} \tilde{E}^{-1}(\mu^0) B(p) u(s) \\
&= \sum_{m=0}^{\infty} [\sigma_1 M_1 + \ldots + \sigma_l M_l]^m \tilde{B}_M u(s) \\
&= \tilde{B}_M u(s) + [\sigma_1 M_1 + \ldots + \sigma_l M_l] \tilde{B}_M u(s) \\
&\quad + [\sigma_1 M_1 + \ldots + \sigma_l M_l]^2 \tilde{B}_M u(s) + \ldots \\
&\quad + [\sigma_1 M_1 + \ldots + \sigma_l M_l]^j \tilde{B}_M u(s) + \ldots
\end{aligned}
\tag{13.7}
$$

where $\sigma_i = \mu_i - \mu_i^0$, $\tilde{E}(\mu^0) = E_0 + \mu_1^0 E_1 + \ldots + \mu_l^0 E_l$, $M_i = -[\tilde{E}(\mu^0)]^{-1} E_i$, $i = 1, 2, \ldots l$, and $\tilde{B}_M = [\tilde{E}(\mu^0)]^{-1} B(p)$. The coefficients in the above series expansion are the moment matrices of the parametrized system. The corresponding multi-moments of the transfer function are those moment matrices multiplied by $C(p)$ from the left.

A matrix sequence based on the moment matrices in (13.7) is defined as below,

$$
\begin{aligned}
R_0 &= B_M, \\
R_1 &= [M_1 R_0, M_2 R_0, \ldots, M_p R_0], \\
R_2 &= [M_1 R_1, M_2 R_1, \ldots, M_p R_1], \\
&\;\;\vdots \\
R_j &= [M_1 R_{j-1}, M_2 R_{j-1}, \ldots, M_p R_{j-1}], \\
&\;\;\vdots
\end{aligned}
\tag{13.8}
$$

where $B_M = \tilde{B}_M$, if $B(p)$ dose not depend on p, i.e. $B(p) = B$. Otherwise, $B_M = [\tilde{B}_{M_1}, \ldots, \tilde{B}_{M_q}]$, if $B(p)$ can be approximated by an affine form, e.g., $B(p) \approx B_1 p_1 + \ldots + B_q p_q$. Here $\tilde{B}_{M_i} = [\tilde{E}(\mu^0)]^{-1} B_i$, $i = 1, \ldots, q$.

Let R be the subspace spanned by the columns of the R_j, $j = 0, 1, \ldots, m$:

$$
R = \mathrm{colspan}\{R_0, \ldots, R_j, \ldots, R_m\}_{\mu^0},
$$

then there exists $z \in \mathbb{R}^r$, such that $x \approx Vz$. Here, the columns in $V \in \mathbb{R}^{n \times r}$ constitute an orthonormal basis of R. The projection matrix V is then used to derive the ROM in (13.5) or (13.6)

For completeness, we show Algorithm 13.1 from [2] for computing $V = [v_1, v_2, \ldots, v_r]$, the algorithm can be seen as repeatedly using the modified Gram-Schmidt process to extract orthogonal vectors $v_1, \ldots, v_r$ from the vectors in $R_j, 1 \leq j \leq m$. It is proved in [2] that the multi-moments included in $R_j, j = 1, \ldots, m$, are matched by the corresponding multi-moments of the ROM.

Algorithm 13.1 Computes $V = [v_1, v_2, ..., v_q]$ for a parametric system (13.2) or (13.3), where $B(p)$ is generally considered as a matrix.

1: Initialize $a_1 = 0$, $a_2 = 0$, $sum = 0$.
2:
3: **if** (multiple input) **then**
4:     Orthogonalize the columns in R_0 using Modified Gram-Schmidt: $[v_1, v_2, ..., v_{q_1}] = $ orthonormalize$\{R_0\}$
5:     $sum = q_1$ (q_1 is the number of remaining columns after orthogonalization.)
6: **else**
7:     Compute the first column in V: $v_1 = R_0 / \|R_0\|_2$
8:     $sum = 1$
9: **end if**
10: compute the orthonormal columns in $R_1, R_2, ..., R_m$ iteratively as below
11:
12: **for** $i = 1, 2, ..., m$ **do**
13:     $a_2 = sum$;
14:     **for** $t = 1, 2, ..., p$ **do**
15:         **if** $a_1 = a_2$ **then**
16:             $stop$
17:         **else**
18:             **for** $j = a_1 + 1, ... a_2$ **do**
19:                 $w = \tilde{E}^{-1} E_t v_j$;
20:                 $col = sum + 1$;
21:                 **for** $k = 1, 2, ..., col - 1$ **do**
22:                     $h = v_k^T w$
23:                     $w = w - h v_k$
24:                 **end for**
25:                 **if** $\|w\|_2 > \varepsilon$ (a small number indicating whether the contributed information from w to V is sufficient to add w to the subspace , e.g. $\varepsilon = 10^{-5}$) **then**
26:                     $v_{col} = \frac{w}{\|w\|_2}$;
27:                     $sum = col$;
28:                 **end if**
29:             **end for**
30:         **end if**
31:     **end for**
32:     $a_1 = a_2$;
33: **end for**
34: Orthogonalize the columns in V by the Modified Gram-Schmidt process.
35:

13.3.2 PMOR for Steady Systems

The above PMOR method computes reduced-order models of the dynamical systems in (13.2) or (13.3). It is easy to see that the method can be naturally applied to steady systems:

$$(E_0 + E_1 p_1 + ... + E_q p_q)x(p, t) = B(p)u(t),$$
$$y(p, t) = C(p)x(p, t). \tag{13.9}$$

Comparing (13.9) with the Laplace transformed system (13.4), we see that they have an identical form. Consequently, Algorithm 13.1 can be used to compute a projection matrix V from the series expansion of x in (13.9). Then the reduced-order model of (13.9) is obtained by Galerkin projection as below,

$$(V^T E_0 V + V^T E_1 V p_1 + \ldots + V^T E_q V p_q) x(p,t) = V^T B(p) u(t). \tag{13.10}$$

13.3.3 PMOR Applied to the ET Models

From the analysis as above, we see that the electrical subsystem in the ET model (13.1) for either the package or the Power-MOS, can be written in the form of the steady system defined in (13.9), so that x_E can be expanded into series as in (13.7). The corresponding projection matrix V_E for reduced order modeling of the electrical subsystem is then obtained from Algorithm 13.1. The ROM of the thermal subsystem can be derived by following the standard steps in subsection 13.3.1.

In the following, we explore the details by taking the package model as an example. The system matrices of the package model are in the form of

$$M(h) = M_0 + h M_1 + \frac{1}{h} M_2,$$

so that the electrical subsystem can be written as

$$\begin{aligned} (A_{E0} + p_1 A_{E1} + p_2 A_{E2}) x_E(p,t) &= -(B_{E0} + p_1 B_{E1} + p_2 B_{E2}) u(t), \\ y_E(p,t) &= C_E(p) x_E(p,t), \end{aligned} \tag{13.11}$$

where $p_1 := h$ and $p_2 := \frac{1}{h}$. It has the same form as the algebraic system in (13.9), so that Algorithm 13.1 is straightforwardly used to compute the projection matrix V_E. The ROM of the electrical subsystem is then given by

$$\begin{aligned} (V_E^T A_{E0} V_E + p_1 V_E^T A_{E1} V_E + p_2 V_E^T A_{E2} V_E^T) z_E &= -V_E^T \tilde{B}_E \tilde{u}(t), \\ \hat{y}_E &= C_E(p) V_E z_E, \end{aligned} \tag{13.12}$$

where $\tilde{B}_E = (B_{E0}, B_{E1}, B_{E2})$, and $\tilde{u} = (u^T, p_1 u^T, p_2 u^T)^T$.

Next we consider reduced order modeling of the thermal subsystem in (13.1). Note that the thermal subsystem is a weakly nonlinear system with quadratic part $F(p) \times_2 x_E(p,s) \times_3 x_E(p,s)$, whereas the proposed PMOR method is only applicable to linear systems. Therefore, the quadratic part is first ignored in order to use Algorithm 13.1 to compute the projection matrix V_T. To this end, we obtain the Laplace transform of the linear thermal subsystem by ignoring the coupling term.

$$sE_T(p)x_T(p,s) = A_T(p)x_T(p,s) + E(p)x_T^0 + B_T(p)u(s),$$
$$y_T(p,s) = C_T(p)x_T(p,s). \tag{13.13}$$

We still take the package model as an example to explain some more details. Using the affine form $M(h) = M_0 + hM_1 + \frac{1}{h}M_2$ of the system matrices, (13.13) can be further written as

$$(sE_{T0} + sp_1 E_{T1} + sp_2 E_{T2}$$
$$-A_{T0} - p_1 A_{T1} - p_2 A_{T2})x_T(p,s) = \tilde{B}_T\tilde{u}(s,p), \tag{13.14}$$
$$y_T = C_T(p)x_T(p,s),$$

where

$$\tilde{B}_T = (E_{T0}x_T^0, p_1 E_{T1}x_T^0, p_2 E_{T2}x_T^0, B_{T0}, p_1 B_{T1}, p_2 B_{T2})$$
$$\tilde{u}(s,p) = (1,1,1,u^T(s), u^T(s), u^T(s))^T.$$

Compared with the system in (13.4), the state vector x in (13.14) can also be expanded into a power series as in (13.7), where $E_0 = -A_{T0}$, $E_i = -A_{Ti}, i = 1,2, E_{j+3} = E_{Tj}, j = 0,1,2, B = \tilde{B}$ and $\mu_i = p_i, i = 1,2, \mu_3 = s$, $\mu_{j+3} = sp_j, j = 1,2$. Therefore, the projection matrix V_T is computed by simply applying Algorithm 13.1 to (13.14).

The corresponding ROM is obtained by applying Galerkin projection to the thermal subsystem, i.e.

$$V_T^T E_T(p)V_T\dot{z}_T = V_T^T A_T(p)V_T z_T + V_T^T B_T(p)u(t)$$
$$+V_T^T F(p) \times_2 V_E z_E \times_3 V_E z_E, \tag{13.15}$$
$$z_T(p,0) = V_T^T x_T^0.$$

where

$$V_T^T E_T(p)V_T = V_T^T E_{T0} V_T + p_1 V_T^T E_{T1} V_T + p_2 V_T^T E_{T2} V_T,$$
$$V_T^T A_T(p)V_T = V_T^T A_{T0} V_T + p_1 V_T^T A_{T1} V_T + p_2 V_T^T A_{T2} V_T,$$
$$V_T^T B_T(p) = V_T^T B_{T0} + p_1 V_T^T B_{T1} + p_2 V_T^T B_{T2}.$$

Based on the affine form of the tensor, $F(p) = F^0 + p_1 F^1 + p_2 F^2$, and after simple calculations, the reduced tensor $V_T^T F(p) \times_2 V_E z_E \times_3 V_E z_E$ can be further written as $\hat{F}(p) \times_2 z_E \times_3 z_E$, where $\hat{F}(p) = (\hat{F}^0 + p_1 \hat{F}^1 + p_2 \hat{F}^2)$, and $\hat{F}_j^k = \sum_{i=1}^{n}(V_T)_{ij}V_E^T F_i^k V_E$. Here $F_i^k \in \mathbb{R}^{n \times n}$ is the i-th slice in the tensor F^k, $\hat{F}_j^k \in \mathbb{R}^{r \times r}$ is the j-th slice in the tensor $\hat{F}^k$, $(V_T)_{ij}$ is the ij-th element in V_T, $i,j = 1,.......,n$, $k = 0,1,2$. Finally, $\hat{F}_j^k$ and subsequently $\hat{F}^k$, needs to be computed only once, and independent of the parameter and time t, when solving the ROM.

Note that the nonzero initial condition is moved to the right-hand side, and is considered as a part of the input, so that system (13.14) is in the same form as (13.4). Another way of dealing with the nonzero initial condition is substituting x_T in the thermal subsystem with another variable $\tilde{x}_T$ given by $\tilde{x}_T = x_T - x_T^0$. As a result, the system with $\tilde{x}_T$ being the state vector has zero initial condition, i.e.

$$E_T(p)\dot{\tilde{x}}_T(p,t) = A_T(p)\tilde{x}_T(p,t) + A_T(p)x_T^0 + B_T(p)u(t) + F(p) \times_2 x_E \times_3 x_E,$$
$$\tilde{x}_T(p,0) = 0.$$
$$(13.16)$$

However, the nonzero initial condition x_T^0 appears in the state equation of $\tilde{x}_T$ because of the variable substitution. Ignoring the quadratic term in (13.16), and applying Laplace transformation, we obtain

$$sE_T(p)\tilde{x}_T(p,s) = A_T(p)\tilde{x}_T(p,s) + \frac{1}{s}A_T(p)x_T^0 + B_T(p)u(s). \qquad (13.17)$$

Using the affine form of the package model, the above system can be written as

$$(sE_{T0} + sp_1 E_{T1} + sp_2 E_{T2} - A_{T0} - p_1 A_{T1} - p_2 A_{T2})\tilde{x}_T(p,s) = \tilde{B}_T(p)\tilde{u}(s),$$
$$(13.18)$$

where

$$\tilde{B}_T(p) = (A_{T0}x_T^0, p_1 A_{T1}x_T^0, p_2 A_{T2}x_T^0, B_{T0}, p_1 B_{T1}, p_2 B_{T2}),$$
$$\tilde{u}(s) = (1, \tfrac{1}{s}, \tfrac{1}{s}, u^T(s), u^T(s), u^T(s))^T.$$

The state vector $\tilde{x}_T(p,s)$ in (13.18) can also be expanded into a power series following the expansion in (13.7). The projection matrix $\tilde{V}_T$ corresponding to (13.18) is derived from Algorithm 13.1. Observing that $\tilde{V}_T$ is from (13.16), therefore Galerkin projection should be applied to (13.16) to obtain the reduced thermal subsystem, i.e.

$$\tilde{V}_T^T E_T \tilde{V}_T(p)\dot{\tilde{z}}_T = \tilde{V}_T^T A_T(p)\tilde{V}_T \tilde{z}_T + \tilde{V}_T^T A_T(p)x_T^0 + \tilde{V}_T^T B_T(p)u(t)$$
$$+ \tilde{V}_T^T F(p) \times_2 \hat{V}_E z_E \times_3 \hat{V}_E z_E,$$
$$\tilde{z}_T(p,0) = 0. \qquad (13.19)$$

The tensor part should be further dealt with using the same technique as discussed before. The original state vector $x_T(p,t)$ is returned back using the relation $x_T(p,t) \approx \tilde{V}_T \tilde{z}_T(p,t) + x_T^0$, since $\tilde{x}_T(p,t)$ is approximated by $\tilde{V}_T \tilde{z}_T(p,t)$. It is easy to see that the initial condition of the original state vector computed from the reduced thermal subsystem (13.19) is exact, because $\tilde{V}_T \tilde{z}_T(p,0) = 0$. For the simulation results in Section 13.5, we use the latter way of dealing with the nonzero initial condition, though both techniques are applicable. The output of (13.19) is $\hat{y}_T = C_T \tilde{V}_T \tilde{z}_T(p,t) + C_T x_T^0$. Finally, the ROM of the ET model is given by

$$V_E^T A_E(p) V_E z_E = -V_E^T B_E(p) u(t)$$
$$\tilde{V}_T^T E_T \tilde{V}_T(p) \dot{\tilde{z}}_T = \tilde{V}_T^T A_T(p) \tilde{V}_T \tilde{z}_T + \tilde{V}_T^T A_T(p) x_T^0$$
$$+ \tilde{V}_T^T B_T(p) u(t) + \tilde{V}_T^T F(p) \times_2 \tilde{V}_E z_E \times_3 \tilde{V}_E z_E,$$
$$\hat{y}(p,t) = C_E(p) V_E z_E + C_T(p) \tilde{V}_T \tilde{z}_T \qquad (13.20)$$
$$+ C_T(p) x_T^0 + D(p) u(t),$$
$$z_E(p,0) = V_E^T x_E^0,$$
$$\tilde{z}_T(p,0) = 0.$$

13.4 Automatically Generating the Reduced ET Models

Obviously, the size of the ROM in (13.5) or (13.6) depends on the number of the columns of the projection matrix V, which is in turn, decided by R_j in (13.8). Unfortunately, the number of columns in R_j increases exponentially. This becomes even worse for multiple input systems, where R_0 is a matrix with at least l_I columns, l_I being the number of inputs.

To avoid exponential increase of the columns of R_j, so as to avoid producing reduced-order models with big sizes, multiple point expansion of the state vector x is preferred. Instead of using a single expansion point μ^0 as in (13.7), one may use several expansion points $\mu^i, i = 1, \ldots d$. One series expansion as in (13.7) is derived for each expansion point, and the corresponding projection matrix V_i is computed using Algorithm 13.1. The final projection matrix V is then derived by orthogonalizing all $V_i, i = 1, \ldots, d$, i.e.

$$V = \text{orth}\{V_1, \ldots, V_d\}.$$

Since multiple expansion points have been used, only a few $R_j's$ need to be taken for each expansion point. For example, only R_0, R_1 are necessary for generating V_i at each expansion point. In this way, the exponential increase of columns might be avoided, and the ROM is kept small.

13.4.1 Adaptively Choosing the Expansion Points

An immediate problem that arises is how to choose the expansion points. In [6], a greedy algorithm for adaptively choosing the expansion points is proposed. It is further applied to reduced-order modelling of the ET models in [7]. The greedy algorithm depends on efficient error estimation of the ROM. One expansion point is selected at every iteration of the algorithm, and corresponds to the point at which the ROM is of biggest error.

Given an error estimation/bound for the ROM, the greedy algorithm is readdressed in Algorithm 13.2. It is clear from Algorithm 13.2 that at each iteration, a new expansion point μ^i corresponding to the biggest error bound

$\Delta(\mu^i)$, is selected from a training set Ξ_{train} of the parameters. Once μ^i is selected, x is expanded into a power series as in (13.7) at μ^i, and the corresponding projection matrix V_i is computed from the R'_js associated with the coefficients of the series expansion. V is then updated by including V_i into the column space of V. The new reduced-order model is generated using the updated V, and the corresponding error bound $\Delta(\mu)$ is again computed to check the accuracy of the new ROM and to select the next expansion point if necessary.

Algorithm 13.2 Adaptively selecting expansion points μ^i, and computing V automatically

---

1: $V = []; V^{du} = [];i=0;$
2: Choose some ε_{tol}; set $\varepsilon(> \varepsilon_{tol})$;
3: Choose Ξ_{train}: a large set of samples of μ, taken over the domain of interest;
4: Choose the initial expansion point: μ^1;
5: **while** $\varepsilon > \varepsilon_{tol}$ **do**
6:     i=i+1;
7:     range$(V_i) = $span$\{R_0, R_1, \ldots, R_{q_r}\}_{\mu^i}$; (Use Algorithm (13.1) to generate V_i);
8:     range$(V_i^{du}) = $span$\{R_0^{du}, R_1^{du}, \ldots, R_{q_r}^{du}\}_{\mu^i}$;
9:                          (Apply Algorithm (13.1) to (13.22) to compute V_i^{du});
10:     $V = $orth$\{V, V_i\}$;
11:     $V^{du} = $orth$\{V, V_i^{du}\}$;
12:     $\mu^i = \arg \max\limits_{\mu \in \Xi_{train}} \Delta(\mu)$;
13:     $\varepsilon = \Delta(\mu^i)$;
14: **end while.**

Algorithm 13.2 relies on the error estimation $\Delta(\mu)$ for the ROM. An error bound for the transfer function of the ROM is proposed in [6], and is applied to the reduced ET models in [7] for automatic model order reduction. In the next subsection, we present the error bound and review its application to the ET models.

13.4.2 Error Bound and its Application to the ET Models

For a parametric Linear Time Invariant (LTI) system in (13.2) or in (13.3), with l_I inputs and l_O outputs, the error bound $\Delta(\mu)$ is defined as

$$\Delta(\mu) = \max_{\substack{1 \le i \le l_O, \\ 1 \le j \le l_I}} \Delta_{ij}(\mu),$$

where $\Delta_{ij}(\mu)$ is the error bound for the (i,j)-th entry of the transfer function matrix of the ROM, i.e.,

$$|H_{ij}(\mu) - \hat{H}_{ij}(\mu)| \le \Delta_{ij}(\mu).$$

$\Delta_{ij}(\mu)$ is defined as [6]:

$$\Delta_{ij}(\mu) = \frac{||r_i^{du}(\mu)||_2 ||r_j^{pr}(\mu)||_2}{\beta(\mu)} + |(\hat{x}_i^{du})^* r_j^{pr}(\mu)|,$$

where

$$r_j^{pr}(\mu) = B_j - E(\mu)\hat{x}_j^{pr},$$

$$\hat{x}_j^{pr} = V[V^T E(\mu)V]^{-1} V^T B_j,$$

$$r_i^{du}(\mu) = -C_i^T - E^*(\mu)\hat{x}_i^{du}.$$

Here, B_j is the j-th column of $B(p)$ and C_i is the i-th row of $C(p)$.

$$E(\mu) = E_0 + \mu_1 E_1 + \ldots \mu_p E_p,$$

$B(p)$, $C(p)$ are defined in (13.4), i.e. $E(\mu)$ is from Laplace transform of (13.2) or (13.3). The variable $\beta(\mu)$ is the smallest singular value of the matrix $E(\mu)$. $\hat{x}_j^{pr}$ is the state vector computed from the ROM of the system

$$E(\mu)x_j^{pr} = B_j, \tag{13.21}$$

and it approximates the exact solution x_j^{pr}. $r_j^{pr}(\mu)$ is the residual of $\hat{x}_j^{pr}$ w.r.t the system in (13.21). The state vector $\hat{x}_i^{du}$ approximates x_i^{du}, the exact solution of the auxiliary system,

$$E^*(\mu)x_i^{du} = -C_i^T, \tag{13.22}$$

where $E^*(\mu)$ is the conjugate transpose of $E(\mu)$. $\hat{x}_i^{du}$ is computed from the ROM of the auxiliary system, i.e.

$$\hat{x}_i^{du} = -V^{du}[(V^{du})^T E^*(\mu)V^{du}]^{-1}(V^{du})^T C_i^T.$$

$r_i^{du}(\mu)$ is the residual of $\hat{x}_i^{du}$ w.r.t. the auxiliary system. In the expression of $\Delta_{ij}(\mu)$, $(\hat{x}_i^{du})^*$ is the conjugate transpose of $\hat{x}_i^{du}$. The matrix V^{du} can be computed, for example, by applying Algorithm 13.1 to the auxiliary system. Note that V^{du} is also updated, once a new expansion point μ^i is selected by the greedy algorithm (see Algorithm 13.2).

It is stated in [6] that the error bound above is also the output error bound for the steady system (13.9), i.e.

$$|y_i - \hat{y}_i| \le \Delta_i(p).$$

$\Delta_i(p)$ is analogously defined by replacing the matrices $E(\mu)$, $B(p)$, $C(p)$ in (13.4) with $E(p) := (E_0 + E_1 p_1 + \ldots + E_q p_q)$, $b(p) := B(p)u$, $C(p)$ in (13.9), respectively. Now $b(p)$ is a vector. Therefore, in (13.21), $B_j = b(p)$, and the index j does not vary. The error bound $\Delta_i(p)$ is the error bound for the i-th entry of the output $\hat{y}$. Finally, $\Delta(p) = \max\limits_{1 \leq i \leq l_O} \Delta_i(p)$.

We use the package model to explain its application to ET models. Since the ET electrical subsystem is in the form of the steady system (13.9), the error bound $\Delta(p)$ for the output of the reduced electrical subsystem is given by replacing the matrices $E(p) := (E_0 + E_1 p_1 + \ldots + E_q p_q)$, $B(p)$ in (13.9) with the corresponding matrices in (13.11), and $C(p)$ is replaced with $C_E(p)$ in (13.1).

The error bound for the transfer function of (the linear part of) the reduced thermal subsystem can also be similarly defined following the definition of $\Delta(\mu)$. It is worth noting that the error bound is only suitable for linear systems. Therefore, for the thermal subsystem, we measure the error between the linear part of the original thermal subsystem and the corresponding linear part of the reduced thermal subsystem. This does not deteriorate our aim, since V_T is actually computed from the linear part, from which we seek adaptively selecting the expansion points and automatically generating V_T. More specifically, the error bound $\Delta(\mu)$ for the reduced thermal subsystem will be computed by replacing the matrices $E(\mu)$, $B(p)$, $C(p)$ with $E(s,p) := s E_T(p) - A_T(p)$, $B_T(s,p) := [A_T(p)x_T^0, B_T(p)]$ from (13.17), and $C_T(p)$ in (13.1), respectively.

Finally, we measure the error of the reduced electrical subsystem and the reduced thermal subsystem using $\Delta(p)$ and $\Delta(\mu)$, respectively. The reduced electrical and thermal subsystems can be separately obtained using Algorithm 13.2.

13.5 Simulation Results

This section shows the results of Algorithm 13.2 applied to the electrothermal simulation of the package model with 34 inputs and 68 outputs. The system is parametrized by the thickness of the top layer and excited by the inputs:

$$u_i = \begin{cases} 1, & i = 1, \\ 0, & 2 \leq i \leq 17, \\ 75 \times 10^8 t + 75, & i = 18, t \in [0, 10^{-8}], \\ 150 & i = 18, t > 10^{-8}, \\ 75 & 19 \leq i \leq 34. \end{cases}$$

The initial condition for all electrical state variables is $0\,\text{V}$, and the initial condition for all thermal state variables is $75\,°\text{C}$. For the electrical sub-system, the training set is $\{1, 2, 5, 8, 10, 20, 30, 40, 50, 60, 70, 80, 90, 100\}$ (in μm), while for the thermal sub-system, the training set contains 20 uniformly distributed frequency samples within $[0\,\mathrm{rad/s}, 100\,\mathrm{rad/s}]$, each of which is paired with a uniformly chosen thickness of the top layer within $[0\mu m, 30\mu m]$. Using the PMOR method proposed, the electrical subsystem is reduced from order 1122 to order 68, the thermal subsystem is reduced from order 8071 to order 606, and the speed-up factor for the electro-thermal simulation is 7.2.

Table 13.1: Convergence behavior of electro-thermal simulation of the package model ($\epsilon_{tol} = 10^{-4}$)

Iteration	Electrical sub-system		Thermal sub-system	
	Selected sample h	Error bound	Selected sample (s, h)	Error bound
1	1	2.1×10^3	$(8.1339, 7.5910)$	7.3×10^6
2	100	3.7×10^0	$(41.065, 29.653)$	2.3×10^1
3	90	6.6×10^{-2}	$(17.494, 15.121)$	1.3×10^{-1}
4	80	6.4×10^{-3}	$(16.455, 4.6942)$	$7.8 \times \times 10^{-5}$
5	70	5.3×10^{-3}	—	—
6	60	4.2×10^{-3}	—	—
7	50	3.1×10^{-3}	—	—
8	40	1.8×10^{-3}	—	—
9	30	8.9×10^{-4}	—	—

The convergence behavior of the adaptive PMOR method is shown in Table 13.1, where the results of Algorithm 13.2 for reduced-order modelling of the electrical and the thermal subsystems are presented, respectively. For the electrical subsystem, samples of the parameter h need to be selected as expansion points for the series expansion of the state vector x_E. It takes 9 iterations to obtain the ROM of the electrical subsystem satisfying the tolerance $\epsilon_{tol} = 10^{-4}$. After the final iteration, the biggest error bound is $\Delta(h) = 8.9 \times 10^{-4}$ at $h = 30$, and is below the tolerance, therefore the algorithm stops, the ROM is then obtained. For the thermal subsystem, Algorithm 13.2 takes 4 iterations to derive the ROM. At the fourth iteration, the biggest error bound is $\Delta(\mu) = 7.8 \times \times 10^{-5}$ at $\mu := (s, h) = (16.455, 4.6942)$, and is below ϵ_{tol}. The thermal flux output y_{36} at the output port 36, and its relative error w.r.t the parameter h and time t are shown in Fig. 13.2. The samples of the parameter in the figure might not be from the training set. It is seen that the error is still below the tolerance.

Note that the ROMs of the electrical and the thermal subsystems obtained by the PMOR method are dense. Furthermore, the package model is a system with multiple inputs and multiple outputs (MIMO). In the companion chapter [1], the proposed automatic PMOR method will be combined with the

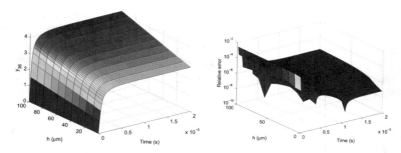

Fig. 13.2: For the package model. **Left**: The thermal flux output y_{36}. **Right**: The relative error of y_{36}.

superposition principle to derive sparse PMOR methods for MIMO systems. The method introduced there is able to obtain ROMs that are block-wise sparse.

The results of Algorithm 13.2 for the Power-MOS model are presented in the companion chapter [11]. The conductivity σ is the parameter considered. It is in fact a random variable, and Uncertainty Quantification (UQ) needs to be done for electro-thermal analysis of the model. In [11], we will show that the automatic PMOR method discussed in this chapter and the stochastic collocation method can be integrated to achieve fast UQ analysis.

13.6 Conclusions

In this chapter, we have discussed a PMOR method based on multi-moment matching for ET coupled models. Thanks to an efficient error bound for the transfer function/output, the electrical and the thermal subsystems can be reduced automatically by applying an adaptive greedy algorithm. Simulation results for a package model have demonstrated the robustness of the adaptive PMOR method.

References

1. Banagaaya, N., Feng, L., and Benner, P.: *Sparse (P)MOR for Electro-Thermal Coupled Problems with Many Inputs.* Chapter in this Book, 2019.
2. Benner, P., and Feng, L.: *A robust algorithm for parametric model order reduction based on implicit moment matching.* In: Quarteroni, A., and Rozza, G. (Eds.): *Reduced Order Methods for Modeling and Computational Reduction.* MS&A - Modeling, Simulation and Applications, 9: 159–185. Springer, Heidelberg, 2014.

3. BENNER, P., GUGERCIN, S., AND WILLCOX, K.: *A Survey of Projection-Based Model Reduction Methods for Parametric Dynamical Systems*. SIAM Review, 57(4): 483–531, 2015.
4. CHEN, Q., AND SCHOENMAKER, W.: *A new tightly-coupled transient electro-thermal Simulation Method for Power Electronics*. In: Proceedings of International Conference On Computer Aided Design (ICCAD), 2016. http://dx.doi.org/10.1145/2966986.2966993.
5. FENG, L., ANTOULAS, A.C., AND BENNER, P.: *Automated Generation of Reduced-Order Models for Linear Parametric Systems*. In: RUSSO, G., CAPASSO, V., NICOSIA, G., AND ROMANO, V. (EDS.): *Progress in Industrial Mathematics at ECMI 2014*, Series Mathematics in Industry Vol. 22, Springer International Publishing, pp. 811–818, 2016.
6. FENG, L., ANTOULAS, A.C., AND BENNER, P.: *Some a posteriori error bounds for reduced order modelling of (non-)parametrized linear systems*. ESAIM: Mathematical Modelling and Numerical Analysis (ESAIM:M2AN), 51:6, pp. 2127–2158, 2017. http://dx.doi.org/10.1051/m2an/2017014.
7. FENG, L., YUE, Y., BANAGAAYA, N., MEURIS, P., SCHOENMAKER, W., AND BENNER, P.: *Parametric Modeling and Model Order Reduction for (Electro-)Thermal Analysis of Nanoelectronic Structures*. Journal of Mathematics in Industry, 6 (10):1–16, 2016. Open access: https://dx.doi.org/10.1186/s13362-016-0030-8.
8. KEYES, D.E., MCINNES, L.C., WOODWARD, C., ET. AL.: *Multiphysics simulations: Challenges and opportunities*. International Journal of High Performance Computing Applications, 27(1):4–83, 2013.
9. KOLDA, T.G., AND BADER, B.W.: *Tensor decompositions and applications*. SIAM Review. 51(3): 455–500, 2009.
10. MEI, Q., SCHOENMAKER, W., WENG, S.H., ZHUANG, H., CHENG, C.K., AND CHEN, Q.: *An efficient transient electro-thermal simulation framework for power integrated circuits*. IEEE Transactions on Computer-Aided Design of Integrated Circuits and Systems, 35(5):832–843, 2016.
11. YUE, Y., FENG, L., BENNER, P., PULCH, R., AND SCHÖPS, S.: *Reduced Models and Uncertainty Quantification*. Chapter in this Book, 2019.

Chapter 14
Sparse (P)MOR for Electro-Thermal Coupled Problems with Many Inputs

Nicodemus Banagaaya, Lihong Feng, Peter Benner

Abstract This chapter reviews the recently proposed (parametric) model order reduction methods for electro-thermal (ET) coupled problems with many inputs, arising from (nano-)microelectronic simulation. For non-parametric problems, we discuss the block-diagonal structured model order reduction (MOR) methods (BDSM-ET) proposed in [6, 7]. For parametric problems, we discuss the parametric MOR method for ET coupled problems with many inputs (IpBDSM-ET) proposed in [3]. By construction, these methods lead to sparse reduced-order models (ROMs) and their efficiency is demonstrated using (non-)parametric ET coupled problems from industrial applications.

14.1 Introduction

Spatial discretization of electro-thermal (ET) coupled problems considering parameter variations leads to a parameterized nonlinear quadratic dynamical system of the following form:

$$\mathbf{E}(\mu)\mathbf{x}'(t) = \mathbf{A}(\mu)\mathbf{x}(t) + \mathbf{x}(t)^T \mathcal{F}(\mu)\mathbf{x}(t) + \mathbf{B}(\mu)\boldsymbol{u}(t), \quad \mathbf{x}(0) = \mathbf{x}_0 \tag{14.1a}$$
$$\boldsymbol{y}(t) = \mathbf{C}(\mu)\mathbf{x}(t) + \mathbf{D}(\mu)\boldsymbol{u}(t), \tag{14.1b}$$

where $\mathbf{x}(t) \in \mathbb{R}^n$ is the state vector and $\mathbf{E}(\mu) \in \mathbb{R}^{n \times n}$ is uniformly singular, indicating that (14.1) is a system of Differential-Algebraic Equations (DAEs) for every parameter μ. Here, $\mathbf{A}(\mu) \in \mathbb{R}^{n \times n}, \mathbf{B}(\mu) \in \mathbb{R}^{n \times m}, \mathbf{C}(\mu) \in \mathbb{R}^{\ell \times n}, \mathbf{D}(\mu) \in \mathbb{R}^{\ell \times m}$, and $\mathcal{F}(\mu) = \left[\mathbf{F}_1(\mu)^T, \ldots, \mathbf{F}_n(\mu)^T \right]^T \in \mathbb{R}^{n \times n \times n}$ is a 3-way tensor of n parametrized matrices $\mathbf{F}_i(\mu) \in \mathbb{R}^{n \times n}$. Each element in $\mathbf{x}(t)^T \mathcal{F}(\mu)\mathbf{x}(t) \in \mathbb{R}^n$ is a scalar $\mathbf{x}(t)^T \mathbf{F}_i(\mu)\mathbf{x}(t) \in \mathbb{R}, i = 1, \ldots, n$. $\boldsymbol{u} = \boldsymbol{u}(t) \in \mathbb{R}^m$

Nicodemus Banagaaya, Lihong Feng, Peter Benner
Max Planck Institute for Dynamics of Complex Technical Systems, Magdeburg, Germany, e-mail: {Banagaaya,Feng,Benner}@mpi-magdeburg.mpg.de

© Springer Nature Switzerland AG 2019
E. J. W. ter Maten et al. (eds.), *Nanoelectronic Coupled Problems Solutions*,
Mathematics in Industry 29, https://doi.org/10.1007/978-3-030-30726-4_14

and $\boldsymbol{y} = \boldsymbol{y}(t, \mu) \in \mathbb{R}^\ell$ are the inputs (excitations) and the desired outputs (observations), respectively. The vector $\mu \in \mathbb{R}^d$ represents the parameter variations which may arise from material properties, system configurations, etc. We assume that the matrices $(\mathbf{E}(\mu), \mathbf{A}(\mu), \mathbf{B}(\mu), \mathbf{C}(\mu), \mathbf{D}(\mu))$ and the tensor $\mathcal{F}(\mu)$ have an affine parameter dependence, i.e.,

$$\mathbf{M}(\mu) = \mathbf{M}_0 + \sum_{i=1}^{m} f_i(\mu)\mathbf{M}_i, \qquad (14.2)$$

where the scalar functions f_i determine the parameter dependency, which can be nonlinear functions of μ, and $\mathbf{M}_i$ can be either a constant matrix or a constant tensor.

If $\mathbf{M}(\mu) = \mathbf{M}_0$ in (14.2) for all matrices and for the tensor in (14.1), (e.g., $\mathbf{E}(\mu) = \mathbf{E}, \mathbf{A}(\mu) = \mathbf{A}, \mathbf{B}(\mu) = \mathbf{B}, \mathbf{C}(\mu) = \mathbf{C}, \mathbf{D}(\mu) = \mathbf{D}$ and $\mathcal{F}(\mu) = \mathcal{F}$), then system (14.1) is considered to be non-parametric leading to

$$\mathbf{E}\mathbf{x}'(t) = \mathbf{A}\mathbf{x}(t) + \mathbf{x}(t)^T \mathcal{F}\mathbf{x}(t) + \mathbf{B}\boldsymbol{u}(t), \quad \mathbf{x}(0) = \mathbf{x}_0, \qquad (14.3\text{a})$$
$$\boldsymbol{y}(t) = \mathbf{C}\mathbf{x}(t) + \mathbf{D}\boldsymbol{u}(t). \qquad (14.3\text{b})$$

In practice, parametric and non-parametric systems have very large dimensions compared to the number of inputs and outputs. Despite the ever increasing computational power, simulation of such systems in acceptable time is very difficult, in particular if multi-query tasks are required. This calls for the application of Parametric Model Order Reduction (PMOR) and non-parametric Model Order Reduction (MOR) to parametric and non-parametric systems, respectively.

PMOR replaces (14.1) by a parametric reduced-order model (PROM):

$$\mathbf{E}_r(\mu)\mathbf{x}'_r(t) = \mathbf{A}_r(\mu)\mathbf{x}_r(t) + \mathbf{x}_r(t)^T \mathcal{F}_r(\mu)\mathbf{x}_r(t) + \mathbf{B}_r(\mu)\mathbf{u}(t), \quad (14.4\text{a})$$
$$\mathbf{y}_r(t) = \mathbf{C}_r(\mu)\mathbf{x}_r(t) + \mathbf{D}(\mu)\mathbf{u}(t), \qquad (14.4\text{b})$$
$$\mathbf{x}_r(0) = \mathbf{x}_{r_0}, \qquad (14.4\text{c})$$

where $\mathbf{E}_r(\mu) = \mathbf{V}^T \mathbf{E}(\mu)\mathbf{V}, \mathbf{A}_r(\mu) = \mathbf{V}^T \mathbf{A}(\mu)\mathbf{V}, \mathbf{B}_r(\mu) = \mathbf{V}^T \mathbf{B}(\mu), \mathbf{C}_r(\mu) = \mathbf{C}(\mu)\mathbf{V}$ and $\mathbf{V} \in \mathbb{R}^{n \times r}$ with $r \ll n$, from which the dimensions of the other reduced matrices follow readily.

Proposition 14.1 ([5]). *Let* $\mathbf{W} = (\mathbf{w}_{ij}) \in \mathbb{R}^{n \times r}$ *be a parameter-independent matrix,* $\mathbf{x}_r \in \mathbb{R}^r$, *and* $\tilde{\mathcal{F}}(\mu) = \left[\tilde{\mathbf{F}}_1(\mu)^T, \dots, \tilde{\mathbf{F}}_n(\mu)^T\right]^T \in \mathbb{R}^{r \times r \times n}$ *a 3-way parameterized tensor, then there exists a 3-way parameterized tensor* $\mathcal{F}_r(\mu) \in \mathbb{R}^{r \times r \times r}$, *such that:*

$$\mathbf{W}^T \left(\mathbf{x}_r^T \tilde{\mathcal{F}}(\mu)\mathbf{x}_r\right) = \mathbf{x}_r^T \mathcal{F}_r(\mu)\mathbf{x}_r, \quad where \quad \mathcal{F}_r(\mu) = \left[\hat{\mathbf{F}}_1(\mu)^T, \dots, \hat{\mathbf{F}}_r(\mu)^T\right]^T$$

with $\hat{\mathbf{F}}_j(\mu) = \sum_{i=1}^{n} \mathbf{w}_{ij}\tilde{\mathbf{F}}_i(\mu) \in \mathbb{R}^{r \times r}, j = 1, \dots, r.$

The proof can be found in [5]. Using Proposition 14.1, the reduced nonlinear term in (14.4a) is obtained by reformulating

$$\mathbf{V}^T\big(\mathbf{x}(t)^T\big(\mathbf{V}^T\mathcal{F}(\mu)\mathbf{V}\big)\mathbf{x}(t)\big) \quad \text{into} \quad \mathbf{x}_r(t)^T\mathcal{F}_r(\mu)\mathbf{x}_r(t),$$

where $\mathcal{F}_r(\mu) = \big[\mathbf{F}_{r_1}(\mu)^T,\ldots,\mathbf{F}_{r_r}(\mu)^T\big]^T \in \mathbb{R}^{r\times r\times r}$ is a 3-way tensor with parametrized matrices $\mathbf{F}_{r_i}(\mu) \in \mathbb{R}^{r\times r}$, and

$$\mathbf{V}^T\mathcal{F}(\mu)\mathbf{V} := \big[\mathbf{V}^T\mathbf{F}_1(\mu)^T\mathbf{V},\ldots,\mathbf{V}^T\mathbf{F}_n(\mu)^T\mathbf{V}\big]^T.$$

The projection matrix $\mathbf{V} \in \mathbb{R}^{n\times r}$ which is valid for all parameters μ in the desired range, and for arbitrary inputs $\mathbf{u(t)}$, can be constructed using, e.g., the implicit moment-matching PMOR method from [9]. However, direct application of this approach to the nonlinear quadratic DAE system (14.1) may produce PROMs which are inaccurate or difficult to simulate due to the underlying hidden constraints. Moreover, if m is large, standard PMOR methods lead to very large and dense ROMs.

For the case of non-parametric systems, MOR replaces (14.3) by a reduced-order model (ROM):

$$\mathbf{E}_r\mathbf{x}_r'(t) = \mathbf{A}_r\mathbf{x}_r(t) + \mathbf{x}_r(t)^T\mathcal{F}_r\mathbf{x}_r(t) + \mathbf{B}_r\mathbf{u}(t), \mathbf{x}_r(0) = \mathbf{x}_{r_0}, \quad (14.5a)$$
$$\mathbf{y}_r(t) = \mathbf{C}_r\mathbf{x}_r(t) + \mathbf{D}\mathbf{u}(t), \quad (14.5b)$$

where $\mathbf{E}_r = \mathbf{V}^T\mathbf{E}\mathbf{V}, \mathbf{A}_r = \mathbf{V}^T\mathbf{A}\mathbf{V}, \mathbf{B}_r = \mathbf{V}^T\mathbf{B}, \mathbf{C}_r = \mathbf{C}\mathbf{V}$. A good ROM should have small approximation error $\|\mathbf{y} - \mathbf{y}_r\|$ in a suitable norm $\|\cdot\|$ for every arbitrary input $\mathbf{u}(t)$. There exist many MOR methods for nonlinear (quadratic) systems such as the snapshot and implicit moment-matching methods, see [8] for a general discussion of MOR methods. The snapshot methods are not flexible for input-dependent systems as considered in this work, hence, we consider the input-independent MOR methods, such as the implicit moment-matching methods [8]. However, it is well known that as the number of inputs increases, the efficiency of moment-matching MOR methods decreases, since the size of the ROM is proportional to the number of inputs. Moreover, they cannot be applied directly to quadratic DAEs [5]. In general, models with numerous inputs and outputs are challenging for MOR, and most MOR methods produce large and dense ROMs for such systems. In [6], the BDSM-ET and SIP-ET methods for non-parameterized ET coupled problems with many inputs are proposed to overcome this problem. The BDSM-ET method is more accurate and leads to much smaller ROMs than the SIP-ET method. This is because the size of the ROM generated by the SIP-ET method depends on the number of zero rows of the input matrices for electrical and thermal subsystems, respectively. However, the BDSM-ET ROMs have dense matrices in the electrical subsystem and a dense 3-way tensor in the thermal subsystem, which restricts their applicability to small and medium sized ET systems. In [7], the modified BDSM-ET method based on the superposition principle was introduced, which leads to sparse ROMs for non-parameterized ET systems. The BDSM-ET methods can naturally be extended to parameterized systems as discussed in [3].

In Sect. 14.2, we review the projection based (p)MOR method for ET coupled problems. In Sect. 14.3 and 14.4, we discuss the BDSM-ET method and its modified invariant, respectively. Sect. 14.5 discusses the modified BDSM-ET method [7] for parametrized ET coupled problems with many inputs. Finally, we present numerical experiments from industrial applications and conclusions. For simplicity, we remove (t) for time dependent variables in the next sections.

14.2 ET Coupled Problems and Projection-Based (P)MOR

In this section, we discuss the projection-based (P)MOR method for ET coupled problems. Recall that discretization of ET coupled problems results in (non-) parametric quadratic DAEs. It is well known that direct simulation and model-order reduction of DAEs is very difficult especially if the index of a given DAE is greater than one. According to [2], the best way to deal with DAEs is first splitting them into underlying differential and algebraic subsystems before simulation or model order reduction.

14.2.1 Decoupling of ET Coupled Problems

Parameterized ET coupled problems arising from nanoelectronics take the form of (14.1) with system matrices and tensor structures as follows:

$$\mathbf{E}(\mu) = \begin{pmatrix} 0 & 0 \\ 0 & \mathbf{E}_T(\mu) \end{pmatrix}, \quad \mathbf{A}(\mu) = \begin{pmatrix} \mathbf{A}_v & 0 \\ 0 & \mathbf{A}_T(\mu) \end{pmatrix}, \quad \mathbf{B}(\mu) = \begin{pmatrix} \mathbf{B}_v(\mu) & 0 \\ 0 & \mathbf{B}_T(\mu) \end{pmatrix},$$

$$\mathbf{C}(\mu) = \begin{pmatrix} \mathbf{C}_v(\mu) & \mathbf{C}_T(\mu) \end{pmatrix}, \quad \mathbf{D}(\mu) = \begin{pmatrix} \mathbf{D}_v(\mu) & \mathbf{D}_T(\mu) \end{pmatrix}, \quad \mathbf{u} = \begin{pmatrix} \mathbf{u}_v^T, \mathbf{u}_T^T \end{pmatrix}^T,$$

$$\mathcal{F}(\mu) = \begin{pmatrix} 0, \ldots, 0, \mathbf{F}_{n_v+1}(\mu)^T, \ldots, \mathbf{F}_n(\mu)^T \end{pmatrix}^T, \tag{14.6}$$

where

$$\mathbf{F}_i(\mu) = \begin{pmatrix} \mathbf{F}_{v_i}(\mu) & 0 \\ 0 & 0 \end{pmatrix} \in \mathbb{R}^{n \times n}, \mathbf{F}_{v_i}(\mu) \in \mathbb{R}^{n_v \times n_v}, i = n_v + 1, \ldots, n,$$

$$\mathbf{A}_v(\mu) \in \mathbb{R}^{n_v \times n_v}, \mathbf{B}_v(\mu) \in \mathbb{R}^{n_v \times \tilde{m}},$$

$$\mathbf{E}_T(\mu) \in \mathbb{R}^{n_T \times n_T}, \mathbf{A}_T(\mu) \in \mathbb{R}^{n_T \times n_T}, \mathbf{B}_T(\mu) \in \mathbb{R}^{n_T \times \tilde{m}},$$

$$\mathbf{C}_v(\mu) \in \mathbb{R}^{\ell \times n_v}, \mathbf{C}_T(\mu) \in \mathbb{R}^{\ell \times n_T},$$

$$\mathbf{D}_v(\mu) \in \mathbb{R}^{\ell \times \tilde{m}}, \mathbf{D}_T(\mu) \in \mathbb{R}^{\ell \times \tilde{m}},$$

and $\mathbf{u}_v, \mathbf{u}_T \in \mathbb{R}^{\tilde{m}}, \tilde{m} = m/2$. Thus, substituting the above matrices and the tensor $\mathcal{F}(\mu)$ into (14.1) leads to an equivalent parametrized decoupled system given by

$$\mathbf{A}_v(\mu)\mathbf{x}_v = -\mathbf{B}_v(\mu)\mathbf{u}_v, \tag{14.7a}$$

$$\mathbf{E}_T(\mu)\mathbf{x}'_T = \mathbf{A}_T(\mu)\mathbf{x}_T + \mathbf{x}_v^T \mathcal{F}_T(\mu)\mathbf{x}_v + \mathbf{B}_T(\mu)\mathbf{u}_T, \quad \mathbf{x}_T(0) = \mathbf{x}_{T_0}, \tag{14.7b}$$

$$\mathbf{y} = \mathbf{C}_v(\mu)\mathbf{x}_v + \mathbf{C}_T(\mu)\mathbf{x}_T + \mathbf{D}_v(\mu)\mathbf{u}_v + \mathbf{D}_T(\mu)\mathbf{u}_T, \tag{14.7c}$$

where $\mathcal{F}_T(\mu) = \left(\mathbf{F}_{T_1}(\mu)^T, \ldots, \mathbf{F}_{T_{n_T}}(\mu)^T\right)^T \in \mathbb{R}^{n_v \times n_v \times n_T}, \mathbf{F}_{T_j}(\mu) = \mathbf{F}_{v_{n_v+j}}(\mu),$ $j = 1, \ldots, n_T$ and $\mathbf{F}_{v_i}(\mu)$ is as defined earlier. Equations (14.7a) and (14.7b) are the electrical and thermal subsystems, respectively. The output solutions can be obtained through (14.7c). We observe that the nonlinear term $\mathbf{x}_v^T \mathcal{F}_T(\mu)\mathbf{x}_v$ can be treated as part of the thermal input, since it is known after first simulating the electrical subsystem.

14.2.2 Projection-Based (P)MOR of ET Coupled Problems

PMOR replaces the decoupled system (14.7) with a reduced-order decoupled system

$$\mathbf{A}_{v_r}(\mu)\mathbf{x}_{v_r} = -\mathbf{B}_{v_r}(\mu)\mathbf{u}_v, \tag{14.8a}$$

$$\mathbf{E}_{T_r}(\mu)\mathbf{x}'_{T_r} = \mathbf{A}_{T_r}(\mu)\mathbf{x}_{T_r} + \mathbf{x}_{v_r}^T \mathcal{F}_{T_r}(\mu)\mathbf{x}_{v_r} + \mathbf{B}_{T_r}(\mu)\mathbf{u}_T, \quad \mathbf{x}_{T_r}(0) = \mathbf{x}_{T_{r_0}}, \tag{14.8b}$$

$$\mathbf{y}_r = \mathbf{C}_{v_r}(\mu)\mathbf{x}_{v_r} + \mathbf{C}_{T_r}(\mu)\mathbf{x}_{T_r} + \mathbf{D}_v(\mu)\mathbf{u}_v + \mathbf{D}_T(\mu)\mathbf{u}_T, \tag{14.8c}$$

where $\mathbf{A}_{v_r}(\mu) \in \mathbb{R}^{r_v \times r_v}, \mathbf{B}_{v_r}(\mu) \in \mathbb{R}^{r_v \times \tilde{m}}, \mathbf{E}_{T_r}(\mu) \in \mathbb{R}^{r_T \times r_T}$ and $\mathbf{A}_{T_r}(\mu) \in \mathbb{R}^{r_T \times r_T}, \mathbf{B}_{T_r}(\mu) \in \mathbb{R}^{r_T \times \tilde{m}}, \mathbf{C}_{v_r}(\mu) \in \mathbb{R}^{\ell \times r_v}, \mathbf{C}_{T_r}(\mu) \in \mathbb{R}^{\ell \times r_T}$ and finally $\mathcal{F}_{T_r}(\mu) \in \mathbb{R}^{r_v \times r_v \times r_T}$, such that the reduced order is $r = r_v + r_T \ll n$. $\tilde{m} = m/2$ as defined earlier. This PMOR method which was proposed in [4] produces simple and accurate ROMs for ET coupled problems. However, if $\tilde{m}$ is large, it may lead to ROMs (14.8) which are very large and dense. In Sec. 14.5, we review the BDSM-ET method for parameterized ET coupled systems with many inputs (IpBDSM-ET) proposed in [3]. Small and sparse PROMs (14.8) can be obtained by the BDSM-ET method.

The non-parameterized decoupled system of (14.3) can be deduced from (14.7) leading to

$$\mathbf{A}_v \mathbf{x}_v = -\mathbf{B}_v \mathbf{u}_v, \tag{14.9a}$$

$$\mathbf{E}_T \mathbf{x}_T' = \mathbf{A}_T \mathbf{x}_T + \mathbf{x}_v^T \mathcal{F}_T \mathbf{x}_v + \mathbf{B}_T \mathbf{u}_T, \quad \mathbf{x}_T(0) = \mathbf{x}_{T_0}, \tag{14.9b}$$

$$\mathbf{y} = \mathbf{C}_v \mathbf{x}_v + \mathbf{C}_T \mathbf{x}_T + \mathbf{D}_v \mathbf{u}_v + \mathbf{D}_T \mathbf{u}_T. \tag{14.9c}$$

MOR replaces the decoupled system (14.9) with a reduced-order decoupled system

$$\mathbf{A}_{v_r} \mathbf{x}_{v_r} = -\mathbf{B}_{v_r} \mathbf{u}_v, \tag{14.10a}$$

$$\mathbf{E}_{T_r} \mathbf{x}_{T_r}' = \mathbf{A}_{T_r} \mathbf{x}_{T_r} + \mathbf{x}_{v_r}^T \mathcal{F}_{T_r} \mathbf{x}_{v_r} + \mathbf{B}_{T_r} \mathbf{u}_T, \quad \mathbf{x}_{T_r}(0) = \mathbf{x}_{T_{r_0}}, \tag{14.10b}$$

$$\mathbf{y}_r = \mathbf{C}_{v_r} \mathbf{x}_{v_r} + \mathbf{C}_{T_r} \mathbf{x}_{T_r} + \mathbf{D}_v \mathbf{u}_v + \mathbf{D}_T \mathbf{u}_T, \tag{14.10c}$$

where $\mathbf{A}_{v_r} \in \mathbb{R}^{r_v \times r_v}$, $\mathbf{B}_{v_r} \in \mathbb{R}^{r_v \times \tilde{m}}$, $\mathbf{E}_{T_r} \in \mathbb{R}^{r_T \times r_T}$, $\mathbf{A}_{T_r} \in \mathbb{R}^{r_T \times r_T}$, $\mathbf{B}_{T_r} \in \mathbb{R}^{r_T \times \tilde{m}}$, $\mathbf{C}_{v_r} \in \mathbb{R}^{\ell \times r_v}$, $\mathbf{C}_{T_r} \in \mathbb{R}^{\ell \times r_T}$, $\mathcal{F}_{T_r} \in \mathbb{R}^{r_v \times r_v \times r_T}$, such that the reduced order $r = r_v + r_T \ll n$. In order to obtain the ROM (14.10), we combine the MOR techniques for algebraic equations and the conventional MOR methods for ODEs to obtain (14.10a) and (14.10b), respectively. MOR methods based on Gaussian elimination could be applied, if the input matrix $\mathbf{B}_v$ has many zero rows, see [6]. Nevertheless, this approach still leads to large and dense ROMs for ET coupled systems with many inputs. In Sec. 14.3 and 14.4, we discuss the BDSM-ET method and its modified version which can be used to reduce non-parameterized ET coupled systems with many inputs efficiently.

14.3 A Block-Diagonal Structured MOR Method for Non-Parametric ET Problems: BDSM-ET Method

In this section, we discuss the BDSM-ET method for ET coupled systems which was proposed in [6]. This method applies the Gaussian elimination based methods, such as SIP [12] and the BDSM method [13], to the electrical (14.9a) and thermal (14.9b) subsystems, respectively, to obtain a reduced-order ET model of the form (14.9). This is briefly described as follows. Assume that $\mathbf{B}_v$ in (14.9a) has many zero rows, then the electrical subsystem (14.9a) can be reformulated and partitioned as

$$\begin{pmatrix} \mathbf{A}_{v_{11}} & \mathbf{A}_{v_{12}} \\ \mathbf{A}_{v_{12}}^T & \mathbf{A}_{v_{22}} \end{pmatrix} \begin{pmatrix} \mathbf{x}_{v_e} \\ \mathbf{x}_{v_I} \end{pmatrix} = - \begin{pmatrix} \mathbf{B}_{v_e} \\ 0 \end{pmatrix} \mathbf{u}_v, \quad \mathbf{y}_v = \begin{pmatrix} \mathbf{C}_{v_e} & 0 \end{pmatrix} \begin{pmatrix} \mathbf{x}_{v_e} \\ \mathbf{x}_{v_I} \end{pmatrix} + \mathbf{D}_v \mathbf{u}_v,$$

$$\tag{14.11}$$

where $\mathbf{x}_{v_e} \in \mathbb{R}^{n_{v_e}}$ and $\mathbf{x}_{v_I} \in \mathbb{R}^{n_{v_I}}$ represent the port and the internal nodal voltages, respectively, and $n_v = n_{v_e} + n_{v_I}$. Eliminating all internal nodes from (14.11) leads to the reduced-order electrical subsystem (14.10a) with matrix coefficients

$$\mathbf{A}_{v_r} = [\mathbf{A}_{v_{11}} - \mathbf{A}_{v_{12}}\mathbf{W}_v] \in \mathbb{R}^{r_v \times r_v}, \mathbf{B}_{v_r} = \mathbf{B}_{v_e} \in \mathbb{R}^{r_v \times \tilde{m}}, \mathbf{C}_{v_r} = \mathbf{C}_{v_e} \in \mathbb{R}^{\ell \times r_v},$$
(14.12)

where $\mathbf{W}_v = \mathbf{A}_{v_{22}}^{-1}\mathbf{A}_{v_{12}}^{T} \in \mathbb{R}^{n_{v_I} \times n_{v_e}}$, $\mathbf{x}_{v_r} = \mathbf{x}_{v_e} \in \mathbb{R}^{r_v}$, and the order of the reduced electrical subsystem becomes $r_v = n_{v_e} \ll n_v$. Clearly, the size of the reduced electrical subsystem (14.12) depends on the number of zero rows in $\mathbf{B}_v$. The reduction in the electrical subsystem induces a reduction in the thermal subsystem through the nonlinear part, leading to

$$\mathbf{E}_T\mathbf{x}_T' = \mathbf{A}_T\mathbf{x}_T + \mathbf{x}_{v_r}^T\tilde{\mathcal{F}}_T\mathbf{x}_{v_r} + \mathbf{B}_T\mathbf{u}_T, \quad \mathbf{x}_T(0) = \mathbf{x}_{T_0},$$
$$\mathbf{y}_T = \mathbf{C}_T\mathbf{x}_T + \mathbf{D}_T\mathbf{u}_T,$$
(14.13)

where $\tilde{\mathcal{F}}_T = \mathcal{F}_{T_{11}} - \mathbf{W}_v^T\mathcal{F}_{T_{21}} - \mathcal{F}_{T_{12}}\mathbf{W}_v + \mathbf{W}_v^T\mathcal{F}_{T_{22}}\mathbf{W}_v \in \mathbb{R}^{r_v \times r_v \times n_T}$ is a 3-way tensor. The 3-way tensors $\mathcal{F}_{T_{11}} \in \mathbb{R}^{n_{v_e} \times n_{v_e} \times n_T}$, $\mathcal{F}_{12} \in \mathbb{R}^{n_{v_e} \times n_{v_I} \times n_T}$ and $\mathcal{F}_{21} \in \mathbb{R}^{n_{v_I} \times n_{v_e} \times n_T}$, $\mathcal{F}_{22} \in \mathbb{R}^{n_{v_I} \times n_{v_I} \times n_T}$ are the partitions of the tensor $\mathcal{F}_T$ corresponding to the partitions in (14.11). The next step is to apply the superposition principle to (14.13). Assume that the thermal input matrix $\mathbf{B}_T$ has no zero columns, so that it can be split into $\mathbf{B}_T = \sum_{i=1}^{\tilde{m}} \mathbf{B}_{T_i}$, where $\mathbf{B}_{T_i} \in \mathbb{R}^{n_T \times \tilde{m}}$ are column rank-1 matrices defined as

$$\mathbf{B}_{T_i}(:,j) = \begin{cases} \mathbf{b}_{T_i} \in \mathbb{R}^{n_T}, & \text{if } j = i, \\ 0, & \text{otherwise}, \end{cases} \quad i = 1,\ldots,\tilde{m}.$$

Here and below, blkdiag denotes the block-diagonal matrix defined by the input arguments. Applying the two-stage superposition principle from [6] to (14.13) leads to a block-diagonal structured system of dimension $\tilde{m}n_T$ given by

$$\mathcal{E}_T\tilde{\mathbf{x}}_T' = \mathcal{A}_T\tilde{\mathbf{x}}_T + \mathbf{x}_{v_r}^T\mathcal{F}_T\mathbf{x}_{v_r} + \mathcal{B}_T\mathbf{u}_T, \quad \tilde{\mathbf{x}}_T(0) = [\mathbf{x}_T(0),0]^T,$$
$$\mathbf{y}_T = \mathcal{C}_T\tilde{\mathbf{x}}_T + \mathbf{D}_T\mathbf{u}_T,$$
(14.14)

where

$$\mathcal{E}_T = \text{blkdiag}(\mathbf{E}_T,\ldots,\mathbf{E}_T) \in \mathbb{R}^{\tilde{m}n_T \times \tilde{m}n_T},$$
$$\mathcal{C}_T = (\mathbf{C}_T,\ldots,\mathbf{C}_T) \in \mathbb{R}^{\ell \times \tilde{m}n_T},$$
$$\mathcal{A}_T = \text{blkdiag}(\mathbf{A}_T,\ldots,\mathbf{A}_T) \in \mathbb{R}^{\tilde{m}n_T \times \tilde{m}n_T},$$
$$\mathcal{B}_T = (\mathbf{B}_{T_1}^{T},\ldots,\mathbf{B}_{T_{\tilde{m}}}^{T})^T \in \mathbb{R}^{\tilde{m}n_T \times \tilde{m}},$$
$$\mathcal{F}_T = \begin{pmatrix} \tilde{\mathcal{F}}_T \\ 0 \end{pmatrix} \in \mathbb{R}^{r_v \times r_v \times \tilde{m}n_T}.$$

The corresponding reduced-order thermal subsystem in the form of (14.10b) has block-diagonal structured matrices given by

$$\mathbf{E}_{T_r} = \mathbf{V}^{\mathrm{T}} \mathcal{E}_T \mathbf{V}, \quad \mathbf{A}_{T_r} = \mathbf{V}^{\mathrm{T}} \mathcal{A}_T \mathbf{V}, \quad \mathbf{B}_{T_r} = \mathbf{V}^{\mathrm{T}} \mathcal{B}_T, \quad \mathbf{C}_{T_r} = \mathcal{C}_T \mathbf{V}, \quad (14.15)$$

where $\mathbf{V} = \mathrm{blkdiag}(\mathbf{V}^{(1)}, \ldots, \mathbf{V}^{(\tilde{m})})$. Consider an arbitrarily chosen expansion point $s_0 \in \mathbb{C}$, the projection matrices $\mathbf{V}^{(i)}$ can be constructed from each subsystem of (14.14) as (see [6] for details)

$$\mathrm{range}(\mathbf{V}^{(i)}) = \mathrm{span}\{\mathbf{R}_i, \mathbf{M}\mathbf{R}_i, \ldots, \mathbf{M}^{r_{T_i}-1}\mathbf{R}_i\}, \quad r_{T_i} \ll n_T,$$

where $\mathbf{M} = (s_0 \mathbf{E}_T - \mathbf{A}_T)^{-1} \mathbf{E}_T \in \mathbb{R}^{n_T \times n_T}$, and $\mathbf{R}_i = (s_0 \mathbf{E}_T - \mathbf{A}_T)^{-1} \mathbf{b}_{T_i} \in \mathbb{R}^{n_T}$, $i = 1, \ldots, \tilde{m}$. The nonlinear term $\mathbf{V}^{\mathrm{T}} \left(\mathbf{x}_{v_r}^T \mathcal{F}_T \mathbf{x}_{v_r} \right)$ can be reformulated as the reduced-order nonlinear term $\mathbf{x}_{v_r}^T \mathcal{F}_{T_r} \mathbf{x}_{v_r}$ using the non-parameterized form of Proposition 14.1. We see that $\mathcal{F}_{T_r}$ in the reduced-order nonlinear term is independent of the time t and can be precomputed before simulating the ROM. Therefore, reformulating the nonlinear term further improves the efficiency of simulating the ROM. It can be seen that $\mathbf{V}^{(i)}$ depends on the single column $\mathbf{b}_{T_i}$, rather than $\mathbf{B}_T$ with many columns, leading to a block-wise sparse ROM as compared with the standard moment-matching methods, such as PRIMA [11]. Finally, the order of the reduced thermal subsystem (14.10b) is $r_T = \sum_{i=1}^{\tilde{m}} r_{T_i}$. From the analysis in [6,13], the block-diagonal system (14.14) yields a system equivalent to (14.13), so that the block-diagonal ROM of (14.14) can be considered as the ROM of (14.13). However, the matrix $\mathbf{A}_{v_r}$ and the tensor $\mathcal{F}_{T_r}$ in the ROM are dense which is still a computational and storage burden. In the next section, we review the modified BDSM-ET method which leads to sparser ROMs also in the electrical subsystem.

14.4 Modified BDSM-ET for Non-Parametric ET Problems

In this section, we discuss the modified BDSM-ET method from [7]. The goal of the modified BDSM-ET method is to reduce the computational and storage burden of simulating the reduced electrical subsystem and the reduced nonlinear term in the thermal subsystem, obtained using the BDSM-ET method discussed in the previous section. The main idea is to first apply the superposition principle to both the electrical and the thermal subsystems before MOR. Assume that the electrical input matrix $\mathbf{B}_v$ has no zero columns, so that it can be split into $\mathbf{B}_v = \sum_{i=1}^{\tilde{m}} \mathbf{B}_{v_i}$, where $\mathbf{B}_{v_i} \in \mathbb{R}^{n_v \times \tilde{m}}$ is a column rank-1 matrix defined as

$$\mathbf{B}_{v_i}(:,j) = \begin{cases} \mathbf{b}_{v_i} \in \mathbb{R}^{n_v}, & \text{if } j = i, \\ 0, & \text{otherwise}, \end{cases} \quad i = 1,\ldots,\tilde{m}.$$

Applying the superposition principle to the electrical subsystem in (14.9) results in an equivalent block-diagonal algebraic system

$$\mathcal{A}_v \xi_v = -\mathcal{B}_v \mathbf{u}_v, \quad \mathbf{y}_v = \mathcal{C}_v \xi_v, \tag{14.16}$$

where $\mathcal{A}_v = \text{blkdiag}(\mathbf{A}_v,\ldots,\mathbf{A}_v)$, $\mathcal{B}_v = (\mathbf{B}_{v_1}^T,\ldots,\mathbf{B}_{v_{\tilde{m}}}^T)^T$, $\mathcal{C}_v = (\mathbf{C}_v,\ldots,\mathbf{C}_v)$ and $\xi_v = (\mathbf{x}_{v_1}^T,\ldots,\mathbf{x}_{v_{\tilde{m}}}^T)^T$. The next step is to reduce the dimension of (14.16). This is done by applying reordering and elimination techniques to each subsystem of (14.16):

$$\mathbf{A}_v \mathbf{x}_{v_i} = -\mathbf{B}_{v_i} \mathbf{u}_v, \quad \mathbf{y}_{v_i} = \mathbf{C}_v \mathbf{x}_{v_i}, i = 1,\ldots,\tilde{m}. \tag{14.17}$$

Assuming each $\mathbf{B}_{v_i}$ has many zero rows, then each subsystem in (14.17) can be reformulated as

$$\begin{pmatrix} \mathbf{A}_{v11}^{(i)} & \mathbf{A}_{v12}^{(i)} \\ \mathbf{A}_{v12}^{(i)T} & \mathbf{A}_{v22}^{(i)} \end{pmatrix} \begin{pmatrix} \mathbf{x}_{ve}^{(i)} \\ \mathbf{x}_{vI}^{(i)} \end{pmatrix} = - \begin{pmatrix} \mathbf{B}_{ve}^{(i)} \\ 0 \end{pmatrix} \mathbf{u}_v, \quad \mathbf{y}_{v_i} = \begin{pmatrix} \mathbf{C}_{ve}^{(i)} & 0 \end{pmatrix} \begin{pmatrix} \mathbf{x}_{ve}^{(i)} \\ \mathbf{x}_{vI}^{(i)} \end{pmatrix}, \tag{14.18}$$

where $\mathbf{x}_{ve}^{(i)} \in \mathbb{R}^{n_{ve}^{(i)}}$ and $\mathbf{x}_{vI}^{(i)} \in \mathbb{R}^{n_{vI}^{(i)}}$ represent the port and the internal nodal voltages, respectively, and $n_v = n_{ve}^{(i)} + n_{vI}^{(i)}$, $i = 1,\ldots,\tilde{m}$. Eliminating all internal nodes from (14.18) leads to the ROM of each subsystem as

$$\mathbf{A}_{v r_i} \mathbf{x}_{v r_i} = \mathbf{B}_{v r_i} \mathbf{u}_v, \quad \mathbf{y}_{v r_i} = \mathbf{C}_{v r_i} \mathbf{x}_{v r_i}, \tag{14.19}$$

where $\mathbf{A}_{v r_i} = [\mathbf{A}_{v11}^{(i)} - \mathbf{A}_{v12}^{(i)} \mathbf{W}_{v_i}] \in \mathbb{R}^{r_{v_i} \times r_{v_i}}$, $\mathbf{B}_{v r_i} = -\mathbf{B}_{ve}^{(i)} \in \mathbb{R}^{r_{v_i} \times \tilde{m}}$, $\mathbf{C}_{v r_i} = \mathbf{C}_{ve}^{(i)} \in \mathbb{R}^{\ell \times r_{v_i}}$, $\mathbf{W}_{v_i} = \mathbf{A}_{v22}^{(i)-1} \mathbf{A}_{v12}^{(i)T} \in \mathbb{R}^{n_{vI}^{(i)} \times n_{ve}^{(i)}}$, $\mathbf{x}_{v r_i} = \mathbf{x}_{ve}^{(i)} \in \mathbb{R}^{r_{v_i}}$, and $r_{v_i} = n_{ve}^{(i)} \ll n_v$. Replacing each $\mathbf{A}_v, \mathbf{B}_{v_i}, \mathbf{C}_v, \mathbf{x}_{v_i}$ in (14.16) with $\mathbf{A}_{v r_i}, \mathbf{B}_{v r_i}, \mathbf{C}_{v r_i}, \mathbf{x}_{v r_i}$ leads to the ROM of (14.16), which is also the ROM of (14.9a) of dimension $r_v = \sum_{i=1}^{\tilde{m}} r_{v_i}$ and with matrices

$$\mathbf{A}_{v r} = \text{blkdiag}(\mathbf{A}_{v r_1},\ldots,\mathbf{A}_{v r_{\tilde{m}}}),$$
$$\mathbf{B}_{v r} = (\mathbf{B}_{v r_1}^T,\ldots,\mathbf{B}_{v r_{\tilde{m}}}^T)^T,$$
$$\mathbf{C}_{v r} = (\mathbf{C}_{v r_1},\ldots,\mathbf{C}_{v r_{\tilde{m}}}).$$

Finally, we reduce the thermal subsystem (14.9b). Here, we discuss an approach which produces a much sparser reduced 3-way tensor than that obtained using the BDSM-ET method. Applying the superposition principle to the algebraic subsystem (14.9a) introduces $\left(\sum_{i=1}^{\tilde{m}} \mathbf{x}_{v_i}^T\right) \mathcal{F}_T \left(\sum_{i=1}^{\tilde{m}} \mathbf{x}_{v_i}\right)$ into the thermal subsystem, i.e. $\mathbf{x}_v$ is replaced by $\sum_{i=1}^{\tilde{m}} \mathbf{x}_{v_i}$ in the nonlinear part. In order to obtain a sparse tensor, we assume small signal modeling, i.e., the approximation $\left(\sum_{i=1}^{\tilde{m}}\mathbf{x}_{v_i}^T\right)\mathcal{F}_T\left(\sum_{i=1}^{\tilde{m}}\mathbf{x}_{v_i}\right) \approx \sum_{i=1}^{\tilde{m}}\mathbf{x}_{v_i}^T\mathcal{F}_T\mathbf{x}_{v_i}$ is introduced into the thermal subsystem. Thus (14.9b) can be approximated as

$$\mathbf{E}_T\mathbf{x}_T' = \mathbf{A}_T\mathbf{x}_T + \xi_v^T\mathcal{F}_T\xi_v, + \mathbf{B}_T\mathbf{u}_T, \quad \mathbf{x}_T(0) = \mathbf{x}_{T_0}, \tag{14.20a}$$
$$\mathbf{y}_T = \mathbf{C}_T\mathbf{x}_T + \mathbf{D}_T\mathbf{u}_T. \tag{14.20b}$$

We have used the equality $\sum_{i=1}^{\tilde{m}}\mathbf{x}_{v_i}^T\mathcal{F}_T\mathbf{x}_{v_i} = \xi_v^T\mathcal{F}_T\xi_v$, where

$$\mathcal{F}_T = \left[\mathcal{F}_{T_1}^T, \ldots, \mathcal{F}_{T_{n_T}}^T\right]^T \in \mathbb{R}^{\tilde{n}_v \times \tilde{n}_v \times n_T}, \tilde{n}_v = \tilde{m}n_v,$$
$$\mathcal{F}_{T_i} = \mathrm{blkdiag}(\mathbf{F}_{T_i}, \ldots, \mathbf{F}_{T_i}) \in \mathbb{R}^{\tilde{n}_v \times \tilde{n}_v},$$
$$\mathbf{F}_{T_i} \in \mathbb{R}^{n_v \times n_v}$$

and ξ_v is defined as in (14.16). We can see that each reduced state in (14.19) induces a reduction in (14.20) leading to

$$\mathbf{E}_T\mathbf{x}_T' = \mathbf{A}_T\mathbf{x}_T + \xi_{v_r}^T\mathcal{F}_{T_r}\xi_{v_r} + \mathbf{B}_T\mathbf{u}_T, \quad \mathbf{x}_T(0) = \mathbf{x}_{T_0}, \tag{14.21a}$$
$$\mathbf{y}_T = \mathbf{C}_T\mathbf{x}_T + \mathbf{D}_T\mathbf{u}_T, \tag{14.21b}$$

where $\xi_{v_r} = (\mathbf{x}_{v_{r_1}}^T, \ldots, \mathbf{x}_{v_{r_{\tilde{m}}}}^T)^T$, $\mathcal{F}_{T_r} = \left[\mathcal{F}_{T_{r_1}}^T, \ldots, \mathcal{F}_{T_{r_{n_T}}}^T\right]^T \in \mathbb{R}^{r_v \times r_v \times n_T}$, with $\mathcal{F}_{T_{r_i}} = \mathrm{blkdiag}(\mathbf{F}_{T_{r_i}}, \ldots, \mathbf{F}_{T_{r_i}}) \in \mathbb{R}^{r_v \times r_v}$, and $\mathbf{F}_{T_{r_i}} = \mathbf{F}_{T_{11}}^{(i)} - \mathbf{W}_{v_i}^T\mathbf{F}_{T_{21}}^{(i)} - \mathbf{F}_{T_{12}}^{(i)}\mathbf{W}_{v_i} + \mathbf{W}_{v_i}^T\mathbf{F}_{T_{22}}^{(i)}\mathbf{W}_{v_i} \in \mathbb{R}^{r_{v_i} \times r_{v_i}}$. Here $\mathbf{F}_{T_{11}}^{(i)}, \mathbf{F}_{T_{12}}^{(i)}, \mathbf{F}_{T_{21}}^{(i)}, \mathbf{F}_{T_{22}}^{(i)}$ are the subblocks of $\mathbf{F}_{T_i}$ partitioned according to the partition of $\mathbf{A}_v$ in (14.18). Since $\sum_{i=1}^{\tilde{m}}\mathbf{x}_{v_i}^T\mathcal{F}_T\mathbf{x}_{v_i}$ can be considered as an extra input for the thermal subsystem, the superposition principle still applies to the thermal subsystem. Therefore, (14.21) can also be split into $\tilde{m}$ subsystems, the thermal state $\mathbf{x}_T$ of (14.21) can be reduced following the steps from (14.14) to (14.15). The reduced thermal system is in the form of (14.10b) with reduced matrices being defined in (14.15). Using the non-parameterized formulation of Proposition 14.1, the nonlinear term $\mathbf{V}^T\left(\xi_{v_r}^T\tilde{\mathcal{F}}_T\xi_{v_r}\right)$, where $\tilde{\mathcal{F}}_T = \begin{pmatrix}\mathcal{F}_{T_r} \\ 0\end{pmatrix} \in \mathbb{R}^{r_v \times r_v \times \tilde{m}n_T}$, $\tilde{\mathcal{F}}_T = \left[\tilde{\mathcal{F}}_{T_1}^T, \ldots, \tilde{\mathcal{F}}_{T_{\tilde{m}n_T}}^T\right]^T$ with $\tilde{\mathcal{F}}_{T_i} \in \mathbb{R}^{r_v \times r_v}$, can also be reformulated as $\xi_{v_r}^T\tilde{\mathcal{F}}_{T_r}\xi_{v_r}$ where $\tilde{\mathcal{F}}_{T_r} = \left[\tilde{\mathcal{F}}_{T_{r_1}}^T, \ldots, \tilde{\mathcal{F}}_{T_{r_{r_T}}}^T\right]^T \in \mathbb{R}^{r_v \times r_v \times r_T}$ with

$$\tilde{\mathcal{F}}_{T_{r_j}} = \sum_{i=1}^{\tilde{m}n_T} v_{ji}\tilde{\mathcal{F}}_{T_i} \in \mathbb{R}^{r_v \times r_v}, j = 1, \ldots, r_T, \mathbf{V} = (\mathbf{v}_{ij}) \in \mathbb{R}^{\tilde{m}n_T \times r_T}.$$

Instead of a dense tensor as in the previous section, here $\tilde{\mathcal{F}}_{T_r}$ is in the block-diagonal form which is sparse. Combining the above block structured reduced electrical and thermal subsystems, we obtain the modified BDSM-ET ROMs of

(14.3) in the form of (14.5) with system matrices

$$\mathbf{E}_r = \begin{pmatrix} 0 & 0 \\ 0 & \mathbf{E}_{T_r} \end{pmatrix}, \mathbf{A}_r = \begin{pmatrix} \mathbf{A}_{v_r} & 0 \\ 0 & \mathbf{A}_{T_r} \end{pmatrix}, \mathbf{B}_r = \begin{pmatrix} \mathbf{B}_{v_r} & 0 \\ 0 & \mathbf{B}_{T_r} \end{pmatrix},$$

$$\mathbf{C}_r = \begin{pmatrix} \mathbf{C}_{v_r} & \mathbf{C}_{T_r} \end{pmatrix}, \mathbf{D} = \begin{pmatrix} \mathbf{D}_v & \mathbf{D}_T \end{pmatrix},$$

$$\mathcal{F}_r = \left(0,\ldots,0,\mathbf{F}_{r_v+1}^T,\ldots,\mathbf{F}_{r_v+r_T}^T\right)^T,$$

$$\mathbf{F}_{r_v+j} = \begin{pmatrix} \tilde{\mathcal{F}}_{T_{r_j}} & 0 \\ 0 & 0 \end{pmatrix} \in \mathbb{R}^{r \times r}, j = 1,\ldots,r_T.$$

Hence, by construction, the modified BDSM-ET method constructs sparser ROMs than the BDSM-ET method discussed in the previous section, since all its reduced matrices and the tensor are block-wise sparse.

14.5 Modified BDSM-ET for Parametric ET coupled Problems

In [3], the modified BDSM-ET method has been extended to derive PROMs of parametrized ET coupled problems. The method is reviewed in this section. It is proposed to apply the superposition principle to the parametrized subsystems (14.7a) and (14.7b) separately. Then, the standard PMOR method from [9] can be applied to each subsystem. Assume that $\mathbf{B}_v(\mu), \mathbf{B}_T(\mu)$ have no zero columns, so that they can be split into $\mathbf{B}_k(\mu) = \sum_{i=1}^{\tilde{m}} \mathbf{B}_{k_i}(\mu), k = v, T$, where $\mathbf{B}_{k_i}(\mu)$ are column rank-1 parametric matrices defined as

$$\mathbf{B}_{k_i}(:,j) = \begin{cases} \mathbf{b}_{k_i}(\mu) \in \mathbb{R}^{n_k}, & \text{if } j = i, \\ 0, & \text{otherwise,} \end{cases} \quad i = 1,\ldots,\tilde{m}.$$

By the superposition principle and using the above splitting of $\mathbf{B}_v(\mu)$, the electrical subsystem (14.7a) can be decomposed into $\tilde{m}$ subsystems

$$\mathbf{A}_v(\mu)\mathbf{x}_{v_i} = -\mathbf{B}_{v_i}(\mu)\mathbf{u}_v, \quad \mathbf{y}_v = \mathbf{C}_v(\mu)\mathbf{x}_{v_i}, \quad i = 1,\ldots,\tilde{m}. \tag{14.22}$$

Then the $\tilde{m}$ parametrized subsystems in (14.22) can be equivalently transformed into a block-diagonal system of dimension $\tilde{n}_v$ given by

$$\mathcal{A}_v(\mu)\xi_v = -\mathcal{B}_v(\mu)\mathbf{u}_v, \quad \mathbf{y}_v = \mathcal{C}_v(\mu)\xi_v + \mathbf{D}(\mu)\mathbf{u}, \tag{14.23}$$

where

$$\xi_v = (\mathbf{x}_{v_1}^T, \ldots, \mathbf{x}_{v_{\tilde{m}}}^T)^T \in \mathbb{R}^{\tilde{n}_v}, \; \tilde{n}_v = \tilde{m} n_v,$$
$$\mathcal{A}_v(\mu) = \text{blkdiag}(\mathbf{A}_v(\mu), \ldots, \mathbf{A}_v(\mu)) \in \mathbb{R}^{\tilde{n}_v \times \tilde{n}_v},$$
$$\mathcal{B}_v(\mu) = (\mathbf{B}_{v_1}(\mu)^T, \ldots, \mathbf{B}_{v_{\tilde{m}}}(\mu)^T)^T \in \mathbb{R}^{\tilde{n}_v \times \tilde{m}},$$
$$\mathcal{C}_v(\mu) = (\mathbf{C}_v(\mu), \ldots, \mathbf{C}_v(\mu)) \in \mathbb{R}^{\ell \times \tilde{n}_v}.$$

The dimension of (14.23) can be reduced by replacing (14.23) with a much smaller ROM

$$\mathcal{A}_{v_r}(\mu)\xi_{v_r} = -\mathcal{B}_{v_r}(\mu)\mathbf{u}_v, \quad \mathbf{y}_{v_r} = \mathcal{C}_{v_r}(\mu)\xi_{v_r} + \mathbf{D}_v(\mu)\mathbf{u}, \qquad (14.24)$$

where $\mathcal{A}_{v_r}(\mu) = \mathbf{V_v}^T \mathcal{A}_v(\mu)\mathbf{V}_v, \mathcal{B}_{v_r}(\mu) = \mathbf{V}_v^T \mathcal{B}_v(\mu), \mathcal{C}_{v_r}(\mu) = \mathcal{C}_v(\mu)\mathbf{V}_v,$
$\xi_v = \mathbf{V}_v \xi_{v_r} \in \mathbb{R}^{r_v}$ and $\mathbf{V}_v = \text{blkdiag}(\mathbf{V}_v^{(1)}, \mathbf{V}_v^{(2)}, \ldots, \mathbf{V}_v^{(m)}) \in \mathbb{R}^{\tilde{m} n \times r_v}, r_v \ll n_v.$
The projection matrices $\mathbf{V}_v^{(i)} \in \mathbb{R}^{n_V \times r_{v_i}}, r_v = \sum r_{v_i}$ can be constructed by applying the existing PMOR methods to each system in (14.22). Then the thermal subsystem (14.7b) can be reduced as follows. Applying the superposition principle to the electrical subsystem (14.7a) introduces

$$\left(\sum_{i=1}^{\tilde{m}} \mathbf{x}_{v_i}^T \right) \mathcal{F}_T(\mu) \left(\sum_{i=1}^{\tilde{m}} \mathbf{x}_{v_i} \right)$$

into the thermal subsystem (14.7b), i.e. $\mathbf{x}_v$ is replaced by $\sum_{i=1}^{\tilde{m}} \mathbf{x}_{v_i}$ in the nonlinear part. In order to obtain a sparse tensor, we assume small signal modeling, i.e., the approximation

$$\left(\sum_{i=1}^{\tilde{m}} \mathbf{x}_{v_i}^T \right) \mathcal{F}_T(\mu) \left(\sum_{i=1}^{\tilde{m}} \mathbf{x}_{v_i} \right) \approx \sum_{i=1}^{\tilde{m}} \mathbf{x}_{v_i}^T \mathcal{F}_T(\mu) \mathbf{x}_{v_i}$$

is introduced into the thermal subsystem.
Thus (14.7b) can be approximated as

$$\mathbf{E}_T(\mu)\mathbf{x}_T' = \mathbf{A}_T(\mu)\mathbf{x}_T + \xi_v^T \mathcal{W}_T(\mu)\xi_v, + \mathbf{B}_T(\mu)\mathbf{u}_T, \quad \mathbf{x}_T(0) = \mathbf{x}_{T_0}, \quad (14.25a)$$
$$\mathbf{y}_T = \mathbf{C}_T(\mu)\mathbf{x}_T + \mathbf{D}_T(\mu)\mathbf{u}_T. \qquad (14.25b)$$

Here, we have used the equality $\sum_{i=1}^{\tilde{m}} \mathbf{x}_{v_i}^T \mathcal{F}_T(\mu)\mathbf{x}_{v_i} = \xi_v^T \mathcal{W}_T(\mu)\xi_v$, where

$$\mathcal{W}_T(\mu) = \left[\mathbf{W}_{T_1}(\mu)^T, \ldots, \mathbf{W}_{T_{n_T}}(\mu)^T \right]^T \in \mathbb{R}^{\tilde{n}_v \times \tilde{n}_v \times n_T},$$
$$\mathbf{W}_{T_i}(\mu) = \text{blkdiag}(\mathbf{F}_{T_i}(\mu), \ldots, \mathbf{F}_{T_i}(\mu)) \in \mathbb{R}^{\tilde{n}_v \times \tilde{n}_v}, \; \mathbf{F}_{T_i}(\mu) \in \mathbb{R}^{n_v \times n_v},$$

and ξ_v is defined as in (14.23). We can see that the reduced state in (14.24) introduces an approximation into (14.25) leading to

$$\mathbf{E}_T(\mu)\mathbf{x}'_T = \mathbf{A}_T(\mu)\mathbf{x}_T + \xi_{v_r}^T \hat{\mathcal{W}}_T(\mu)\xi_{v_r} + \mathbf{B}_T(\mu)\mathbf{u}_T, \quad \mathbf{x}_T(0) = \mathbf{x}_{T_0}, \quad (14.26a)$$
$$\mathbf{y}_T = \mathbf{C}_T(\mu)\mathbf{x}_T + \mathbf{D}_T(\mu)\mathbf{u}_T, \quad (14.26b)$$

where $\hat{\mathcal{W}}_T(\mu) = \mathbf{V_v}^T \mathcal{W}_T(\mu)\mathbf{V}_v := \sum_{i=1}^{\tilde{m}} \mathbf{V}_v^{(i)^T} \mathcal{F}_T(\mu)\mathbf{V}_v^{(i)} \in \mathbb{R}^{r_v \times r_v \times n_T}$. Here,

$$\mathbf{V}_v^{(i)^T} \mathcal{F}_T(\mu)\mathbf{V}_v^{(i)} := \left[\mathbf{V}_v^{(i)^T} \mathbf{F}_{T_1}(\mu)^T \mathbf{V}_v^{(i)}, \dots, \mathbf{V}_v^{(i)^T} \mathbf{F}_{T_{n_T}}(\mu)^T \mathbf{V}_v^{(i)}\right]^T.$$

Since $\xi_{v_r}^T \hat{\mathcal{W}}_T(\mu)\xi_{v_r}$ can be considered as an extra input for the thermal subsystem, the superposition principle still applies to the thermal subsystem. By the superposition principle and using the splitting of $\mathbf{B}_T(\mu)$, the thermal subsystem (14.26) can also be split into $\tilde{m}$ subsystems

$$\mathbf{E}_T(\mu)\mathbf{x}'_{T_1} = \mathbf{A}_T(\mu)\mathbf{x}_{T_1} + \xi_{v_r}^T \hat{\mathcal{W}}_T(\mu)\xi_{v_r} + \mathbf{B}_{T_1}(\mu)\mathbf{u}_T, \quad \mathbf{x}_{T_1}(0) = \mathbf{x}_{T_0},$$
$$(14.27a)$$

$$\mathbf{E}_T(\mu)\mathbf{x}'_{T_i} = \mathbf{A}_T(\mu)\mathbf{x}_{T_i} + \mathbf{B}_{T_i}(\mu)\mathbf{u}_T, \quad \mathbf{x}_{T_i}(0) = 0, \quad i = 2, \dots, \tilde{m}, \quad (14.27b)$$
$$\mathbf{y}_T = (\mathbf{C}(\mu), \dots, \mathbf{C}(\mu))\xi_T + \mathbf{D}_T(\mu)\mathbf{u}_T, \quad (14.27c)$$

where $\xi_T = (\mathbf{x}_{T_1}^T, \dots, \mathbf{x}_{T_{\tilde{m}}}^T)^T \in \mathbb{R}^{\tilde{n}_T}$, $\tilde{n}_T = \tilde{m}n_T$. System (14.27) can be reformulated as a block-diagonal structured system of dimension $\tilde{n}_T$ given by

$$\mathcal{E}_T(\mu)\xi'_T = \mathcal{A}_T(\mu)\xi_T + \xi_{v_r}^T \tilde{\mathcal{W}}_T(\mu)\xi_{v_r} + \mathcal{B}_T(\mu)\mathbf{u}_T,$$
$$\mathbf{y}_T = \mathcal{C}_T(\mu)\xi_T + \mathbf{D}_T(\mu)\mathbf{u}_T, \quad \xi_T(0) = (\mathbf{x}_{T_0}^T, 0, \dots, 0)^T, \quad (14.28)$$

where

$$\mathcal{E}_T(\mu) = \text{blkdiag}(\mathbf{E}_T(\mu), \dots, \mathbf{E}_T(\mu)),$$
$$\mathcal{A}_T(\mu) = \text{blkdiag}(\mathbf{A}_T(\mu), \dots, \mathbf{A}_T(\mu)) \in \mathbb{R}^{\tilde{n}_T \times \tilde{n}_T},$$
$$\tilde{\mathcal{W}}_T(\mu) = \begin{pmatrix} \hat{\mathcal{W}}_T(\mu) \\ 0 \end{pmatrix} \in \mathbb{R}^{r_v \times r_v \times \tilde{n}_T},$$
$$\mathcal{B}_T(\mu) = (\mathbf{B}_{T_1}^T(\mu), \dots, \mathbf{B}_{T_{\tilde{m}}}^T(\mu))^T \in \mathbb{R}^{\tilde{n}_T \times \tilde{m}},$$
$$\mathcal{C}_T(\mu) = (\mathbf{C}_T(\mu), \dots, \mathbf{C}_T(\mu)) \in \mathbb{R}^{\ell \times \tilde{n}_T}.$$

The corresponding ROM of (14.28) is given by

$$\mathcal{E}_{T_r}(\mu)\xi'_{T_r} = \mathcal{A}_{T_r}(\mu)\xi_T + \xi_{v_r}^T \mathcal{W}_{T_r}(\mu)\xi_{v_r} + \mathcal{B}_{T_r}(\mu)\mathbf{u}_T,$$
$$\mathbf{y}_{T_r} = \mathcal{C}_{T_r}(\mu)\xi_{T_r} + \mathbf{D}_T(\mu)\mathbf{u}_T, \quad \xi_{T_r}(0) = (\mathbf{x}_{T_{r_0}}^T, 0, \dots, 0)^T, \quad (14.29)$$

where

$$\mathcal{E}_{T_r}(\mu) = \mathbf{V}_T^T \mathcal{E}_T(\mu) \mathbf{V}_T \in \mathbb{R}^{r_T \times r_T},$$
$$\mathcal{A}_{T_r}(\mu) = \mathbf{V}_T^T \mathcal{A}_T(\mu) \mathbf{V} \in \mathbb{R}^{r_T \times r_T},$$
$$\mathcal{B}_{T_r}(\mu) = \mathbf{V}_T^T \mathcal{B}_T(\mu) \in \mathbb{R}^{r_T \times \tilde{m}},$$
$$\mathcal{C}_{T_r}(\mu) = \mathcal{C}_T(\mu) \mathbf{V}_T \in \mathbb{R}^{\ell \times r_T},$$
$$\mathcal{W}_{T_r}(\mu) = \left[\hat{\mathbf{W}}_{T_1}(\mu)^T, \dots, \hat{\mathbf{W}}_{T_{r_T}}(\mu)^T \right]^T \in \mathbb{R}^{r_v \times r_v \times r_T}$$
$$\hat{\mathbf{W}}_{T_i}(\mu) \in \mathbb{R}^{r_v \times r_v}, \ i = 1, \dots, r_T, \ \text{block-diagonal matrices},$$
$$\mathbf{V}_T = \mathrm{blkdiag}(\mathbf{V}_T^{(1)}, \dots, \mathbf{V}_T^{(\tilde{m})}) \in \mathbb{R}^{\tilde{n}_T \times r_T}.$$

Using Proposition 14.1, the reduced nonlinear term $\xi_{v_r}^T \mathcal{W}_{T_r}(\mu) \xi_{v_r}$, can be reformulated using $\mathbf{V}_T^T \left(\xi_{v_r}^T \tilde{\mathcal{W}}_T \xi_{v_r} \right)$. Note that by construction, $\mathcal{W}_{T_r}(\mu)$ is a sparse tensor since it has block-wise sparse $\hat{\mathbf{W}}_{T_i}(\mu)$ slices. The projection matrices $\mathbf{V}_T^{(1)}$ and $\mathbf{V}_T^{(i)}$, $i = 2, \dots, \tilde{m}$, can be constructed by applying existing PMOR methods to the subsystem (14.27a) and each subsystem in (14.27b), respectively. In this work, we use the implicit moment-matching PMOR method proposed in [9]. This implies that $\mathbf{V}_T^{(1)}$ can be constructed by ignoring the nonlinear part of (14.27a). This approach allows us to automatically construct $\mathbf{V}_v$ and $\mathbf{V}_T$, respectively, using the global a posteriori error bound [5], and it is used to produce the simulation results in the subsection 14.6.2.

14.6 Simulation Results

In this section, we consider (non-)parameterized numerical examples. Simulation on the large-scale model ($n = 925,286$) is done on a Unix computing server with 1TB main memory while the other simulations were done using MATLAB®Version 2012b on a Laptop with 6GB RAM, CPU@ 2.00 GHz. The sparse tensor computations were performed with the MATLAB Tensor Toolbox [1]. All used (P)MOR methods are defined as follows: BDSM-ET method denotes the MOR method for ET coupled problems with many inputs discussed in Sec. 14.3; modified BDSM-ET (sparse MOR method) denotes the MOR method for ET coupled problems with many inputs discussed in Sec. 14.4; IpMOR-ET denotes the standard PMOR method for ET coupled problems proposed in [5] and IpBDSM-ET (sparse PMOR method) denotes the modified BDSM-ET method for parametric systems discussed in Sect. 14.5.

14.6.1 Non-Parameterized Numerical Examples

In this subsection, we compare the BDSM-ET method discussed in Sec. 14.3 with the modified BDSM-ET (sparse MOR method) discussed in Sec. 14.4 using non-parameterized ET coupled problems.

Example 14.1. Here, we consider four non-parameterized ET coupled models from industrial applications, namely, a package model ($n = 9,193, m = 34, \ell = 68$), a power-MOS model ($n = 13,216, m = 6, \ell = 12$), a MOS block model ($n = 17,147, m = 260, \ell = 130,$) and a power cell model ($n = 925,286, m = 408, \ell = 816$) as shown in Table 14.1.

Table 14.1: Electro-thermal coupled models.

Model (n)	n_v	n_T	ℓ	m
9,193	8,071	1,122	68	34
13,216	11,556	1,660	12	6
17,147	5,833	11,314	260	130
925,286	392,773	532,513	408	816

As discussed in subsection 14.2.1, we first decouple the ET coupled systems into electrical and thermal subsystems of dimensions n_v and n_T, respectively as shown in Table 14.1. We can observe that $n = n_V + n_T$ for each ET coupled system as expected. m and ℓ are the number of inputs and outputs, respectively. Obviously, the models have many inputs.

Next, we compare the ROMs obtained using both methods for the models as illustrated in Tables 14.2 and 14.3 for linear and nonlinear systems, respectively, under the same conditions. In Table 14.2, we show the results for the linear case, i.e., $\mathcal{F} = 0$, while in Table 14.3, we consider the nonlinear case, i.e., $\mathcal{F} \neq 0$.

We can observe that both methods produced computationally efficient ROMs in the linear case. However, the BDSM-ET method failed to reduce the linear ET model of original size $n = 925,286$ because of the memory limitations while the modified BDSM-ET method has successfully reduced it and has achieved a speed-up of $1,172$.

Table 14.2: Linear case ($\mathcal{F} = 0$), $r = r_1 + r_2$.

	BDSM-ET ROM						Modified BDSM-ET ROM				
Model (n)	r_1	r_2	r	% Red	Speed-Up	Error	r_1	r_2	r	% Red	Speed-Up	Error
9,193	188	72	260	97.2	209.6	5.2×10^{-4}	238	72	310	96.6	331.5	5.2×10^{-4}
13,216	160	63	223	98.3	678.7	3.9×10^{-8}	160	63	223	98.3	523.4	2.2×10^{-8}
17,147	2,102	198	2,300	86.6	18.6	2.3×10^{-4}	2,158	198	2,356	86.3	25.3	2.3×10^{-4}
925,286	–	–	–	–	–	–	9,396	4,305	13,596	98.5	1,172	6.3×10^{-8}

It is seen from Table 14.3 that the BDSM-ET method failed to reduce the ET coupled model of original size $n = 17,147$ because of the memory limitations while the modified BDSM-ET method reduced it with a speed-up of 9.4.

Table 14.3: Nonlinear case ($\mathcal{F} \neq 0$), $r = r_1 + r_2$.

	BDSM-ET ROM					Modified BDSM-ET ROM					
Model (n)	r_1	r_2	r	% Red	Speed-Up	Error	r_1	r_2	r	% Red	Speed-Up	Error
9,193	188	72	260	97.2	38.9	5.2×10^{-4}	238	72	310	96.6	82.5	5.2×10^{-4}
13,216	160	63	223	98.3	63.1	3.5×10^{-5}	160	63	223	98.3	177	3.5×10^{-5}
17,147	-	-	-	-	-	-	2,158	198	2,356	86.3	9.4	3.2×10^{-4}

From Table 14.2 and 14.3, we can see that the modified BDSM-ET method produces computationally efficient ROMs for ET coupled problems with many inputs. Hence, the modified BDSM-ET method is reliable and robust compared to the BDSM-ET method.

14.6.2 Parameterized Numerical Examples

In this subsection, we compare the BDSM-ET method for parameterized ET coupled problems discussed in Sec. 14.5 with the standard PMOR for ET coupled problems (IpMOR-ET) proposed in [5].

Example 14.2. We consider electro-thermal simulation of a power-MOS device model. It is a parameterized quadratic system described in (14.1) with matrices and tensor in the form of (14.6) and $n = 13216$ state variables. There are $m = 6$ inputs and $\ell = 12$ outputs. The matrices $\mathbf{E}(\mu), \mathbf{A}(\mu), \mathbf{B}(\mu), \mathbf{C}(\mu), \mathbf{D}(\mu)$ and the tensor $\mathbf{F}(\mu)$ exhibit a parameter dependence of an affine form (14.2) given by $\mathbf{M}(\mu) = \mathbf{M}_0 + \mu \mathbf{M}_1$, where $\mathbf{M}(\mu)$ indicates any parameterized matrix or tensor in (14.1) and $\mathbf{M}_i$ ($i = 0, 1$) are parameter-independent coefficients. For this example, $\mu = \sigma$ describes the electrical conductivity of the power-MOS device. This system is decoupled into the form (14.7) with $n_v = 1,660$ electrical equations, and $n_T = 11,556$ thermal equations.

Table 14.4: Simulation efficiency comparison.

	Original model			Reduced order models			
pMOR methods	n	m	ℓ	r	Speed-up	Reduction rate (%)	Error
IpMOR-ET	13,216	6	12	56	1365.1	99.6	1.4×10^{-6}
IpBDSM-ET	13,216	6	12	56	1376.2	99.6	3.3×10^{-5}

We used the IpMOR-ET and the modified IpBDSM-ET methods to reduce the decoupled system leading to pROMs with 4 electrical and 52 thermal equations, respectively, as shown in Table 14.4. Hence, the power-MOS device model is reduced from $13,216$ to 56 leading to a reduction rate of 99.6. We simulated both ROMs within the time interval $t \in [0, 2 \times 10^{-6}]$ and chose the electrical conductivity $\sigma \in [10^7, 10^8]$ leading to excellent speed-ups. However, the advantage of the IpBDSM-ET method should be significant if there are many inputs, which has been shown in the previous example. Furthermore, the modified IpBDSM-ET leads to sparser ROMs than the IpMOR method as illustrated in [3].

14.7 Conclusions

Sparse (P)MOR for ET coupled problems with numerous inputs and outputs is computationally cheaper since it leads to a block diagonal structure. By construction, these methods produce sparse yet accurate ROMs as compared to the standard (P)MOR methods. However, the size of the ROMs depend on the number of inputs, which could be avoided by grouping the inputs. Future work will include the extension of the method to more general coupling structures.

References

1. BADER, B.W., AND KOLDA, T.G.: *Matlab tensor toolbox*, version 2.6 (2015) Available via DIALOG. http://www.sandia.gov/~tgkolda/TensorToolbox. Cited 03 May 2017.
2. BANAGAAYA, N.: *Index-aware model order reduction methods*, Ph.D.-Thesis, Eindhoven University of Technology, Eindhoven, Netherlands, 2014.
3. BANAGAAYA, N., BENNER, P., AND FENG, L.: *Parametric Model Order Reduction for Electro-Thermal Coupled Problems with Many Inputs*. In: QUINTELA, P., BARRAL, P., GÓMEZ, D., PENA, F.J., RODRIGUEZ, J., SALGADO, P., AND VÁZQUEZ-MÉNDEZ, M.E. (EDS): *Progress in Industrial Mathematics at ECMI 2016*. Springer International Publishing AG, Series Mathematics in Industry, Vol. 26, pp. 263-270, 2018. http://dx.doi.org/10.1007/978-3-319-63082-3_41.
4. BANAGAAYA, N., FENG, L., SCHOENMAKER, W., MEURIS, P., AND BENNER, P.: *Model order reduction of an electro-thermal package model*. IFAC-PapersOnLine. 8 (1) pp. 934–935, 2015.
5. BANAGAAYA, N., FENG, L., SCHOENMAKER, W., MEURIS, P., AND BENNER, P.: *An Index-aware Parametric Model Order Reduction for Parametrized Quadratic DAEs*. Applied Mathematics and Computation, 319, pp. 409–424, 2018.
6. BANAGAAYA, N., FENG, L., SCHOENMAKER, W., MEURIS, P., GILLON, R., AND BENNER, P.: *Model order reduction for nanoelectronics coupled problems with many inputs*. In: Proceedings 2016 Design, Automation & Test in Europe Conference & Exhibition, DATE 2016, Dresden, pp. 313–318, 2016.

7. BANAGAAYA, N., FENG, L., SCHOENMAKER, W., MEURIS, P., GILLON, R., AND BENNER, P.: *Sparse Model Order Reduction for Electro-Thermal Problems with Many Inputs*. In: LANGER, U., AMRHEIN, W., AND ZULEHNER, W. (EDS): *Scientific Computing in Electrical Engineering - SCEE 2016*. Springer International Publishing AG, Series Mathematics in Industry, Vol. 28, pp. 189–202, 2018. https://dx.doi.org/10.1007/978-3-319-75538-0_18.

8. BAUR, U., BENNER, P., AND FENG, L.: *Model order reduction for linear and nonlinear systems: A system-theoretic perspective*. Arch. Comput. Methods Eng., 21(4), pp. 331–358, 2014.

9. FENG, L., AND BENNER, P.: *A robust algorithm for parametric model order reduction based on implicit moment matching*. In: ALFIO, Q., AND GIANLUIGI, R. (EDS.): *Reduced Order Methods for modeling and computational reduction*, Springer, Heidelberg, pp. 159–186, 2014.

10. FENG, L., YUE, Y., BANAGAAYA, N., MEURIS, P., SCHOENMAKER, W., AND BENNER, P.: *Parametric modeling and model order reduction for (electro-)thermal analysis of nanoelectronic structures*, J. Mathematics in Industry 6 (10), pp. 1–16, 2016. Open access: https://dx.doi.org/10.1186/s13362-016-0030-8.

11. ODABASIOGLU, A., CELIK, M., AND PILEGGI. L.: *PRIMA: passive reduced-order interconnect macromodeling algorithm*. IEEE Trans. Comput.-Aided Design Integr. Circuits Syst., vol. 17, no. 8, pp. 645–654, 1997.

12. YE, Z., VASILYEV, D., ZHU, Z., AND PHILLIPS, J.: *Sparse Implicit Projection (SIP) for reduction of general many-terminal networks*. In Proc. IEEE/ACM ICCAD, pp. 736–743, 2008.

13. ZHANG, Z., HU, C.C. X., AND WONG, N.: *A Block-Diagonal Structured Model Reduction Scheme for Power Grid Networks*. In Proc. Design, Automation & Test in Europe Conference & Exhibition, DATE 2011, Grenoble, France, pp. 44–49, 2011

Chapter 15
Reduced Models and Uncertainty Quantification

Yao Yue, Lihong Feng, Peter Benner, Roland Pulch, Sebastian Schöps

Abstract Uncertainty quantification analyses the variability of system outputs with respect to process variations and thus represents a useful tool for robust design. Often statistics of a dynamical system's outputs are quantities of interest. A sampling of the outputs requires many transient simulations. Due to the complexity of systems in nanoelectronics, methods of model order reduction (MOR) are applied for accelerating the uncertainty quantification. We consider coupled problems or multiphysics systems. We employ parametric MOR techniques to build parameter-dependent reduced-order models, which can be used for fast computations at all parameter samples. Sampling-based techniques like the Latin hypercube method, for example, or quadrature rules yield the parameter values. We apply this approach to an electro-thermal coupled system. Furthermore, we illustrate a co-simulation technique with different quadrature grids for the subsystems. Now just some parts of a coupled problem are substituted by parametric MOR, if the others cannot be reduced efficiently. This method is applied to a circuit-electromagnetic coupled system.

15.1 Parametric MOR and Uncertainty Quantification

In the first part of this chapter, we use parametric model order reduction to construct a surrogate model of a quantity of interest. Consequently, the

Yao Yue, Lihong Feng, Peter Benner
Max Planck Institute for Dynamics of Complex Technical Systems, Magdeburg. e-mail: {Yue,Feng,Benner}@mpi-magdeburg.mpg.de

Roland Pulch
University of Greifswald,Germany. e-mail: Roland.Pulch@uni-greifswald.de

Sebastian Schöps
Technical University of Darmstadt, Germany. e-mail: Schoeps@gsc.tu-darmstadt.de

© Springer Nature Switzerland AG 2019   329
E. J. W. ter Maten et al. (eds.), *Nanoelectronic Coupled Problems Solutions*,
Mathematics in Industry 29, https://doi.org/10.1007/978-3-030-30726-4_15

reduced-order model can be sampled with low computational effort to obtain statistics. We demonstrate the approach using the mathematical model of a power-MOS device.

15.1.1 Mathematical Modeling of the Power-MOS Device

The model of interest in this section is a Power-MOS device, shown in Figure 15.1(a). Power-MOS devices are commonly used in energy harvesting, where energy from external sources like light and environmental heat is collected in order to power small devices such as implanted biosensors [18, 21]. Therefore, electro-thermal co-simulation is crucial in the analysis of Power-MOS devices. The device shown in Figure 15.1(a) has three contacts: the drain, the source, and the back contact. In our model, the inputs are the voltages and temperatures on these contacts, while the currents and thermal fluxes on these contacts are chosen as the outputs.

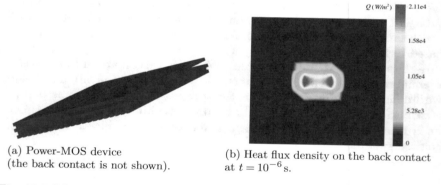

(a) Power-MOS device
(the back contact is not shown).

(b) Heat flux density on the back contact
at $t = 10^{-6}$ s.

Fig. 15.1: The Power-MOS device and its heat flux density on the back contact at $t = 10^{-6}$ s.

To illustrate the system behavior, we now excite the device with a voltage on the source. As time progresses within the interval $[0\,\mathrm{s}, 10^{-6}\,\mathrm{s}]$, the voltage on the source rises linearly, and the chip is heated up. The heat flux density at the time $t = 10^{-6}$ s on the back contact is shown in Figure 15.1(b).

To construct the electro-thermal coupled model, we first ignore the coupling and consider the electrical sub-system and the thermal sub-system separately. Ignoring the local charging, the electrical part is governed by a linear Poisson equation:

$$\nabla \cdot (\sigma \cdot \nabla U) = 0,$$

where U is the electrical potential and σ is the electrical conductivity. Ignoring the dependence of temperature on the thermal conductivity and thermal capacitance, the thermal sub-system is governed by:

$$\nabla \cdot \boldsymbol{\phi}_q + \frac{\partial w(T)}{\partial t} = Q,$$

$$\boldsymbol{\phi}_q = -\kappa \nabla T,$$

$$w(T) = C_T(T - T_{\mathrm{ref}}),$$

where $\boldsymbol{\phi}_q$ is the heat flux, w is the local energy storage, κ is the thermal conductivity, C_T is the thermal capacitance and T_{ref} is the reference temperature.

Then we consider the coupling between the electrical part and the thermal part. The two sub-systems are coupled in several ways. First and foremost, the Joule self-heating

$$Q_{\mathrm{SH}} = \mathbf{E} \cdot \mathbf{J},$$

which is generated when electric current passes through a conductor, is the most important coupling from the electrical sub-system to the thermal sub-system. Additionally, the temperature influences the electrical conductivity. In this section, the temperature influence is ignored and we focus on the Joule self-heating. Therefore, the system is one-way coupled: only the coupling from the electrical sub-system to the thermal subsystem is considered.

After spatial discretization, the electro-thermal model of the power-MOS device that we consider is

$$\begin{cases} \mathbf{A}_{\mathrm{E}}(p)\mathbf{x}_{\mathrm{E}}(p,t) = -\mathbf{B}_{\mathrm{E}}(p)\mathbf{u}(t), & \text{(Electrical Part),} & (15.1\mathrm{a}) \\[4pt] \mathbf{E}_{\mathrm{T}}(p)\dot{\mathbf{x}}_{\mathrm{T}}(p,t) = \mathbf{A}_{\mathrm{T}}(p)\mathbf{x}_{\mathrm{T}}(p,t) + \mathbf{B}_{\mathrm{T}}(p)\mathbf{u}(t) \\[2pt] \qquad\qquad\quad +\mathbf{F}(p) \times_2 \mathbf{x}_{\mathrm{E}}(p) \times_3 \mathbf{x}_{\mathrm{E}}(p), & \text{(Thermal Part),} & (15.1\mathrm{b}) \\[4pt] \mathbf{x}_{\mathrm{T}}(p,0) = \mathbf{x}_{\mathrm{T}}^0, \; \mathbf{x}_{\mathrm{E}}(p,0) = \mathbf{x}_{\mathrm{E}}^0, & \text{(Initial Conditions)} & (15.1\mathrm{c}) \\[4pt] \mathbf{y}(p,t) = \mathbf{C}_{\mathrm{E}}(p)\mathbf{x}_{\mathrm{E}}(p,t) \\[2pt] \qquad\quad +\mathbf{C}_{\mathrm{T}}(p)\mathbf{x}_{\mathrm{T}}(p,t) + \mathbf{D}(p)\mathbf{u}(t), & \text{(Output).} & (15.1\mathrm{d}) \end{cases}$$

The following comments on the model (15.1) are in order:

- For the whole system, p represents parameter(s), which is the conductivity of the third metal layer in our case. $\mathbf{u}(t) \in \mathbb{R}^l$ is the input vector representing the voltages and temperatures of the contacts, and $\mathbf{y}(p) \in \mathbb{R}^m$ is the output vector representing the currents and heat fluxes of the contacts.
- In the electrical part (15.1a), $\mathbf{A}_{\mathrm{E}}(p) \in \mathbb{R}^{n_{\mathrm{E}} \times n_{\mathrm{E}}}$ is the system matrix, $\mathbf{B}_{\mathrm{E}}(p) \in \mathbb{R}^{n_{\mathrm{E}} \times l}$ is the input matrix and $\mathbf{x}_{\mathrm{E}} \in \mathbb{R}^{n_{\mathrm{E}}}$ is the state vector. It is described by linear algebraic equations that do not depend on the thermal sub-system.

- In the thermal part governed by the system of ordinary differential equations (15.1b), $\mathbf{B}_T(p) \in \mathbb{R}^{n_T \times l}$ is the input matrix, $\mathbf{x}_T(p) \in \mathbb{R}^{n_T}$ is the state vector, $\mathbf{A}_T(p), \mathbf{E}_T(p) \in \mathbb{R}^{n_T \times n_T}$ are system matrices.

  – The tensor $\mathbf{F}(p) \in \mathbb{R}^{n_T \times n_E \times n_E}$, which can be considered as n_T slices of n_E by n_E matrices $\mathbf{F}_i(p) \in \mathbb{R}^{n_E \times n_E}$, $i = 1, \ldots, n_T$, represents the nonlinear coupling from the electrical part to the thermal part. Denoting the i-mode tensor-matrix product by $\times_i$ (for details, please refer to references [2, 11]), the product $\mathbf{F}(p) \times_2 \mathbf{x}_E(p) \times_3 \mathbf{x}_E(p)$ is a vector of length n_T, whose i-th component is the standard vector-matrix-vector product $\mathbf{x}_E(p)^\top \mathbf{F}_i(p) \mathbf{x}_E(p)$. The term $\mathbf{F}(p) \times_2 \mathbf{x}_E(p) \times_3 \mathbf{x}_E(p)$ models the Joule self-heating. Actually this is an approximation since the full Joule self-heating is described by $\mathbf{F}(p) \times_2 \mathbf{x}_E(p) \times_3 \mathbf{x}_E(p) + \mathbf{G}(p) \times_2 \mathbf{x}_E(p) \times_3 \mathbf{u}$, which also models the Joule self-heating related to the input $\mathbf{u}$ by the tensor $\mathbf{G}(p)$. Since the influence of the second therm is quite limited, it is ignored in the model (15.1).

- In the output part (15.1d), $\mathbf{D}(p) \in \mathbb{R}^{m \times l}$ represents the feedthrough, which is nonzero and parameter-dependent in our model, and $\mathbf{C}_E(p) \in \mathbb{R}^{m \times n_E}$ and $\mathbf{C}_T(p) \in \mathbb{R}^{m \times n_T}$ represent the output matrices corresponding to the electrical part and the thermal part, respectively.

In our numerical tests, p represents a single parameter σ, the conductivity of the third metal layer. In this case, all parametric matrices in system (15.1) are of the affine form

$$\mathbf{Y}(\sigma) = \mathbf{Y}_c + \sigma \mathbf{Y}_v, \qquad \mathbf{Y} \in \{\mathbf{A}_E, \mathbf{A}_T, \mathbf{B}_E, \mathbf{B}_T, \mathbf{C}_E, \mathbf{C}_T, \mathbf{D}, \mathbf{E}_T, \mathbf{F}\}. \quad (15.2)$$

15.1.2 Parametric MOR of the Power-MOS Model

Several approaches can be employed to reduce the electro-thermal coupled system (15.1). For example, we can combine the electrical sub-system (15.1a) and the thermal sub-system (15.1b) to form a model described by Differential-Algebraic Equations (DAEs) and apply Parametric Model Order Reduction (PMOR) to these DAEs directly. However, this approach is quite inefficient. In this section, we will exploit the one-way coupling structure to develop a fast PMOR method for the Power-MOS model.

In general, we use the PMOR method developed in [8, 9]. We start from the electrical sub-system (15.1a) since it is an algebraic equation with no coupling from the thermal sub-system, which is easy to reduce. When the parameter dependency is described by (15.2), the electrical sub-system can be written as:

$$\left(\dot{\mathbf{A}}_{Ec} + \sigma \mathbf{A}_{Ev} \right) \mathbf{x}_E(\sigma, t) = - \begin{bmatrix} \mathbf{B}_{Ec} & \mathbf{B}_{Ev} \end{bmatrix} \begin{bmatrix} \mathbf{u}(t) \\ \sigma \mathbf{u}(t) \end{bmatrix}.$$

Actually, it is a special form of [8, equations (4)] by assigning $\mathbf{E}_0 = \mathbf{A}_{\mathrm{Ec}}$, $\mathbf{E}_1 = \mathbf{A}_{\mathrm{Ev}}$, $\mu_1 = \sigma$, $\mathbf{B}(p) = [\mathbf{B}_{\mathrm{Ec}} \ \mathbf{B}_{\mathrm{Ev}}]$, $s = t$, $\mathbf{u}(s) = [\mathbf{u}(t) \ \sigma\mathbf{u}(t)]^\top$. Therefore, the electrical sub-system (15.1a) can be readily reduced by using the PMOR method developed in [8,9], i.e., the multi-moment-matching scheme and the adaptive selection of expansion points based on the error bound proposed. We denote the basis matrix generated by such a method as $\mathbf{V}_{\mathrm{E}}$.

To reduce the thermal sub-system (15.1b), however, we must deal with the nonlinear coupling from the electrical sub-system. Our approach is to first ignore the coupling and reduce the linear system. Then we use the generated bases to reduce the original *nonlinear* system using the projection framework.

- **Step 1:** Use a multi-moment-matching method to reduce the linear part. We first ignore the nonlinear part in system (15.1b) and conduct a shift

$$\widetilde{\mathbf{x}}(t) = \mathbf{x}_{\mathrm{T}}(t) - \mathbf{x}_{\mathrm{T}}^0 \tag{15.3}$$

to get an equivalent system:

$$\mathbf{E}_{\mathrm{T}}(p)\dot{\widetilde{\mathbf{x}}} = \mathbf{A}_{\mathrm{T}}(p)\widetilde{\mathbf{x}} + \mathbf{A}_{\mathrm{T}}(p)\mathbf{x}_{\mathrm{T}}^0 + \mathbf{B}_{\mathrm{T}}(p)\mathbf{u}(t). \tag{15.4}$$

Then, we apply the Laplace transform on (15.4) to obtain its frequency domain representation:

$$
(\mathbf{A}_{\mathrm{Tc}} + \sigma\mathbf{A}_{\mathrm{Tv}} - s\mathbf{E}_{\mathrm{Tc}} - (\sigma s)\mathbf{E}_{\mathrm{Tv}})\mathbf{X}
$$
$$
= [\mathbf{B}_{\mathrm{Tc}} \ \mathbf{B}_{\mathrm{Tv}} \ \mathbf{A}_{\mathrm{Tc}}\mathbf{x}_{\mathrm{T}}^0 \ \mathbf{A}_{\mathrm{Tv}}\mathbf{x}_{\mathrm{T}}^0]\left[-\mathbf{U}^\top \ -\sigma\mathbf{U}^\top \ \frac{-1}{s} \ \frac{-\sigma}{s}\right]^\top, \tag{15.5}
$$

where $\mathbf{X}$ and $\mathbf{U}$ represent the Laplace transforms of the state vector $\widetilde{\mathbf{x}}_{\mathrm{T}}(p,t)$ and the input vector $\mathbf{u}$, respectively, and s and σ denote the radial frequency and the conductivity, respectively. The system above is also in the form of [8, equation (4)] by assigning

$$\mathbf{E}_0 = \mathbf{A}_{\mathrm{Tc}}, \ \mu_1 = \sigma, \ \mathbf{E}_1 = \mathbf{A}_{\mathrm{Tv}}, \ \mu_2 = -s, \ \mathbf{E}_2 = \mathbf{E}_{\mathrm{Tc}}, \ \mu_3 = -\sigma s, \ \mathbf{E}_3 = \mathbf{E}_{\mathrm{Tv}},$$

$$\mathbf{B}(p) = [\mathbf{B}_{\mathrm{Tc}} \ \mathbf{B}_{\mathrm{Tv}} \ \mathbf{A}_{\mathrm{Tc}}\mathbf{x}_{\mathrm{T}}^0 \ \mathbf{A}_{\mathrm{Tv}}\mathbf{x}_{\mathrm{T}}^0], \ \mathbf{u}(s) = \left[-\mathbf{U}^\top \ -\sigma\mathbf{U}^\top \ \frac{-1}{s} \ \frac{-\sigma}{s}\right]^\top.$$

We denote the basis matrix built for (15.5) by $\mathbf{V}_{\mathrm{T}}$.

- **Step 2:** Use the bases to reduce the nonlinear coupled model. To obtain a pROM for (15.1b), we approximate $\mathbf{x}_{\mathrm{E}}$ by $\mathbf{V}_{\mathrm{E}}\widehat{\mathbf{x}}_{\mathrm{E}}$ and $\mathbf{x}_{\mathrm{T}}$ by $\mathbf{V}_{\mathrm{T}}\widehat{\mathbf{x}}_{\mathrm{T}}$, and then force the approximation error to be orthogonal to the range of $\mathbf{V}_{\mathrm{T}}$. The resulting pROM is

$$\widehat{\mathbf{E}}_{\mathrm{T}}(p)\dot{\widehat{\mathbf{x}}}_{\mathrm{T}}(p,t) = \widehat{\mathbf{A}}_{\mathrm{T}}(p)\widehat{\mathbf{x}}_{\mathrm{T}}(p,t) + \widehat{\mathbf{B}}_{\mathrm{T}}(p)\mathbf{u} + \widehat{\mathbf{F}}(p) \times_2 \widehat{\mathbf{x}}_{\mathrm{E}}(p) \times_3 \widehat{\mathbf{x}}_{\mathrm{E}}(p), \tag{15.6}$$

where

$$\widehat{\mathbf{E}}_{\mathrm{T}}(p) = \mathbf{V}_{\mathrm{T}}^{\top}\mathbf{E}_{\mathrm{T}}(p)\mathbf{V}_{\mathrm{T}},$$
$$\widehat{\mathbf{A}}_{\mathrm{T}}(p) = \mathbf{V}_{\mathrm{T}}^{\top}\mathbf{A}_{\mathrm{T}}(p)\mathbf{V}_{\mathrm{T}},$$
$$\widehat{\mathbf{B}}_{\mathrm{T}}(p) = \mathbf{V}_{\mathrm{T}}^{\top}\mathbf{B}_{\mathrm{T}}(p),$$
$$\widehat{\mathbf{F}}(p) = \mathbf{F}(p) \times_1 \mathbf{V}_{\mathrm{T}} \times_2 \mathbf{V}_{\mathrm{E}} \times_3 \mathbf{V}_{\mathrm{E}}.$$

To obtain the reduced tensor $\widehat{\mathbf{F}}(p)$, we first approximate $\mathbf{x}_{\mathrm{E}}(p)$ in the range of $\mathbf{V}_{\mathrm{E}}$, and then project the approximation onto the test subspace $\mathbf{V}_{\mathrm{T}}$, i.e., the tensor product

$$\widehat{\mathbf{F}}(p) \times_2 \widehat{\mathbf{x}}_{\mathrm{E}}(p) \times_3 \widehat{\mathbf{x}}_{\mathrm{E}}(p)$$

is approximated by

$$\mathbf{V}_{\mathrm{T}}^{\top} \left[\mathbf{F}(p) \times_2 \left(\mathbf{V}_{\mathrm{E}}\widehat{\mathbf{x}}_{\mathrm{E}}(p) \right) \times_3 \left(\mathbf{V}_{\mathrm{E}}\widehat{\mathbf{x}}_{\mathrm{E}}(p) \right) \right].$$

Thanks to the reduced tensor, evaluating the ROM does not require computations with quantities of the order of the FOM. In our actual computations, the parametric matrices in the ROM are computed by

$$\widehat{\mathbf{Y}}(\sigma) = \widehat{\mathbf{Y}}_{\mathrm{c}} + \sigma\widehat{\mathbf{Y}}_{\mathrm{v}}, \qquad \widehat{\mathbf{Y}} \in \{\widehat{\mathbf{A}}_{\mathrm{T}}, \widehat{\mathbf{B}}_{\mathrm{T}}, \widehat{\mathbf{C}}_{\mathrm{T}}, \widehat{\mathbf{E}}_{\mathrm{T}}, \widehat{\mathbf{F}}\}, \tag{15.7}$$

where $\widehat{\mathbf{Y}}_{\mathrm{c}}$ and $\widehat{\mathbf{Y}}_{\mathrm{v}}$ are pre-computed at the construction of the pROM.

15.1.3 PMOR-based UQ

Uncertainty Quantification (UQ) methods can be categorized into non-intrusive methods and intrusive methods [12,20]. Non-intrusive methods conduct UQ by solving the original deterministic system, e.g., system (15.1), at various parameter points. Intrusive methods, however, require building a large-scale coupled system, which is often of a much higher order than the original deterministic system. In this section, we focus on non-intrusive methods since the parametric ROM of the original deterministic model (FOM) can be directly used to replace the FOM in UQ. We embed our parametric ROMs into two UQ methods, namely the Latin Hypercube Sampling (LHS) method and the Stochastic Collocation (SC) method.

- LHS [10]. The LHS method is an improvement of the traditional Monte-Carlo sampling. To obtain n samples, LHS divides the input distribution into n intervals of equal probability, and selects one sample randomly in each interval. In this way, LHS ensures that the ensemble of the samples represents the real variability well. The mean and standard deviation of the samples are used to approximate those of the original continuous model. Compared to the standard Monte-Carlo sampling, LHS ensures a set of evenly distributed samples and a faster convergence is usually observed.

- SC [12]. Since the mean and the standard deviation can be computed via probabilistic integration, SC uses a quadrature rule to present the relevant integrals as a weighted sum of the corresponding function values at the collocation points. As we conduct UQ on a single normally-distributed random variable, we use the Gauss-Hermite quadrature rule. The efficiency of SC is guaranteed by the quick convergence of the quadrature rules used [1]. Normally, SC converges much faster than LHS.

The computationally dominant part of both LHS and SC is the simulation of the high-order FOM at all sampled points p_i. Since our pROMs are highly accurate for these simulations as we will show in Sect. 15.1.4, PMOR-based UQ, which replaces the FOM (15.1) with pROMs for these simulations, achieves a significant speedup. Similar ideas have been proposed in [7] and [6], which employed the proper orthogonal decomposition (POD) method and the reduced basis method for (P)MOR, respectively.

15.1.4 Numerical Results

In this section, we apply the proposed methods to the Power-MOS circuit model, which has 6 inputs and 12 outputs. The system is parameterized by the conductivity of the third metal layer and excited by the inputs:

$$u_i = \begin{cases} 0, & i = 1,2, \\ 10^7 t, & i = 3, t \in [0, 10^{-6}], \\ 10, & i = 3, t > 10^{-6}, \\ 26.85, & i = 4,5,6. \end{cases}$$

The outputs we consider here include the currents (I) and thermal fluxes (ϕ) of the three contacts, namely the source, the drain, and the back contact. The initial condition for all electrical state variables is $0\,\mathrm{V}$, and the initial condition for all thermal state variables is $26.85\,^\circ\mathrm{C}$. For the electrical sub-system, the training set is $\sigma \in \{10^7\,\mathrm{S/m}, 3 \times 10^7\,\mathrm{S/m}, 5 \times 10^7\,\mathrm{S/m}\}$, while for the thermal sub-system, the training set contains 20 uniformly distributed frequency samples within $[0\,\mathrm{rad/s}, 10^6\,\mathrm{rad/s}]$, each of which is paired with a randomly chosen conductivity $\sigma \in [10^7\,\mathrm{S/m}, 5 \times 10^7\,\mathrm{S/m}]$.

15.1.4.1 Assessing the Quality of the Reduced Model

Using the PMOR methods proposed, the electrical sub-system is reduced from order 1160 to order 2, the thermal sub-system is reduced from order

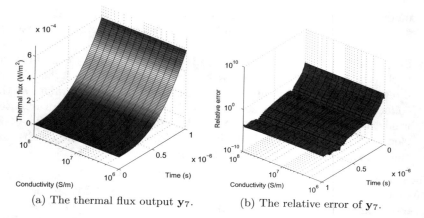

(a) The thermal flux output $\mathbf{y}_7$.   (b) The relative error of $\mathbf{y}_7$.

Fig. 15.2: The thermal flux output $\mathbf{y}_7$ and its relative error for the Power-MOS model.

11556 to order 35, and the speed-up factor for the electro-thermal simulation is 65.93. The convergence behavior of the adaptive PMOR method is shown in Table 15.1 and the thermal flux output $\mathbf{y}_7$, along with its relative error, is shown in Figure 15.2.

Figure 15.2(b) shows that the relative error is large when t is small. The reason is that the device is still hardly heated – the thermal flux is still very close to zero, and therefore, the numerical noise resulting from the numerical error of the finite element discretization dominates the output of the FOM since the true physical dynamics is still small. As Figure 15.2(b) shows, the ROM approximates the thermal flux accurately after the thermal flux dominates the numerical error $(t > 2 \times 10^{-7})$. Therefore, the ROM approximates the true dynamics accurately and, in addition, the ROM is also robust to

Table 15.1: Convergence behavior of electro-thermal simulation of the Power-MOS model (with the error tolerance $\epsilon_{tol} = 10^{-12}$ for the method presented in [8]).

	Electrical sub-system		Thermal sub-system	
Iteration	Selected sample σ	Error bound	Selected sample (s,σ)	Error bound
1	10^7	7.165399×10^{-24}	$(0, 2.736 \times 10^7)$	43.73
2	—	—	$(10^6, 2.537 \times 10^7)$	4.225×10^{-4}
3	—	—	$(2.632 \times 10^5, 1.694 \times 10^7)$	4.345×10^{-8}
4	—	—	$(5.790 \times 10^5, 2.687 \times 10^7)$	9.774×10^{-11}
5	—	—	$(5.263 \times 10^4, 2.836 \times 10^7)$	4.041×10^{-13}

Table 15.2: UQ results for the outputs at $t = 10^{-6}$ s. ($E(\cdot)$ and std($\cdot$) represent the mean and standard deviation, respectively).

	LHS using FOM	LHS using ROM	SC using FOM	SC using ROM
$E(I_{\text{drain}})$	7.4621e-04	7.4621e-04	7.4602e-04	7.4602e-04
std(I_{drain})	2.4794e-04	2.4794e-04	2.4867e-04	2.4867e-04
$E(I_{\text{source}})$	-7.4621e-04	-7.4621e-04	-7.4602e-04	-7.4602e-04
std(I_{source})	2.4794e-04	2.4794e-04	2.4867e-04	2.4867e-04
$E(I_{\text{back}})$	0	0	0	0
std(I_{back})	0	0	0	0
$E(\phi_{\text{drain}})$	5.8479e-04	5.8478e-04	5.8479e-04	5.8479e-04
std(ϕ_{drain})	1.5838e-10	1.5677e-10	1.5985e-10	1.5719e-10
$E(\phi_{\text{source}})$	4.1977e-04	4.1975e-04	4.1977e-04	4.1977e-04
std(ϕ_{source})	1.8528e-10	9.1986e-11	4.6370e-11	9.2124e-11
$E(\phi_{\text{back}})$	6.6781e-07	6.6773e-07	6.6781e-07	6.6781e-07
std(ϕ_{back})	1.5682e-14	1.7778e-14	1.1199e-14	1.6189e-14
Number of sampled points	100	100	11	11
CPU time	6001.14 s	94.19 s	733.64 s	30.51 s

the numerical error present in the FOM due to discretization. Furthermore, although the samples are selected within the range $[10^7 \text{ S/m}, 5 \times 10^7 \text{ S/m}]$, Figure 15.2(b) shows that the parametric ROM is valid in a much wider range.

15.1.4.2 UQ based on the Reduced Model

Now we apply the pROMs to the UQ analysis of the electro-thermal system (15.1). Here we conduct UQ on the outputs at $t = 10^{-6}$ s. We assume that the conductivity obeys the normal distribution $\mathcal{N}(3 \times 10^7, (10^7)^2)$. The numerical results in Table 15.2 show that for both UQ methods, pROM-based UQ computes highly accurate means ($E(\cdot)$). For the nontrivial electrical outputs I_{drain} and I_{source}, which are sensitive to the change in the conductivity with the same coefficient of variation (CV) of 33.23%, the standard deviations (std($\cdot$)) are also computed with high accuracy. The thermal outputs ϕ_{drain}, ϕ_{source} and ϕ_{back}, however, are insensitive to the change in the conductivity with CV's of 3.77e-07, 4.4138e-07 and 2.3483e-08, respectively. Although the standard deviations to these insensitive thermal outputs are not resolved with high accuracy, the relative orders are correct, which gives sufficient information to engineers that the thermal outputs are highly insensitive to the electrical conductivity of the third metal layer.

15.2 Parametric MOR and Co-Simulation

In the second part of this chapter, we investigate coupled problems, which consist of two or more subsystems. Co-simulation or dynamic iteration yields a numerical solution in the time domain. Now parametric MOR, see [5], is applied only to a strict subset of the partial systems.

15.2.1 Problem Definition

We consider coupled problems or multiphysics problems with weakly coupled parts. Without loss of generality, we assume that the coupled problem consists of two parts, since modifications for the case of more subsystems are straightforward. Physical parameters of the problem are replaced by random variables to perform an uncertainty quantification. Statistics of the solution can be computed by a sampling method or a quadrature formula. A monolithic integration approach yields the solution of the coupled problem in the time domain for each node of the quadrature scheme separately.

Alternatively, we apply a co-simulation in the sense of a dynamic iteration for a transient simulation, see [3]. The dynamic iteration allows for a decoupling of the parts such that multiscale or multirate approaches can be used, where different time step sizes are involved for each subsystem. Moreover, the decoupling enables to apply an own quadrature formula for each part of the coupled problem. Thus a local refinement of the accuracy can be done. For example, a part, which is more sensitive with respect to parameter variations, is solved using a quadrature with more nodes. A transfer of information between two different quadrature grids is required just in the communication time points of the dynamic iteration. Global approximations yield this transition in the parameter domain. However, we expect this approach to be efficient only if the original deterministic problem can be solved efficiently by a dynamic iteration. The algorithm of this strategy was outlined and used successfully in [14], where a circuit-heat problem with two parts was considered without using an MOR.

Now we combine methods based on polynomial chaos, see [20], and reduced-order models (ROMs) in a co-simulation. The substitution of a full-order model (FOM) by an ROM introduces an additional error. The idea of the approach using different quadrature grids is to compensate this error partly by a reduction of the error from the quadrature. Since the ROM can be evaluated cheaply, a much finer quadrature grid is applicable, which results in a negligible quadrature error for the ROM part. If all parts of the coupled problem are replaced by ROMs, then this approach is not reasonable any more, because the complete coupled problem can be solved with a low computational effort and thus a single fine quadrature grid can be applied for all subsystems. We require parametric MOR for this purpose, see [5, 17], because the ROM is solved for a large number of grid points in the parameter space.

In this chapter, we illustrate the algorithm for the dynamic iteration of a coupled problem with two different parts, where each part is solved on an own quadrature grid. We verify the applicability of the algorithm by the simulation of a circuit-electromagnetic problem, which is also employed as a test example in [15, 16] and other chapters of this book.

15.2.2 Description of the Method

The algorithm is formulated for time-dependent coupled problems consisting of two parts, i.e.,

$$
\begin{aligned}
\mathbf{F}_1\left(\mathbf{y}_1(t,\mathbf{p}), \mathbf{y}_2^{\mathrm{cpl}}(t,\mathbf{p}), t, \mathbf{p}\right) &= 0, \\
\mathbf{F}_2\left(\mathbf{y}_2(t,\mathbf{p}), \mathbf{y}_1^{\mathrm{cpl}}(t,\mathbf{p}), t, \mathbf{p}\right) &= 0,
\end{aligned}
\tag{15.8}
$$

where independent random variables $\mathbf{p}: \Omega \to \Pi \subseteq \mathbb{R}^Q$ are included from some probability space $(\Omega, \mathcal{A}, \mu)$. The operators $\mathbf{F}_1, \mathbf{F}_2$ represent systems of ordinary differential equations or differential algebraic equations typically stemming from the semi-discretization of partial differential equations. Hence time derivatives are involved in each part. The operators $\mathbf{F}_i$ comprise n_i equations and the solution of the system (15.8) is $\mathbf{y}_i : [t_0, t_{\mathrm{end}}] \times \Pi \to \mathbb{R}^{N_i}$ for $i = 1, 2$, where initial values are given pointwise for all $\mathbf{p}$. The coupling variables are defined as $\mathbf{y}_i^{\mathrm{cpl}} := \mathbf{B}_i \mathbf{y}_i$ with constant matrices $\mathbf{B}_i \in \{0,1\}^{R_i \times N_i}$ such that the coupling variables include just a subset of $\mathbf{y}_i$ for each $i = 1, 2$. Typically, it holds that $R_1 \ll N_1$ and $R_2 \ll N_2$, i.e., the coupling variables represent just a small portion of the solution. Furthermore, it is allowed that just one of the two subsystems in (15.8) includes all the parameters.

To solve the coupled problem (15.8) for a fixed $\mathbf{p} \in \Pi$, we consider a dynamic iteration, where the total time span is split into windows with a first window $[t_0, t_{\mathrm{win}}]$. The resulting iteration of Gauss-Seidel type reads as

$$
\begin{aligned}
\mathbf{F}_1\left(\mathbf{y}_1^{(\nu+1)}(t,\mathbf{p}), \mathbf{y}_2^{\mathrm{cpl}\,(\nu)}(t,\mathbf{p}), t, \mathbf{p}\right) &= 0, \\
\mathbf{F}_2\left(\mathbf{y}_2^{(\nu+1)}(t,\mathbf{p}), \mathbf{y}_1^{\mathrm{cpl}\,(\nu+1)}(t,\mathbf{p}), t, \mathbf{p}\right) &= 0,
\end{aligned}
\qquad \text{for } \nu = 0, 1, 2, \dots
\tag{15.9}
$$

and $t \in [t_0, t_{\mathrm{win}}]$ using the starting values $\mathbf{y}_2^{(0)}(t,\mathbf{p}) \equiv \mathbf{y}_2(t_0,\mathbf{p})$. A numerical method yields the solutions $\mathbf{y}_1, \mathbf{y}_2$ only on a discrete set of time points, which may also differ for the two subsystems. We assume that all coupling variables are interchanged in a few communication time points $\bar{t}_j$ with

$$
t_0 \le \bar{t}_1 < \bar{t}_2 < \cdots < \bar{t}_J = t_{\mathrm{win}}.
\tag{15.10}
$$

Interpolation in time yields approximations of the coupling variables $\mathbf{y}_i^{\mathrm{cpl}}(t,\mathbf{p})$ for $t \in [t_0, t_{\mathrm{win}}]$ and $i = 1, 2$.

Statistical information for a function $g : \Pi \to \mathbb{R}$ depending on the random parameters is obtained by probabilistic integrals

$$\mathrm{E}(g) := \int_\Omega g(\mathbf{p}(\omega))\, \mathrm{d}\mu(\omega) = \int_\Pi g(\mathbf{p})\,\rho(\mathbf{p})\, \mathrm{d}\mathbf{p} \tag{15.11}$$

provided that the integral is finite, where the existence of a joint probability density function $\rho : \Pi \to \mathbb{R}$ is assumed. For example, probabilistic integration can be applied to the solution of (15.8) component-wise. The expected value as well as the variance represent elementary statistics. Our aim is to compute statistics of the solution $\mathbf{y}_1, \mathbf{y}_2$ for either the complete time interval or just at a final time.

A quadrature scheme or a sampling method yields an approximation of a probabilistic integral (15.11), see [60] and the references therein. We obtain a finite sum of the form

$$\mathrm{E}(g) \approx S(g) = \sum_{k=1}^{K} w_k g(\mathbf{p}^{(k)})$$

with grid points $\mathbf{p}^{(1)}, \ldots, \mathbf{p}^{(K)} \in \Pi$ and weights $w_1, \ldots, w_K \in \mathbb{R}$. For a quantity of interest

$$g(\mathbf{p}) = \tilde{g}(\mathbf{y}_1(t_{\mathrm{end}}, \mathbf{p}), \mathbf{y}_2(t_{\mathrm{end}}, \mathbf{p}))$$

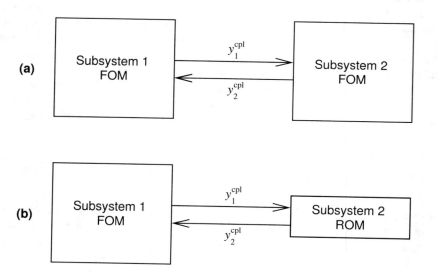

Fig. 15.3: Original setting with full-order models (a) and inclusion of a reduced-order model (b).

at some final time, it follows that an initial value problem of the system (15.8) has to be resolved K times for the different realizations of the parameters.

If the solutions of the subsystems in the coupled problem (15.8) behave differently with respect to the random parameters, then the application of different quadrature formulas might become advantageous. A higher variance within a subsystem often indicates that a higher accuracy of the quadrature is required. Now a parametric ROM can be solved on a much finer quadrature grid, since its evaluation requires just a low computational effort. Figure 15.3 illustrates the inclusion of a reduced-order model. If sufficiently accurate ROMs are available for both systems, then the complete system can be sampled with a low computational effort on a fine grid and the usage of different quadrature grids is not relevant. Thus the interesting situation is that one submodel cannot be reduced (efficiently).

We introduce two grids

$$\mathcal{G}_i := \left\{ \mathbf{p}_i^{(1)}, \ldots, \mathbf{p}_i^{(K_i)} \right\} \quad \text{with } \mathbf{p}_i^{(k)} \in \Pi \tag{15.12}$$

for $i = 1, 2$ dedicated to the two parts of the coupled problem (15.8). The numbers of grid points K_1, K_2 may differ significantly. The subsystem for $\mathbf{F}_i$ together with its solution $\mathbf{y}_i$ is integrated in time for the grid points in $\mathcal{G}_i$ and each $i = 1, 2$.

Following (15.9), we have to solve the problems

$$\mathbf{F}_1\left(\mathbf{y}_1^{(\nu+1)}(t, \mathbf{p}_1^{(k)}), \mathbf{y}_2^{\text{cpl}(\nu)}(t, \mathbf{p}_1^{(k)}), t, \mathbf{p}_1^{(k)}\right) = 0 \quad \text{for } k = 1, \ldots, K_1,$$
$$\mathbf{F}_2\left(\mathbf{y}_2^{(\nu+1)}(t, \mathbf{p}_2^{(k)}), \mathbf{y}_1^{\text{cpl}(\nu+1)}(t, \mathbf{p}_2^{(k)}), t, \mathbf{p}_2^{(k)}\right) = 0 \quad \text{for } k = 1, \ldots, K_2,$$
$$\tag{15.13}$$

in each step of the dynamic iteration. The first iteration step $\nu = 0$ in (15.13) for $\mathbf{F}_1$ can be computed directly using the globally defined initial values. The output is $\mathbf{y}_1^{(1)}(\bar{t}_j, \mathbf{p}_1^{(k)})$ for $k = 1, \ldots, K_1$ in the communication time points $\bar{t}_1, \ldots, \bar{t}_J$ from (15.10). To this end, we need the coupling variables $\mathbf{y}_1^{\text{cpl}(1)}(\bar{t}_j, \mathbf{p}_2^{(k)})$ for $k = 1, \ldots, K_2$ and $j = 1, \ldots, J$. Likewise, the output of $\mathbf{F}_2$ is the solution $\mathbf{y}_2^{(1)}(\bar{t}_j, \mathbf{p}_2^{(k)})$ for $k = 1, \ldots, K_2$ and has to be transformed into the coupling variables $\mathbf{y}_2^{\text{cpl}(1)}(\bar{t}_j, \mathbf{p}_1^{(k)})$ for $k = 1, \ldots, K_1$, i.e., the evaluation on the other quadrature grid is required. This strategy repeats in each iteration step. Hence transitions between the two grids have to be defined for a fixed time point.

For the interchange of information between the two grids, we consider global approximations of the coupling variables in the parameter space Π. If a global approximation is available, we can evaluate at any point $\mathbf{p} \in \Pi$. There are two techniques to construct a global approximation:

1. **Best-approximation in integral norm.**
   Orthogonal basis polynomials are available with respect to the L^2-inner product of the probability space induced by the integral (15.11). Hence

a truncated sum of the polynomial chaos expansion is used, see [60]. Let the time $\bar{t}$ be fixed from the set (15.10). The global approximation reads as

$$\tilde{\mathbf{y}}_i^{\mathrm{cpl}}(\bar{t}, \mathbf{p}) := \sum_{m=0}^{M_i} \mathbf{u}_{i,m}(\bar{t}) \Phi_m(\mathbf{p}) \qquad (15.14)$$

for $i = 1, 2$ with known basis polynomials $\Phi_m : \Pi \to \mathbb{R}$ satisfying the orthonormality condition $\mathrm{E}(\Phi_m \Phi_n) = \delta_{mn}$. In general, all polynomials up to a certain degree are involved. The coefficient functions in (15.14) are determined approximately by

$$
\begin{aligned}
\mathbf{u}_{i,m}(\bar{t}) &:= \int_{\Pi} \mathbf{y}_i^{\mathrm{cpl}}(\bar{t}, \mathbf{p}) \Phi_m(\mathbf{p}) \rho(\mathbf{p}) \, \mathrm{d}\mathbf{p} \\
&\approx \sum_{k=1}^{K_i} w_i^{(k)} \mathbf{y}_i^{\mathrm{cpl}}(\bar{t}, \mathbf{p}_i^{(k)}) \Phi_m(\mathbf{p}_i^{(k)})
\end{aligned}
\qquad (15.15)
$$

for $i = 1, 2$, where the values $w_i^{(k)} \in \mathbb{R}$ represent the weights of quadrature formulas on the grids $\mathcal{G}_i$. Thus the sums (15.14) can be evaluated for an arbitrary $\mathbf{p} \in \Pi$.

2. **Interpolation scheme in parameter space.**
The coupling variables are approximated by an interpolating function. For fixed time $\bar{t}$, it holds that

$$\tilde{\mathbf{y}}_i^{\mathrm{cpl}}(\bar{t}, \mathbf{p}) := \sum_{k=1}^{K_i} \Psi_{i,k}(\mathbf{p}) \mathbf{y}_i^{\mathrm{cpl}}(\bar{t}, \mathbf{p}_i^{(k)}) \qquad (15.16)$$

for $i = 1, 2$ with the functions $\Psi_{i,k} : \Pi \to \mathbb{R}$ satisfying

$$\Psi_{i,k}(\mathbf{p}_i^{(\ell)}) = \begin{cases} 0 \text{ for } k \neq \ell, \\ 1 \text{ for } k = \ell, \end{cases}$$

which are independent of time. For example, Lagrange polynomials are applicable in the case of uniform grids. Obviously, the approximation coincides with the exact coupling values at the grid points. Again the formula (15.16) can be evaluated at an arbitrary parameter value.

Since the number of coupling variables is relatively low in comparison to the dimension of the coupled problem (15.8), the computational effort for the global approximation is usually negligible compared to the time integration.

After the convergence of the dynamic iteration in a time window, the same approach is repeated in the next time window. Therein, initial values can be transformed between the two grids again by the above procedure. If the approximations have been computed at the final time t_{end}, then we reconstruct statistical data by quadrature formulas using the same grid points.

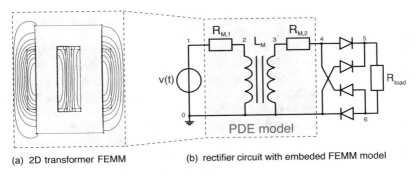

(a) 2D transformer FEMM   (b) rectifier circuit with embeded FEMM model

Fig. 15.4: Diagram of circuit-electromagnetic problem.

15.2.3 Simulation of Circuit-EM Problem

For a verification of the algorithm from Section 15.2.2, we simulate a test example. The numerical computations are done within the software package MATLAB. We consider a coupled circuit-electromagnetic (or circuit-field) problem, see [3, 15, 16]. Figure 15.4 illustrates the problem. The electric circuit includes four diodes and a transformer. This circuit represents a rectifier, where a sinusoidal input voltage is converted into a rectified output voltage at a load resistance. For the transformer, a refined modeling is applied, where a quasilinear parabolic form of Maxwell's equations describes the device in two space dimensions. The software package FEMM [13] yields a space-discretization of the partial differential equations by a finite element method. Thus we obtain a coupled problem of the form (15.8) consisting of a nonlinear circuit part and a linear device part. We arrange the amplitude of the input voltage in the circuit part and the conductance in the device part as independent random variables. Uniform probability distributions are applied with a range of 15% around the mean values.

Now the FOM of the device part is replaced by an ROM. Therein, parametric model order reduction (pMOR) preserves the conductance as an independent variable. An adaptive Krylov-space method from [4] is employed for the pMOR. This method uses an error bound to automatically select the interpolation points to build a low-order ROM with a guaranteed high accuracy.

For the grids (15.12) in the parameter space, the tensor-product Gauss-Legendre quadrature formula is applied. The first grid of the circuit part includes $2 \times 2 = 4$ points, whereas the second grid of the device part with the ROM is refined to $4 \times 4 = 16$ points. The exchange of information between the two grids is done by the best approximation in the L^2-norm as outlined in Section 15.2.2. In (15.14) and (15.15), the multivariate Legendre polynomials are included up to the total degree two. We apply the first grid to calculate the approximations of the expected value and the standard deviation for the output voltage in the complete time interval.

The transient simulation is done in a time interval with the length of about one period of the input voltage. In the dynamic iteration, relatively small time windows are included. Just a single step of the dynamic iteration (15.13) is done in each time window. The implicit Euler method performs the time integration in both parts of the coupled system.

For comparison, a monolithic time integration yields a reference solution, where still the FOM is included and a single quadrature grid with 9 nodes from the Stroud formula of order five, see [19], is used. Figure 15.5 depicts the expected value as well as the standard deviation of the output voltage for both numerical simulations. We observe a good agreement between the FOM variant and the ROM variant. Furthermore, the absolute difference between these statistics is shown in Figure 15.6.

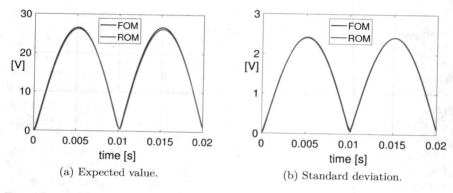

(a) Expected value.   (b) Standard deviation.

Fig. 15.5: Statistics of the output voltage computed by the full order model (FOM) variant and by the reduced order model (ROM) variant.

15.3 Conclusions

In this chapter, we developed two types of methods that use (parametric) ROMs to accelerate UQ of large-scale couples systems.

- In the first part, we developed a PMOR-based UQ framework for an electro-thermal one-way coupled system. We applied an adaptive multi-moment-matching PMOR method on both the electrical sub-system and the linear part of the thermal sub-system. With the computed bases, we reduced the system including the nonlinear coupling term. We used the

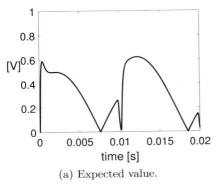
(a) Expected value.

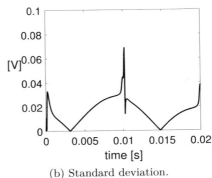
(b) Standard deviation.

Fig. 15.6: Modulus of the difference between the full order model (FOM) variant and the reduced order model (ROM) variant for the statistics of the output voltage.

parametric ROM to accelerate two types of sampling-based UQ methods, namely the Latin hypercube sampling method and the stochastic collocation method. Both methods gave accurate UQ results with significant speedups.

- In the second part, we derived an algorithm for solving a coupled problem with random parameters by a dynamic iteration including different quadrature grids for the separate parts of the problem. The inclusion of reduced order models is feasible in this technique and allows for a refinement of the respective grid to increase the accuracy. We confirmed the feasibility of the method by the simulation of a test example. The efficiency of this approach still has to be investigated further by both theoretical analysis and numerical experiments.

References

1. BABUŠKA, I., NOBILE, F., AND TEMPONE, R.: *A stochastic collocation method for elliptic partial differential equations with random input data.* SIAM Review, 52, pp. 317–355, 2010.
2. BADER, B.W., KOLDA, T.G., ET AL.: *Matlab tensor toolbox version 2.6.* http:// www.sandia.gov/~tgkolda/TensorToolbox/, Sandia National Laboratories, February 2015. Online: http://www.tensortoolbox.org/.
3. BARTEL, A., BRUNK, M., GÜNTHER, M., AND SCHÖPS, S.: *Dynamic iteration for coupled problems of electric circuits and distributed devices,* SIAM J. Sci. Comput., 35, pp. B315–B335, 2013.
4. BENNER, P., AND FENG, L.: *A robust algorithm for parametric model order reduction based on implicit moment matching.* In: QUARTERONI, A., AND ROZZA, G.

(EDS.): *Reduced Order Methods for Modeling and Computational Reduction.* MS&A - Modeling, Simulation and Applications, 9: 159–185. Springer, Heidelberg, 2014.

5. BENNER, P., GUGERCIN, S., AND WILLCOX, K.: *A Survey of Projection-Based Model Reduction Methods for Parametric Dynamical Systems.* SIAM Review, 57(4): 483–531, 2015.

6. BENNER, P., AND HESS, M.W.: *Reduced basis modeling for uncertainty quantification of electromagnetic problems in stochastically varying domains.* In: BARTEL, A., CLEMENS, M., GÜNTHER, M, AND TER MATEN, E.J.W. (EDS.): *Scientific Computing in Electrical Engineering SCEE 2014,* Series Mathematics in Industry, Vol. 23, Springer International Publishing, pp. 215–222, 2016.

7. BENNER, P., AND SCHNEIDER, J.: *Uncertainty quantification for Maxwell's equations using stochastic collocation and model order reduction.* International Journal for Uncertainty Quantification, 5, pp. 195–208, 2015.

8. FENG, L., AND BENNER, P.: *Parametric Model Order Reduction for Electro-Thermal Coupled Problems.* Chapter in this Book, 2019.

9. FENG, L., YUE, Y., BANAGAAYA, N., MEURIS, P., SCHOENMAKER, W., AND BENNER, P.: *Parametric Modeling and Model Order Reduction for (Electro-)Thermal Analysis of Nanoelectronic Structures.* Journal of Mathematics in Industry, 6 (10):1–16, 2016. Open access: https://dx.doi.org/10.1186/s13362-016-0030-8.

10. HELTON, J., AND DAVIS, F.: *Latin hypercube sampling and the propagation of uncertainty in analyses of complex systems.* Reliability Engineering & System Safety, 81, pp. 23–69, 2003.

11. KOLDA, T.G., AND BADER, B.W.: *Tensor decompositions and applications,* SIAM Review, 51, pp. 455–500, 2009.

12. LE MAÎTRE, O., AND KNIO, O.: *Spectral Methods for Uncertainty Quantification: With Applications to Computational Fluid Dynamics,* Scientific Computation, Springer Netherlands, 2010.

13. MEEKER, D.: *FEMM - Finite Element Method Magnetics.* User's Manual version 4.2, 2010. http://www.femm.info/wiki/HomePage.

14. PULCH, R., BARTEL, A., AND SCHÖPS, S.: *Quadrature methods with adjusted grids for stochastic models of coupled problems.* In: RUSSO, G., CAPASSO, V., NICOSIA, G., AND ROMANO, V. (EDS.): *Progress in Industrial Mathematics at ECMI 2014,* Series Mathematics in Industry, Vol. 22, Springer, pp. 377–384, 2018.

15. SCHÖPS, S., DE GERSEM, H., AND SCHÖPS, S.: *A cosimulation framework for multirate time integration of field/circuit coupled problems.* IEEE Trans. Magn., 46, pp. 3233–3236, 2010.

16. SCHÖPS, S., DE GERSEM, H., AND WEILAND, T.: *Winding functions in transient magnetoquasistatic field-circuit coupled simulations.* COMPEL, 32, pp. 2063–2083, 2013.

17. SOLL, T., AND PULCH, R.: *Sample selection based on sensitivity analysis in parameterized model order reduction.* J. Comput. Appl. Math., 316, pp. 369–379, 2017.

18. SPIRITO, P., BREGLIO, G., D'ALESSANDRO, V., AND RINALDI, N.: *Thermal instabilities in high current power MOS devices: experimental evidence, electro-thermal simulations and analytical modeling.* In: 23rd International Conference on Microelectronics, vol. 1, MIEL, pp. 23–30, 2002.

19. STROUD, A.: *Approximate Calculation of Multiple Integrals,* Prentice Hall, 1971.

20. XIU, D.: *Numerical Methods for Stochastic Computations: a Spectral Method Approach.* Princeton University Press, 2010.

21. YUE, Y., FENG, L., MEURIS, P., SCHOENMAKER, W., AND BENNER, P.: *Application of Krylov-type parametric model order reduction in efficient uncertainty quantification of electro-thermal circuit models.* In: PIERS Proceedings, Prague, 6–9, July 2015, pp. 379–384.

Part V
Robustness, Reliability, Ageing

Part V covers the following topics.

- **Estimating Failure Probabilities**
  Authors:
  E. Jan W. ter Maten (Bergische Universität Wuppertal, Germany),
  Theo G.J. Beelen (Eindhoven University of Technology, the Netherlands),
  Roland Pulch (Universität Greifswald, Germany),
  Ulrich Römer (Technische Universität Braunschweig, Germany),
  Herbert De Gersem (Technische Universität Darmstadt, Germany),
  Rick Janssen, Jos J. Dohmen, Bratislav Tasić (NXP Semiconductors, Eindhoven, the Netherlands),
  Renaud Gillon, Aarnout Wieers, Frederik Deleu (ON Semiconductor, Oudenaarde, Belgium).
- **Fast Fault Simulation for Detecting Erroneous Connections in ICs**
  Authors:
  Jos J. Dohmen, Bratislav Tasić, Rick Janssen (NXP Semiconductors, Eindhoven, the Netherlands),
  E. Jan W. ter Maten (Bergische Universität Wuppertal, Germany),
  Theo G.J. Beelen (Eindhoven University of Technology, the Netherlands),
  Roland Pulch (Universität Greifswald, Germany),
  Michael Günther (Bergische Universität Wuppertal, Germany).
- **Calibration of Probability Density Function**
  Authors:
  Jos J. Dohmen (NXP Semiconductors, Eindhoven, the Netherlands),
  Theo G.J. Beelen, Oryna Dvortsova (Eindhoven University of Technology, the Netherlands),
  E. Jan W. ter Maten (Bergische Universität Wuppertal, Germany),
  Bratislav Tasić, Rick Janssen (NXP Semiconductors, Eindhoven, the Netherlands).
- **Ageing Models and Reliability Prediction**
  Authors:
  Renaud Gillon, Aarnout Wieers, Frederik Deleu (ON Semiconductor, Oudenaarde, Belgium),
  Tomas Gotthans (Brno University of Technology, Czech Republic),
  Rick Janssen (NXP Semiconductors, Eindhoven, the Netherlands),
  Wim Schoenmaker (MAGWEL, Leuven, Belgium),
  E. Jan W. ter Maten (Bergische Universität Wuppertal, Germany).

Chapter 16
Estimating Failure Probabilities

E. Jan W. ter Maten, Theo G.J. Beelen, Alessandro Di Bucchianico,
Roland Pulch, Ulrich Römer, Herbert De Gersem, Rick Janssen,
Jos J. Dohmen, Bratislav Tasić, Renaud Gillon, Aarnout Wieers,
Frederik Deleu

Abstract System failure describes an undesired configuration of an engineering device, possibly leading to the destruction of material or a significant
loss of performance and a consequent loss of yield. For systems subject to
uncertainties, failure probabilities express the probability of this undesired
configuration to take place. The accurate computation of failure probabilities, however, can be very difficult in practice. It may also become very costly,
because of the many Monte Carlo samples that have to be taken, which may
involve time consuming evaluations. In this chapter we present an overview of
techniques to realistically estimate the amount of Monte Carlo runs that are

E. Jan W. ter Maten
Bergische Universität Wuppertal, Germany,
e-mail: terMaten@math.uni-wuppertal.de

Theo G.J. Beelen, Alessandro Di Bucchianico, E. Jan W. ter Maten
Eindhoven University of Technology, the Netherlands,
e-mail: {A.D.Bucchianico,E.J.W.ter.Maten}@tue.nl;Th.Beelen@gmail.com

Roland Pulch
Universität Greifswald, Germany,
e-mail: Roland.Pulch@uni-greifswald.de

Ulrich Römer
Technische Universität Braunschweig, Germany,
e-mail: U.Roemer@tu-braunschweig.de

Herbert De Gersem
Technische Universität Darmstadt, Germany,
e-mail: DeGersem@temf.tu-darmstadt.de

Rick Janssen, Jos J. Dohmen, Bratislav Tasić
NXP Semiconductors, Eindhoven, the Netherlands,
e-mail: {Rick.Janssen,Jos.J.Dohmen,Bratislav.Tasic}@nxp.com

Renaud Gillon, Aarnout Wieers, Frederik Deleu
ON Semiconductor, Oudenaarde, Belgium,
e-mail: {Renaud.Gillon,Aarnout.Wieers,Frederik.Deleu}@onsemi.com

© Springer Nature Switzerland AG 2019
E. J. W. ter Maten et al. (eds.), *Nanoelectronic Coupled Problems Solutions*,
Mathematics in Industry 29, https://doi.org/10.1007/978-3-030-30726-4_16

needed to guarantee sharp bounds for relative errors of failure probabilities. They are presented for Monte Carlo sampling and for Importance Sampling. These error estimates apply to both non-parametric and parametric sampling. In the case of parametric sampling we propose a hybrid algorithm that combines simulations of full models and approximating response surface models. We illustrate this hybrid algorithm with a computation of bond wire fusing probabilities.

16.1 Motivation of Failure Analysis

The determination of failure probabilities directly addresses the reliability of electronic devices in view of undesired variations in an industrial production or construction [6,11,27,30,32,33]. Assuming that most of the produced devices exhibit a good or acceptable functionality, only a small number of devices have to be rejected. Thus a numerical simulation should determine small failure probabilities to confirm that an industrial production is acceptable and efficient. A strong criterion is the six sigma (6σ) concept, where variations of six standard deviations under the assumption of a normal distribution for an input parameter still results in acceptable outputs. The associated failure probability is $10^{-7}\%$ (or a probability of 10^{-9}) for a left, one-sided tail, or twice this amount for two-sided failure tails, which in both cases is extremely small.

16.2 Mathematical Formulation

The mathematical formulation of a failure probability is as follows. Let $\mathbf{p} \in \Gamma \subseteq \mathbb{R}^{n_p}$ be the parameters for state variables $\mathbf{y}(\mathbf{p})$ (when being time-independent), or $\mathbf{y}(t,\mathbf{p})$ (when time-dependent) of a mathematical model such as e.g. a system of ordinary differential equations or partial differential equations, for example. The failure of the system is described by a "performance" function

$$h : \Gamma \to \mathbb{R}, \qquad \text{with} \qquad \begin{array}{l} h(\mathbf{p}) \geq 0 \text{ for acceptance,} \\ h(\mathbf{p}) < 0 \text{ for failure.} \end{array} \qquad (16.1)$$

In general, for a given $\mathbf{p}$, the value $h(\mathbf{p}) = \hat{h} \circ \mathbf{y}(\mathbf{p})$ is the result of a post-processing function $\hat{h}$ on $\mathbf{y}(\mathbf{p})$ when applied to a steady-state solution, or, on a time-dependent solution $\mathbf{y}(t,\mathbf{p})$, when applied at a specific time measurement moment. It also may depend on a time integration result of $\mathbf{y}(t,\mathbf{p})$ (e.g., a total power loss of voltages times currents through a specific resistor). In a coupled electromagnetic-heat problem too strong currents may cause damage because of heating.

Due to (16.1), the *region of failure in the parameter space* becomes $\Gamma_F = \{\mathbf{p} \in \Gamma : h(\mathbf{p}) < 0\}$. When we have a measure μ on Γ and assume that Γ_F is measurable, then $\mu(\Gamma_F)/\mu(\Gamma)$ can be taken as failure probability.

We now assume that the parameters $\mathbf{p}$ are subject to uncertainty. By this, we mean thatthe parameters $\mathbf{p}$ become random variables $\mathbf{p} : \omega \in \Omega \to \Gamma$, defined on some probability space $(\Omega, \mathcal{A}, \mu)$ with event space Ω, sigma-algebra $\mathcal{A}$ and probability measure μ. Under the assumption that a joint probability density function $\rho : \Gamma \to \mathbb{R}$ of the random variables exists, the failure probability reads as

$$P_F = \int_\Omega \chi_{\Gamma_F}(\mathbf{p}(\omega)) \, d\mu(\omega) = \int_\Gamma \chi_{\Gamma_F}(\mathbf{p})\rho(\mathbf{p}) \, d\mathbf{p}. \qquad (16.2)$$

Therein, for an arbitrary subset $A \subseteq \Gamma$, the characteristic function $\chi_A(\mathbf{p})$ is defined by

$$\chi_A(\mathbf{p}) := \begin{cases} 1 & \text{for } \mathbf{p} \in A, \\ 0 & \text{for } \mathbf{p} \notin A. \end{cases} \qquad (16.3)$$

Each evaluation of the integrand in (16.2) requires to solve the model equations for $\mathbf{y}(\mathbf{p})$. We notice that $A = \Gamma_F$ is defined indirectly via the output space of $\mathbf{y}$. This makes the identification of Γ_F and quantification of P_F for a given h not trivial. The inverse problem, determining the optimal (largest) part of the parameter space that guarantees a given small failure probability, is even much harder.

In the next sections we present an overview of techniques to estimate the amount of Monte Carlo samples or of samples in Importance Sampling to guarantee sharp bounds on relative errors in the failure probabilities. Because of simplified notation they are presented for non-parametric sampling. However they apply to parametric problems as well.

For parametric problems also exploiting a response surface model is discussed. A hybrid algorithm that combines full model simulation with results from a response surface model is presented and is applied to a bond wire fusing problem.

16.3 Tail Probabilities for the Normal Distribution

We start by considering a real-valued random variable Y that has a normal probability density function $N(0,1)$ [26]. For the moment we do not consider parameters. Hence, we directly treat the *region of failure in the output space* and assume it to be a set $A = (-\infty, x)$ (say for some $x < 0$). Then the failure probability is defined as

$$P(Y \in A) = \Phi(x) = \frac{1}{\sqrt{2\pi}} \int_{-\infty}^{x} e^{-z^2/2} dz, \qquad (16.4)$$

where $\Phi(x)$ denotes the cumulative distribution function (cdf) of the normal, Gaussian, distribution $N(\mu, \sigma^2)$ with mean $\mu = 0$ and standard deviation $\sigma = 1$. Hence the failure probability becomes a so-called tail probability α, that is the outcome $\Phi(x)$, for some x, of the integral in (16.4). For each $\alpha \in (0, 1)$ we can define the *quantiles* z_α by $\Phi(-z_\alpha) = \alpha$ (here we exploit the monotonicity of the cdf). Clearly α is the (left) tail probability for $x = -z_\alpha$. With the error function $\mathrm{erf}(z) = \frac{2}{\sqrt{\pi}} \int_{-\infty}^{z} e^{-t^2/2} dt$ and its inverse $\mathrm{erfinv}(z)$ we find

$$\alpha = \frac{1}{2}\left[1 + \mathrm{erf}(\frac{-z_\alpha}{\sqrt{2}})\right], \quad \text{or} \quad z_\alpha = -\sqrt{2}\,\mathrm{erfinv}(2\alpha - 1). \tag{16.5}$$

Fig. 16.1 shows the decimal powers of the left tail probability, $\log_{10}(\alpha)$, versus the quantiles z_α of the normal distribution with $\mu = 0$ and $\sigma = 1$. Actually the vertical axis provides z_α along a σ- scale. Our interest goes to variations up to 6σ. For tail probabilities $\alpha = 10^{-1}$, 10^{-3}, 10^{-9}, 10^{-10} we find $z_\alpha \equiv 1.28\sigma$, 3.09σ, 6σ, 6.36σ, respectively. For more values, see Table 16.1.

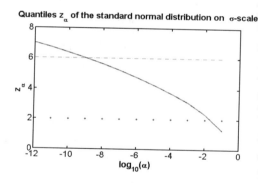

Fig. 16.1: Powers of a left-sided tail accuracy, $\log_{10}(\alpha)$, versus the quantiles z_α of the normal distribution along a σ-scale. Our interest goes to quantiles up to 6σ where the tail probability is in the order of 10^{-9}. See also [8].

Table 16.1: Typical values of quantiles z_α of the standard normal distribution along a σ-scale [27].

α	10^{-12}	10^{-11}	10^{-10}	10^{-9}	10^{-8}	10^{-7}	10^{-6}	10^{-5}	10^{-4}	10^{-3}	10^{-2}	10^{-1}
z_α	7.03	6.71	6.36	6.00	5.61	5.20	4.75	4.26	3.72	3.09	2.33	1.28

16.4 Approximating Tail Probabilities by Monte Carlo

We now take Y to be a real-valued random variable with probability density function f, that is not necessarily the normal distribution $N(0,1)$. We still assume that $A = (-\infty, x)$ is the region of failure. The probability failure becomes

$$P(Y \in A) = p \equiv \int_{-\infty}^{x} f(z)dz. \tag{16.6}$$

For varying x, $p = p(x)$ becomes the cdf that is implied by f. However, we can not directly calculate p in (16.6). We can approximate p by use of Monte-Carlo (MC) samples Y_i of Y [4, 22, 23, 26]. We assume that $N = N_{\mathrm{MC}}$ independent random observations Y_i $(i = 1, \ldots, N)$ of Y are taken and we define, for the given set $A = (-\infty, x)$, the event indicator $X_i = I_A(Y_i)$, where $I_A(Y_i) = 1$, if $Y_i \in A$ and 0 otherwise. By this, each X_i is Bernoulli distributed with success probability p. The individual means and variances are $\mathrm{E}[X_i] = p$ and $\mathrm{Var}[X_i] = p(1-p)$. By taking the sample mean of the X_i we can estimate p better with a smaller variance. Hence we define $p_f^{\mathrm{MC}}(A) \equiv p_N(A) = \frac{1}{N}\sum_{i=1}^{N} X_i$. Then $Np_{\mathrm{MC}} \sim \mathrm{Bin}(N,p)$ is binomially distributed (N samples, each with success probability p), and thus for the expectation one has $\mathrm{E}[p_f^{\mathrm{MC}}] = \mathrm{E}[p_N(A)] = \frac{1}{N}Np = p$ and for the variance $\sigma^2[p_f^{\mathrm{MC}}] = \sigma^2[p_N(A)] = \frac{p(1-p)}{N}$.

Below, we collect some approaches for estimating the required simulation sample size $N = N_{\mathrm{MC}}$ in order to guarantee a prescribed accuracy in estimating p.

A rule-of-thumb may be to continue simulation until 10 failures appear. Then N becomes $N \approx 10/p$, which causes a large number of simulations for small p. However, as we will see in Section 16.5.2, a realistic estimate is even in the order of $100/p$.

From the Central Limit Theorem (CLT) [12, p. 244], we derive that *for each fixed $\beta \in \mathbb{R}$ and $N = N_{\mathrm{MC}}$ large enough*

$$P\left(\frac{p_f^{\mathrm{MC}} - p}{\sigma[p_f^{\mathrm{MC}}]} < \beta\right) = P\left(\frac{Np_f^{\mathrm{MC}} - Np}{\sigma[X]\sqrt{N}} < \beta\right) \longrightarrow \Phi(\beta). \tag{16.7}$$

Here X stands for a Bernoulli distributed random variable with success probability p. Thus the (cumulative) *distribution function* of the scaled (normalized) error $(p_{\mathrm{MC}} - p)/\sigma[p_f^{\mathrm{MC}}]$ converges to a normal distribution.

Hence, for given $\varepsilon > 0$ and $N = N_{\mathrm{MC}}$ large enough such that the scaled $z \equiv \varepsilon/\sqrt{p(1-p)/N} > z_{\alpha/2}$, we obtain

$$P(|p_f^{\mathrm{MC}} - p| > \varepsilon) = P\left(\frac{|p_f^{\mathrm{MC}} - p|}{\sigma[p_f^{\mathrm{MC}}]} > z\right) \leq 2\Phi(-z_{\alpha/2}) = \alpha, \tag{16.8}$$

where $\sigma[p_f^{MC}] = \sqrt{\frac{p(1-p)}{N}}$, $z = \varepsilon/\sqrt{p(1-p)/N}$ (the factor 2 comes from the fact that the CLT considers two tails) [8, 27, 29]. The convergence holds for all points z in (16.8) for which the distribution is continuous. We note that, after scaling the error $|p_f^{MC} - p|$ with $\sigma[p_f^{MC}]$, z involves a factor $\sqrt{N}$. In our case it allows to derive an error estimate for a particular value of z, which leads to estimate the number of samples we have to take. We deduce

$$N_{MC} \geq p(1-p)\left(\frac{z_{\alpha/2}}{\varepsilon}\right)^2 = \frac{1-p}{p}\left(\frac{z_{\alpha/2}}{\nu}\right)^2, \quad \text{for } \varepsilon = \nu p. \qquad (16.9)$$

Then (16.8) effectively considers a probability for a relative error ν of p. Let us assume $\nu = 0.1$ and $p = 10^{-10}$. Now for $\alpha = 0.02$ (which is quite moderate) we have $z_{\alpha/2} \approx 2$ (see also Fig. 16.1) and (16.9) gives $N_{MC} \geq 4 \cdot 10^{12}$. This is large, but it looks reasonable if we compare it to the small value of p. We see that N_{MC} may grow with $1/p$ and not necessarily with $1/p^2$.

A problem arises if we do not know p. Then we can use $p(1-p) \geq 1/4$ for all p, yielding $N_{MC} \geq \frac{1}{4}z_{\alpha/2}^2/\varepsilon^2 = 10^{22}$, which is much larger than the previous bound found just before. This new larger lower bound indeed even grows with $1/p^2$ [8, 27, 29].

Even more general, if N_{MC} is not large enough to apply the CLT, we always can apply Chebyshev's inequality (16.10). This inequality is valid for any random variable U with finite mean μ and variance σ^2

$$P(|U - \mu| > \varepsilon) \leq \frac{\sigma^2}{\varepsilon^2}. \qquad (16.10)$$

However, the Chebyshev inequality (16.10) is very conservative. In our extreme example above ($U = p_f^{MC}$, $\mu = p$, $\varepsilon = \nu p$, $\sigma^2[p_f^{MC}] = \frac{p(1-p)}{N}$, with $\nu = 0.1$, $p = 10^{-10}$), it requires that $N \geq 10^{24}$. Also this lower bound grows with $1/p^2$. These last two, more general, lower bounds are much too pessimistic and there is a big gap to the situations where we know p, or where we have a good estimate for it. Large Deviations Theory (LDT), see [3, 13, 16], bridges this gap and results in a sharp upper bound for the (non-scaled) relative error. The expression for the upper bound also nicely involves N_{MC}, which allows to estimate N_{MC} to obtain a guaranteed accuracy for the tail probability.

16.4.1 Large Deviations Theory for Tail Probabilities by MC

In the previous section A was $A = (-\infty, x)$, for which we defined $p_N(A)$ using estimates X_i in which $X_i = I_A(Y_i)$ of Monte Carlo samples Y_i and where the I_A is the event indicator that flags if $Y_i \in A$. In a similar way we can define $P_N(S)$ for any set $S \subset \mathbb{R}$. Clearly, for A and S one uses different event indicators.

The key ingredient is the *Large-Deviation Principle* in Large Deviations Theory (LDT) [3, 13, 16] that we assume and state as follows.

Theorem 16.1. *The sequences of the Monte Carlo results $P_N(S)$ satisfy a "Large-Deviation Principle", meaning that there is some "rate function" I from $\mathbb{R} \to \mathbb{R} \cup \{-\infty, \infty\}$ such that*

(i) $\limsup_{N \to \infty} \frac{1}{N} \ln P_N(C) \leq -\inf_{x \in C} I(x)$, *for all* **closed** *subsets $C \subset \mathbb{R}$,*

(ii) $\liminf_{N \to \infty} \frac{1}{N} \ln P_N(G) \geq -\inf_{x \in G} I(x)$, *for all* **open** *subsets $G \subset \mathbb{R}$.*

In our case for the set $A = (-\infty, x)$, each X_i is a Bernoulli variable X with success probability p. The *logarithmic moment generating function* for X is given by $\ln(\mathrm{E}[e^{\lambda X}]) = \ln(q + e^{\lambda} p)$, where $q = 1 - p$. We define the following function [27, 29]

$$J(x, \lambda) = \lambda x - \ln\left(\mathrm{E}[e^{\lambda X}]\right) \tag{16.11}$$

$$= \lambda x - \ln(q + e^{\lambda} p), \tag{16.12}$$

where $x, \lambda \in \mathbb{R}$. For x outside $[0,1]$, $J(x, \lambda)$ is unbounded. For $0 < x < 1$ we can find a maximum value λ^* that satisfies

$$0 = \frac{\partial J}{\partial \lambda} = x - \frac{p e^{\lambda^*}}{q + p e^{\lambda^*}}, \quad \text{hence}$$

$$\lambda^* = \ln\left(\frac{qx}{p(1 - x)}\right), \quad \text{and}$$

$$p e^{\lambda^*} = \frac{qx}{1 - x}, \quad \text{and}$$

$$q + p e^{\lambda^*} = \frac{q}{1 - x}.$$

The rate function I of Theorem 16.1 can be derived and, for $0 < x < 1$, shown (see [16]) to be equal to

$$I(x) = \sup_{\lambda \in \mathbb{R}} J(x, \lambda) \tag{16.13}$$

$$= x \ln\left(\frac{qx}{p(1 - x)}\right) - \ln\left(\frac{q}{1 - x}\right). \tag{16.14}$$

From (16.14) we can calculate $I'(x)$ and $I''(x)$ explicitly [27, 29]

$$I'(x) = \ln(\frac{q}{p}) + \ln(x) - \ln(1 - x), \tag{16.15}$$

$$I''(x) = \frac{1}{x(1 - x)}. \tag{16.16}$$

Hence, for $x \in (0,1)$, we have $I''(x) > 0$, implying that I' is increasing and that I is *convex*. Also $I(0^+) = -\ln(q) > 0$ and $I(1^-) = \ln(q/p) \in \mathbb{R}$. Thus I can be extended continuously at both $x = 0$ and at $x = 1$.

Theorem 16.2. *Let $B \equiv (p - \nu p, p + \nu p) \subset (0,1)$. For $N = N_{MC}$ large enough, the Monte Carlo results $P_N := p_f^{MC}$ approximate p with a relative precision ν such that*

$$P(|p_f^{MC} - p| \geq \nu p) \leq \exp\left(-N \inf_{|x-p| \geq \nu p} I(x)\right) \approx \exp\left(-\frac{N_{MC}}{2} \frac{p}{1-p} \nu^2\right).$$
(16.17)

The exponential type of bound in (16.17) is also valid from below and thus is sharp.

Proof. We summarize results of [27, 29]. The rate function function $I(x)$ is continuous on the interval $(0,1)$. With $B \equiv (p - \nu p, p + \nu p) \subset (0,1)$, we take $C = [0,1] \backslash B$ (closed) and $G = \mathbb{R} \backslash C$ (open). The Large-Deviation Principle of Theorem 16.1 for the upper bound on the limsup implies

$$\lim_{N \to \infty} \frac{1}{N} \ln\left(\left|\frac{1}{N} \sum_{k=1}^{N} X_k - p\right| \geq \nu p\right) \leq - \inf_{|x-p| \geq \nu p} I(x).$$

At p, $I(p) = 0$ is a global minimum. This implies that actually the infimum of I on $\{x : |x - p| \geq \nu p\}$ is assumed at $x = p \pm \nu p$, i.e., on the boundary of C. Hence

$$\inf_{|x-p| \geq \nu p} I(x) = \min\{I(p - \nu p), I(p + \nu p)\}.$$
(16.18)

This can be analysed further using Taylor expansion at p, where we notice that

$$I'''(p) = \frac{1}{p(1-p)} = \frac{1}{pq} = \frac{N}{\text{Var}[P_N(C)]}.$$
(16.19)

We obtain

$$I(p \pm \nu p) = \frac{Np^2}{2\text{Var}[P_N(C)]} \nu^2 + \mathcal{O}(\nu^3) = \frac{p}{2q} \nu^2 + \mathcal{O}(\nu^3).$$

Thus from part (i) of the Large Deviation Principle in Theorem 16.1, we obtain the so-called binomial case of the *Cramér bound*

$$P\left(\left|\frac{1}{N} \sum_{k=1}^{N} X_k - p\right| \geq \nu p\right) \leq \exp\left(-N \inf_{|x-p| \geq \nu p} I(x)\right)$$

$$\approx \exp\left(-\frac{N^2 p^2}{2\text{Var}[P_N(C)]} \nu^2\right) = \exp\left(-\frac{Np}{2q} \nu^2\right),$$

for all N with a possible exception of finitely many values of N. Part (ii) of Theorem 16.1 implies that the exponential bound in (16.17) is also valid from below and thus is sharp. ∎

16.4.2 Approximating Tail Probabilities by LDT

The main remark to (16.17) is that it is a point-wise statement for each p and not a convergence result for a distribution function, like in the CLT convergence (16.8). And since it concerns a tail probability it even is a global result. By this the remarkable exponential could show up. However, similar as for the CLT it assumes that N_{MC} is "large enough". We now apply (16.17) to our example (see also [8, 27, 29]). For $\nu = 0.1$, $p = 10^{10}$ and $\alpha = 0.02$, as above, we find $N_{\mathrm{MC}} \geq 8\,10^{12}$. Indeed this lower bound for N_{MC} is close to the optimal one found with the CLT ($4\,10^{12}$), but now without using special knowledge. As mentioned before, the LDT result is sharp. It means that in general one really needs $\mathcal{O}(10^{12})$ samples, and, more generally, $N_{\mathrm{MC}} = \mathcal{O}(1/p)$ (but with a factor not that close to 1). We finally notice that an extra k-th decimal in ν increases N_{MC} with a factor k^2.

16.5 Approximating Tail Probabilities by Importance Sampling

With Importance Sampling (IS) [3,18,31,35] one samples the Y_i according to a distribution function g (called design distribution) that is different from f. We assume that $g(z) \neq 0$ on $A = (-\infty, x)$ and observe that $p_f(A) = \int_{-\infty}^{x} f(z)dz = \int_{-\infty}^{x} \frac{f(z)}{g(z)} g(z)dz$. For an arbitrary Y with probability density function g we define a *weighted success indicator* $V = V(A) = I_A(Y)f(Y)/g(Y)$. Then, with the g-distribution we have for the expectation

$$E_g[V] = \int_{-\infty}^{\infty} I_A(y)\frac{f(y)}{g(y)}g(y)dy = \int_{-\infty}^{x} f(z)dz = p_f(A). \qquad (16.20)$$

Hence, if we determine $V_i = I_A(Y_i)f(Y_i)/g(Y_i)$ from g-distributed Y_i, we can define

$$p_g^{\mathrm{IS}} = p_g^{\mathrm{IS}}(A) = \frac{1}{N}\sum_{i=1}^{N} V_i. \qquad (16.21)$$

Its expectation becomes $E_g[p_g^{\mathrm{IS}}] = \frac{1}{N}\sum_{i=1}^{N} E_g(V_i) = \frac{1}{N} N p_f(A) = p_f(A)$, which is unbiased. Notice that this g-sampling may already be a benefit: sampling according to a known and simple g may be more efficient than sampling according to a more general density f that involves more calculations.

The expression(s) for the variance $\mathrm{Var}[p_g^{\mathrm{IS}}]$ are more involved. Under restrictions on g, when compared to f, one can obtain guarantees for variance reduction.

16.5.1 Variance Reduction and Efficiency by Importance Sampling

In order to consider the required simulation size for Importance Sampling, we will need an expression for the variance. This will also show that we can obtain variance reduction compared to MC that is based on f, by appropriately choosing g.

Assuming that Y_i and Y_j are independent for $i \neq j$, we also have that $I_A(Y_i)$ and $I_A(Y_j)$, and thus V_i and V_j, are independent. Hence

$$N\,\mathrm{Var}_g[p_g^{\mathrm{IS}}] = N\,\mathrm{Var}_g\left[\frac{1}{N}\sum_{i=1}^{N} V_i\right] = \mathrm{Var}[V]$$

$$= \int \left(I_A(y)\frac{f(y)}{g(y)} - p_f(A)\right)^2 g(y)\,d(y) \tag{16.22}$$

$$= \int I_A(y)\frac{f^2(y)}{g(y)}\,dy - p_f^2(A). \tag{16.23}$$

We also have

$$N\,\mathrm{Var}_g[p_g^{\mathrm{IS}}] = \int_{-\infty}^{x} \frac{f^2(y)}{g(y)}\,dy - p_f^2(A)\left(\int_{-\infty}^{x} g(y)\,dy + \int_{x}^{\infty} g(y)\,dy\right)$$

$$= \int_{-\infty}^{x} \left(\frac{f(y)}{g(y)} - p_f(A)\right)^2 g(y)\,dy + 2p_f(A)\int_{-\infty}^{x} f(y)\,dy$$

$$-2p_f^2(A)\int_{-\infty}^{x} g(y)\,dy - p_f^2(A)\int_{x}^{\infty} g(y)\,dy$$

$$= \int_{-\infty}^{x} \left(\frac{f(y)}{g(y)} - p_f(A)\right)^2 g(y)\,dy + p_f^2(A)\int_{x}^{\infty} g(y)\,dy. \tag{16.24}$$

Here (16.22)-(16.24) are three equivalent formulations. It follows from (16.24) that if one could choose $g(y) = 0$ for $y > x$ and $f(y)g(y) = p_f(A)$ for $y < x$ (note that this choice of g indeed yields a density), the variance of the estimator p_g^{IS} would be zero. This is not surprising, since then the estimator is constant and hence, its variance is zero. In practice we cannot implement this perfect choice, since it requires knowledge of the quantity $p_f(A)$ that we are trying to estimate. So preferably one should have $g(y) \approx 0$ for $y > x$, and $f(y)/g(y) \approx p_f(A)$ for $y < x$ (i.e., p_g^{IS} nearly constant). In order to approximate this one usually applies an estimate $\tilde{x}$ for x and restricts oneself to $\tilde{x} \approx \mathrm{E}[g(Y)]$, or one minimizes the normalized standard deviation $\mathrm{Var}[p_g^{\mathrm{IS}}]/\mathrm{E}[p_g^{\mathrm{IS}}]$. Theorem 16.3 states a condition for Variance Reduction (see also [27, 29]).

Theorem 16.3. *We take the same samples $N_{\mathrm{IS}} = N_{\mathrm{MC}}$ for the Importance Sampling with distribution density g and for the Monte Carlo Sampling using the distribution function f, resulting in estimates p_g^{IS} and p_f^{MC}, respectively. Let g dominate f on $A = (-\infty, x)$*

$$\frac{f(z)}{g(z)} \le \kappa \le 1 \text{ on } A. \tag{16.25}$$

Then we obtain Variance Reduction

$$\mathrm{Var}_g[p_g^{\mathrm{IS}}] \le \kappa \mathrm{Var}_f[p_f^{\mathrm{MC}}] - \frac{1-\kappa}{N}.p^2 \tag{16.26}$$

Proof. From (16.23), we immediately derive for $\kappa = 1$

$$\frac{f(z)}{g(z)} \le 1 \text{ on } A \implies \mathrm{Var}_g[p_g^{\mathrm{IS}}] \le \mathrm{Var}_f[p_f^{\mathrm{MC}}]. \tag{16.27}$$

For $\kappa < 1$ the same expression (16.23) also easily leads to (16.26). Thus we obtain *variance reduction* for the Importance Sampling estimates using the same number of samples as for obtaining estimates by Monte Carlo sampling. ∎

We remark that when (16.25) is violated, such that $\frac{f(z)}{g(z)} > 1$ for some $z \in A$, one can obtain estimates by Importance Sampling that are much less accurate than those obtained by Monte Carlo estimates.

The variance reduction does not yet imply more efficiency. However, similar to (16.17), we generalize the LDT result for Importance Sampling.

Theorem 16.4. *Let Y be distributed according to g. We assume that there is no $y \in \mathbb{R}$ such that $P(Y = y) = 1$ (Y is not supported by a single point). Let $V = V(Y) = I_{(-\infty,x)}(Y) f(Y)/g(Y)$ and $v(y) = I_{(-\infty,x)}(y) f(y)/g(y)$. We assume some restrictions on the ratios f/g.*

1. *Let $\mathrm{E}_g[e^{\lambda V}] = \int_{-\infty}^{\infty} g(y) e^{\lambda I_{(-\infty,x)}(y) f(y)/g(y)} dy$ be the moment generating function of V w.r.t. g. We assume that $0 < \mathrm{E}_g[e^{\lambda V}] < \infty$, for all $\lambda \in \mathbb{R}$.*
2. *Let $\rho_\lambda(y)$ be the density function*

$$\rho_\lambda(y) = \frac{e^{\lambda v(y)} g(y)}{\mathrm{E}_g[e^{\lambda V}]}, \tag{16.28}$$

which is well-defined for all $\lambda \in \mathbb{R}$ and satisfies $\int \rho_\lambda(y) dy = 1$. Let Y_λ be a random variable distributed according to ρ_λ. Then the main assumption is that the expectation and the variance of Y_λ are finite, i.e., we assume that for all $\lambda \in \mathbb{R}$

$$\mathrm{E}_{\rho_\lambda}[Y_\lambda] = \int y_\lambda \rho_\lambda(y_\lambda) dy_\lambda = \int y \frac{e^{\lambda v(y)} g(y)}{\mathrm{E}_g[e^{\lambda V}]} dy < \infty,$$

$$\mathrm{Var}_{\rho_\lambda}[Y_\lambda] = \mathrm{E}_{\rho_\lambda}[Y_\lambda^2] - \mathrm{E}_{\rho_\lambda}^2[Y_\lambda] < \infty. \tag{16.29}$$

Let $B \equiv (p - \nu p, p + \nu p) \subset (0, 1)$. For $N = N_{\mathrm{IS}}$ large enough, the Importance Sampling results $P_N := p_f^{\mathrm{IS}}$ approximate p with a relative precision ν such that

$$P(|p_f^{\mathrm{IS}} - p| \geq \nu p) \leq \exp\left(-N \inf_{|x-p| \geq \nu p} I_{\mathrm{IS}}(x)\right) \approx \exp\left(-\frac{N_{\mathrm{IS}}\, p^2}{2\mathrm{Var}_g[V]} \nu^2\right). \tag{16.30}$$

The exponential type of bound in (16.17) is also valid from below and thus is sharp.

Proof. For a complete proof, which is more technical than for the regular Monte Carlo case, we refer to [27, 29]. Basically we want to follow the arguments in the proof of Theorem (16.2). When comparing to (16.11) we like to replace the Bernoulli distributed function X in (16.11) by a V that is defined by the weighted random variable $V = V(Y) = I_{(-\infty, x)}(Y) f(Y)/g(Y)$. More precisely, we have to consider the supremum of the rate function $I(x) = \sup_{\lambda \in \mathbb{R}} J(x, \lambda)$ [16], where

$$J(x, \lambda) = \lambda x - \varphi(\lambda), \tag{16.31}$$

for $x, \lambda \in \mathbb{R}$ and in which $\varphi(\lambda) = \ln\left(\mathrm{E}_g[e^{\lambda V}]\right)$. The assumptions assure that $\varphi(\lambda)$ is a well-defined, strictly convex, two times differentiable function with first and second order derivatives that are equal to the finite values of the mean and variance of a suitable random variable Y_λ.

$$\varphi'(\lambda) = \frac{\mathrm{E}_g\left[V e^{\lambda V}\right]}{\mathrm{E}_g\left[e^{\lambda V}\right]} = \mathrm{E}_{\rho_\lambda}(Y_\lambda), \tag{16.32}$$

$$\varphi''(\lambda) = \frac{\mathrm{E}_g\left[V^2 e^{\lambda V}\right]}{\mathrm{E}_g\left[e^{\lambda V}\right]} - \frac{\mathrm{E}_g^2\left[V e^{\lambda V}\right]}{\mathrm{E}_g^2\left[e^{\lambda V}\right]} = \mathrm{Var}_{\rho_\lambda}(Y_\lambda). \tag{16.33}$$

In [27, 29] it is derived that for x, λ such that $\frac{d}{d\lambda} J(x, \lambda) = 0$, we can express

$$x = \Psi(\lambda), \quad \text{where } \Psi(\lambda) = \frac{\int y e^{\lambda v(y)} g(y)\, dy}{\int e^{\lambda v(y)} g(y)\, dy} = \mathrm{E}_{\rho_\lambda}[Y_\lambda], \tag{16.34}$$

is an invertible function Ψ and thus $\lambda = \lambda(x) = \Psi^{-1}(x)$. By this the rate function $I(x)$ becomes

$$I(x) = J(x, \lambda(x)). \tag{16.35}$$

Clearly $\rho_{\lambda=0}(y) = g(y)$. Further, to calculate the first (total) derivative of $I(x)$, we differentiate with respect to x and substitute (16.34) to obtain

$$I'(x) = \lambda(x) + x\lambda'(x) - \lambda'(x)\frac{\mathrm{E}_g[Ve^{\lambda(x)V}]}{\mathrm{E}_g[e^{\lambda(x)V}]}$$

$$= \lambda(x) + \lambda'(x)(x - \mathrm{E}_{\rho_\lambda}[Y_\lambda])$$

$$= \lambda(x). \tag{16.36}$$

For the second derivative of $I(x)$, we first implicitly differentiate (16.34) with respect to x, which yields $1 = \frac{\partial}{\partial\lambda}(\mathrm{E}_{\rho_\lambda}[Y_\lambda])\,\lambda'(x)$. By (16.32)-(16.33) the derivative of the expectation in this expression can be rewritten as

$$\frac{\partial}{\partial\lambda}(\mathrm{E}_{\rho_\lambda}[Y_\lambda]) = \frac{\partial}{\partial\lambda}\frac{\mathrm{E}_g[Ve^{\lambda V}]}{\mathrm{E}_g[e^{\lambda V}]} = \frac{\mathrm{E}_g[Ve^{\lambda V}]}{\mathrm{E}_g[e^{\lambda V}]} - \frac{\mathrm{E}_g^2[Ve^{\lambda V}]}{\mathrm{E}_g^2[e^{\lambda V}]}$$

$$= \mathrm{E}_{\rho_\lambda}[V] - \mathrm{E}_{\rho_\lambda}^2[V] = \mathrm{Var}_{\rho_\lambda}[Y_\lambda].$$

Substituting these expressions when differentiating (16.36) with respect to x, we obtain [27, 29]

$$I''(x) = \lambda'(x) = \frac{1}{\frac{\partial x}{\partial\lambda}} = \frac{1}{\frac{\partial}{\partial\lambda}(\mathrm{E}_{\rho_\lambda}[Y_\lambda])} = \frac{1}{\mathrm{Var}_{\rho_\lambda}[Y_\lambda]}.$$

By [16, Lemma I.4, p. 8], $I(x)$ is strictly (proper) convex which means that the minimizer of I is unique. Next we consider $x = p$. The Strong Law of Large Numbers implies that the empirical measure of every neighbourhood of p tends to one. Hence, p is the unique minimizer of I and $I'(p) = 0$. Since p is also an internal point, we obtain that $0 = I'(p) = \lambda(p)$. Hence

$$I''(p) = \frac{1}{\mathrm{Var}_{\rho_{\lambda=0}}[Y_\lambda]} = \frac{1}{\mathrm{Var}_g[V]}.$$

Finally, by Taylor expansion,

$$I(p \pm \nu p) = \frac{1}{2}\nu^2 p^2 I''(p) + \mathcal{O}(\nu^3 p^3) \approx \frac{1}{2}\frac{\nu^2 p^2}{\mathrm{Var}_g[V]}.$$

Thus, after applying the Large-Deviation Principle of Theorem 16.1 we derive

$$P\left(\left|\frac{1}{N}\sum_{k=1}^N V_k - p\right| \geq \nu p\right) \leq \exp\left(-N\inf_{|x-p|\geq\nu p} I(x)\right)$$

$$\approx \exp\left(-\frac{Np^2}{2\mathrm{Var}_g[V]}\nu^2\right),$$

for all sufficiently large N.
Also this result is sharp as it was for (16.17). This completes the proof. ∎

We are now able to compare the efficiency between Importance Sampling and Monte Carlo Sampling.

Theorem 16.5. *We assume that* (16.25) *holds. Then*

$$\frac{N_{\mathrm{IS}}}{N_{\mathrm{MC}}} \leq \kappa(1+\zeta), \quad \text{for } |(1-\frac{1}{\kappa}+\mathcal{O}(p^2)| \leq \zeta, \qquad (16.37)$$

which for $\kappa = 0.1$ *and* $p = 10^{-10}$ *means that* $\zeta \leq 10^{-9}$.

Proof. We will assume that (16.25) holds. Then the assumptions of Theorem 16.4 on the ratios f/g are satisfied.

Comparing (16.17) and (16.30) we see the same type of exponential decay as a function of N. Hence an improvement for Importance Sampling should come from a proper choice of the distribution function g. Assuming the same upper bounds values in (16.17) and (16.30), comparing them gives [8, 27, 29]

$$\frac{N_{\mathrm{IS}}}{N_{\mathrm{MC}}} = \frac{\mathrm{Var}_g[V]}{p(1-p)} = \frac{E_g[V^2]-p^2}{p(1-p)} = \frac{E_g[V^2]}{pq} - \frac{p}{q} \leq \frac{\kappa}{q} - \frac{p}{q} \leq \kappa(1+\zeta), \quad (16.38)$$

where $q = 1-p$ and when $|(1 - \frac{1}{\kappa} + \mathcal{O}(p^2)| \leq \zeta$. The remaining part follows easily.

This completes the proof.   ∎

The upper bound (16.37) can also be obtained by equating the normalized standard deviations $\sigma_f[p_f^{\mathrm{MC}}]/E_f[p_f^{\mathrm{MC}}]$ and $\sigma_g[p_g^{\mathrm{IS}}]/E_g[p_g^{\mathrm{IS}}]$. However, the approach via (16.17) and (16.30) indicates the sharpness.

We conclude that, for $\kappa = 0.1$ and $p = 10^{-10}$, we can take an *order less samples* with Importance Sampling to get the same accuracy as with regular Monte Carlo. This even becomes better with smaller κ. The main message is that by Importance Sampling we gain *efficiency*. Also the asymptotic accuracy improves when compared to regular Monte Carlo, but the improvement is less impressive than for the efficiency. By (16.26) we can derive *variance reduction* $\sigma_g[p_g^{\mathrm{IS}}] \leq \sqrt{\kappa}\,\sigma_f[p_f^{\mathrm{MC}}]$, which for $\kappa = 0.1$ means that here not an order is gained, but a factor $\sqrt{\kappa} \approx 0.316$ (see also [8, 29]).

16.5.2 First examples on choices for g

We first note that, if $g(x) \equiv 1$, we have $\mathrm{Var}_g[V] = \mathrm{Var}_f[V] = pq$. Then (16.17) and (16.30) show the same upper bounds.

As a next example we consider $f \sim N(\mu, \sigma^2)$ with $\mu = 0$, $\sigma = 1$ and the quantile set to $z_\alpha = \mu + 3.5\sigma$. Then the failure probability $p \approx 2.3\,10^{-4}$. For IS we take $g \sim N(\mu, 4\sigma^2)$ (thus with a double size of standard deviation when compared to f), see Fig. 16.2.

For $N = 10^5$ samples we get $N_{\mathrm{MC}} \leq 20$ and $N_{\mathrm{IS}} \approx 4\,10^3$ successful samples in the Region-of-Interest. Here IS already gives a good approximation, while MC does not.

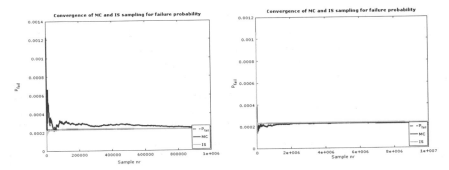

Fig. 16.2: For a Gaussian distribution $f \sim N(\mu, \sigma^2)$ with $\mu = 0$, $\sigma = 1$ we consider the failure probability $p = 2.3263 \, 10^{-4}$ for the (right tail) Region-of-Interest $A = (z_\alpha, \infty)$ with quantile $z_\alpha = \mu + 3.5\sigma$. The graphs show the convergence of Monte Carlo sampling and of sampling by IS, where we have taken $g \sim N(\mu, 4\sigma^2)$ (thus with a double size of standard deviation when compared to f). **Left:** Results after 10^6 samples. **Right:** Results after 10^7 samples. Clearly, IS converges much faster than MC. IS is already accurate with 10^5 samples, but MC is not.

With $N = 10^6$ samples we get $N_{MC} \leq 250$ and $N_{IS} \approx 4 \, 10^4$ successful samples in the Region-of-Interest (the, with 'randn' (Matlab) generated, samples $\tilde{x}_i$ provide for f: $|x_i = \sigma * \tilde{x}_i| < 5\sigma$ and for g: $|x_i = 2 * \sigma * \tilde{x}_i| < 10\sigma$). Now also MC is able to provide a moderately reasonable approximation for p. The outcomes confirm that for MC one can not simply take $N = 10/p$ as rule-of-thumb for the number of simulations. Here $N = 100/p$ is more realistic. Actually, with $N \geq 2 * 10^6$ samples, MC has sufficient samples generated in the Region-of-Interest to provide a good approximation for p.
For $N = 10^7$ samples MC looks stable (for f: $|x_i| < 5.2\sigma$ and for g: $|x_i| < 10.5\sigma$). Clearly, a 6σ variation for the x_i for f is difficult to achieve by using randn.
Next we consider some particular choices for Gaussian distributions.

16.5.2.1 Exploiting a broader or a shifted Gaussian distribution

The condition (16.25) is easily satisfied if $f \sim N(0, 1)$ is a Gaussian probability density distribution and $g = g_\sigma \sim N(0, \sigma^2)$, with $\sigma \geq 1$, or $g = g_\mu \sim N(-\mu, 1)$ has a broader or a shifted Gaussian distribution, with enough density on the Region-of-Interest $A = (-\infty, x)$. In both cases, similar to (16.38), we derive

$$\frac{N_{IS}}{N_{MC}} \leq \frac{\frac{f(x)}{g(x)} - p}{1 - p}, \text{ in which } x = -z_\alpha \text{ (quantile) for } \alpha = p. \qquad (16.39)$$

For instance, if $f(z) \sim N(0,1)$ and the broader Gaussian distribution $g(z) = g_\sigma(z) \sim N(0,\sigma^2)$, with $\sigma \geq 1$, already a moderately larger σ provides a speed-up with Importance Sampling. Fig. 16.3 (Top) [8] shows the speed-up that one can obtain. Along the vertical axis the $\log_{10}$ of the right-hand side of (16.39) is calculated for $x = -z_\alpha$ for $\alpha = p$. In all cases we have $(-\infty, -z_\alpha) \subset (-\infty, x_1)$ where $x_1 < 0$ is such that for $x < x_1$ one has $f(x)/g_\sigma(x) < 1$. The figure also indicates convergence with respect to increasing σ. Indeed $f(x)/g_\sigma(x)$ is minimal when $\sigma_{\text{opt}}^2 = x^2 = z_\alpha^2$. For $\alpha = p = 10^{-10}$, we find $\log_{10}(N_{\text{IS}}/N_{\text{MC}}) \approx -7.87$ and thus an optimal speed up of approximately $7.4\ 10^7$.

A shifted Gaussian distribution for g can be used as well, but now we need to assure that the mean of g remains outside the Region-of-Interest A. Let $f(z) \sim N(0,1)$ and $g(z) \equiv g_\mu \sim N(-\mu, 1)$. Then $z \leq x_\mu \equiv \frac{2\ln(\kappa) - \mu^2}{2\mu}$ provides $f(z)/g_\mu(z) \leq \kappa \leq 1$. Fig. 16.3 (Bottom) [8] shows the speed-up that can be obtained in this case. Also here we can determine an optimum value $\mu_{\text{opt}} = -x = z_\alpha$. Comparing the two figures in Fig. 16.3, the figure at the Bottom indicates a higher speed-up than what is indicated by the figure at the Top. For $\alpha = p = 10^{-10}$ the optimal upper bound for the shifted Gaussian distribution g is ≈ 0.094 times smaller than the optimal upper bound for the broader Gaussian distribution g.

By construction, for the shifted Gaussian distribution g, a fraction

$$\frac{1}{\sqrt{2\pi}} \int_{-z_\alpha}^\infty e^{-\frac{(x+z_\alpha)^2}{2}} dx = \frac{1}{\sqrt{2\pi}} \int_0^\infty e^{-\frac{x^2}{2}} dx = 0.5$$

is sampled outside $A = (-\infty, x)$, which is better than in the previous case of a sampling with a broader Gaussian distribution, where we can derive 0.8413 for the portion that is sampled outside the Region-of-Interest. However, in the case of shifting the mean, one has to be more careful in order to guarantee that $f(x)/g_\mu < 1$ on the Region-of-Interest A.

16.5.2.2 Exploiting a uniform distribution

Finally we consider a violating choice for g. Let $a \leq x \leq b \leq 0$ and $A' = (-\infty, a)$. Thus $A' \subset A = (-\infty, x)$. We define a uniform distribution $g(y) = 1/(b-a)$ for $y \in [a, b]$ and 0 elsewhere. This g violates the assumption that $g(y) \neq 0$ on A, but it is easy to guarantee that $g(y) \neq 0$ on a significant portion of A. Then, at best, by using g, we estimate a biased result

$$\int_a^x \frac{f(z)}{g(z)} g(z)\, dz = \int_a^x f(z)\, dz = p_f(A) - p_f(A') \leq p_f(A). \qquad (16.40)$$

If $p_f(A') \ll p_f(A)$ then we can still obtain a good result. Notice that $p_f(A) - p_f(A') = p^{-\alpha} - p^{-\alpha'}$, using the quantiles z_α seen before. The estimates for the variance receive a similar biasing effect. Notice again, that, for

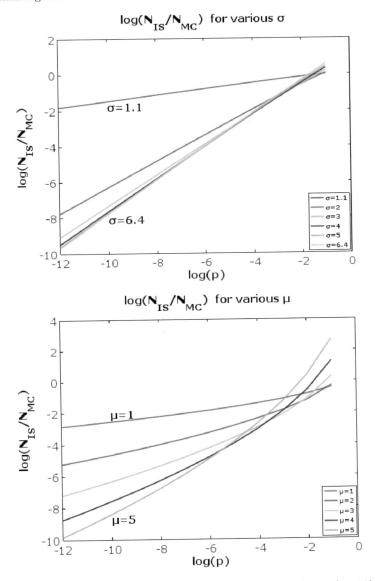

Fig. 16.3: **Top:** The upper bound of $\log_{10}(N_{\mathrm{IS}}/N_{\mathrm{MC}})$ in (16.39) versus $\log_{10}(p)$ for $f(z) \sim N(0,1)$ and broader density distributions $g(z) \sim N(0,\sigma^2)$. The curve at the top is for $\sigma = 1.1$, the one at the bottom is for $\sigma = 6.4$. **Bottom:** Similarly for $\log_{10}(N_{\mathrm{IS}}/N_{\mathrm{MC}})$ versus $\log_{10}(p)$ for $f(z) \sim N(0,1)$ and shifted density distributions $g(z) \sim N(-\mu,1)$. The curve on the top at the left is for $\mu = 1$, the one at the bottom is for $\mu = 5$. See also [8].

a normal distribution $f \sim N(0,1)$, due to monotonicity on $[a,x]$, $f(y)/g(y) \leq (b-a)\exp(-\frac{1}{2}x^2)$, where $x = -z_\alpha$.

An alternative is to use

$$\tilde{g}(x) = 0.9\,g(x) + 0.1\,f(x+x_\beta), \qquad (16.41)$$

for a suitable x_β and to guarantee $f(x)/\tilde{g}(x) < 1$ on A.

16.5.2.3 Additional Remarks

In [20] an adaptive Non-parametric Adaptive Importance Sampling (NAIS) procedure was proposed, that starts from some initial set of rare event samples and then, within an iterative loop, uses a sum of kernel density estimates as distribution g to find more rare events that are added to the set found so far, after which g can be updated for the next iteration. In [27] an adaptive procedure is given (in the programming language R [2,39]) that builds g by kernel density estimates as well. It maintains a distance between sample points and is adaptive in the sense that when the levels are monotonically adapted (and thus the tail probability being successfully decreased) the distribution function g is easily adapted. In practice, this is especially appreciated when constructing a cumulative distribution function needed for the re-sampling. In [32] a good overview is given of several further adaptive approaches to estimate failure probabilities. In [31] several practical details on Importance Sampling are found, including details for higher dimensions. We notice that all adaptive methods aim to find more samples in the Region-of-Interest for failures A. Less or no emphasis is made to also obtain variance reduction. We have seen before that in order to obtain variance reduction it is beneficial when $f(x)/g(x) < 1$ on the failure region A. This means that in all algorithms that adaptively determine g one should not aim to detect points that are too deep in the interior of A. On the contrary, we should try to give preference to (internal) points close to the boundary of A may be given.

Finally, we mention two variants of IS: a Weighted IS and a Regression IS. In (16.21) the estimate $p_g^{\text{IS}} = p_g^{\text{IS}}(A) = \frac{1}{N}\sum_{i=1}^N V_i = \overline{V}$ was introduced. Here $V_i = V(Y_i) = I_A(Y_i)W(Y_i)$ for $V(Y) = I_A(Y)W(Y)$ and $W(Y) = f(Y)/g(Y)$. After introducing $\overline{W} = \frac{1}{N}\sum_{i=1}^N W_i$, with $W_i = W(Y_i)$, we observe that $\overline{W}$ is not necessarily bounded. However, a Weighted or Normalized IS can be defined by $p_g^{\text{WIS}} = \overline{V}/\overline{W}$ [14,15,31]. We notice that $E_g[W(Y)] = 1$, which is a good result as seen from the point of view of numerical stability. In [14] it was shown that $p_g^{\text{WIS}} = p + \sum_{k=1}^\infty a_k$, in which the first term of the expansion series a_1 has $E_g[a_1] = 0$ and variance $\sigma_g^2[a_1] = \mathcal{O}(1/N)$.

In [15] a regression variant was considered on the pairs (V_i, W_i), which leads to a line $Z = \gamma^\star W + \delta^\star$, for optimal values $\gamma^\star$ and $\delta^\star$. Because for all i $E_g[W_i] = 1$, the estimate $p_g^{\text{RIS}} = Z(W = 1) = \gamma^\star + \delta^\star$ is taken as estimate for p. Let $\mathbf{1} = (1,1,\ldots,1)^T \in \mathbb{R}^N$, $\mathbf{v} = (V_1,\ldots,V_N)^T$, $\mathbf{w} = (W_1,\ldots,W_N)^T$, $\mathbf{A} = (\mathbf{w},\mathbf{1})$, $\mathbf{x} = (\gamma,\delta)^T$ and $\mathbf{x}^\star = (\gamma^\star,\delta^\star)^T$. Then $\mathbf{x}^\star = (\mathbf{A}^T\mathbf{A})^{-1}\mathbf{A}^T\mathbf{v}$ is the solution such that $||\mathbf{v} - \mathbf{A}\mathbf{x}||$ is minimal; notice that $(\mathbf{A}^T\mathbf{v})^T = (\mathbf{w}^T\mathbf{v}, N\overline{V})^T$. It was proven [15] that $\mathbf{E}_g[p_g^{\mathrm{RIS}}] = p - \mathcal{O}(1/N)$ and $\sigma_g^2[p_g^{\mathrm{RIS}}] = \mathcal{O}(1/N)$. We notice that for both the WIS and RIS variants of Importance Sampling further focussing on variance reduction still has to be made.

16.6 Parametric Problems

The previous sections basically did consider non-parametric sampling of a random variable Y that has a probability density function f. In general f will not be known. At best one may have obtained some impression. This makes estimating failure probabilities for non-parametric problems not that straight forward, despite the fact that in Importance Sampling we may sample according to some preferred density function g. For the weightings f/g we still need to be able to evaluate f.

Contrarily, we observe that in several simulations the nonlinear output response $\mathbf{y}(\mathbf{p})$, for parameters $\mathbf{p} = (p_1,\ldots,p_m)^T$, depends on independent *input parameters* p_i *with known density distribution functions* f_i (in most cases a normal distribution). For these parameters, $f(\mathbf{p}) = \prod_{i=1}^m f_i(p_i)$ and, for instance, $\mathrm{E}[\mathbf{y}] = \int \mathbf{y}(\mathbf{p})\, f(\mathbf{p})\, d\mathbf{p}$ can be determined.

In this case the ratio $f(\mathbf{p})/g(\mathbf{p})$, with $g(\mathbf{p}) = \prod_{i=1}^m g_i(p_i)$, is considered in $\mathbf{p}$-space, where f is known and thus the ratio can easily be calculated. Of course, in a more dimensional parameter space the definition of $g(\mathbf{p})$ that should cover the area of parameters for the rare events of interest, requires more attention. Notice, however, that the product reveals a weakness. For a large parameter space in the explicit calculation of the product probability densities f and g underflow may occur (when not taking special measures). In Section 16.6.1 we consider a problem with a moderate number of parameters. In the next sections we consider two techniques that are beneficial for parametric problems: stratification (Section 16.6.2) and exploiting a response-surface model for sampling (Section 16.6.3).

16.6.1 Multivariate Sampling for SRAM

Integrated circuits like Static Random Access Memories (SRAM) can provide an extremely high bandwidth combined with high read- and write speeds, SRAM memories are the first choice for fast cache memories used in personal computers and microcontrollers in mobile handsets. To guarantee sufficient yield for a 10 Mb SRAM, a failure probability of $p_f = 10^{-10}$ is taken into account in SRAM bitcell design for all relevant performance indicators.

SRAM Performance Indicators (PI) include Static Noise Margin (SNM), Read Current, Write Margin and Leakage Currents, which are post-processing functions imposed on the solution of an electronic circuit. For each PI only a very moderate number of different parameters are considered. E.g., each of the six transistors in an SRAM depends on a Gaussian distributed parameter $p = V_t$ (threshold voltage). The V_t's are responsible for the extremes in the distributions of the output PIs. This makes the application suitable for Importance Sampling (IS). Here we will only give a result for the SNM. The SNM is a measure for the read stability of the cell. The SNM is the amount of noise that can be imposed on the internal nodes of the SRAM cell before it changes its state. It is determined by plotting the voltage transfer curve of one half of the SRAM cell together with the inverse of the voltage transfer curve of the other half of the cell. This gives two areas (so-called 'eyes') of which the areas are estimated by the largest internal square. The smallest of the two is taken as SNM [9, 27].

For each V_t a Gaussian distribution $N(\mu, \sigma_{V_t}^2)$ was used for Monte Carlo (MC) sampling. Only $N = 10^5$ samples SNM$(v_t^{1,i}, \ldots, V_t^{6,i})$, $i = 1, \ldots, N$, were taken, which is much less than 10^{12} samples needed for a tail probability of 10^{-10} of the cumulative distribution function of the SNM according to Theorem 16.2. The IS simulations used broad uniform distributions g with a $\sigma_g = 6\sigma[V_t]$ for the V_t's of each transistor in the SRAM cell. Hence, actually, these g-distributions violate the assumptions in Theorem 16.4 - the contribution from the area outside this distribution range (outback) is not taken into account for a correct result. Furthermore, also here for g, only $N = 10^5$ samples SNM$(v_t^{1,i}, \ldots, V_t^{6,i})$, $i = 1, \ldots, N$ were taken.

Fig. 16.4 shows the cumulative distribution functions p_f of an SRAM cell [9, 27]. Standard MC (blue) can only simulate down to $p_f \approx 10^{-5}$. Statistical noise becomes apparent below $p_f \approx 10^{-4}$. Using IS we are able to simulate until $p_f \approx 10^{-10}$.

The (smooth) red curve shows the extrapolation of a Gaussian curve with estimated values for μ and σ by taking the sample approximations for these after 1000 MC simulations. It under estimates p_f.

The other two curves in Fig. 16.4 were obtained after appropriate binning and applying the Trapezoidal Rule for integration.

For use of stochastic yield constraints in further optimization see [10, 29]. In [19] also the Weighted Importance Sampling method [14] was applied to SRAM problems.

16.6.2 Stratification

Stratification combines randomness with the benefits of sampling on sub-regions. Good overviews are given in [4, 31]. The basic idea is to partition the (input) parameter space Γ in M blocks Γ_m, with $m = 1, \ldots, M$. We assume

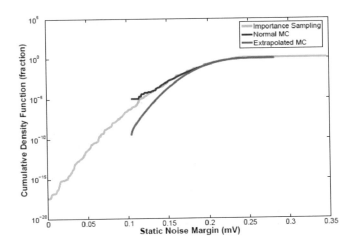

Fig. 16.4: SNM cumulative distribution functions of an SRAM cell using a Gaussian sampling function f (blue), an extrapolated sampling function (red) and the SNM distribution obtained by IS (green). The simulations used 50k trials. See also [27, 29].

that the parameters p have a global probability density function $f(p)$ on Γ, with mean $\mu_p = \int p f(p) dp$ and variance $\sigma_p^2 = \int (p - \mu_p)^2 f(p) dp$. Let $\mu_{p,m}$ be the local, weighted, mean on Γ_m, given by

$$\mu_{p,m} = \frac{1}{|\Gamma_m|} \int_{\Gamma_m} p f(p) dp, \text{ where } |\Gamma_m| = \int_{\Gamma_m} f(p) dp.$$

The variance of a stochastic quantity is the minimum over all constants of the mean quadratic error of this stochastic quantity to such a constant. Hence, applying this locally on each Γ_m, the 'summed variance' $\sigma_{p,s}^2$ gives

$$\sigma_{p,s}^2 = \sum_m \frac{1}{|\Gamma_m|} \int_{\Gamma_m} (p - \mu_{p,m})^2 f(p) dp \leq \sum_m \frac{1}{|\Gamma_m|} \int_{\Gamma_m} (p - \mu_p)^2 f(p) dp = \sigma_p^2.$$

Similarly, for a function $y(p)$ with global mean $\mu_y = \int y(p) f(p) dp$, global variance $\sigma_y^2 = \int (y(p) - \mu_y)^2 f(p) dp$ and local, weighted means

$$\mu_{y,m} = \frac{1}{|\Gamma_m|} \int_{\Gamma_m} y(p) f(p) dp,$$

we find for the summed variance $\sigma_{y,s}^2$

$$\sigma_{y,s}^2 = \sum_m \frac{1}{|\Gamma_m|} \int_{\Gamma_m} (y(p) - \mu_{y,m})^2 f(p) dp$$

$$\leq \sum_m \frac{1}{|\Gamma_m|} \int_{\Gamma_m} (y(p) - \mu_y)^2 f(p) dp = \sigma_y^2.$$

For M blocks Γ_m and a large $N = \sum_{m=1}^M n_m$ we may obtain a cheap *impression* of the distribution of sampled input parameter values P_k^m, with $P_k^m \in \Gamma_m$, and $k = 1,\ldots,n_m$, in advance, before really evaluating $y(P_k^m)$. Note that also after obtaining the sampled P_k^m values we may re-define the M blocks Γ_m to our convenience (for instance to ensure that we have no empty blocks). The local fraction $f_m = n_m/N$ gives a local density on Γ_m. We can now actually re-sample only k_m points $\widetilde{P}_k^m$ in each Γ_m, with $K = \sum_m k_m \ll N$ (but not necessarily $k_m \leq n_m$), evaluate $\widetilde{Y}_k^m = y(\widetilde{P}_k^m)$ and determine the global sample mean

$$\hat{\mu}_y = \sum_m \hat{\mu}_{y,m} = \sum_m \left(\frac{1}{k_m} \sum_k \widetilde{Y}_k^m \right) f_m = \sum_m \sum_k \widetilde{Y}_k^m \frac{f_m}{k_m}, \quad (16.42)$$

where $\hat{\mu}_{y,m} = \sum_m \left(\frac{1}{k_m} \sum_k \widetilde{Y}_k^m \right) f_m$ is the local, weighted, sample mean on Γ_m. For the summed sample variance $\hat{\sigma}_{y,s}^2$ we find for the summed variance $\hat{\sigma}_{y,s}^2$

$$\hat{\sigma}_{y,s}^2 = \sum_m \left(\frac{1}{k_m} \sum_k (\widetilde{Y}_k^m - \hat{\mu}_{y,m})^2 \right) f_m \leq \sum_m \left(\frac{1}{k_m} \sum_k (\widetilde{Y}_k^m - \hat{\mu}_y)^2 \right) f_m = \hat{\sigma}_y^2,$$

where $\hat{\sigma}_y^2$ is the global sample variance.
The result is that

- originally less populated blocks are sampled more;
- originally more populated blocks are sampled less.

For costly evaluations $y(p)$ stratification can improve efficiency because of a serious reduction in actual evaluations as long as $K \ll N$. Stratification can easily be combined with Importance Sampling. The procedure also offers options for refinement (hierarchically, or by Kriging, etc) [37,38].
Finally, we remark that in the circuit simulator Spectre the sampled parameters $p_{k'}^m$ can be generated with the 'iterVsValue' command: in fact this is a parameter distribution scan before doing the actual simulations. By this we can get a cheap impression of the distribution of the sampled input parameter values.

16.6.3 Hybrid Technique with Surrogate-Model Computation

In this section we consider the use of response surface models (also called surrogate models). These models may be derived for $\mathbf{y}(\mathbf{p})$, e.g., by Parametric Model Order Reduction (PMOR), or directly for $h(\mathbf{p})$ in (16.1), both providing relatively cheap approximations $\tilde{h}(\mathbf{p})$ for $h(\mathbf{p})$ [11, 24, 25, 28, 30]. Automated procedures can be found at the website in [36].

As example for a function $h(\mathbf{p})$, with respect to the bond wire fusing application, one can use

$$h(\mathbf{p}) = - \max_{i=1,\dots,L} \max_{t \in I} T_i(t, \mathbf{p}) + T_{\text{fs}}, \qquad (16.43)$$

where T_{fs} refers to the fusion temperature, I to the time domain, and T_i denotes the temperature of the i-th bond wire, $i = 1, \dots, L$. These temperatures depend on a parameter vector $\mathbf{p} = (p_1, \dots, p_L)^T$, containing the different bond wire lengths p_i [6, 7, 34].

For evaluating h at $\mathbf{p}$, a system of partial differential equations (PDEs) needs to be solved in general (e.g., an electro-thermal system). Here, we adopt the approach to obtain a response surface model approximation $\tilde{h} = h_N$ for h by applying a collocation procedure [1], which is non-intrusive and hence readily applicable to an existing PDE (electro-thermal) solver. The full tensor collocation method approximates the performance function as

$$h(\mathbf{p}) \approx \tilde{h}(\mathbf{p}) \equiv h_N(\mathbf{p}) = \sum_{k=1}^{N} h(\mathbf{p}^k) l_k(\mathbf{p}), \qquad (16.44)$$

where the $\mathbf{p}^k$ refer to well-chosen collocation points and the functions l_k (tensor-product) to Lagrange polynomials, respectively. The collocation points in each j-th parameter coordinate $(\mathbf{p}^k)_j$ of $\mathbf{p}^i$ are chosen as the roots of orthogonal polynomials with respect to $L_{\rho_j}^2(\Gamma_j)$ (equipped with the L^2-integral inner-product with density function ρ_j). As the number of input parameters n_p can be large in practice, the tensor collocation approach suffers from the curse-of-dimensionality and we employed isotropic sparse grids instead, see [1] for details.

Another surrogate model, based on Polynomial-Chaos Expansion (PCE) and numerical quadrature has been used in [33] for the computation of failure probabilities in oscillators.

We consider the failure indicator (16.1) and the failure probability P_F, see (16.2), by using M samples $\mathbf{p}^i$ for determining P_F

$$P_\text{F} = \frac{1}{M} \sum_{i=1}^{M} \chi_A(\mathbf{p}^i), \qquad (16.45)$$

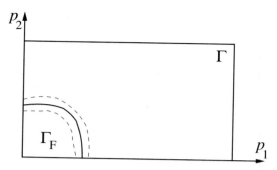

Fig. 16.5: Two-dimensional parameter domain Γ with failure region Γ_{F} and a neighborhood of the boundary of Γ_{F} indicated by dashed lines. The last is controlled by a threshold value γ.

where $A \equiv \Gamma_{\mathrm{F}} = \{\mathbf{p} \in \Gamma : h(\mathbf{p}) < 0\}$ (see also Fig. 16.5). The response surface model (16.44) can approximate the failure probability (16.45) by

$$P_{\mathrm{F}} \approx \frac{1}{M} \sum_{i=1}^{M} \chi_{\tilde{A}}(\mathbf{p}^i), \qquad (16.46)$$

with $\tilde{A} \equiv \widetilde{\Gamma}_{\mathrm{F}} = \{\mathbf{p} \in \Gamma : \tilde{h}(\mathbf{p}) < 0\}$. Sampling the surrogate model only requires polynomial evaluations and can be achieved for a large M at moderate cost. The main effort actually consists in computing the surrogate model itself. Unfortunately, it has been observed, that failure probabilities based on surrogate models can be inaccurate [25], even for large N and M, which is due to the convergence in L_{ρ}^2 (where $\rho(\mathbf{p}) = \prod_j \rho_j(p_j)$). In the areas for failure probability the density $\rho(\mathbf{p})$ usually does not contribute that much to the accuracy of the approximations. Here even Gibbs' phenomena can be observed for the approximating functions. This motivates the use of a hybrid scheme as presented in this section.

It should be noted that these errors are in addition to inaccuracies due to discretization errors and due to time integration procedures. In several cases the function h will depend on some post-processing procedures (e.g., by calculating stress from temperatures). Also modeling errors in libraries can affect the estimate of failure probabilities. E.g., compact models of transistor devices assume some range for involved parameters (here one also considers temperature as a parameter). Additionally, outside some range, simulations may become more expensive.

In [25] a hybrid method was proposed, where the concepts of full order model evaluations (leading to $h(\mathbf{p})$ evaluations) and surrogate model evaluations (leading to $\tilde{h}(\mathbf{p})$ approximations) are combined. This technique was applied successfully to academic test examples. An improved variant of this method was constructed in [24]. The crucial idea of this approach is to partition the domain of the parameters using a neighborhood of the boundary of the failure region. Figure 16.5 sketches this strategy. A threshold value γ is used to identify a neighbourhood (and actually its size along the normal to the boundary) of Γ_F. When $|\tilde{h}(\mathbf{p})| \geq \gamma$ we assume that we can trust the accuracy of $\tilde{h}(\mathbf{p})$ in properly signalling failure or proper behavior. This only involves cheap evaluations. For those $\mathbf{p}$ for which $|\tilde{h}(\mathbf{p}| < \gamma$ we need to rely on the more accurate full model and have to evaluate $h(\mathbf{p})$. More precisely, we consider

$$\chi_{\Gamma_F^\gamma} = \chi_{\{\tilde{h} \leq -\gamma\}} + \chi_{\{|\tilde{h}| < \gamma\} \cap \{h < 0\}}, \tag{16.47}$$

involving some threshold value $\gamma > 0$. This leads to

$$P_F \approx P_F^{\gamma,M} \equiv \frac{1}{M} \sum_{i=1}^{M} \chi_{\Gamma_F^\gamma}(\mathbf{p}^i). \tag{16.48}$$

For given γ one can derive error estimates for P_F, see [24, 28].

The basic hybrid method is provided by Algorithm 16.1, given below. In the discussion above, the parameter γ was chosen in advance as a function of the accuracy of the surrogate model. This accuracy is hard to estimate in practice, but it can be iteratively approximated in an implicit way. This has been outlined in [25, Section 4.4], is summarized in Algorithm 16.1 and will be explained below. As inputs for Algorithm 16.1, the full model h, the sur-

---

Algorithm 16.1 Basic Iterative Failure Probability Algorithm

1: **procedure** P_F=**HYBRID1**$(h, \tilde{h}, S_\mathbf{p}, M, \delta M, \eta)$
2:     Set $k = 0, M^{(k)} = 0$
3:     Evaluate P_F^0 using (16.46), i.e., the complete surrogate estimate with $\tilde{A}$, involving the samples $\mathbf{p}^i \in S_\mathbf{p}$ $(i = 1, \ldots, M)$
4:     Find permutation π of set $[1 : M]$, with $\pi_i = \pi(i)$, such that $\{|\tilde{h}(\mathbf{p}^{\pi_i})|\}_{i=1}^{M}$ is ascending
5:     **while** $k < \lceil M/\delta M \rceil$ **do**
6:         Define $S_z^{(k)} = \pi_{[M_1:M_2]}$, where $M_1 = M^{(k)} + 1, M_2 = M^{(k)} + \delta M$
7:         **for each** $\pi_i \in S_z^{(k)}$ **do**
8:             Evaluate $h(\mathbf{p}^{\pi_i})$
9:             Determine $\chi_A(\mathbf{p}^{\pi_i})$
10:         **end for**
11:         Evaluate $P_F^{(k)} = P_F^{(k-1)} + 1/M \sum_{\pi_i \in S_z^{(k)}} \left(-\chi_{\tilde{A}}(\mathbf{p}^{\pi_i}) + \chi_A(\mathbf{p}^{\pi_i}) \right)$
12:         **if** $|P_F^{(k)} - P_F^{(k-1)}| \leq \eta$ **then**
13:             **return** $P_F = P_F^k$
14:         **end if**
15:         update $k = k + 1, M^{(k)} = M^{(k-1)} + \delta M$
16:     **end while**
17:     **return** $P_F = P_F^k$
18: **end procedure**

Algorithm 16.2 Failure Probability Algorithm with Neighborhood Tolerance

---

1: **procedure** P_F=**HYBRID2**$(h, \tilde{h}, S_{\mathbf{p}}, M, \gamma)$
2:      Evaluate P_F^0 using (16.46), i.e., the full surrogate estimate with $\tilde{A}$
3:      Find permutation π of set $[1:M]$, with $\pi_k = \pi(k)$, such that $\{|\tilde{h}(\mathbf{p}^{\pi_k})|\}_{k=1}^M$ is ascending
4:      Define $S_{\mathbf{p}}^{\gamma} = \{\pi_k \mid k \in [1:M] \text{ and } |\tilde{h}(\mathbf{p}^{\pi_k})| < \gamma\}$
5:      **for each** $\pi_k \in S_{\mathbf{p}}^{\gamma}$ **do**
6:          Evaluate $h(\mathbf{p}^{\pi_k})$
7:          **if** $h(\mathbf{p}^{\pi_k}) < \gamma$ **then**
8:              Determine $\chi_A(\mathbf{p}^{\pi_k})$
9:              Evaluate $P_F^{(k)} = P_F^{(k-1)} + 1/M\left(-\chi_{\tilde{A}}(\mathbf{p}^{\pi_k}) + \chi_A(\mathbf{p}^{\pi_k})\right)$
10:          **end if**
11:      **end for**
12:      **return** $P_F = P_F^k$
13: **end procedure**

---

rogate model $\tilde{h} = h_N$, the set of random samples $S_{\mathbf{p}} = \{\mathbf{p}^i\}_{i=1}^M$, a "stepsize" δM (a size for a portion of sorted Monte Carlo results) and a tolerance η for the failure probability estimate are required.

The Algorithm first approximates the failure probability using (16.46) (Step 3). In the while loop starting at 5 the smallest absolute values of $\tilde{h}$ are updated by evaluating the full model h. The correction to the failure probability is made in Step 11 (actually the subtraction needs a more stable formulation). In Step 12 the tolerance η implicitly determines γ.

The next Algorithm 16.2 involves the parameter γ explicitly, rather than η. However now one needs to determine a proper value for γ. The equality (16.47) is often fulfilled even for a moderate accuracy of the surrogate model [24]. We conclude with a useful error estimate from [25, Theorem 4.1, p. 8971] that estimates the error between the failure probability $P_F^{\gamma,M}$ in (16.48) obtained by the hybrid method and the failure probability P_F of the full model in (16.45).

Theorem 16.6. *Let $r > 0$ and $L_{\rho}^r(\Gamma)$ be the r-integral norm involving the density ρ. We define $\theta = ||h - \tilde{h}||_{L_{\rho}^r(\Gamma)}$.*

For $\varepsilon > 0$ and $\gamma = \theta / \sqrt[r]{\varepsilon}$ the error between the failure approximation $P_F^{\gamma,M}$ by the hybrid method in (16.48) and P_F of the full model in (16.45) has the bound $|P_F - P_F^{\gamma,M}| \leq \varepsilon$.

Clearly, for estimating small failure probabilities, rigorous error control has to be taken into account to determine θ. Section 16.7 provides an approach based on dual systems.

16.7 Error control for an Electro-Thermal Problem

The error control in Theorem 16.6 is based on the accuracy θ in which $\tilde{h}$ approximates h. We take $r = 2$ and we assume that $\tilde{h}$ is built from polynomial approximations in all coordinates of $\mathbf{p}$. Hence θ can then only be influenced by the degree d of the used approximations.

Below we indicate how θ may be estimated for a linear failure indicator function h of an electro-thermal problem using an adjoint system of equations. On a domain D and time interval I, we are concerned with a nonlinear, coupled, electro-thermal problem for the electro-magnetic potential $\varphi(t,x)$ and temperature $T(t,x)$, where $(t,x) \in I \times D$,

$$-\nabla \cdot (\sigma(T)\nabla\varphi) = 0, \qquad \text{in } I \times D, \qquad (16.49a)$$

$$\rho c \partial_t T - \nabla \cdot (\lambda(T)\nabla T) = \sigma|\nabla\varphi|^2, \qquad \text{in } I \times D, \qquad (16.49b)$$

endowed with suitable initial and mixed boundary conditions. Here $\sigma(T)$ and $\lambda(T)$ are the temperature-dependent conductivity and diffusion densities. Furthermore, ρ and c are the mass density and the specific heat capacity of the material, respectively. The equations are discretized using a Finite Element Method (FEM).

Bond wires are a cheap way to connect an IC to its surrounding packaging. Fusing needs to be avoided during operation. The bond wires are incorporated in (16.49) using lumped elements after discretization with FEM [6, 7, 34]

$$\mathbf{K}_\alpha(\mathbf{p},T) = \mathbf{K}_\alpha^{\mathrm{FE}}(T) + \mathbf{K}_\alpha^{\mathrm{lump}}(\mathbf{p},T), \quad \alpha \in \{\lambda, \sigma\}.$$

Therein, $\mathbf{K}_\alpha^{\mathrm{FE}}$ denotes the standard Finite Element stiffness matrix, $\mathbf{K}_\alpha^{\mathrm{lump}}$ includes the lumped bond wire contribution and $\mathbf{p}$ are the bond wire lengths as random parameters. With $\mathbf{y} = (\varphi, T)$ we obtain a parametric index-1 differential algebraic equation

$$\mathbf{M}\dot{\mathbf{y}} + \mathbf{K}(\mathbf{p},\mathbf{y})\mathbf{y} = \mathbf{Q}(\mathbf{p},\mathbf{y}),$$

where $\mathbf{M}$ is independent of $\mathbf{p}$ and of $\mathbf{y}$. Now we apply the hybrid technique outlined in Section 16.6.3, assuming a *linear* failure indicator function (16.1), that reads as

$$h = \int_I \mathbf{f}^\top \mathbf{y} \, \mathrm{d}t,$$

for some given vector $\mathbf{f}$, which, here, is independent of $\mathbf{p}$ and $\mathbf{y}$. We notice that h implicitly depends on $\mathbf{p}$.

In [5, 17] the solution $\mathbf{z}$ of the adjoint system

$$-\mathbf{M}\dot{\mathbf{z}} + \overline{\mathbf{K}}(\mathbf{p})\mathbf{z} = \mathbf{f},$$

in which $\overline{\mathbf{K}}(\mathbf{p}) = \frac{\partial \mathbf{K}(\mathbf{p},\mathbf{y})}{\partial \mathbf{y}}\mathbf{y} + \mathbf{K}(\mathbf{p},\mathbf{y}) - \frac{\partial \mathbf{Q}}{\partial \mathbf{y}}(\mathbf{p},\mathbf{y})$, is solved backward in time (starting with initial value $\mathbf{0}$, which is allowed for DAEs of index up to 1). The adjoint solution $\mathbf{z}$ can be used to efficiently determine the sensitivity of $\mathrm{d}h/\mathrm{d}\mathbf{p}$.

Things simplify if we freeze $\mathbf{y}$ to some $\bar{\mathbf{y}}$ in $\mathbf{K}$ and in $\mathbf{Q}$. Then $\overline{\mathbf{K}}(\mathbf{p}) = \mathbf{K}(\mathbf{p},\bar{\mathbf{y}})$. We have

$$h = \int_I \mathbf{f}^\top \mathbf{y} \ \mathrm{d}t = \int_I \mathbf{z}^\top \left(\mathbf{M}\dot{\mathbf{y}} + \mathbf{K}(\mathbf{p},\bar{\mathbf{y}})\mathbf{y}\right) \ \mathrm{d}t = \int_I \mathbf{z}^\top \mathbf{Q}(\mathbf{p},\bar{\mathbf{y}}) \ \mathrm{d}t.$$

The surrogate model $\tilde{h}$ is constructed by a Stochastic Collocation (SC) approach as in [1], based on a tensor product grid or a sparse grid. With the stochastic collocation approximation $\tilde{\mathbf{y}}$, we obtain

$$\tilde{h} = \int_I \mathbf{f}^\top \tilde{\mathbf{y}} \ \mathrm{d}t = \int_I \mathbf{z}^\top \left(\mathbf{M}(\tilde{\mathbf{y}})' + \mathbf{K}(\mathbf{p},\bar{\mathbf{y}})\tilde{\mathbf{y}}\right) \ \mathrm{d}t.$$

Combining we find the estimate

$$\theta = \|h - \tilde{h}\|_{L^2_\rho(\Gamma)} = \left\| \int_I \mathbf{z}^\top \left(\mathbf{Q}(\cdot,\bar{\mathbf{y}}) - [\mathbf{M}(\tilde{\mathbf{y}})' + \mathbf{K}(\cdot,\bar{\mathbf{y}})\tilde{\mathbf{y}}]\right) \ \mathrm{d}t \right\|_{L^2_\rho(\Gamma)}.$$

Notice that we may take as freezing value $\bar{\mathbf{y}} = \tilde{\mathbf{y}}$ in $\mathbf{K}$ and $\mathbf{Q}$. In the numerical computation, we consider two bond wires with uncertainties in the length of the wires. The length of a wire is $l_i = l_{i,0}/(1 - p_i)$ with random variables p_i uniformly distributed in the interval $(0.17, 0.048)$. In the stochastic collocation method, we use a tensor product grid of degree d [1,34] in the domain of the parameters. The computational effort of the hybrid method is dominated by the number of evaluations of the full order model. Table 16.2 shows the required number of full model evaluations and the resulting thresholds γ for different degrees d.

Table 16.2: Computational effort in hybrid approach for calculating the failure probability of bond wire fusion for different degrees d of the collocation grid [34].

degree d	threshold γ	no. full evaluations
1	4.31×10^{-2}	768
2	4.60×10^{-4}	12
3	1.31×10^{-6}	0

16.8 Enhancement to Ageing

The hybrid technique can be extended to simulate ageing. In addition to uncertainties due to manufacturing imperfections, ageing may cause parameter variations. Modeling of ageing can be achieved by introducing a random process $\tilde{Y} : \tilde{I} \times \Gamma \to \mathbb{R}$, where $\tilde{I} \subset \mathbb{R}$ contains the time-scale of the ageing process. We can then seek for an approximation in separated form

$$\tilde{Y}_{n_y}(\tau,\omega) = \sum_{i=1}^{n_y} f_i(\tau) Y_i(\omega), \tag{16.50}$$

e.g., by truncating the Karhunen-Loève expansion. In this way we identify again a set of uncorrelated random variables. If necessary, independence can be achieved by using an additional Hermite polynomial chaos expansion. Now, also the system's response depends on the time τ, i.e., $\tilde{g} : \tilde{I} \times \Gamma \to \mathbb{R}$ and one might define the performance function as

$$g = \min_{\tau \in \tilde{I}} \tilde{g}(\tau,\cdot). \tag{16.51}$$

Provided that (16.50) and (16.51) can be computed efficiently, failure probabilities can be evaluated in the presence of ageing effects using the hybrid method described above.

As an example, consider the threshold voltage of transistors shifted by a negative bias temperature instability [21]. There, the ageing effect is modeled as

$$\Delta V_{\text{th}}(\tau,\omega) = Z_1(\omega) \log \left(Z_2(\omega) + Z_3(\omega)\tau \right), \tag{16.52}$$

where the Z_i, $i = 1,2,3$, denote random variables with statistics inferred from measurement data. If the Z_i from (16.52) are independent, we can directly identify $Y_i = Z_i$, $i = 1,2,3$. If the data is correlated, the Karhunen-Loève expansion (16.50) can be used instead.

References

1. BABUSKA, I., NOBILE, F., AND TEMPONE, R.: *A stochastic collocation method for elliptic partial differential equations with random input data*, SIAM Review, 52.2, pp. 317–355, 2010.
2. BRAUN, W.J., AND MURDOCH, D.J.: *A First Course in Statistical Programming with R*, Cambridge University Press, 2007.
3. BUCKLEW, J.A.: *Introduction to Rare Event Simulation*, Springer, 2004.
4. CAFLISCH, R.E.: *Monte Carlo and quasi-Monte Carlo methods*, Acta Numerica, 1–49, 1998.
5. CAO, Y., LI, S., PETZOLD, L., AND SERBAN, R.: *Adjoint sensitivity for differential-algebraic equations: the adjoint DAE system and its numerical solution*, SIAM J. Sci. Comput., Vol. 24-3, pp. 1076–1089, 2002.

6. CASPER, T., DE GERSEM, H., GILLON, R., GOTTHANS, T., KRATOCHVÍL, T., MEURIS, P., AND SCHÖPS, S.: *Electrothermal Simulation of Bonding Wire Degradation under Uncertain Geometries*. Proceedings 2016 Design, Automation & Test in Europe Conference & Exhibition (DATE), Paper 0776, pp. 1297–1302, 2016.
7. CASPER, T., RÖMER, U., AND SCHÖPS, S.: *Efficient evaluation of bond wire fusing probabilities*. Presentation at SIAM Conference on Uncertainty Quantification, April 5-8, 2016, Lausanne, Switzerland.
8. DI BUCCIANICO, A., TER MATEN, J., PULCH, R., JANSSEN, R., NIEHOF, J., HANSSEN, M., AND KAPORA, S.: *Robust and efficient uncertainty quantification and validation of RFIC isolation*, Radioengineering 23:1, 308–318, 2014.
9. DOORN, T.S., TER MATEN, E.J.W., CROON, J.A., DI BUCCIANICO, A., WITTICH, O.: *Importance Sampling Monte Carlo simulation for accurate estimation of SRAM yield*. In: Proc. IEEE ESSCIRC'08, 34th Eur. Solid-State Circuits Conf., Edinburgh, Scotland, pp. 230–233, 2008.
10. DOORN, T.S., CROON, J.A., TER MATEN, E.J.W., DI BUCCIANICO, A.: *A yield statistical centric design method for optimization of the SRAM active column*. In: Proc. IEEE ESSCIRC'09, 35th Eur. Solid-State Circuits Conf., Athens, Greece, pp. 352–355, 2009.
11. DUBOURG, V.: *Adaptive Surrogate Models for Reliability Analysis and Reliability-Based Design Optimization*, PhD-Thesis, Univ. Blaise Pascal, Clermont II, Clermont-Ferrand, France, 2011.
12. FELLER, W.: *An Introduction to Probability Theory and Its Applications*, Vol. 1, Third Edition. John Wiley & Sons, Inc, New York, USA„ 1968. Online, https://archive.org/details/ AnIntroductionToProbabilityTheoryAndItsApplicationsVolume1.
13. DE HAAN, L., FERREIRA, A.: *Extreme Value Theory*. Springer, 2006.
14. HESTERBERG, T.: *Weighted average importance sampling and defensive mixture distributions*, Technometrics, 37(2):185–194, 1995.
15. HESTERBERG, T.C.: *Advances in Importance Sampling*, PhD-Thesis Stanford Univ., 1988; online available at http://citeseerx.ist.psu.edu/viewdoc/download?doi= 10.1.1.136.5735&rep=rep1&type=pdf, including added remarks, 2003.
16. DEN HOLLANDER, F.: *Large Deviations*. Fields Institute Monographs 14, The Fields Institute for Research in Math. Sc. and AMS, Providence, R.I., 2000.
17. ILIEVSKI, Z., XU, H., VERHOEVEN, A., TER MATEN, E.J.W., SCHILDERS, W.H.A., AND MATTHEIJ, R.M.M.: *Adjoint transient sensitivity analysis in circuit simulation*. In: CIUPRINA, G., AND IOAN, D. (EDS.): *Scientific Computing in Electrical Engineering SCEE 2006*, Series Mathematics in Industry Vol. 11, Springer, pp. 183–189, 2007.
18. JOURDAIN, B., AND LELONG, J.: *Robust adaptive importance sampling for normal random vectors*, The Annals of Appl. Prob., 19.5, pp. 1687–1718, 2009.
19. KANJ, R., JOSHI, R., NASSIF, S.: *Mixture Importance Sampling and its application to the analysis of SRAM designs in the presence of rare failure events*, Proc. DAC 2006, 5.3, pp. 69–72, 2006.
20. KIM, Y.B., ROH, D.S., AND LEE, M.Y.: *Nonparametric adaptive importance sampling for rare event simulation*. In: JONES, J.A., BARTON, R.R., KANG, K., AND FISHWICK, P.A. (EDS.): *Proceedings of the 2000 Winter Simulation Conference*, pp. 767–772, 2000.
21. KLEEBERGER, V.B., BARKE, M., WERNER, C., SCHMITT-LANDSIEDEL, D., AND SCHLICHTMANN, U.: *A compact model for NBTI degradation and recovery under use-profile variations and its application to aging analysis of digital integrated circuits*, Microelectronics Reliability, 54.6, pp. 1083–1089, 2014.
22. LAPEYRE, B., AND LELONG, J.: *A framework for adaptive Monte-Carlo procedures*, Monte Carlo Methods and Applications, Walter de Gruyter, 17.1, pp. 77–98, 2011.

23. LI, X., LE, J., GOPALAKRISHNAN, P., AND PILEGGI, L.T.: *Asymptotic Probability Extraction for Nonnormal Performance Distributions*, IEEE Trans. on Comp.-Aided Design of Integrated Circuits and Systems (TCAD), Vol. 26, No. 1, 16–37, 2007.
24. LI, J., LI, J., AND XIU, D.: *An efficient surrogate-based method for computing rare failure probability*, J. Comput. Phys. 230:24, 8683–8697, 2011.
25. LI, J., AND XIU, D.: *Evaluation of failure probability via surrogate models*, J. Comput. Phys. 229:23, 8966–8980, 2010.
26. MARTINEZ, W.L., MARTINEZ, A.R.: *Computational Statistics Handbook with Matlab*, Chapman & Hall/CRC Press LLC, Boca Raton, FL, USA, 2002.
27. TER MATEN, E.J.W., DOORN, T.S., CROON, J.A., BARGAGLI, A., DI BUCCHIANICO, A., WITTICH, O.: *Importance Sampling for high speed statistical Monte-Carlo simulations – Designing very high yield SRAM for nanometer technologies with high variability*. TUE-CASA Report 2009-37, TU Eindhoven, 2009. http://www.win.tue.nl/analysis/reports/rana09-37.pdf
28. TER MATEN, E.J.W., PULCH, R., SCHILDERS, W.H.A., AND JANSSEN, H.H.J.M.: *Efficient calculation of Uncertainty Quantification*. In: FONTES, M., GÜNTHER, M., AND MARHEINEKE, N. (EDS): *Progress in Industrial Mathematics at ECMI 2012*, Series Mathematics in Industry Vol. 19, Springer, pp. 361–370, 2014. https://dx.doi.org/10.1007/978-3-319-05365-3_50.
29. TER MATEN, E.J.W., WITTICH, O., DI BUCCHIANICO, A., DOORN, T.S., AND BEELEN, T.G.J.: *Importance sampling for determining SRAM yield and optimization with statistical constraint*. In: MICHIELSEN, B., AND POIRIER, J.-R. (EDS.): *Scientific Computing in Electrical Engineering SCEE 2010*. Mathematics in Industry Vol. 16, Springer, pp. 39–48, 2012. https://dx.doi.org/10.1007/978-3-642-22453-9_5.
30. MCCONAGHY, T., AND GIELEN, G.G.E.: *Globally Reliable Variation-Aware Sizing of Analog Integrated Circuits via Response Surfaces and Structural Homotopy*, IEEE Trans. on Design of Integr. Circuits and Systems (TCAD), 28.11, pp. 1627–1640, 2009.
31. OWEN, A.B.: *Monte Carlo Theory, Methods and Examples*, Online: http://statweb.stanford.edu/~owen/mc/, 2013.
32. PROPPE, C.: *Estimation of failure probabilities by local approximation of the limit state function*. Structural Safety 30, pp. 277–290, 2008.
33. PULCH, R.: *Polynomial chaos for the computation of failure probabilities in periodic problems*. In: ROOS, J., AND COSTA, L. (EDS.): *Scientific Computing in Electrical Engineering SCEE 2008*. Mathematics in Industry Vol. 14, Springer, Berlin, pp. 191–198, 2010.
34. RÖMER, U., SCHÖPS, S., AND CASPER, T.: *On the Evaluation of Failure Probabilities for Nanoelectronic Applications*, Poster, GAMM-2016 Annual Meeting, March 7-11, 2016, Braunschweig, Germany.
35. SRINIVASAN, R.: *Importance Sampling - Applications in Communications and Detection*, Springer Verlag, Berlin, 2002.
36. SUMO: *SUrrogate MOdeling Lab*, Ghent University, Belgium, 2018. http://sumo.intec.ugent.be/.
37. TYAGI, A.K.: *Speeding up Rare-Event Simulations in Electronic Circuit Design by Using Surrogate Models*. PhD-Thesis, TU Eindhoven, 2018. https://research.tue.nl/files/106593385/20181009_Tyagi.pdf (Open Access).
38. TYAGI, A.K., JONSSON, X., BEELEN, T.G.J, AND SCHILDERS, W.H.A.: *Hybrid Importance Sampling Monte Carlo Approach for Yield Optimization in Circuit Design*. Journal Mathematics in Industry 8:11, 2018. https://doi.org/10.1186/s13362-018-0053-4 (Open Access).
39. VENABLES, W.N., SMITH, D.M., AND THE R CORE TEAM: *An Introduction to R - Notes on R: A Programming Environment for Data Analysis and Graphics*, Version 3.4.3 (2017-11-30), https://cran.r-project.org/doc/manuals/r-release/R-intro.pdf.

Chapter 17
Fast Fault Simulation for Detecting Erroneous Connections in ICs

Jos J. Dohmen, Bratislav Tasić, Rick Janssen, E. Jan W. ter Maten,
Theo G.J. Beelen, Roland Pulch, Michael Günther

Abstract Imperfections in manufacturing processes can be modelled as unwanted connections (defects, or faults) that are added to the nominal, "golden", fault-free design of an electronic circuit to study their impact. Testing in a structured way using fault simulation techniques to obtain information on the impact of faults and guaranteeing defect coverage and test quality is not a common practice during the design of analog or mixed signal ICs. Fault simulation involves defect extraction and injection of defects into the netlist of the analog or mixed signal circuit and performing analogue simulation (DC, AC, or Transient) of the tests. The major drawback is the long CPU time associated with the many analogue simulations. For example, if simulation of the test suite takes one hour, it may take several years to perform all simulations for more than 10,000 defects (when not exploiting parallelism).

In the transient simulation the solution due to an inserted fault is compared to a golden, fault-free, solution. A strategy is developed to efficiently simulate the faulty solutions until their moment of detection. We obtain a significant speed-up of over 100x over sequential approaches, while a useful estimate

Jos J. Dohmen, Bratislav Tasić, Rick Janssen
NXP Semiconductors, Eindhoven, the Netherlands,
e-mail: {Jos.J.Dohmen,Bratislav.Tasic,Rick.Janssen}@nxp.com

E. Jan W. ter Maten, Michael Günther
Bergische Universität Wuppertal, Germany,
e-mail: {terMaten,Guenther}@math.uni-wuppertal.de

E. Jan W. ter Maten, Theo G.J. Beelen
Eindhoven University of Technology, the Netherlands,
e-mail: {E.J.W.ter.Maten,T.G.J.Beelen}@tue.nl;Th.Beelen@gmail.com

Roland Pulch
Universität Greifswald, Germany,
e-mail: Roland.Pulch@uni-greifswald.de

© Springer Nature Switzerland AG 2019
E. J. W. ter Maten et al. (eds.), *Nanoelectronic Coupled Problems Solutions*,
Mathematics in Industry 29, https://doi.org/10.1007/978-3-030-30726-4_17

of the detection status and the defect coverage can still be ensured. Our strategy can also be used when exploiting parallelism.

17.1 Introduction

Defects in the manufacturing processes may cause unwanted connections in the nominal, "golden", fault-free design of an electronic circuit. The majority of these defects can be modelled by linear conductivities that are inserted in the golden design. There are two aspects here to consider:

- The connection can take place between any two locations of the circuit. Hence any pair of nodes in the circuit may have to be considered. Such a new connection can be parallel to an existing direct or indirect connection between the nodes. But it can also be a completely new connection.
- The amount of the conductivity (a $1/R$ value, where R is a resistor value) can vary over a large range: from being effectively close to be a short-cut up to be close to be an "open". In practise one may observe a bias to the last situation (i.e., the values can be assumed to have an exponential distribution).

Of course the unexpected connection may also be an additional linear capacitor or inductor. A (nearly) broken connection can be modelled by an additional voltage source. However, in practice, most faults have a dominant effect due to the conductivity in the connection. Hence, in this chapter we will focus on the case of conductivities. We notice that our approaches can be extended to also deal with the other types of connections [17].

We can consider the problem also from a different point of view. Essentially one is looking for the spots in the circuit that are weakest in allowing changes of the functionality of the circuit due to external causes. In our application they are due to the manufacturing process. However the fault can also show up later, due to effects of ageing of the design, or by stress effects due to heating. The problem is also related to other types of networks, e.g., in analyzing traffic behaviour in a city when suddenly a road is blocked, or when a new connection pops up. Our approach can be extended to energy distribution networks, sewage systems, and even to networks that are not constant of size in time.

17.2 Circuit Equations, DC and Transient Simulation

The electronic circuit equations can be written as [6]

$$\frac{\mathrm{d}}{\mathrm{d}t}\mathbf{q}(\mathbf{x}) + \mathbf{j}(\mathbf{x}) = \mathbf{s}(\mathbf{x}, t). \tag{17.1}$$

It involves equations for Kirchhoff's Current Law and Voltage Law as formulated in the Modified Nodal Analysis (MNA). Here $\mathbf{x}(t)$ represents the time-varying vector of unknowns, involving nodal voltages and currents through voltage-defined elements. Furthermore, $\mathbf{q}$ and $\mathbf{j}$ model capacitors and conductances, respectively. Inductors contribute to $\mathbf{q}$ and $\mathbf{j}$, both. At the right-hand side $\mathbf{s}(\mathbf{x}, t)$ represents the specifications of the sources. Current sources can directly be found here. Voltage sources contribute partly to $\mathbf{j}$ and to $\mathbf{s}(\mathbf{x}, t)$, both. We assume that $\mathbf{q}(0) = \mathbf{0}$ and $\mathbf{j}(0) = \mathbf{0}$. The steady-state solution, which is called the direct current (DC)-solution, $\mathbf{x}_{\mathbf{DC}}$, satisfies:

$$\mathbf{j}(\mathbf{x}_{\mathbf{DC}}) = \mathbf{s}(\mathbf{x}_{\mathbf{DC}}, \mathbf{0}). \tag{17.2}$$

The importance of solving the DC-problem lies in the fact that the DC-solution is crucial as starting solution for a number of next analyses (transient analysis, AC analysis, harmonic balance analysis, periodic steady-state analysis). In general, the problem (17.2) is nonlinear and in many cases it is hard to solve. When adding a new element to the circuit the resulting modified circuit may be even harder to simulate in DC (but possibly also during transient simulation). In [18] we gave an overview of several approaches to solve (17.2) and introduced a novel, robust and efficient Source Stepping by Pseudo Transient (SSPT) method, that modifies already existing Pseudo Transient methods in the case of controlled elements (that may involve recursion). The SSPT preserves the structure of the matrices derived from $\mathbf{q}$, $\mathbf{j}$ and $\mathbf{s}$. For details and for references to other methods we refer to [18].

In general, (17.1) forms a system of Ordinary Differential Equations (ODEs) or a system of Differential-Algebraic Equations (DAEs). In the latter case we assume that the DAE has differential index 1 [6, p. 547]. We will exploit arguments from sensitivity analysis of parametric DAEs [11, 15].

For time integration in circuit simulation one usually applies the Backward-Differentiation Formulas (BDF) methods (BDF1 and BDF2 being well-known first and second order variants), or the Trapezodial Rule (TR), or a combination of BDF2 and TR [4, 6]. For simplicity, we just apply BDF1, being Euler Backward. We introduce some notation and assume some well-known approaches in solving the discretized equations as well as some stepsize control. Assuming time points $t_{k+1} = t_k + h_k$ $(k \geq 0)$ with stepsizes h_k and approximation $\mathbf{x}^n$ at t_n, BDF1 calculates $\mathbf{x}^{n+1}$ at t_{k+1} by

$$\frac{\mathbf{q}^{n+1} - \mathbf{q}^n}{h_n} + \mathbf{j}^{n+1} = \mathbf{s}^{n+1}. \tag{17.3}$$

Here $\mathbf{q}^k = \mathbf{q}(\mathbf{x}^k)$, $\mathbf{j}^k = \mathbf{j}(\mathbf{x}^k)$, for $k = n, n+1$ and $\mathbf{s}^{n+1} = \mathbf{s}(\mathbf{x}^{n+1}, t_{n+1})$. The system is solved by a Newton-Raphson procedure. One may choose between exploiting a fixed Jacobian, or preferring new LU-decompositions, or re-using of subspaces in case of iterative methods, like Krylov-methods. One can also benefit from hierarchical bypassing [4]. A fast and general approach for time integration is found in [1].

17.3 Fast Fault Simulation: Algorithmic Details

Recently, testing of analog and mixed signal ICs received a lot of attention [3, 7, 8, 12, 20]. Fault simulation is part of analog testing and emphasis is on balancing speed-up (fast fault simulation (FFS)) against coverage in detection of faults. Imperfections in manufacturing processes may cause unwanted connections (faults) that are added to the nominal, "golden", design of an electronic circuit. By fault analysis these faults can be detected [9, 16–19]. In [18] the faulty elements are represented by adding linear conductivities to the circuit. All linear elements contribute to the system of equations by a linear term in the solution $\mathbf{x}$ with as local coefficient matrix a rank-1 matrix [4]. A conductivity, seen as a resistor $R(a,b)$ between two nodes a and b in the circuit, with value R, contributes to the system of equations by the expression for the current $i = \frac{\mathbf{x}_b - \mathbf{x}_a}{R}$ (Ohm's law) in which $\mathbf{x}_b = \mathbf{e}_b^T \mathbf{x}$ is the voltage value of $\mathbf{x}$ at node b and with $\mathbf{x}_a = \mathbf{e}_a^T \mathbf{x}$ defined similarly. Here $\mathbf{e}_a$ and $\mathbf{e}_b$ are appropriate euclidean unit vectors. Hence, we get a contributing term to the system of equations of the form $p\mathbf{u}\mathbf{v}^T\mathbf{x}$, in which $p = 1/R$ and $\mathbf{u} = \mathbf{e}_b - \mathbf{e}_a$ and $\mathbf{v}^T = \mathbf{e}_b^T - \mathbf{e}_a^T$ (the order of b and a is just by convention). Clearly, $\mathbf{u}$ and $\mathbf{v}$ address the location of the nodes a and b in the circuit and the parameter p represents the amount of conductivity.

We start analysing a faulty circuit by assuming that $\mathbf{u}$ and $\mathbf{v}$ are given. Then we only have to deal with varying values of p. The faulty solution $\mathbf{x}(t,p)$ satisfies (17.1) in which $\mathbf{j}$ is given by

$$\mathbf{j}(\mathbf{x}(t,p),p) = \mathbf{j}_0(\mathbf{x}(t,p)) + p\mathbf{u}\mathbf{v}^T\mathbf{x}(t,p). \tag{17.4}$$

The golden solution of the fault-free circuit corresponds with $p = 0$ and will be denoted by $\mathbf{x}(t) \equiv \mathbf{x}(t,0)$. It uses $\mathbf{j}(\mathbf{x}(t,0),0) = \mathbf{j}_0(\mathbf{x}(t,0)) = \mathbf{j}_0(\mathbf{x}(t))$ in (17.1). Of course, the golden solution $\mathbf{x}(t)$ does not depend on $\mathbf{u}$ and $\mathbf{v}$.

We now introduce notations for discrete values of solution and of derivatives. Let $\mathbf{x}_p^k \equiv \mathbf{x}^k(p) \approx \mathbf{x}(t_k,p)$, for $k = n, n+1$, be the numerical approximations of the faulty system at time t_k and $\mathbf{x}^k \equiv \mathbf{x}_0^k$ ($k = n, n+1$) the corresponding ones for the golden, fault-free solution. We define the derivative matrices $\mathbf{C}_p^k \equiv \frac{\partial \mathbf{q}(\mathbf{x}_p^k)}{\partial \mathbf{x}}$, $\mathbf{G}_p^k \equiv \frac{\partial \mathbf{j}(\mathbf{x}_p^k)}{\partial \mathbf{x}}$ for $\mathbf{j}$ as defined in (17.4) and $\mathbf{S}_p^k \equiv \frac{\partial \mathbf{s}(\mathbf{x}_p^k, t_k)}{\partial \mathbf{x}}$. With these the Jacobian matrix for $\mathbf{x}_p^k$ of (17.3) becomes

$$\mathbf{A}_p^{n+1} = \frac{1}{h_n}\mathbf{C}_p^{n+1} + \mathbf{G}_p^{n+1} - \mathbf{S}_p^{n+1}. \tag{17.5}$$

For $p = 0$ we get the corresponding matrices $\mathbf{C}^k = \mathbf{C}_0^k$ ($k = n, n+1$), $\mathbf{G}^{n+1} = \mathbf{G}_0^{n+1}$ and $\mathbf{S}^{n+1} = \mathbf{S}_0^{n+1}$ and the Jacobian matrix $\mathbf{A}^{n+1} = \mathbf{A}_0^{n+1}$. With $\mathbf{s}^{n+1} = \mathbf{s}(\mathbf{x}^{n+1}, t^{n+1})$, the golden solution $\mathbf{x}^k$ satisfies

$$\mathbf{A}^{n+1}\mathbf{x}^{n+1} = \mathbf{s}^{n+1} + \frac{1}{h_n}\mathbf{C}^n\mathbf{x}^n + \mathbf{R}, \tag{17.6}$$

where $\mathbf{R}$ abbreviates a deviation from linearity (note that in (17.1) we did assume that $\mathbf{q}(0) = \mathbf{0}$ and $\mathbf{j}(0) = \mathbf{0}$). For our purposes we will denote the right-hand side of (17.6) by $\mathbf{r}(t_{n+1}, \mathbf{x}^n, \mathbf{x}^{n+1}) = \mathbf{s}^{n+1} + \frac{1}{h_n}\mathbf{C}^n\mathbf{x}^n + \mathbf{R}$ and (17.6) simply becomes

$$\mathbf{A}^{n+1}\mathbf{x}^{n+1} = \mathbf{r}(t_{n+1}, \mathbf{x}^n, \mathbf{x}^{n+1}). \tag{17.7}$$

Let $\hat{\mathbf{x}}_p^k \equiv \frac{\partial \mathbf{x}_p^k}{\partial p}$ be the p-sensitivity of the faulty solution $\mathbf{x}_p^k$ ($k = n, n+1$) and let $\hat{\mathbf{x}}^k \equiv \hat{\mathbf{x}}_0^k = \lim_{p\downarrow 0}\frac{\partial \mathbf{x}_p^k}{\partial p}$ be the corresponding *limit sensitivity* for the golden solution. By Taylor expansion we have

$$\mathbf{x}_p^k = \mathbf{x}^k + p\hat{\mathbf{x}}^k + \mathcal{O}(p^2) \quad (k = n, n+1). \tag{17.8}$$

Differentiating (17.3) and (17.4) with respect to p gives for the p-sensitivities

$$\mathbf{A}_p^{n+1}\hat{\mathbf{x}}_p^{n+1} = -\mathbf{u}\mathbf{v}^T\mathbf{x}_p^{n+1} + \frac{1}{h_n}\mathbf{C}_p^n\hat{\mathbf{x}}_p^n \quad (p \neq 0), \tag{17.9}$$

$$\mathbf{A}^{n+1}\hat{\mathbf{x}}^{n+1} = -\mathbf{u}\mathbf{v}^T\mathbf{x}^{n+1} + \frac{1}{h_n}\mathbf{C}^n\hat{\mathbf{x}}^n \quad (p \downarrow 0). \tag{17.10}$$

Combining (17.8), (17.9) and (17.10) yields [17–19]

$$[\mathbf{A}^{n+1} + p\mathbf{u}\mathbf{v}^T]\mathbf{x}_p^{n+1}$$
$$= \mathbf{r}(t_{n+1}, \mathbf{x}^n, \mathbf{x}^{n+1}) + \frac{1}{h_n}\mathbf{C}^n(\mathbf{x}_p^n - \mathbf{x}^n) + \mathcal{O}(p^2 + \frac{p^2}{h_n}), \tag{17.11}$$

$$= \mathbf{r}(t_{n+1}, \mathbf{x}^n, \mathbf{x}^{n+1}) + \mathcal{O}(p^2 + \frac{p^2}{h_n} + \frac{p}{h_n}). \tag{17.12}$$

Here the left-hand side involves the matrix $\mathbf{A}^{n+1}$ with a rank-one update. This invites for applying the Sherman-Morrison-Woodbury formula [5, 10]. In both right-hand sides (17.11) and (17.12) we can ignore the $\mathcal{O}$-terms. Then suitable *predictions* for $\mathbf{x}_p^{n+1}$ become

$$\tilde{\mathbf{x}}_p^{n+1} = \mathbf{x}^{n+1} - \frac{\mathbf{v}^T\mathbf{x}^{n+1}}{\frac{1}{p} + \mathbf{v}^T\mathbf{w}}\mathbf{w}, \text{ with } \mathbf{w} \text{ the solution of } \mathbf{A}^{n+1}\mathbf{w} = \mathbf{u}, \text{ or} \tag{17.13}$$

$$\tilde{\mathbf{x}}_p^{n+1} = (\mathbf{x}^{n+1} + \mathbf{y}) - \frac{\mathbf{v}^T(\mathbf{x}^{n+1} + \mathbf{y})}{\frac{1}{p} + \mathbf{v}^T\mathbf{w}}\mathbf{w}, \text{ where } \mathbf{w}, \mathbf{y}, \text{ are the solutions of}$$

$$\mathbf{A}^{n+1}\mathbf{w} = \mathbf{u}, \quad \mathbf{A}^{n+1}\mathbf{y} = \frac{p}{h_n}\mathbf{C}^n\hat{\mathbf{x}}^n. \tag{17.14}$$

Both predictions can be improved by subsequent Newton-Raphson iterations (see also Section 17.3.1). Prediction (17.13) is based on (17.12), which implicitly uses arguments from sensitivity analysis. The prediction (17.14) is based on (17.11), which explicitly determines the effects of the sensitivity quantities by solving for $\mathbf{y}$. Here also $\hat{\mathbf{x}}^n$ has to be determined by additional integration in time, starting from the sensitivity at DC when $t = 0$. Because p is scalar this is just a vector, but one that is different for different pairs of $\mathbf{u}$, $\mathbf{v}$. Notice that (17.13) may be a more accurate alternative than (17.14). However, for simplicity, we will use (17.13).

A simplification arises at $t = 0$ for the DC solutions

$$\mathbf{j}(\mathbf{x}_{DC,p}) = \mathbf{j}_0(\mathbf{x}_{DC,p}) + p\mathbf{u}\mathbf{v}^T\mathbf{x}_{DC,p} = \mathbf{s}(\mathbf{x}_{DC,p}), \tag{17.15}$$

$$\mathbf{j}_0(\mathbf{x}_{DC}) = \mathbf{s}(\mathbf{x}_{DC}), \tag{17.16}$$

where in both cases $\mathbf{s}$ is evaluated at $t = 0$. The Jacobian matrices are $\mathbf{A}_p \equiv \mathbf{G}_p - \mathbf{S}_p$ and, for $p = 0$, $\mathbf{A} \equiv \mathbf{A}_0$, which involves $\mathbf{G} = \mathbf{G}_0$ and $\mathbf{S} = \mathbf{S}_0$. Thus the corresponding equations for (17.7), (17.9) and (17.10) become

$$\mathbf{A}\mathbf{x}_{DC} = \mathbf{s}(\mathbf{x}_{DC}) + \mathbf{R}, \tag{17.17}$$

$$\mathbf{A}_p\hat{\mathbf{x}}_{DC,p} = -\mathbf{u}\mathbf{v}^T\mathbf{x}_{DC,p}, \tag{17.18}$$

$$\mathbf{A}\hat{\mathbf{x}}_{DC} = -\mathbf{u}\mathbf{v}^T\mathbf{x}_{DC}. \tag{17.19}$$

Similar to (17.13) we deduce

$$[\mathbf{A} + p\mathbf{u}\mathbf{v}^T]\mathbf{x}_{DC,p} = \mathbf{s}(\mathbf{x}_{DC}) + \mathbf{R} + \mathcal{O}(p^2). \tag{17.20}$$

We find as *prediction* (see also [16])

$$\tilde{\mathbf{x}}_{DC,p} = \mathbf{x}_{DC} - \frac{\mathbf{v}^T\mathbf{x}_{DC}}{\frac{1}{p} + \mathbf{v}^T\mathbf{w}} \mathbf{w}, \text{ where } \mathbf{w} \text{ is defined by } \mathbf{A}\mathbf{w} = \mathbf{u}. \tag{17.21}$$

We outline the basic algorithm in Alg. 17.1. Here the input consists of $\mathcal{P}$, being the list of triples of faults $(\mathbf{u}, \mathbf{v}, p)$; $\mathcal{M}$ being the list of time points θ_m for checking deviations between the faulty solutions and the golden solution; T being the end point of the time integration; τ defining the threshold for checking the deviations. The output $\mathcal{D}$ is the list of all $(\mathbf{u}, \mathbf{v}, p), \theta_m, \mathbf{x}_p^{k+1}, \mathbf{x}^{k+1}$ for which a defect has been detected.

Remarks

- One important observation is that the loop of faulty connections is treated within the overall time integration. This is a main difference compared to implementations of parametric problems, where usually the parameter loop is outside the time loop. In Alg. 17.1 (steps 15-22) all faulty solutions basically use the same time step as used for the golden solution ("integration of ensemble"). However, when needed, but not shown in Alg. 17.1, a faulty solution may take smaller time steps until synchronisation at t_{k+1}. Notice that all steps (8-12, 15-22) for the faulty solutions can be treated in parallel.

Algorithm 17.1 Basic Algorithm Fast Fault Simulation

1: **procedure** $\mathcal{D}$=FFS($\mathcal{P} = \{(\mathbf{u},\mathbf{v},p),...\}, \mathcal{M} = \{\theta_m,...\}, T, \Delta t_{\text{start}}, \tau$)
2:     $k = 0$; $t_k = 0$; $\Delta t = \Delta t_{\text{start}}$; $\theta_m = \min\{\theta \in \mathcal{M}\}$; $\mathcal{D} = \emptyset$;
3:     **while** $t_k \leq \min(T, \theta_m)$ **do**
4:         **while** $\mathcal{P} \neq \emptyset$ **do**
5:             Select $(\mathbf{u},\mathbf{v},p) \in \mathcal{P}$
6:             **if** $t_k = 0$ **then**
7:                 Solve DC problem (17.16) for $\mathbf{x}^0 = \mathbf{x}_{\text{DC}}$
8:                 Predict $\tilde{\mathbf{x}}_p^0$ by (17.21)
9:                 Solve DC problem (17.15) for $\mathbf{x}_p^0 = \mathbf{x}_{\text{DC},p}$ $\triangleright$ Includes bypassing
10:                 **if** $\|\mathbf{x}_p^0 - \mathbf{x}^0\| > \tau$ **then**
11:                     $\mathcal{D} = \mathcal{D} \cup \{(\mathbf{u},\mathbf{v},p), 0, \mathbf{x}_p^0, \mathbf{x}^0\}$
12:                 **end if**
13:             **end if**
14:             Determine golden solution $\mathbf{x}^{k+1}$ at $t_{k+1} = t_k + \Delta t$
15:             Predict $\tilde{\mathbf{x}}_p^{k+1}$ by (17.13)
16:             Determine faulty solution $\mathbf{x}_p^{k+1}$ of (17.3) $\triangleright$ Includes bypassing
17:             **if** $t_{k+1} = \theta_m$ **then**
18:                 **if** $\|\mathbf{x}_p^{k+1} - \mathbf{x}^{k+1}\| > \tau$ **then**
19:                     $\mathcal{D} = \mathcal{D} \cup \{(\mathbf{u},\mathbf{v},p), \theta_m, \mathbf{x}_p^{k+1}, \mathbf{x}^{k+1}\}$
20:                     $\mathcal{P} = \mathcal{P} \setminus \{(\mathbf{u},\mathbf{v},p)\}$ $\triangleright$ Abandon continuing for this fault
21:                 **end if**
22:             **end if**
23:         **end while**
24:         **if** $t_{k+1} = \theta_m$ **then**
25:             $\mathcal{M} = \mathcal{M} \setminus \{\theta_m\}$, $\theta_m = \min\{\theta \in \mathcal{M}\}$
26:         **end if**
27:         Estimate Δt for $\mathbf{x}^{k+2}$ at $t_{k+2} = t_{k+1} + \Delta t$ such that $t_{k+2} \leq \min(T, \theta_m)$
28:         $k = k + 1$
29:     **end while**
30:     **return** Output Database $\mathcal{D}$
31: **end procedure**

- In (17.13), (17.14) and (17.21), the vector $\mathbf{w}$ that satisfies $\mathbf{Aw} = \mathbf{u}$ can be re-used for several triples $(\mathbf{u},\mathbf{v},p) \in \mathcal{P}$ that use the same $\mathbf{u}$.
- We note that the time integration of a faulty solution is stopped after detection of the defect: the list $\mathcal{P}$ becomes smaller. This "Abandoning" is shown in Fig. 17.1.
  Additionally, solving for $\mathbf{x}_p^{k+1}$ can be simplified by reuse on a submodel of values of $\mathbf{x}^{k+1}$ when $\mathbf{x}_p^{k+1}$ and $\mathbf{x}^{k+1}$ do not defer that much at the submodel's terminals. By this an hierarchical branch can be completely bypassed, which reduces computational effort. See again Fig. 17.1.
- A special attention has to be given to connections between internal nodes of two submodels $\mathcal{S}_1$ and $\mathcal{S}_2$. In principle this violates the hierarchical structure. To accommodate with this we did extend all submodels with an extra terminal. By default the corresponding voltage is set to the ground. However, when $R = R(a,b)$ is such that $\mathbf{x}_a \in \mathcal{S}_1^{\text{int}}$ and $\mathbf{x}_b \in \mathcal{S}_2^{\text{int}}$

we introduce $R_1 = R_1(a,c)$ and $R_2 = R_1(c,b)$ with $R_1 = R_2 = 0.5R$ and include $\mathbf{x}_c$ as the extra terminal unknown in $\mathcal{S}_1$ and $\mathcal{S}_2$ and in the parent submodels until the two hierarchical branches meet. By this the overall systems for the golden solution and for the faulty ones all have the same size.

- Notice that the detection criterion in the steps 10 and 18 is formulated for the norm, but for measurements this can be reduced to a specific nodal voltage.

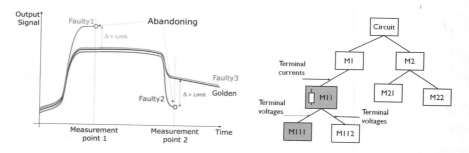

Fig. 17.1: **Left**: Abandoning of detected faults from the fault list: we stop their time integration. **Right**: An hierarchical tree of submodels is shown, with qat th top level the circuit model. In Submodel M11 a fault is represented by a symbol for a resistor. By this fault submodel M11 may affect submodel M111, but perhaps not M112. Even when M1 is affected, M2 may be not. In such cases the branches starting in M112 and in M2 can be bypassed [4, 18].

17.3.1 Non-Linear Solver for the Circuit with Rank-1 Modifications

The Sherman-Morrison-Woodbury formula [5, 10] for expressing the inverse of a matrix $\mathbf{M}$ with a rank-1 update in terms of the inverse of the original matrix $\mathbf{M}$ is given by the following expression

$$(\mathbf{M} + p\mathbf{u}\mathbf{v}^T)^{-1} = \mathbf{M}^{-1} - \frac{\mathbf{M}^{-1}\mathbf{u}\mathbf{v}^T\mathbf{M}^{-1}}{\frac{1}{p} + \mathbf{v}^T\mathbf{M}^{-1}\mathbf{u}}. \tag{17.22}$$

We consider a Newton process to determine a root of

$$\mathbf{F}(\mathbf{x}) + p\mathbf{u}\mathbf{v}^T\mathbf{x} = \mathbf{0}. \tag{17.23}$$

Let $\mathbf{x}^{(k-1)}$ be the $(k-1)$-th iterand, $\mathbf{F}^{(k-1)} = \mathbf{F}(\mathbf{x}^{(k-1)})$ and $\mathbf{A}^{(k-1)}$ be the Jacobian of $\mathbf{F}^{(k-1)}$, evaluated at $\mathbf{x}^{(k-1)}$. Then the Newton process to determine the root of (17.23) can be formulated as

$$\mathbf{y} = \mathbf{x}^{(k-1)} - \mathbf{M}^{-1}\mathbf{F}^{(k-1)}, \text{ where } \mathbf{M} \equiv \mathbf{A}^{(k-1)} \tag{17.24}$$

$$\mathbf{x}^{(k)} = \mathbf{y} - \mathbf{z}\frac{\mathbf{v}^T\mathbf{y}}{\frac{1}{p} + \mathbf{v}^T\mathbf{z}}, \text{ where } \mathbf{z} = \mathbf{M}^{-1}\mathbf{u}. \tag{17.25}$$

For (17.25) a similar remarks applies as for (17.22) when we include a parameter p as multiple for $\mathbf{u}$. Notice that $\mathbf{y}$ is the Newton correction for $\mathbf{F}(\mathbf{x}) = \mathbf{0}$, when starting from $\mathbf{x}^{(k-1)}$. If $\mathbf{y} \perp \mathbf{v}$ the correction in (17.25) is $\mathbf{0}$ and $\mathbf{z}$ is not needed at all. If $\mathbf{v}$ is sparse (which happens in case of a conductor) this is easily checked.

To continue, one has two options concerning the matrix $\mathbf{A}^{(k-1)}$:

- One can take the exact Jacobian $\mathbf{A}^{(k-1)}$ of $\mathbf{F}(\mathbf{x}^{(k-1)})$ for the golden circuit, evaluated at $\mathbf{x}^{(k-1)}$. Notice that $\mathbf{A}^{(k-1)}$ will use the same AMD-order as for the matrix of the golden circuit, which is very convenient during debugging. This is an implementation that facilitates easy comparison of matrix entries during testing. Alternatively one can simply include the faulty element in the list of elements and perform a conventional Newton process for (17.23).
- One can re-use the final Jacobian of the golden solution as fixed Jacobian for all k. This may be beneficial when $|p| \ll 1$.

17.3.2 Hierarchical Matrix and Unknowns

In [4] an hierarchical matrix-vector system was described that fits the definition of part of the matrix on each submodel and device (the latter is a library block for a transistor, say). At each level we partition the unknowns $\mathbf{x} = (\mathbf{x}_e^T, \mathbf{x}_i^T)^T$, where $\mathbf{x}_e$ corresponds to the terminals and $\mathbf{x}_i$ involves the internal unknowns. The matrix $\mathbf{M}$ has corresponding blocks $\mathbf{M}_{ee}$, $\mathbf{M}_{ei}$, $\mathbf{M}_{ie}$ and $\mathbf{M}_{ii}$ and right-hand side vector $\mathbf{b} = (\mathbf{b}_e^T, \mathbf{b}_i^T)^T$. This order of unknowns fits an hierarchical top-down definition with at the top level the governing circuit model (where the ground serves as terminal). We assume that a system $\mathbf{Mx} = \mathbf{b}$ has to be solved on each level in which $\mathbf{M}$ and $\mathbf{b}$ first have to be assembled (like in a Newton iteration).

The hierarchical system is accessed from the top. A recursion asks for the matrix and right-hand side assembly from the involved lower level submodels and combines that with the contributions from the current level. Hence, effectively, one starts assembling at the lowest level in each hierarchical branch (bottom-up recursion for each sub tree). Here also a partial $\mathbf{M}_{ii} = \mathbf{U}_{ii}\mathbf{L}_{ii}$ is performed, starting at the last equation and treating elimination upwards in the last column. Notice that with a reverse order of the unknowns this corresponds with an ordinary **LU** decomposition. However the current order more naturally fits an hierarchical top-down definition.

As a result the equations for the internal unknowns become $\mathbf{Lx}_i = \mathbf{b}'_i - \mathbf{M}'_{ie}\mathbf{x}_e$, with $\mathbf{b}'_i := \mathbf{U}^{-1}\mathbf{b}_i$ and $\mathbf{M}'_{ie} := \mathbf{U}^{-1}\mathbf{M}_{ie}$.

All internal unknowns are eliminated from the terminal equations. This can be done by extending the bottom up row-loop for determining **U** to also cover the lower numbered rows. Notice that, effectively, one determines the Schur complement $\mathbf{M}'_{ee} = \mathbf{M}_{ee} - \mathbf{M}_{ei}\mathbf{L}^{-1}\mathbf{U}^{-1}\mathbf{M}_{ie}$ and $\mathbf{b}'_e = \mathbf{b}_e - \mathbf{M}_{ei}\mathbf{L}^{-1}\mathbf{U}^{-1}\mathbf{b}_i$. Clearly $\mathbf{M}'_{ei} := \mathbf{0}$. After this, $\mathbf{M}'_{ee}$ and $\mathbf{b}'_e$ are added to the equations in the parent model. This decomposition process is shown in Fig. 17.2.

After this all what remains for the solution process is an hierarchical top-down recursion with a lower triangular system. On each submodel level the terminal values $\mathbf{x}_e$ are assumed to be determined earlier at the parent model level and to be passed to the current submodel. Here the remaining internal $\mathbf{x}_i$ are determined by solving $\mathbf{Lx}_i = \mathbf{b}'_i - \mathbf{M}'_{ie}\mathbf{x}_e$. Next all necessary terminal values for the child submodels are passed to these. Unused terminals are indirectly set to a (global) ground value via a terminal connection at the parent level.

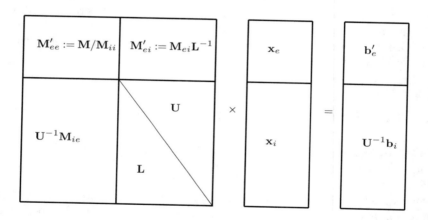

Fig. 17.2: Decomposition of the matrix **M** on a submodel involves a partial decomposition of the lower right part of the matrix $\mathbf{M}_{ii} = \mathbf{UL}$. The top-left matrix becomes the Schur complement $\mathbf{M}'_{ee} := \mathbf{M}/\mathbf{M}_{ii} = \mathbf{M}_{ee} - \mathbf{M}_{ei}\mathbf{L}^{-1}\mathbf{U}^{-1}\mathbf{M}_{ie}$. The top part of the right-hand side is $\mathbf{b}'_e = \mathbf{b}_e - \mathbf{M}_{ei}\mathbf{L}^{-1}\mathbf{U}^{-1}\mathbf{b}_i$. After the partial elimination effectively $\mathbf{M}'_{ei} = \mathbf{0}$, but for re-use options (like for solving additional systems based on the Sherman-Morrison-Woodbury formula) we store $\mathbf{M}_{ei}\mathbf{L}^{-1}$ here. Then for a new right-hand side **b** the part $\mathbf{b}'_e$ can be determined easily.

Remark

We remark that a circuit partitioner may generate a hierarchy that is different than one that is obtained from an input list description of the circuit design. For the new hierarchy the goal is to balance the sizes of the hierarchical blocks for further improving cpu performance. In the reversed order of unknowns one can exploit a Bordered Block Diagonal (BBD) linear solver together with parallelism [1].

17.3.3 Hierarchical Inner Product

The inner product of unknowns $\mathbf{x}$ and $\mathbf{y}$ can be defined in an hierarchical way:

$$\mathbf{x}^T \mathbf{y} = \sum_S \mathbf{x}_i^T \mathbf{y}_i, \text{ where } \mathbf{x}_i, \mathbf{y}_i \text{ are the internal unknowns on } S. \quad (17.26)$$

On a submodel level, a faulty conductivity connects an internal unknown $(\mathbf{x}_i)_k$ to either and internal unknown, or to a terminal of the same submodel. Thist means that the involved $\mathbf{u}$ and $\mathbf{v}$ are sparse vectors. By this and due to the hierarchical structure of the matrix $\mathbf{A}^{n+1}$, the correction due to the fault in equation (17.22) is applied to the local matrix $\mathbf{M}$ of the model were the fault is introduced. Not all matrices of the other sub-models do not need a correction. The fault only affects the matrix of the parent model through the Schur complement (Definition 17.1), which contributes to the matrix of the parent model.

17.3.4 Hierarchical Sherman-Morrison-Woodbury Formula

Definition 17.1. Suppose the matrix $\mathbf{M}$ has the following block form

$$\mathbf{M} = \begin{bmatrix} \mathbf{M}_{ee} & \mathbf{M}_{ei} \\ \mathbf{M}_{ie} & \mathbf{M}_{ii} \end{bmatrix} \quad (17.27)$$

where $\mathbf{M}_{ii}$ is a non-singular matrix. The Schur complement $\mathbf{M}/\mathbf{M}_{ii}$ of the matrix $\mathbf{M}$ with respect to $\mathbf{M}_{ii}$ is defined to be

$$\mathbf{M}/\mathbf{M}_{ii} = \mathbf{M}_{ee} - \mathbf{M}_{ei}\mathbf{M}_{ii}^{-1}\mathbf{M}_{ie}. \quad (17.28)$$

This means that the matrices of all parent models will be modified due to the updates in their Schur complements caused by the fault in a submodel.

The Schur complement is modified by a well-defined rank-one update, which is a result that serves debugging of code.

Theorem 17.1. *Let* $\mathbf{u} = (\mathbf{u}_e^T, \mathbf{u}_i^T)^T$ *and* $\mathbf{v} = (\mathbf{v}_e^T, \mathbf{v}_i^T)^T$ *be local vectors according to the partitioning of* $\mathbf{M}$. *Let the matrix*

$$\hat{\mathbf{M}} = \mathbf{M} + \mathbf{u}\mathbf{v}^T \qquad (17.29)$$

have a same partitioning as $\mathbf{M}$. *Then the Schur complement of* $\hat{\mathbf{M}}$ *with respect to* $\hat{\mathbf{M}}_{ii}$ *can be expressed as*

$$\hat{\mathbf{M}}/\hat{\mathbf{M}}_{ii} = \mathbf{M}/\mathbf{M}_{ii} + \frac{1}{1+\alpha}\hat{\mathbf{u}}_e\hat{\mathbf{v}}_e^T, \qquad (17.30)$$

where $\hat{\mathbf{u}}_e = \mathbf{u}_e - \mathbf{M}_{ei}\mathbf{M}_{ii}^{-1}\mathbf{u}_i$ *and* $\hat{\mathbf{v}}_e^T = \mathbf{v}_e^T - \mathbf{v}_i^T\mathbf{M}_{ii}^{-1}\mathbf{M}_{ie}$ *and* $\alpha = \mathbf{v}_i^T\mathbf{M}_{ii}^{-1}\mathbf{u}_i$.

Proof. The rank-1 matrix $\mathbf{u}\mathbf{v}^T$ of equation (17.29) is partitioned according to the same block form as equation (17.27). Hence

$$\mathbf{u}\mathbf{v}^T = \begin{bmatrix} \mathbf{u}_e \\ \mathbf{u}_i \end{bmatrix} \begin{bmatrix} \mathbf{v}_e^T, \mathbf{v}_i^T \end{bmatrix} = \begin{bmatrix} \mathbf{u}_e\mathbf{v}_e^T & \mathbf{u}_e\mathbf{v}_i^T \\ \mathbf{u}_i\mathbf{v}_e^T & \mathbf{u}_i\mathbf{v}_i^T \end{bmatrix}. \qquad (17.31)$$

Equation (17.29) can now be written as

$$\hat{\mathbf{M}} = \mathbf{M} + \mathbf{u}\mathbf{v}^T = \begin{bmatrix} \mathbf{M}_{ee} + \mathbf{u}_e\mathbf{v}_e^T & \mathbf{M}_{ei} + \mathbf{u}_e\mathbf{v}_i^T \\ \mathbf{M}_{ie} + \mathbf{u}_i\mathbf{v}_e^T & \mathbf{M}_{ii} + \mathbf{u}_i\mathbf{v}_i^T \end{bmatrix}. \qquad (17.32)$$

The Schur complement of matrix $\hat{\mathbf{M}}$ with respect to $\hat{\mathbf{M}}_{ii} := \mathbf{M}_{ee} + \mathbf{u}_e\mathbf{v}_e^T$ is given by the matrix

$$\hat{\mathbf{M}}/\hat{\mathbf{M}}_{ii} = \mathbf{M}_{ee} + \mathbf{u}_e\mathbf{v}_e^T - (\mathbf{M}_{ei} + \mathbf{u}_e\mathbf{v}_i^T)(\mathbf{M}_{ii} + \mathbf{u}_i\mathbf{v}_i^T)^{-1}(\mathbf{M}_{ie} + \mathbf{u}_i\mathbf{v}_e^T).$$

$$(17.33)$$

Applying the identity (17.22) to the factor $(\mathbf{M}_{ii} + \mathbf{u}_i\mathbf{v}_i^T)^{-1}$ gives

$$\hat{\mathbf{M}}/\hat{\mathbf{M}}_{ii} = \mathbf{M}_{ee} + \mathbf{u}_e\mathbf{v}_e^T - (\mathbf{M}_{ei} + \mathbf{u}_e\mathbf{v}_i^T)(\mathbf{M}_{ii}^{-1} - \frac{\hat{\mathbf{u}}_i\hat{\mathbf{v}}_i^T}{1+\alpha})(\mathbf{M}_{ie} + \mathbf{u}_i\mathbf{v}_e^T), \quad (17.34)$$

where $\hat{\mathbf{u}}_i := \mathbf{M}_{ii}^{-1}\mathbf{u}_i$, $\hat{\mathbf{v}}_i^T := \mathbf{v}_i^T\mathbf{M}_{ii}^{-1}$ and $\alpha := \mathbf{v}_i^T\mathbf{M}_{ii}^{-1}\mathbf{u}_i = \hat{\mathbf{v}}_i^T\mathbf{u}_i = \mathbf{v}_i^T\hat{\mathbf{u}}_i$. Hence

$$\hat{\mathbf{M}}/\hat{\mathbf{M}}_{ii} = \mathbf{M}_{ee} + \mathbf{u}_e\mathbf{v}_e^T -$$
$$(\mathbf{M}_{ei}\mathbf{M}_{ii}^{-1} + \mathbf{u}_e\mathbf{v}_i^T\mathbf{M}_{ii}^{-1} - \frac{\mathbf{M}_{ei}\hat{\mathbf{u}}_i\hat{\mathbf{v}}_i^T}{1+\alpha} - \frac{\mathbf{u}_e\mathbf{v}_i^T\hat{\mathbf{u}}_i\hat{\mathbf{v}}_i^T}{1+\alpha})(\mathbf{M}_{ie} + \mathbf{u}_i\mathbf{v}_e^T)$$
$$= \mathbf{M}_{ee} + \mathbf{u}_e\mathbf{v}_e^T -$$
$$(\mathbf{M}_{ei}\mathbf{M}_{ii}^{-1} + \mathbf{u}_e\hat{\mathbf{v}}_i^T - \frac{\mathbf{M}_{ei}\hat{\mathbf{u}}_i\hat{\mathbf{v}}_i^T}{1+\alpha} - \frac{\alpha\mathbf{u}_e\hat{\mathbf{v}}_i^T}{1+\alpha})(\mathbf{M}_{ie} + \mathbf{u}_i\mathbf{v}_e^T).$$

By this

$$\hat{\mathbf{M}}/\hat{\mathbf{M}}_{ii} = \mathbf{M}/\mathbf{M}_{ii} + \mathbf{u}_e \mathbf{v}_e^T$$

$$-\mathbf{u}_e \hat{\mathbf{v}}_i^T \mathbf{M}_{ie} + \frac{\mathbf{M}_{ei} \hat{\mathbf{u}}_i \hat{\mathbf{v}}_i^T \mathbf{M}_{ie}}{1+\alpha} + \frac{\alpha \mathbf{u}_e \hat{\mathbf{v}}_i^T \mathbf{M}_{ie}}{1+\alpha}$$

$$-\mathbf{M}_{ei} \hat{\mathbf{u}}_i \mathbf{v}_e^T - \mathbf{u}_e \hat{\mathbf{v}}_i^T \mathbf{u}_i \mathbf{v}_e^T + \frac{\mathbf{M}_{ei} \hat{\mathbf{u}}_i \hat{\mathbf{v}}_i^T \mathbf{u}_i \mathbf{v}_e^T}{1+\alpha} + \frac{\alpha \mathbf{u}_e \hat{\mathbf{v}}_i^T \mathbf{u}_i \mathbf{v}_e^T}{1+\alpha}$$

$$= \mathbf{M}/\mathbf{M}_{ii} + \mathbf{u}_e \mathbf{v}_e^T$$

$$-\mathbf{u}_e \hat{\mathbf{v}}_i^T \mathbf{M}_{ie} + \frac{\mathbf{M}_{ei} \hat{\mathbf{u}}_i \hat{\mathbf{v}}_i^T \mathbf{M}_{ie}}{1+\alpha} + \frac{\alpha \mathbf{u}_e \hat{\mathbf{v}}_i^T \mathbf{M}_{ie}}{1+\alpha}$$

$$-\mathbf{M}_{ei} \hat{\mathbf{u}}_i \mathbf{v}_e^T - \alpha \mathbf{u}_e \mathbf{v}_e^T + \frac{\alpha \mathbf{M}_{ei} \hat{\mathbf{u}}_i \mathbf{v}_e^T}{1+\alpha} + \frac{\alpha^2 \mathbf{u}_e \mathbf{v}_e^T}{1+\alpha}$$

$$= \mathbf{M}/\mathbf{M}_{ii} + \mathbf{u}_e \mathbf{v}_e^T \left(1 - \alpha + \frac{\alpha^2}{1+\alpha}\right)$$

$$+\mathbf{u}_e \hat{\mathbf{v}}_i^T \mathbf{M}_{ie}\left(-1 + \frac{\alpha}{1+\alpha}\right) + \frac{\mathbf{M}_{ei} \hat{\mathbf{u}}_i \hat{\mathbf{v}}_i^T \mathbf{M}_{ie}}{1+\alpha} + \mathbf{M}_{ei} \hat{\mathbf{u}}_i \mathbf{v}_e^T\left(-1 + \frac{\alpha}{1+\alpha}\right)$$

$$= \mathbf{M}/\mathbf{M}_{ii} + \frac{\mathbf{u}_e \mathbf{v}_e^T}{1+\alpha} - \frac{\mathbf{u}_e \hat{\mathbf{v}}_i^T \mathbf{M}_{ie}}{1+\alpha} + \frac{\mathbf{M}_{ei} \hat{\mathbf{u}}_i \hat{\mathbf{v}}_i^T \mathbf{M}_{ie}}{1+\alpha} - \frac{\mathbf{M}_{ei} \hat{\mathbf{u}}_i \mathbf{v}_e^T}{1+\alpha}$$

$$= \mathbf{M}/\mathbf{M}_{ii} + \frac{(\mathbf{u}_e - \mathbf{M}_{ei} \hat{\mathbf{u}}_i)(\mathbf{v}_e^T - \hat{\mathbf{v}}_i^T \mathbf{M}_{ie})}{1+\alpha}, \qquad (17.35)$$

from which (17.30) follows. □

The above theorem shows that each submodel with a fault generates a factor $1/(1+\alpha)$ in which $\alpha = \mathbf{v}_i^T \mathbf{M}_{ii}^{-1} \mathbf{u}_i$. If $\mathbf{v}^T = p\mathbf{u}^T$ we will have $\mathbf{v}_i^T = p\mathbf{u}_i^T$ and thus, when $\mathbf{M}_{ii}$ is positive-definite, $\alpha \geq 0$. When additionally $\mathbf{M}_{ei} = \mathbf{M}_{ie}^T$ one has $\hat{\mathbf{v}}_e^T = p\hat{\mathbf{u}}_e^T$ that is passed to the parent model level.

17.4 Simulation Results

In [3] it was demonstrated that exploiting the golden solution at each time time step as prediction for the Newton iterations to solve the faulty solutions at the next time point greatly reduced the number of Newton iterations for the faulty ones. Additional reduction of the percentage of cpu showed the benefit of bypassing hierarchical branches when terminal values did not change that much.

We used CAN (Controller Area Network) Tranceivers for testing performance and obtained speed-ups from 3.2 to 400 (when compared to a conventional, sequential run of all faults). For CDACs (Control Digital to Analog Converter, an IP block in several products of NXP Semiconductors) the speed-ups were between 3.9 and 11.

Further increase of speed-up factors was due to exploiting prediction by implicit sensitivity and was demonstrated in [17, 18]: 58 for the CDAC in case of just using the prediction and still more than 10 in case of up to 5 nonlinear iterations. For a LIN (Local Interconnect Network) Converter IP Block the corresponding speed-ups were 219 (just prediction) and 43 (up to 5 nonlinear iterations), respectively. This clearly emphasizes the benefits in predicting the initial solutions by implicit sensitivity (17.13), exploiting the Sherman-Morrison-Woodbury formula.

Fig. 17.3 overviews speed-up obtained for the TJA2021 chip [14, 19]. The cpu time needed for a conventional run is simply estimated as the number of faults times the cpu time for the golden solution. No parallelism was used for the faulty runs inside Fast Fault Simulation (FFS). The detected faults covered 90% of the total set, being is a good coverage result.

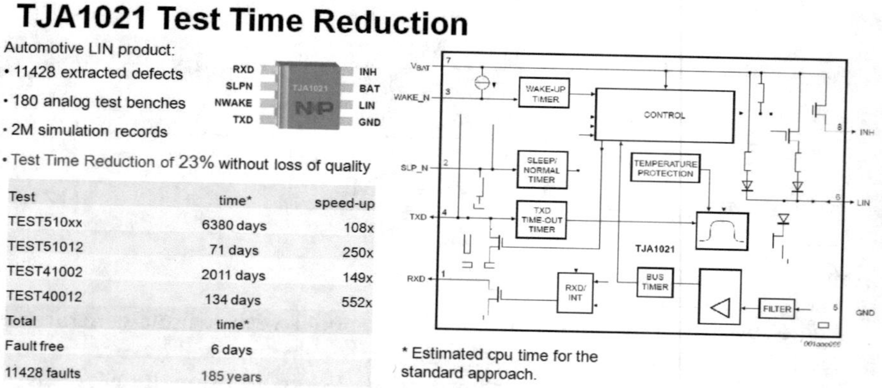

TJA1021 Test Time Reduction

Automotive LIN product:

- 11428 extracted defects
- 180 analog test benches
- 2M simulation records
- Test Time Reduction of 23% without loss of quality

Test	time*	speed-up
TEST510xx	6380 days	108x
TEST51012	71 days	250x
TEST41002	2011 days	149x
TEST40012	134 days	552x
Total	time*	
Fault free	6 days	
11428 faults	185 years	

* Estimated cpu time for the standard approach.

Fig. 17.3: Speed-up in fast fault simulation for the TJA1021 chip of NXP Semiconductors, see also [14, 19]. The time⋆ for the convential sequential simulations of the faulty solutions was estimated by the number of faults times the cpu time needed for the golden solution.

At the end of the time integration in Alg. 17.1 the remaining list of faults $\mathcal{P}$ mostly consists of non-detectable faults ("the untouchables") and during that integration phase they are responsible for contributing most to the cpu time effort that is spent here. In [19] a novel approach was described: only during a time window before a measurement time point $\theta_m \in \mathcal{M}$ an approximation for each faulty solution was determined, by adding the rank-one update to the "golden" system and starting up from the golden solution. Note that the rank-one update can be invoked smoothly by increasing p with time (thus effectively introducing a time-dependent conductance, similar as done for sources in the Source Stepping by Pseudo Transient (SSPT) approach in [18]). This time integration for the faulty system is stopped after passing the time point θ_m. The process of invoking a fault is restarted during a time window before the next time point for measurement. See Fig 17.4.

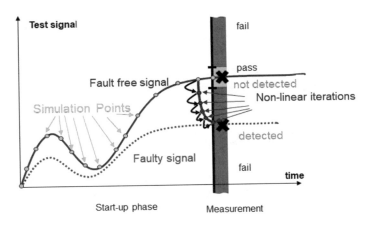

Fig. 17.4: [19] it was proposed to invoke a fault shortly before a time point θ_m for measurement and start from the golden (fault-free) solution, perform a prediction based on implicit sensitivity and determine nonlinear corrections to approximate the corresponding faulty solution. When the fault is detected at θ_m the fault is removed from the fault list. When the fault is not detected the list remains unchanged. After passing θ_m we continue with the time integration of the golden solution until we reach a next time point θ_m for measurement.

This approach in waking up faults gave an additional order of magnitude in speed-up. By this reduction of simulation time candidate faults could be detected that would have been impossible otherwise because of excessive CPU time [19]. Also the really non-detectable faults can be identified before starting a more accurate FFS process as described in Alg. 17.1 in which these faults are yet excluded from $\mathcal{P}$.

17.5 Sensitivity and UQ

We observed that predictions for DC did result in higher speed-ups for FFS in DC than during Transient Simulation. Indeed the prediction based on (17.21) is more accurate than the one using implicit sensitivity (17.13) during time integration. In [19] this benefit was demonstrated for ordinary Monte Carlo simulations in DC for a CAN Tranceiver. This indicates potential for explicitly exploiting sensitivity by (17.14) during time integration. We noticed before that the sensitivity $\hat{\mathbf{x}}$ of the golden solution with respect to p depends on the choice of the vectors $\mathbf{u}$, $\mathbf{v}$ and has to be determined by additional time integration.

Alternatively, sensitivity can be provided by Uncertainty Quantification (UQ). When only sensitivity is needed for one value of p then a transient simulation of $\hat{\mathbf{x}}$ at p is the cheapest approach. When sensitivity is required at more different values of p then alternative procedures may be beneficial such as by UQ [13, 21] and which is detailed in the next sections.

17.5.1 UQ for Fast Fault Simulation

For a given pair $\mathbf{u}$, $\mathbf{v}$ we can exploit UQ for determining sensitivity with respect to p Assume that the parameter values p actually are random according to some probability density function (say an exponential one, like the Laguerre distribution). Then we can expand the p-parameterized ensemble of solutions $\mathbf{x}(t,p) = \sum_k \mathbf{z}_k(t)\phi_k(p)$ in a Polynomial-Chaos Expansion (PCE) [13,21], in which the $\phi_k(\cdot)$ are (scalar) polynomials that are evaluated at p. They are orthonormal with respect to the L_2-inner-product $< ., . >_p$, provided by the probability density function (integration over p). The vector coefficients $\mathbf{z}_k(t)$ depend on $\mathbf{u}$ and $\mathbf{v}$ and satisfy

$$\mathbf{z}_k(t) \equiv \mathbf{z}_k(t \mid \mathbf{u}, \mathbf{v}) = < \mathbf{x}(t,p), \phi_k(p) >_p .$$

Stochastic Collocation (SC) [13, 21] approximates the $\mathbf{z}_k(t)$ by a quadrature

$$\mathbf{z}_k(t) = < \mathbf{x}(t,p), \phi_k(p) >_p \approx \sum_{q=1}^{Q} w_q \mathbf{x}(t,p_q)\,\phi_k(p_q), \qquad (17.36)$$

using (deterministic) quadrature points p_q and weights w_q. The calculation of the $\mathbf{x}(t,p_q)$ does not need modifications of a time integration procedure. Hence, SC is said to be non-intrusive.

Now the golden solution $\mathbf{x}(t,0)$ and the faulty solutions $\mathbf{x}(t,p)$ can be easily compared for any value of p at any moment of measurement in time. We also notice that the sensitivity with respect to p can easily be obtained from (17.36).

We remark that the PCE expansion has to be changed when the pair $\mathbf{u}$, $\mathbf{v}$ changes. However, for each situation, one can obtain the deterministic solutions $\mathbf{x}(t,p_q)$ using the non-linear solver described in Section 17.3.1.

17.5.2 UQ for the Time Integration of a p-Parameterized Ensemble

The main drawback of SC is that it returns the benefits of the series expansion only after the complete time-integration has been finished for all deterministic parameter values p_q. We now consider the joint time integration of the p-ensemble of solutions at values p_q, similar as we have adopted for Fast Fault Simulation. By integrating the ensembleintegration of ensemble one can

- at each new time level predict the solution for a new parameter value p_q from some solution at a parameter value p'_q calculated before ("re-use").
- exploit sensitivity since, actually, it is immediately available at the previous time level from the series expansion that was valid until that moment (however this opportunity is not used in the default SC implementation).

To be a bit more precise, integrating the ensemble means

$$\forall n \in \{0,\ldots,N\} \ \forall q \ \text{Solve} \ \mathbf{x}(t_{n+1},p_q) = \text{Integrate}(t_n,t_{n+1},\mathbf{x}(t_n,p_q)).$$
$$(17.37)$$

Here, parallelism could be exploited, but in a sequential approach the main benefits of (17.37) are that at t_{n+1}, for p_q, one can immediately use results from neighbouring $p_{q'}$ with $q' < q$, in addition to sensitivity information from the previous time level. For example, one can predict $\mathbf{x}(t_{n+1},p_q)$ from $\mathbf{x}(t_{n+1},p_0)$, for some selected p_0, as follows

$$\mathbf{x}(t_{n+1},p_q) = \mathbf{x}(t_{n+1},p_0) + \frac{\partial \mathbf{x}}{\partial p}(t_{l+1},p_0)(p_q-p_0) + \mathcal{O}(|p_q-p_0|^2),$$

$$= \mathbf{x}(t_{n+1},p_0) + \frac{\partial \mathbf{x}}{\partial p}(t_n,p_0)\,(p_q-p_0) + \mathbf{R}_1, \quad \text{where} \quad (17.38)$$

$$\frac{\partial \mathbf{x}}{\partial p}(t_n,p_0) \approx \sum_{k=0}^{m} \mathbf{z}_k(t_n)\,\frac{\partial \phi_k}{\partial p}(p_0)$$

$$= \sum_{k=0}^{m}[\sum_{q=1}^{Q} w_q\,\mathbf{x}(t_n,p_q)\,\phi_i(p_q)]\,\frac{\partial \phi_k}{\partial p}(p_0) + \mathbf{R}_2,$$

$$\mathbf{R}_1 = \mathcal{O}(|p_q-p_0|\,\Delta t) + \mathcal{O}(|p_q-p_0|^2),$$

$$\mathbf{R}_2 = \text{Quadrature error}.$$

We remark that $\mathbf{R}_2$ is not affected by the difference $p_q - p_0$. If p_q is close to p_0 we can neglect $\mathbf{R}_1$ and still have a prediction that will reduce Newton-Raphson iterations compared to when starting from $\mathbf{x}(t_n,p_q)$ (we simply assumed Euler Backward time integration here).

Notice that $\mathbf{R}_2$ is the quadrature error in the sensitivity expression. Here one has some freedom to use a different quadrature formula than used for $\mathbf{x}(t_{n+1},p_q)$. However, for efficiency reasons it would be attractive if the quadrature points are a subset of those used for $\mathbf{x}(t_{n+1}, p_q)$. Using a subset also reduces the need for storage.

17.5.2.1 Stochastic Collocation with Coarse and Fine Quadrature

In [19] an alternative approach is described that maintains the non-intrusive character as long as possible by using two SC sweeps: a first sweep is applied that uses a coarse quadratue rule. This is a traditional non-intrusive implementation. It is just intended to have sensitivity available for the second sweep in which a higher accurate quadrature formula is used. Now, during the second time integration one has sensitivity results available from the expansion resulting as postprocessing after the first sweep and which is provided by just a call to a procedure in the UQ-Library Dakota [2]. This minimizes the intrusive implementation and makes sensitivity available where it can be exploited.

17.6 Conclusion

We outlined details of a successful algorithm for Fast Fault Simulationfast fault simulation (FFS). The speed-ups were obtained by a combination of different techniques (hierarchical simulation, bypassing, on-the-fly-reduction of the list of faults, prediction of neighbouring problems by implicit sensitivity) together with elegant enhancement of the hierarchical modeling using extra ports. Up to our knowledge, NXP Semiconductors' algorithm for Fast Fault Simulation is the best in the world in this area [18,19]. NXP Semiconductors can identify locations on a chip that are probably affected by very tiny manufacturing inaccuracies and may cause faulty behaviour in time. This can then be detected at predefined time points for measurements.

References

1. BENK, J., DENK, G., KOWITZ, C., AND WALDHERR, K.: *A holistic approach for fast and accurate transient simulations of analog circuits.* Journal Mathematics in Industry 7:12, 2017. Open access: https://mathematicsinindustry.springeropen.com/articles/10.1186/s13362-017-0042-z.
2. DAKOTA 6.6: *Algorithms for design exploration and simulation*, https://dakota.sandia.gov/, Sandia National Laboratories, Albuquerque, NM, USA, 2017.
3. DE JONGHE, D., MARICAU, E., GIELEN, G., McCONAGHY, T., TASIĆ, B., AND STRATIGOPOULOS, H.: *Advances in variation-aware modeling, verification, and testing of analog ICs.* Proceedings of Design, Automation and Test in Europe (DATE) 2012, pp. 1615–1620, 2012.

4. FIJNVANDRAAT, J.G., HOUBEN, S.H.M.J., TER MATEN, E.J.W., AND PETERS, J.M.F.: *Time domain analog circuit simulation*, J. Comput. Appl. Math., Vol. 185, No. 2, pp. 441–459, 2006.

5. GOLUB, G.H., AND VAN LOAN, C.F.: *Matrix Computations, Third edition*. The Johns Hopkins University Press, Baltimore, MD, 1996.

6. GÜNTHER, M., FELDMANN, U., AND TER MATEN, J.: *Modelling and discretization of circuit problems*. In: SCHILDERS, W.H.A. AND TER MATEN, E.J.W. (EDS): Handbook of Numerical Analysis, Vol. XIII, Special Volume on Numerical Methods in Electromagnetics, Elsevier BV, North-Holland, Amsterdam, pp. 523–659, 2005.

7. HAPKE, F., REDEMUND, W., SCHLOEFFEL, J., KRENZ-BAATH, R., GLOWATZ, A., WITTKE, M., HASHEMPOUR, H., AND EICHENBERGER, S.: *Defect-oriented cell-internal testing*, Proc. 2010 IEEE Int. Test Conference (ITC 2010), Austin, TX, paper 10.1 (10 pages), 2010.

8. HASHEMPOUR, H., DOHMEN, J., TASIĆ, B., KRUSEMAN, B., HORA, C., VAN BEURDEN, M., AND XING, Y.: *Test time reduction in analogue/mixed-signal devices by defect oriented testing: an industrial example*, Proceedings of Design, Automation, and Test in Europe (DATE) 2011, Grenoble, doi: 10.1109/DATE.2011.5763065 (6 pages), 2011.

9. HOU, J., AND CHATTERJEE, A. *Concurrent transient fault simulation for analog circuits*, IEEE Trans. on Comp.-Aided Design of Integr. Circuits and Systems (TCAD), Vol. 22, No. 10, pp. 1385–1398, 2003.

10. HOUSEHOLDER, A.S. *A survey of some closed methods for inverting matrices*, SIAM J. Appl. Math, Vol. 5, No. 3, pp. 155–169, 1957.

11. ILIEVSKI, Z., XU, H., VERHOEVEN, A., TER MATEN, E.J.W., SCHILDERS, W.H.A., AND MATTHEIJ, R.M.M.: *Adjoint transient sensitivity analysis in circuit simulation*. In: CIUPRINA, G. AND IOAN D. (EDS): Scientific Computing in Electrical Engineering SCEE 2006, Series Mathematics in Industry 11, Springer, pp. 183–189, 2007.

12. KRUSEMAN, B., TASIĆ, B., HORA, C., DOHMEN, J., HASHEMPOUR, H., VAN BEURDEN, M., AND, XING, Y.: *Defect oriented testing for analog/mixed-signal devices*, Proc. 2011 IEEE Int. Test Conference (ITC 2011), Anaheim, CA, paper 1.1 (8 pages), 2011.

13. LE MAÎTRE, O.P., AND KNIO, O.M.: *Spectral methods for uncertainty quantification, with applications to computational fluid dynamics*, Springer, Science+Business Media B.V., Dordrecht, 2010.

14. TER MATEN, E.J.W., PUTEK, P., GÜNTHER, M., PULCH, R., TISCHENDORF, C., STROHM, C., SCHOENMAKER, W., MEURIS, P., DE SMEDT, B., BENNER, P., FENG, L., BANAGAAYA, N., YUE, Y., JANSSEN, R., DOHMEN, J.J., TASIĆ, B., DELEU, F., GILLON, R., WIEERS, A., BRACHTENDORF, H.-G., BITTNER, K., KRATOCHVÍL, T., PETŘZELA, J., ŠOTNER, R., GÖTTHANS, T., DŘÍNOVSKÝ, J., SCHÖPS, S., DUQUE GUERRA, D.J., CASPER, T., DE GERSEM, H., RÖMER, U., REYNIER, P., BARROUL, P., MASLIAH, D., AND ROUSSEAU, B.: *Nanoelectronic COupled Problems Solutions – nanoCOPS: Modelling, Multirate, Model Order Reduction, Uncertainty Quantification, Fast Fault Simulation*. Journal Mathematics in Industry 7:2, 2016. Open access: http://dx.doi.org/10.1186/s13362-016-0025-5.

15. PETZOLD, L., LI, Y., CAO, S.T., AND SERBAN, R.: *Sensitivity analysis of differential-algebraic equations and partial differential equations*, Computers and Chemical Engineering, Vol. 30, Nr. 10-12, pp. 1553–1559, 2006.

16. SHI, C.-J.R., TIAN, M.W., AND SHI, G.: *Efficient DC fault simulation of nonlinear analog circuits: one-step relaxation and adaptive simulation continuation*, IEEE Trans. on Comp.-Aided Design of Integr. Circuits and Systems (TCAD), Vol. 25, No. 27, pp. 1392–1400, 2006.

17. TASIĆ, B., DOHMEN, J.J., TER MATEN, E.J.W., BEELEN, T.G.J., JANSSEN, H.H.J.M, SCHILDERS, W.H.A., AND GÜNTHER, M.: *Fast Fault Simulation to identify subcircuits involving faulty components*. In: G. RUSSO, V. CAPASSO, G. NICOSIA,

V. ROMANO (EDS.): *Progress in Industrial Mathematics at ECMI 2014*, Mathematics in Industry, 22, Springer, pp. 369–376, 2017.

18. TASIĆ, B., DOHMEN, J.J., TER MATEN, E.J.W., BEELEN, T.G.J, SCHILDERS, W.H.A., DE VRIES, A., AND VAN BEURDEN, M.: *Robust DC and efficient time-domain fast fault simulation*. COMPEL, Vol. 33, Nr. 4, pp. 1161–1174, 2014.

19. TASIĆ, B., DOHMEN, J.J., JANSSEN, R., TER MATEN, E.J.W., BEELEN, T.G.J., AND PULCH, R.: *Fast Time-Domain Simulation for Reliable Fault Detection*. Proceedings of Design, Automation and Test in Europe (DATE) 2016, Paper 0994, pp. 301–306, 2016.

20. XING, Y. (1998): *Defect-oriented testing of mixed-signal ICs: some industrial experience*, Proc. IEEE Int. Test Conference ITC'98, Washington, DC, pp. 678–687, 1998.

21. XIU, D.: *Numerical methods for stochastic computations - A spectral method approach*, Princeton Univ. Press, Princeton, NJ, USA, 2010.

Chapter 18
Calibration of Probability Density Function

Jos J. Dohmen, Theo G.J. Beelen, Oryna Dvortsova, E. Jan W. ter Maten, Bratislav Tasić, Rick Janssen

Abstract The capability performance index (C_{pk}) is often used to measure the capability of the production process and to predict yield. However, this C_{pk} is only defined for the Gaussian distribution. At NXP Semiconductors an on-chip calibration technique is frequently used to reduce the effect of process variations. The resulting distribution has a much flatter peak than a Gaussian density and consequently the C_{pk} is significantly underestimated. In this chapter we propose two possible approaches to address accurate C_{pk} calculation for non-normal distributions. One approach is to use the so-called Generalized Gaussian distribution function and to estimate its defining parameters. We propose a numerical fast and reliable method for computing these parameters and a simple formula to calculate the C_{pk} value from these defining parameters. Another approach is to transform data as a way to deal with non-normal distributions. We show that both approaches significantly outperform the standard C_{pk} calculation for the non-normal distributions of interest.

Jos J. Dohmen, Bratislav Tasić, Rick Janssen
NXP Semiconductors, Eindhoven, the Netherlands,
e-mail: {Jos.J.Dohmen,Bratislav.Tasic,Rick.Janssen}@nxp.com

Theo G.J. Beelen, Oryna Dvortsova, E. Jan W. ter Maten
Eindhoven University of Technology, the Netherlands,
e-mail: {T.G.J.Beelen,E.J.W.ter.Maten}@tue.nl
e-mail: {Th.Beelen,Oryna.Dvortsova}@gmail.com

E. Jan W. ter Maten
Bergische Universität Wuppertal, Germany,
e-mail: Jan.ter.Maten@math.uni-wuppertal.de

© Springer Nature Switzerland AG 2019
E. J. W. ter Maten et al. (eds.), *Nanoelectronic Coupled Problems Solutions*,
Mathematics in Industry 29, https://doi.org/10.1007/978-3-030-30726-4_18

18.1 Introduction

Many applications in the semiconductor industry typically have challenging yield targets. The yield prediction of the process is based on the customer requirements, specifications, tolerances and the model of the process variation. As a measure for the capability of the process to produce output within the specification limits, the capability performance index (C_{pk}) is used at NXP Semiconductors (NXP). This C_{pk} is only defined for the Gaussian distribution, and no suitable generalization of the C_{pk} exists for other distributions. As a consequence, when the probability distribution is non-normal and a C_{pk} index is calculated using conventional methods then this leads to erroneous interpretations of the process capability.

In circuit design one aims to increase robustness and yield [11]. Specially added electronic control is applied to obtain narrow tails in empirical probability density functions and thus tiny tail probability. This process is called by engineers (electronic) *'trimming'*. It has no relation to statistical techniques like *Winsoring* (when outliers are clipped to a boundary percentile), or *Trimming* (when outliers are simply neglected). Here it is an an automated calibration process where for example a variable resistor is tuned to obtain the desired output values. To avoid confusion, henceforth we will call this trimming process "*calibration*". At NXP an on-chip calibration technique is frequently used to reduce the effect of process variations. The resulting distribution has a much flatter peak than a Gaussian density and consequently, the C_{pk} is significantly underestimated by the conventional methods.

An example is shown in Fig. 18.1, where a histogram of the output voltage of a low-dropout regulator is plotted before and after the on-chip calibration technique was applied.

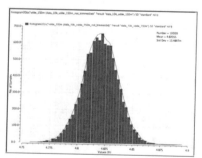 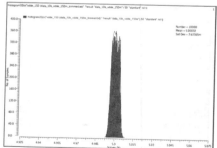

Fig. 18.1: **Left:** Simulated voltage of a low-dropout regulator before calibration (mean $\mu = 4.82V$ and standard deviation $\sigma = 15.5mV$).
Right: Simulated voltage of a low-dropout regulator after calibration ($\mu = 5.00V$ and $\sigma = 2.63mV$).

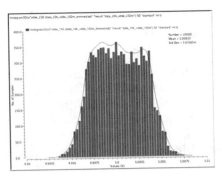

Fig. 18.2: Zoom-in of the simulated voltage of a low-dropout regulator after calibration (resulting in $\mu = 5.00V$ and $\sigma = 2.63mV$) [2].

After chip fabrication, this calibration is used to adjust the operating parameters of the circuit to a certain target value. In the example for a low-dropout regulator of Fig. 18.1 the target of the output voltage value is 5.0V. We see that the calibration has shifted the mean value from 4.82V to the target of 5.0V. At the same time the standard deviation is reduced from 15.5mV to 2.63mV. Fig. 18.2 shows a zoom-in of the figure at the right in Fig. 18.1.

In practice, electronic calibration is often realized with a small digital circuit consisting of a small nonvolatile memory, coupled with an array of FET switches (involving Field Effect Transistors), resistors and capacitors acting as the equivalent of a trim pot or trim capacitor. During the wafer test or final test a digital calibration code (for example a 12 bit number) is written to the internal programmable memory which then leads to the adjustment or correction of the output voltage. Fig. 18.2 clearly shows the shape of the distribution after calibration. As expected, the resulting distribution is roughly symmetric, but the peak is much flatter than a Gaussian density. The step size of the digital calibration codes determines the width of the resulting distribution. The ideal calibration process will change all output values into the exact target value and the distribution into a delta function, but due to the steps in the calibration codes we will have a finite region of non-zero density and zero density elsewhere. However, we do not see this in Fig. 18.2 and instead the distribution clearly has a long tail. There are several reasons why a tail exists after calibration:

- During calibration the output voltage is measured and the optimal calibration code is determined. Random measurement errors introduce spread in the output voltage and this results in a tail.
- The digital calibration code has a minimum and maximum value. Due to this limitation the original tail cannot completely be shifted to the target value.

- The calibration circuit itself is influenced by process variation.
- Since the calibration is done under certain conditions (e.g. temperature conditions) the tail can still exist for other conditions.

Assume that at some measurement point a circuit has a DC solution $V(C, R, p)$, that depends on environmental conditions C (e.g. ambient temperature, supply voltage, external load, etc.), a resistor R and an uncertain parameter p. The resistor can have only discrete values $R = R_0 + R_1 n$, where the conversion of the calibration codes a resistance R that is linear with offset R_0 and with a factor R_1 that determines the step size of additional resistance values with multiples n. The calibration code n is a non-negative integer with $n \in [0, 2^m - 1]$ where m is the number of bits used in the digital calibration circuit.

The circuit design aims to satisfy a performance criterion $V_{\text{Low}} \leq V \leq V_{\text{Up}}$. Now for certain conditions C and for each p we may determine how V depends on R. An optimal $R(C, p)$ assures that $V(R(C, p), p) = V_{\text{Ref}} \in [V_{\text{Low}}, V_{\text{Up}}]$, but there is no guarantee that this can be achieved for all C and p. If $R(C, p)$ exists, $R(C, p)$ can be determined by some nonlinear solution technique, involving solving the circuit equations several times. More general, we determine $R(C, p)$ such that $|V(C, R(C, p), p) - v|$ for $v \in [V_{\text{Low}}, V_{\text{Up}}]$ is minimum. The effect is that a probability density function (pdf) becomes more concentrated around a mean value and tails become more narrow.

In literature, various methods have been proposed to calculate a C_{pk} like index for non-normal distributions. However, these methods mainly focus on skewed distributions and no suitable methods are offered for symmetric non-normal distributions under severe departures from normality. In the following sections, we propose two possible approaches to address accurate C_{pk} calculation for symmetric distributions. The first approach (in section 18.2) is to use the so-called Generalized Gaussian Distribution function and to estimate its defining parameters. We propose a numerical fast and reliable method for computing these parameters. The second approach in section (18.3) is to transform data as a way to deal with non-normal distributions. We compare the performance of both methods and with the conventional C_{pk} calculation and we discuss the limitations of the approaches. The comparison is carried out by sampling data from a known non-normal distribution and also by simulating a circuit including the calibration process. We show that the performance is dependent on the ability to capture tail behavior of the underlying distribution.

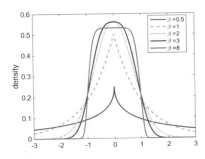
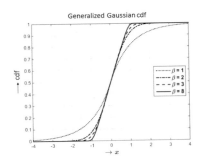

Fig. 18.3: **Left**: Generalized Gaussian probability density function $f(x)$ as in (18.1) for varying β and with $\mu = 0$ and $\alpha = 1$ [2].
Right: Generalized Gaussian cumulative distribution function $F(x)$ as in (18.4) [2].

18.2 Fitting Generalized Gaussian Distributions for Process Capability Index

We assume N independent samples x_i, obtained from an electronic calibration process in some given interval $[A, B]$, and based on some empirical density function [10]. To define a quality measure index (in Section 18.2.2) we are first interested in the 'best' fitting function within the family of Generalized Gaussian Density (GGD) distributions, as shown in Fig.18.3 and given by the expression [2]

$$f(x) = \frac{\beta}{2\alpha\,\Gamma(1/\beta)} \exp\left(-\left(\frac{|x-\mu|}{\alpha}\right)^{\beta}\right), \tag{18.1}$$

where $\alpha, \beta > 0$, $\mu \in \mathbb{R}$ and $\Gamma(z) = \int_0^\infty t^{z-1} e^{-t} dt$, for $z > 0$, is the Gamma function [10, 15]. The mean and the variance of the GGD (18.1) are given by μ and $\alpha^2 \Gamma(3/\beta)/\Gamma(1/\beta)$, respectively. Hence, after expressing

$$\alpha = \sigma\sqrt{\Gamma(1/\beta)/\Gamma(3/\beta)}, \tag{18.2}$$

we get that, for all β, the variance is σ^2. In the case of a normal distribution (where $\beta = 2$) we have that $\alpha = \sigma\sqrt{2}$ since $\Gamma(\frac{1}{2}) = \sqrt{\pi}$ and $\Gamma(\frac{3}{2}) = \frac{1}{2}\sqrt{\pi}$. The cumulative distribution function (cdf) $F(x)$ corresponding to the GGD (18.1) is given by

$$F(x) = \int_{-\infty}^{x} f(y)\, dy$$

$$= \frac{\beta}{2\alpha\Gamma(1/\beta)} \int_{-\infty}^{x} \exp\left(-\left(\frac{|y-\mu|}{\alpha}\right)^{\beta}\right) dy \qquad (18.3)$$

The substitution $z = \left(\frac{|y-\mu|}{\alpha}\right)^{\beta}$ yields $y = \mu + sgn(y-\mu)\,\alpha z^{\frac{1}{\beta}}$ and the corresponding $dy = sgn(y-\mu)\frac{\alpha}{\beta} z^{(1/\beta)-1}\, dz$. Using this substitution for dy in (18.3) we get

$$F(x) = \frac{1}{2\Gamma(1/\beta)} \left(-\int_{\infty}^{0} + sgn(x-\mu)\int_{0}^{(|\mu-x|/\alpha)^{\beta}}\right) z^{(1/\beta)-1} \exp(-z)\, dz$$

$$= \frac{1}{2} + sgn(x-\mu)\frac{1}{2\Gamma(1/\beta)}\int_{0}^{(|\mu-x|/\alpha)^{\beta}} z^{(1/\beta)-1}\exp(-z)\, dz.$$

$$(18.4)$$

where

$$sgn(x) = \begin{cases} 1 & x \geq 0; \\ -1 & x < 0. \end{cases} \qquad (18.5)$$

By using the Lower Incomplete Gamma function (of the 1st kind), defined by

$$\gamma(a, x) = \int_{0}^{x} t^{a-1} e^{-t} dt, \qquad (18.6)$$

we can rewrite (18.4) as

$$F(x) = \frac{1}{2} + sgn(x-\mu)\frac{\gamma\left(1/\beta, \left(\frac{|\mu-x|}{\alpha}\right)^{\beta}\right)}{2\Gamma(1/\beta)}. \qquad (18.7)$$

Graphical impressions of $f(x)$ and $F(x)$ are given in Fig. 18.3. For the Lower Incomplete Gamma function [14,15] there is standard software available. The parameter β determines the shape. For $\beta = 2$ one has the Gaussian distribution and for $\beta = 1$ the GGD corresponds to a Laplacian distribution; for $\beta \to +\infty$ the probability density function (pdf) in (18.1) converges to a uniform distribution in $(\mu - \sqrt{3}\sigma, \mu + \sqrt{3}\sigma)$, and when $\beta \downarrow 0$ we get a degenerate distribution in $x = \mu$ (but with a finite variance) [2]. In our applications we are facing broad pdfs with relatively long tails and where the center-part of the distribution has steep slopes. We want to estimate the tails accurately, so we are interested in the cases when $\beta \geq 2$.

Despite the fact that several distributions of output results will not be symmetrical, we restrict ourselves here to the family of GGD (18.1).

The parameters of the 'best' fitting distribution function can be found by maximizing the logarithm of the likelihood function $L = \ln(\mathcal{L}) = \sum_{i=1}^{N} f(x_i)$.

The necessary conditions are $[1,2]$

$$\frac{\partial L}{\partial \alpha} = 0 : \alpha = \left(\frac{\beta}{N} \sum_{i=1}^{N} |x_i - \mu|^\beta \right)^{1/\beta}, \tag{18.8}$$

$$\frac{\partial L}{\partial \beta} = 0 : \frac{1}{\beta} + \frac{\Psi(1/\beta)}{\beta^2} - \frac{1}{N} \sum_{i=1}^{N} \left| \frac{x_i - \mu}{\alpha} \right|^\beta \ln \left| \frac{x_i - \mu}{\alpha} \right| = 0, \tag{18.9}$$

$$\frac{\partial L}{\partial \mu} = 0 : \sum_{x_i \geq \mu} |x_i - \mu|^{\beta-1} - \sum_{x_i < \mu} |x_i - \mu|^{\beta-1} = 0. \tag{18.10}$$

Here Ψ is the Digamma function $\Psi(x) = \frac{d}{dx} \ln(\Gamma(x)) = \Gamma'(x)/\Gamma(x)$, see [15]. Several papers $[3,5,7]$ consider estimates for α and β to solve the equations (18.8)–(18.9), but they assume that the sample size is large enough and/or that $\beta \leq 3$, motivated by the various application areas. We note that $[8,9]$ also consider the case for a small sample size.

We exploit the explicit elimination of α in (18.9) after which only two additional equation remain

$$g(\beta,\mu) = 0, \tag{18.11}$$
$$h(\beta,\mu) = 0, \tag{18.12}$$

in which β and μ are unknown and α follows from equation (18.8). The analytical formulae for $g(\beta,\mu)$ and $\frac{\partial g(\beta,\mu)}{\partial \beta}$ are given by (see also $[1,2,7,9,15]$)

$$g(\beta,\mu) = 1 + \frac{\Psi(1/\beta)}{\beta} - \frac{\sum_{i=1}^{N} |x_i - \mu|^\beta \ln|x_i - \mu|}{\sum_{i=1}^{N} |x_i - \mu|^\beta} + \frac{\ln\left(\frac{\beta}{N} \sum_{i=1}^{N} |x_i - \mu|^\beta \right)}{\beta},$$

$$\frac{\partial g(\beta,\mu)}{\partial \beta} = -\frac{\Psi(1/\beta)}{\beta^2} - \frac{\Psi'(1/\beta)}{\beta^3} + \frac{1}{\beta^2}$$

$$- \frac{\sum_{i=1}^{N} |x_i - \mu|^\beta (\ln|x_i - \mu|)^2}{\sum_{i=1}^{N} |x_i - \mu|^\beta} + \left(\frac{\sum_{i=1}^{N} |x_i - \mu|^\beta \ln|x_i - \mu|}{\sum_{i=1}^{N} |x_i - \mu|^\beta} \right)^2$$

$$+ \frac{\sum_{i=1}^{N} |x_i - \mu|^\beta \ln|x_i - \mu|}{\beta \sum_{i=1}^{N} |x_i - \mu|^\beta} - \frac{\ln\left(\frac{\beta}{N} \sum_{i=1}^{N} |x_i - \mu|^\beta \right)}{\beta^2}. \tag{18.13}$$

The analytical formulae for $h(\beta,\mu)$ and $\frac{\partial h(\beta,\mu)}{\partial \mu}$ are given by

$$h(\beta,\mu) = \sum_{x_i \geq \mu} |x_i - \mu|^{\beta-1} - \sum_{x_i < \mu} |x_i - \mu|^{\beta-1},$$

$$\frac{\partial h(\beta,\mu)}{\partial \mu} = -(\beta-1) \sum_{i=1}^{N} |x_i - \mu|^{\beta-2}. \tag{18.14}$$

In addition we can derive the expressions for $\frac{\partial g(\beta,\mu)}{\partial \mu}$ and $\frac{\partial h(\beta,\mu)}{\partial \beta}$ and solve (18.11) and (18.12) by Newton's method. When we simply ignore $\frac{\partial g(\beta,\mu)}{\partial \mu}$ and $\frac{\partial h(\beta,\mu)}{\partial \beta}$, we will still find the correct solution, but we lose quadratic convergence. On the other hand this decouples both nonlinear equations and we do not have to evaluate the expressions for $\frac{\partial g(\beta,\mu)}{\partial \mu}$ and $\frac{\partial h(\beta,\mu)}{\partial \beta}$. Hence, both equations can now be solved by Newton's method independently.

As an initial estimator μ_0 for μ we use the sample average $\mu_0 = \frac{1}{N}\sum_{i=1}^{N} x_i$ and use (18.13) to solve for β by Newton's method. This β allows us to calculate the next value μ_1 for μ from (18.14). If the new value of μ is not sufficiently close to the previous value then the process is repeated with the new value of μ. We outline our algorithm in Alg. 18.1.

After having computed the parameters $\hat{\mu}, \hat{\alpha}, \hat{\beta}$ we consider the resulting density function $f(x; \hat{\mu}, \hat{\alpha}, \hat{\beta})$ as best fit to the measured data (see [1]). We make the following observation [1,2]. We introduced M-times the steps 5–11 within a loop and taking averages, see Alg. 18.1. One can choose $M = 1$ and N sufficiently large (usually $N \gg 1000$). In our case, $M = 50$, $N = 200$. So, N can be taken smaller.

In our numerical experiments in the next section we observed that due to the large value of $\partial\alpha/\partial\beta$ averaging the $\hat{\alpha}_k$ gives better results for α than by using (18.8) on $\hat{\beta}$. Finally, to also conveniently deal with $\beta \to \infty$, one can introduce $\tilde{\beta} = 1/\beta$ as unknown (but this is not shown in Alg. 18.1).

Algorithm 18.1 Averaged Generalized Gaussian Distribution Fit [1,2]

1: **procedure** AGGDF(**X**, N, M)
2:     Determine the empirical pdf [10] $\hat{f}(x)$ from the data **X**.   ▷ See Fig. 18.4
3:     Compute the cumulative distribution function $\hat{F}(x) = \int_{-\infty}^{x} \hat{f}(t)dt$.
4:     **for** $k = 1, ..., M$ **do**
5:         Generate random values $\{x_i^k \,|\, i = 1, ..., N\}$ using $\hat{F}^{-1}$.
6:         Set initial guess $\hat{\beta}_k = 2$, $\hat{\mu}_k = \frac{1}{N}\sum_{i=1}^{N} x_i^k$
7:         **while** not converged **do**
8:             Compute root $\hat{\beta}_k$ of $g(\beta, \hat{\mu}_k) = 0$, using these x_i^k-values.   ▷ See (18.11)
9:             Compute root $\hat{\mu}_k$ of $h(\hat{\beta}_k, \mu) = 0$, using these x_i^k-values.   ▷ See (18.12)
10:        **end while**
11:        Compute $\hat{\alpha}_k$.   ▷ See (18.8)
12:    **end for**
13:    Average $\hat{\alpha} = \frac{1}{M}\sum_{i=1}^{M}\hat{\alpha}_k$, $\hat{\beta} = \frac{1}{M}\sum_{i=1}^{M}\hat{\beta}_k$, $\hat{\mu} = \frac{1}{M}\sum_{i=1}^{M}\hat{\mu}_k$.
14:    **return** $\hat{\alpha}$, $\hat{\beta}$, $\hat{\mu}$.
15: **end procedure**

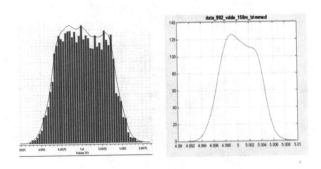

Fig. 18.4: **Left**: Measured data with associated empirical probability density function $\hat{f}$. **Right**: The empirical probability density function $\hat{f}$ for a different set of data used in our calculations [1, 2].

18.2.1 Numerical Results for Determining the GGD Parameters

We tested Alg. 18.1 on several data sets and we found that for small sample sizes (e.g., $N = 200$) the iterations of step 7-9 can be omitted. For such small sample sizes, the variation in α, β and μ dominates the inaccuracy of solving the non-linear equations, because the initial guess of μ as sample average is already sufficiently close to the converged solution. We also applied Alg. 18.1 (with $M = 50$ and $N = 200$) to the 'calibrated' data from NXP IC-measurements (Fig. 18.4). The computed values β_k and their mean $\hat{\beta}$ are shown in Fig. 18.5-(left). The computed density function f as well as the initially fitted (non-symmetrical) density function $\hat{f}$ are given in Fig. 18.6. Note that the tails are very well approximated in Fig. 18.5-(right). To get an impression of the sensitivity of the computed density w.r.t. $\hat{\alpha}$ we varied the computed value of $\hat{\alpha}$ with +/- 10%, plotted the corresponding densities and computed the Mean Squared Error (MSE), see Fig. 18.6 and [1]. Notice that in Fig. 18.6 a 10% variation in α has a large effect on the pdf f. Clearly, the approximation of f around its top using $\hat{\alpha}$ is better than the pdfs with $\hat{\alpha} \pm 10\%$. A similar observation holds for the slopes of pdf f. The best fit $\hat{\alpha}$ was obtained by the mean as in step 9 in Alg. 18.1. So, we consider $\hat{\alpha}$ as best fit.

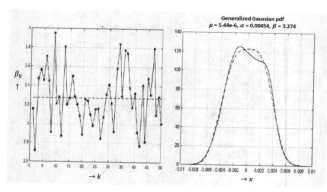

Fig. 18.5:
Left: The computed β_k with mean $\hat{\beta} = 3.27$ and $|\beta_k - \hat{\beta}| < 20\%$ [1,2].
Right: The empirical probability function [10] (solid) and the final fitted GGD (dashed) [1,2].

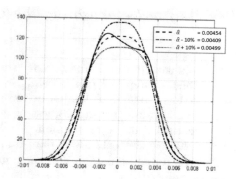

Fig. 18.6: Sensitivity of the density function f w.r.t. $\hat{\alpha}$.
MSE= $(14.31, 56.94, 91.95)$ for $\hat{\alpha} = (454, 499, 409) * 10^{-5}$ [1,2].

18.2.2 A Quality Measure Index for the GGD

Assuming an underlying distribution being standard Gaussian, the capability of a manufacturing process can be measured using some process capability indices like [1,2]

$$C_p = \frac{U - L}{6\sigma} \quad \text{and} \quad C_{pk} = \frac{\min (U - \mu, \mu - L)}{3\sigma}, \tag{18.15}$$

where $[L, U]$ is the specification interval, μ is the process mean and σ is the process standard deviation. A process is said to be capable if the process capability index exceeds an a priori chosen value $k \geq 1$. Usually $k = 4/3$ is taken.

In [2] we introduced a process capability index C_{pkg} for the GGD, but we can improve this definition by generalizing equation (18.15) for all non-normal distributions. We can express this general C_{pk} as

$$C_{pk} = \frac{\min(\Phi^{-1}(F_Y(U)), -\Phi^{-1}(F_Y(L)))}{3}, \tag{18.16}$$

where Φ is the cumulative distribution function (cdf) of the standard normal distribution (i.e. $\mu = 0$ and $\sigma = 1$) and F_Y is the cdf of a random variable Y for which we want to determine the C_{pk}. Notice that when a random variable Y has a known cdf F_Y, then

$$X = \Phi^{-1}(F_Y(Y)) \tag{18.17}$$

has a standard normal distribution. Indeed, the cdf $F_X(x)$ of X equals

$$\begin{aligned}
F_X(x) &= P(X \leq x) \\
&= P(\Phi^{-1}(F_Y(Y)) \leq x) \\
&= P(Y \leq F_Y^{-1}(\Phi(x))) \\
&= F_Y(F_Y^{-1}(\Phi(x))) \\
&= \Phi(x).
\end{aligned}$$

Hence, equation (18.17) simply transforms the specification limits L and U to the standard normal distribution and thus equation (18.16) is identical to the standard C_{pk} formula (18.15) applied to the transformed limits.

In case of a GGD (18.1) we can introduce a process capability index C_{pkg} as quality indicator, similar to the standard Gaussian case as

$$C_{pkg} = \frac{\sqrt{2}}{3} \min\left(sgn(U - \mu) \left(\frac{|U - \mu|}{\alpha}\right)^{\beta/2}, \ sgn(\mu - L) \left(\frac{|\mu - L|}{\alpha}\right)^{\beta/2} \right), \tag{18.18}$$

where L and U are the lower and upper tolerance levels, respectively.

Equation (18.18) approximates equation (18.16) for the GGD assuming that the specification limits L and U are in the tails of the distribution (see section 18.6 appendix). Since our typical yield targets are $C_{pk} \geq 4/3$, this assumption is justified (i.e. (18.15) implies that $|limit - \mu| > 4\sigma$). In Fig. 18.7 we compare the exact C_{pk} values of equation (18.16) with the approximation of equation (18.18). Since μ is a location parameter and α is a scale parameter, we only study the influence of the shape parameter β. In Fig. 18.7 we see that the approximation (C_{pkg}) follows the exact curve quite well. The error is smaller than 1%, except in the region of low β and low C_{pk} where the errors are more

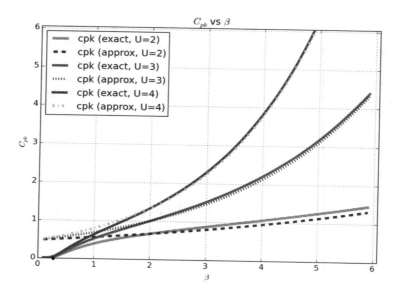

Fig. 18.7: Exact C_{pk} and the approximation C_{pkg} as a function of β for several values of the upper specification limit U. The values of the remaining GGD parameters are chosen as $\mu = 0$ and $\alpha = \sqrt{2}$.

significant, but this is still acceptable as long as the C_{pkg} remains sufficiently low. If the C_{pk} is below target then the exact value is not important.

The C_{pkg} simply results as a post processing facility of Alg. 18.1, after having determined α, β and μ. In practice, formula (18.18) can be applied in two ways. First, a specific value $C_{pkg} = c$ can be given in order to meet certain yield requirements. Assuming $|\mu - L| < |U - \mu|$, then L can be computed via $c = C_{pkg} = (\mu - L)/\sigma$. On the other hand, if L and U are known from product specs, then C_{pkg} can be determined as a measure for the yield.

18.3 The Modulus Transform

Instead of fitting a distribution of a specific type to the data, a different approach is to transform the data to a normal distribution. Box and Cox [4] introduced a family of power transformations to map non-normal data to a data set that is reasonably normal. The Box-Cox Transform does not allow data being negative or zero valued. However, a uniform offset can be added to the data set such that all values are positive. Unfortunately, when applying a power transformation like the Box-Cox Transform some degree of skewness will be introduced to a symmetric data set. To avoid this problem we use a modification of the Box-Cox Transform to allow negative values. This modified transformation is called the Modulus Transform [6] and is given by

$$
T_\lambda(x) = \begin{cases} sgn(x) \, \frac{(|x|+1)^\lambda - 1}{\lambda} & \lambda \neq 0 \\ sgn(x) \, \ln(|x|+1) & \lambda = 0 \end{cases} , \tag{18.19}
$$

where the sign function $sgn(x)$ was defined by (18.5). Notice that the skewness introduced by this transform can be minimized by shifting the median of the data set to zero before applying the transform.

18.3.1 Maximum Likelihood Estimation of the Parameters

The problem now is to find the parameter λ of the Modulus Transform that maps the data best to a normal distribution. Assume we have a λ and a data set that can be transformed to a Gaussian distribution, then the pdf of the transformed data is given by

$$
f(\mu, \sigma, y_\lambda) = \frac{1}{2\pi\sigma^2} \exp\left(-\frac{(y_\lambda - \mu)^2}{2\sigma^2} \right), \tag{18.20}
$$

where $y_\lambda = T_\lambda(x)$ is the transformed variable and x is the original variable. We can express equation (18.20) in terms of the original variable x through integration by substitution of $y_\lambda = T_\lambda(x)$. Then $\int_{-\infty}^{\infty} f(\mu, \sigma, y_\lambda) \, dy_\lambda = 1$ transforms into $\int_{-\infty}^{\infty} f(\mu, \sigma, y_\lambda) \frac{dT_\lambda(x)}{dx} \, dx = 1$. Thus $f(\mu, \sigma, x)$ can be written as

$$
\tilde{f}(\mu, \sigma, x) = \frac{(|x|+1)^{\lambda-1}}{2\pi\sigma^2} \exp\left(-\frac{(T_\lambda(x) - \mu)^2}{2\sigma^2} \right). \tag{18.21}
$$

The likelihood function in relation to the original observations x_i is given by

$$
L(\mu, \sigma^2, \lambda | x_i) = \frac{1}{(2\pi\sigma^2)^{N/2}} \exp\left(-\sum_{i=1}^{N} \frac{(T_\lambda(x_i) - \mu)^2}{2\sigma^2} \right) J, \tag{18.22}
$$

where N is the sample size and the Jacobian J is

$$
J = \Pi_{i=1}^{N} |\partial T_\lambda(x_i)/\partial x_i|. \tag{18.23}
$$

Consequently, the log likelihood function is

$$\ln(L(\mu,\sigma^2,\lambda|x_i)) = -(N/2)\ln(2\pi\sigma^2) - \sum_{i=1}^{N} \frac{(T_\lambda(x_i)-\mu)^2}{2\sigma^2} + (\lambda-1)\sum_{i=1}^{N}\ln(|x_i|+1)$$

(18.24)

Estimates of μ, σ and λ are found by maximizing the log likelihood function. This can be done by using the classical Newton-CG algorithm.

18.3.2 Skewed Density Distributions

The transformation in the previous section works well for symmetric distributions. If the distribution is skewed then the transformation is not optimal for reducing both the skewness and the excess kurtosis.
We will show that a family of skewed distribution exists that allows the C_{pk} to

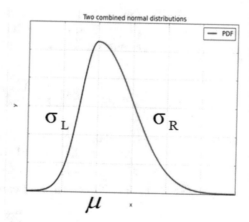

Fig. 18.8: Probability density function consisting of two parts of a normal distribution with different σ values. For $x < \mu$ we have $\sigma = \sigma_L$ and $\sigma = \sigma_R$ for $x \geq \mu$. The median of the combined distribution is μ.

be calculated from a simple formula. Also, the Gaussian distribution belongs to this family as a special case. Hence, if we transform the data set to this skewed distribution instead of the normal distribution, we can handle both skewed and symmetric distributions. This skewed distribution consists of two different normal distributions. Fig. 18.8 shows an impression of such a distribution. For $x < \mu$ we have a part that is a normal distribution with $\sigma = \sigma_L$ and for $x \geq \mu$ we have $\sigma = \sigma_R$, where μ is the median of the distribution. The pdf is given by

$$f(x, \mu, \sigma_L, \sigma_R) = \begin{cases} \frac{1}{\sigma_L \sqrt{2\pi}} \exp -\frac{(x-\mu)^2}{2\sigma_L^2} & \text{for } x < \mu \\ \frac{1}{\sigma_R \sqrt{2\pi}} \exp -\frac{(x-\mu)^2}{2\sigma_R^2} & \text{for } x \geq \mu \end{cases}. \tag{18.25}$$

Notice that when $\sigma_L \neq \sigma_R$ the density function has a discontinuity point at $x = \mu$, but we do not care about the center of the distribution since large C_{pk} values are determined by the tails of the distribution.

Obviously, the C_{pk} for this distribution can be calculated as

$$C_{pk} = \min \left(\frac{U - \mu}{3\sigma_R}, \frac{\mu - L}{3\sigma_L} \right). \tag{18.26}$$

The parameters μ, σ_L and σ_R can be estimated from the 5th, 50th and 95th percentile of the data set, which are easy to calculate. For a normal distribution we have that $\mu = P50$ and $\sigma = (P50 - P05)/z_{90} = (P95 - P50)/z_{90}$, where $z_{90} = 1.645$ is the scaling factor to have tails of 5% each for a 90% confidence interval. Hence, for the distribution of eq. (18.25), we find that $\mu = P50$, $\sigma_L = (P50 - P05)/z_{90}$ and $\sigma_R = (P95 - P50)/z_{90}$.

If we transform the data to distribution (18.25) instead of to a normal distribution then the likelihood function can be given analogously to (18.22). For convenience we introduce the factor $l_\chi(\mu, \sigma_\chi^2, \lambda | x_i)$ where $\chi \in \{L, R\}$ and the sets $\mathcal{S}_\chi$ with $\mathcal{S}_L = \{x_i | T_\lambda(x_i) < \mu\}$ and $\mathcal{S}_R = \{x_i | T_\lambda(x_i) \geq \mu\}$ where $T_\lambda(x_i)$ is the Modulus Transform of (18.19). The cardinality of $\mathcal{S}_\chi$ is $N_\chi = |\mathcal{S}_\chi|$. Obviously, the sample size is $N = N_L + N_R$. The factor $l_\chi(\mu, \sigma_\chi^2, \lambda | x_i)$ is defined by

$$l_\chi(\mu, \sigma_\chi^2, \lambda | x_i) = \frac{1}{(2\pi\sigma_\chi^2)^{N_\chi/2}} \exp \left(-\sum_{x \in \mathcal{S}_\chi} \frac{(T_\lambda(x) - \mu)^2}{2\sigma_\chi^2} \right).$$

The likelihood function can now be given by

$$L(\mu, \sigma_L^2, \sigma_R^2, \lambda | x_i) = \left(\Pi_{\chi \in \{L, R\}} \, l_\chi(\mu, \sigma_\chi^2, \lambda | x_i) \right) J, \tag{18.27}$$

where the Jacobian J is defined as in (18.23). The log likelihood function is

$$\ln(L(\mu, \sigma_L^2, \sigma_R^2, \lambda | x_i)) = \sum_{\chi \in \{L, R\}} \ln(l_\chi(\mu, \sigma_\chi^2, \lambda | x_i)) + (\lambda - 1) \sum_{i=1}^{N} \ln(|x_i| + 1). \tag{18.28}$$

The log of the factor $l_\chi(\mu, \sigma_\chi^2, \lambda | x_i)$ can be expressed as

$$\ln(l_\chi(\mu, \sigma_\chi^2, \lambda | x_i)) = -\frac{N_\chi}{2} \ln(2\pi\sigma_\chi^2) - \sum_{x \in \mathcal{S}_\chi} \frac{(T_\lambda(x) - \mu)^2}{2\sigma_\chi^2}.$$

The maximum likelihood estimates for λ, μ, σ_L and σ_R follow from maximizing the log likelihood function (18.28). Again, the maximum can be found using the Newton-CG algorithm.

18.3.3 The Distribution after the Modulus Transform

In Section 18.3.1 we explained how we transformed data to a normal distribution. However, this transformation is not capable to transform any arbitrary data set to a normal distribution. In this section we will show which distributions can be transformed to a normal distribution. We have seen that the transformed distribution of the original data is given by (18.21). Hence, the Modulus Transform to normality is equivalent to modeling the distribution of the data set as the pdf of equation (18.21).

Graphs of this probability density function are shown in Fig. 18.9 for dif-

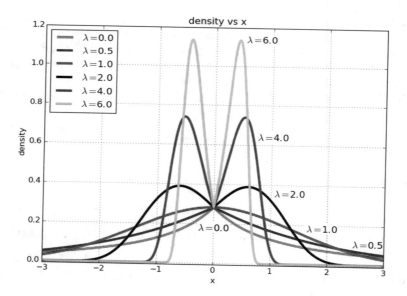

Fig. 18.9: The probability density function associated with the Modulus Transform for different values of λ. The values of the remaining parameters are $\mu = 0$ and $\sigma = \sqrt{2}$.

ferent values of λ. For $\lambda = 1$ we have a Gaussian distribution and for $\lambda > 1$ the distribution has two peaks, which is different from the plateau shape of the GGD for $\beta > 2$. If we compare the tail of (18.21) with the tail of the

GGD (both distributions have similar shape parameters and scale factors for $2\lambda = \beta$ and $\alpha^{\beta} = 2\sigma^2\lambda^2$) then we see that the tail of (18.21) is longer than the tail of the GGD due to the additional factor $(|x|+1)^{\lambda-1}$ in equation (18.21). For the accuracy of the C_{pk} calculation the tail behaviour is more important than the bulk of the distribution. We can calculate the C_{pk} from the normal distribution of equation (18.20) if we transform the specification limits according to the transformation of equation (18.19) and the result is

$$C_{pk} = \frac{\min\left(T_\lambda(U) - \mu, \mu - T_\lambda(L)\right)}{3\sigma}. \tag{18.29}$$

The maximum likelihood estimate will find the parameters λ, μ and σ that give the best fit, but this does not mean that the transformed data is normal. If the transformed data is not normal, then we cannot calculate the C_{pk} from equation (18.29), because it assumes normal data. Therefore we use the Shapiro-Wilk Test of Normality [13] to find evidence whether the transformed data is close to normal or not.

18.4 Comparison of Different C_{pk} Calculation Methods

In this section we compare the results of three different C_{pk} calculation methods: the standard C_{pk} of equation (18.15), the Generalized Gaussian Distribution fit of equation (18.18) and the Modulus Transform approach of equation (18.29). In Fig. 18.10 we sampled from a distribution that consists of a combination of a uniform distribution and tails of a Gaussian distribution. This

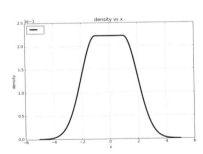 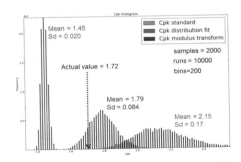

Fig. 18.10: **Left:** Distribution consisting of a combination of a uniform distribution and tails of a Gaussian distribution used to generate the samples. **Right:** Histograms of C_{pk} values for the three C_{pk} calculation methods (the standard C_{pk} (mean=1.45), the Modulus Transform to normality (mean=1.79) and fitting the GGD to the data (mean=2.15)).

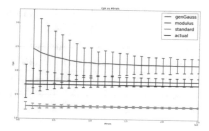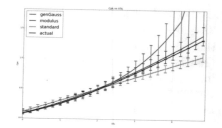

Fig. 18.11: **Left:** C_{pk} value as a function of the sample size (number of trials) for three calculation methods (the standard C_{pk}, fitting the GGD to the data, and the Modulus Transform to normality) and the actual value. The error bars indicate the standard deviation of the C_{pk} value. **Right:** C_{pk} values as a function of the Upper Specification Limit (USL) for the three calculation methods and the actual value.

is a distribution that resembles the measured IC chip production data after calibration. The limits were chosen such that the $C_{pk} = 1.72$. The C_{pk} was calculated 10000 times from different sets of 2000 samples. In Fig. 18.10 we see that the standard C_{pk} significantly underestimates the actual C_{pk} value. The GGD fit gives a C_{pk} value that is far too optimistic and the Modulus Transform is quite close. The GGD has tails that are much shorter than Gaussian, which explains the optimistic C_{pk} values. We also see that the spread in C_{pk} values is much larger for both the GGD fit and the Modulus Transform than for the standard C_{pk} values. For the standard C_{pk} we only need to estimate 2 parameters, while for the other two methods 3 parameters need to be estimated from the data set. Since C_{pk} depends exponentially on λ (in case of the Modulus Transform) or β (in case of the GGD), the C_{pk} value is more sensitive to uncertainty in these parameters. In Fig. 18.11 we sample from the same distribution as shown in Fig. 18.10. We vary the sample size and the specification limits to study the influence on the C_{pk} value. The standard deviation of the C_{pk} is shown as an error bar in Fig. 18.11. We see that the standard deviation of the C_{pk} value decreases with increasing sample size. As our sample size increases, the uncertainty in our parameter estimate decreases and we have greater precision. The disadvantage of the GGD fit and the modulus transform is the relatively large confidence intervals compared with the standard C_{pk}. At the right side of Fig. 18.11 we see that the deviation from the actual value increases with increasing Upper Specification Limit (USL). However, the Modulus Transform follows the actual C_{pk} curve quite well.

In Fig. 18.12 we sample from a Generalized Gaussian Distribution with shape parameter $\beta = 1$. This is a distribution with a sharper peak and longer tails than normal. The standard C_{pk} overestimates the actual value, because it assumes that the tails are normal, while the tails of the GGD are much longer than normal. Both methods, the GGD fit, and the Modulus Transform to normality, perform very well on this data set. Obviously, the GGD fit works very well and the only source of errors is caused by the finite sample size. In Fig. 18.13 we sample from a Generalized Gaussian Distribution with shape parameter $\beta = 4$. In this case the distribution has a flatter peak and shorter tails than normal. We see that the Modulus Transform cannot follow the actual C_{pk} curve. However, it still performs much better than the standard C_{pk} calculation. Again, fitting a GGD to the data sampled from a GGD works well as could be expected. We also consider the C_{pk} calculation for a *skewed*

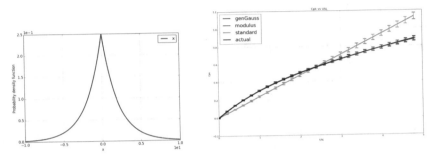

Fig. 18.12: **Left:** Generalized Gaussian Distribution with shape parameter $\beta = 1$ is used to generate samples. **Right:** C_{pk} values as a function of the Upper Specification Limit (USL) for the three calculation methods (the standard C_{pk}, fitting the GGD to the data, and the Modulus Transform to normality) and the actual value.

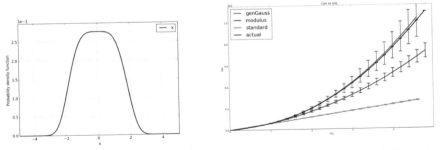

Fig. 18.13: **Left:** Distribution used to sample from: a Generalized Gaussian Distribution with shape parameter $\beta = 4$. **Right:** C_{pk} values as a function of the Upper Specification Limit (USL) for the three calculation methods (the standard C_{pk}, fitting the GGD to the data, and the Modulus Transform to normality) and the actual value.

distribution. For skewed distributions we use the C_{pk} formula of equation (18.26). We do not try to fit the GGD since it is symmetric and therefore does not make sense. We estimate μ, σ_L and σ_R from the median and the 5th and 95th percentile as we explained in Section 18.3.2.

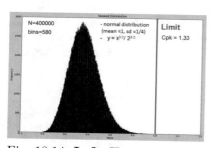 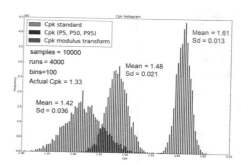

Fig. 18.14: **Left:** Histogram of the data set used to calculate the C_{pk}. It is obtained from samples of a normal distribution with $\mu = 1$ and $\sigma = \frac{1}{4}$ and then applying the formula $y = \left(\frac{x}{2}\right)^{\frac{3}{2}}$ to the samples. **Right:** Histograms of the C_{pk} values for three calculation methods (the Modulus Transform to normality (mean = 1.42), using two halves of a normal distribution with different σ (mean = 1.48) and the standard C_{pk} (mean = 1.61)) and the actual value.

The skewed distribution is obtained from the samples of a normal distribution with with $\mu = 1$ and $\sigma = \frac{1}{4}$. These samples are then transformed according to $y = \left(\frac{x}{2}\right)^{\frac{3}{2}}$ which introduces the skewness. The Upper Specification Limit is chosen as USL = 1 which results to $C_{pk} = \frac{4}{3}$. In Fig. 18.14 we notice that all methods give values that are too optimistic. However, the percentile formula is a significant improvement compared to the standard C_{pk}. The Modulus Transform is closest to the actual value.

18.5 Conclusions

We developed a new robust numerical procedure for computing the parameters of a GGD. The GGD did fit very accurately. Using the GGD a quality measure can be defined analogously to the C_{pk} index for the standard Gaussian distributions. Additionally, we proposed a normalizing data transformation technique by a Modulus Transform that allows for both symmetric and skewed data sets. Also for this normalizing transformation a quality measure was defined based on the C_{pk} index for standard Gaussian distributions. We also compared our transformation method with our modeling approach of the data set as a GGD and with the standard C_{pk} calculation. We found that both methods give much better results than the standard C_{pk} calculation for non-normal distributions. However, the results strongly depend on the behaviour of the tail of the distribution. We expect that the distribution resulting from electronic calibration resembles a distribution consisting of a combination of a uniform distribution with tails of a Gaussian distribution. We have seen that for such distributions the Modulus Transform outperforms the GGD modeling approach.

18.6 Appendix

In Section 18.2.2 we introduced a generalized C_{pk} formula for non-normal distributions of equation (18.16). For convenience, we recall formula (18.16)

$$C_{pk} = \frac{\min(\Phi^{-1}(F_Y(U)), -\Phi^{-1}(F_Y(L)))}{3}, \qquad (18.30)$$

where U is the upper limit, L is the lower limit, Φ is the cdf of the standard normal distribution. We assume that F_Y is the cdf of the GGD.

In this section we will show that this equation (18.16) can be approximated as equation (18.18) for the GGD if we assume that the specification limits are located in the tail of the distribution. The cumulative standard normal distribution (i.e. $\mu = 0$, $\sigma = 1$) is given by

$$\Phi(x) = \frac{1}{2}\left[1 + \text{erf}(\frac{x}{\sqrt{2}})\right], \qquad (18.31)$$

where erf is the so-called error function. An expansion of the error function, which converges rapidly for all real values of z, is given in [12] and if we keep the lowest order term, then the error function is approximated by

$$\text{erf}(z) \approx sgn(z)\sqrt{1 - e^{-z^2}}, \qquad (18.32)$$

where the sign function $sgn(x)$ was defined by equation (18.5). By (18.32) the inverse error function can be approximated by

$$\text{erf}^{-1}(y) \approx sgn(y)\sqrt{-\ln(1 - y^2)}. \qquad (18.33)$$

The inverse $\Phi^{-1}(x)$ is obtained from equation (18.31). Taking the inverse erf^{-1} on both sides of $2\Phi(x) - 1 = \text{erf}(\frac{x}{\sqrt{2}})$ leads to $\sqrt{2}\text{erf}^{-1}(2\Phi(x) - 1) = x$, or, equivalently, $\sqrt{2}\text{erf}^{-1}(2\Phi(x) - 1) = \Phi^{-1}(\Phi(x))$. Thus when we replace $\Phi(x)$ with x, we have $\Phi^{-1}(x) = \sqrt{2}\text{erf}^{-1}(2x - 1)$. Using the approximation for $\text{erf}^{-1}(y)$ of equation (18.33) gives, after some rearrangements, that $\Phi^{-1}(x)$

is approximately given by

$$\Phi^{-1}(x) \approx sgn(2x-1)\sqrt{-2\ln(4x(1-x))}. \tag{18.34}$$

From Section 18.2 we recall that the cdf of the GGD was given by equation (18.7)

$$F(x) = \frac{1}{2} + sgn(x-\mu)\frac{\gamma\left(1/\beta, \left(\frac{|\mu-x|}{\alpha}\right)^{\beta}\right)}{2\Gamma(1/\beta)}.$$

By definition, the Lower Incomplete Gamma function $\gamma(a,z) = \int_0^z t^{a-1}e^{-t}dt$ and Upper Incomplete Gamma function $\Gamma(a,z) = \int_z^{\infty} t^{a-1}e^{-t}dt$ satisfy

$$\gamma(a,z) + \Gamma(a,z) = \Gamma(a), \tag{18.35}$$

since the ordinary gamma function is defined as $\Gamma(a) = \int_0^{\infty} t^{a-1}e^{-t}dt$. The asymptotic expansion of the Upper Incomplete Gamma function for large $|z|$ is (e.g., [16])

$$\Gamma(a,z) = z^{a-1}e^{-z}\sum_{n=0}^{\infty}\frac{\Gamma(a)}{\Gamma(a-n)z^n} \tag{18.36}$$

$$\approx z^{a-1}e^{-z}\left(1+O(\frac{1}{z}).\right)$$

Using the lowest order term of equation (18.36) we find the tail approximation of the cdf

$$F_Y(x) = \begin{cases} 1 - \dfrac{\left(\frac{|x-\mu|}{\alpha}\right)^{1-\beta}\exp\left(-\left(\frac{|x-\mu|}{\alpha}\right)^{\beta}\right)}{2\Gamma(\frac{1}{\beta})}, & x \gg \mu \\[3mm] \dfrac{\left(\frac{|x-\mu|}{\alpha}\right)^{1-\beta}\exp\left(-\left(\frac{|x-\mu|}{\alpha}\right)^{\beta}\right)}{2\Gamma(\frac{1}{\beta})}, & x \ll \mu \end{cases}. \tag{18.37}$$

Combining equation (18.37) with equation (18.34) gives

$$\Phi^{-1}(F_Y(Y)) \approx sgn(x-\mu)\sqrt{-2\ln\left(2\frac{\left(\frac{|x-\mu|}{\alpha}\right)^{1-\beta}\exp\left(-\left(\frac{|x-\mu|}{\alpha}\right)^{\beta}\right)}{\Gamma(\frac{1}{\beta})}\right)}, \tag{18.38}$$

which can be rewritten as

$$\Phi^{-1}(F_Y(Y)) \approx sgn(x-\mu)\sqrt{2\left(\frac{|x-\mu|}{\alpha}\right)^{\beta} - 2\ln\left(2\frac{\left(\frac{|x-\mu|}{\alpha}\right)^{1-\beta}}{\Gamma(\frac{1}{\beta})}\right)}. \tag{18.39}$$

Let us now assume that $|x - \mu|$ is sufficiently large such that

$$\left| \ln \left(2 \frac{\left(\frac{|x-\mu|}{\alpha} \right)^{1-\beta}}{\Gamma(\frac{1}{\beta})} \right) \right| \ll \left(\frac{|x-\mu|}{\alpha} \right)^{\beta}.$$

Taking this into account we can approximate equation (18.39) as

$$\Phi^{-1}(F_Y(x)) \approx sgn(x-\mu)\sqrt{2} \left(\frac{|x-\mu|}{\alpha} \right)^{\beta/2}. \tag{18.40}$$

Combining equation (18.40) and equation (18.30) gives the C_{pk} as

$$C_{pk} = \frac{\sqrt{2}}{3} \min \left(sgn(U-\mu) \left(\frac{|U-\mu|}{\alpha} \right)^{\beta/2}, \; sgn(\mu-L) \left(\frac{|\mu-L|}{\alpha} \right)^{\beta/2} \right), \tag{18.41}$$

which is indeed the same expression as equation (18.18).

References

1. BEELEN, T.G.J., AND DOHMEN, J.J.: *Parameter estimation for a generalized Gaussian distribution*. CASA Report 15-40, TU Eindhoven, 2015. Online: http://www.win.tue.nl/analysis/reports/rana15-40.pdf.
2. BEELEN, T.G.J., DOHMEN, J.J., TER MATEN, E.J.W., AND TASIĆ, B.: *Fitting generalized Gaussian distributions for process capability index*. In: LANGER, U., AMRHEIN, W., AND ZULEHNER, W. (EDS.): *Scientific Computing in Electrical Engineering – SCEE 2016*. Springer International Publishing AG, Series Mathematics in Industry, Vol. 28, pp. 169–176, 2018. http://dx.doi.org/10.1007/978-3-319-75538-0_16.
3. BOMBRUN, L., PASCAL, F., TOURNERET, J.-Y., AND BERTHOUMIEU, Y.: *Performance of the maximum likelihood estimators for the parameters of multivariate generalized Gaussian distributions*. Proc. ICASSP-2012, IEEE. Int. Conf. on Acoustics, Speech, and Signal Processing, Kyoto, Japan, 3525-3528, 2012. https://hal.archives-ouvertes.fr/hal-00744600.
4. BOX, G.E.P., AND COX, D.R.: *An Analysis of Transformations*. Journal of the Royal Statistical Society. Series B (Methodological), Vol. 26, No. 2, pp. 211–252, 1964.
5. GONZÁLEZ-FARÍAS, G., DOMÍNGUEZ-MOLINA, J.A., AND RODRÍGUEZ-DAGNINO, R.M.: *Efficiency of the approximated shape parameter estimator in the generalized Gaussian distribution*. IEEE Trans. on Vehicular Technology 58:8, 4214–4223, 2009.
6. JOHN, J.A., AND DRAPER, N.R.: *An Alternative Family of Transformations*. Journal of the Royal Statistical Society. Series C (Applied Statistics) Vol. 29, No. 2, pp. 190–197, 1980.
7. KOKKINAKIS, K., AND NANDI, A.K.: *Exponent parameter estimation for generalized Gaussian probability density functions with application to speech modeling*. Signal Processing 85, 1852–1858, 2005.

8. KRUPIŃSKI, R.: *Approximated fast estimator for the shape parameter of generalized Gaussian distribution for a small sample size.* Bull. of the Polish Academy of Sciences, Technical Sciences 63:2, 405–411, 2015.

9. KRUPIŃSKI, R., AND PURCZYŃSKI, J.: *Approximated fast estimator for the shape parameter of generalized Gaussian distribution.* Signal processing 86, 205–211, 2006.

10. MARTINEZ, W.L., AND MARTINEZ, A.R.: *Computational statistics handbook with Matlab.* Chapman & Hall/CRC, 2002.

11. TER MATEN, E.J.W., WITTICH, O., DI BUCCHIANICO, A., DOORN, T.S., AND BEELEN, T.G.J.: *Importance sampling for determining SRAM yield and optimization with statistical constraint.* In: MICHIELSEN, B., AND POIRIER, J.-R. (EDS.): *Scientific Computing in Electrical Engineering SCEE 2010.* Series Mathematics in Industry Vol. 16, Springer, 39–48, 2012.

12. SCHÖPF, H.M., AND SUPANCIC, P.H.: *On Bürmann's Theorem and Its Application to Problems of Linear and Nonlinear Heat Transfer and Diffusion.* The Mathematica Journal, 16, 2014. Online: http://www.mathematica-journal.com/data/uploads/2014/11/Schoepf.pdf

13. DE SMITH, M.J.: *STATSREF: Statistical Analysis Handbook.* http://www.statsref.com/, 2015. For Shapiro-Wilk Test, see http://www.statsref.com/HTML/index.html?shapiro-wilk.html.

14. TEMME, N.M.: *Computational aspects of Incomplete Gamma Functions with large complex parameters.* Int. Series of numerical Mathematics 119, 551–562, 1994.

15. WIKIPEDIA: *Generalized Normal Distribution.* https://en.wikipedia.org/wiki/Generalized_normal_distribution; *Gamma function.* https://en.wikipedia.org/wiki/Gamma_function; *Digamma function.* https://en.wikipedia.org/wiki/Digamma_function; *Incomplete Gamma Function.* https://en.wikipedia.org/wiki/Incomplete_gamma_function. 2016.

16. WINITZKI, S.: *Computing the incomplete gamma function to arbitrary precision.* In: KUMAR, V., GAVRILOVA, M.L., TAN, C.J.K., AND L'ECUYER, P. (EDS.): *Computational Science and Its Applications — ICCSA 2003.* Lecture Notes in Computer Science, Vol. 2667. Springer, Berlin, Heidelberg, pp. 790–798, 2003.

Chapter 19
Ageing Models and Reliability Prediction

Renaud Gillon, Aarnout Wieers, Frederik Deleu, Tomas Gotthans,
Rick Janssen, Wim Schoenmaker, E. Jan W. ter Maten

Abstract We overview reliability related activities like ageing and life time prediction. We want to predict the number of thermal stress cycles an IC can handle before showing passivation cracks. For this a sufficient model for electro-migration was used that can be applied to an IC with multiple drivers and knowing a required thermal profile. At first state-of-the-art reliability concepts are reviewed. Next a new framework is introduced that aims at simplifying and speeding-up the process of assessing the reliability of complex application profiles.

19.1 introduction

We derive a predictive model for predicting thermally induced cracks in passivation, by simulating electrical, thermal and stress properties, and making correlations between them. The objective was to predict the number of thermal stress cycles an IC can handle before showing passivation cracks. We exploit a model for electro-migration that can be applied to an integrated

Renaud Gillon, Aarnout Wieers, Frederik Deleu
ON Semiconductor Belgium BVBA, Oudenaarde, Belgium, e-mail: {Renaud.Gillon, Aarnout.Wieers,Frederik.Deleu}@onsemi.com

Tomas Gotthans (Tomáš Götthans)
Brno University of Technology, Czech Republic, e-mail: Gotthans@feec.vutbr.cz

Rick Janssen
NXP Semiconductors, Eindhoven, the Netherlands, e-mail: Rick.Janssen@nxp.com

Wim Schoenmaker
MAGWEL NV, Leuven, Belgium, e-mail: Wim.Schoenmaker@magwel.com

E. Jan W. ter Maten
Bergische Universität Wuppertal, Germany, e-mail: Jan.ter.Maten@math.uni-wuppertal.de

© Springer Nature Switzerland AG 2019
E. J. W. ter Maten et al. (eds.), *Nanoelectronic Coupled Problems Solutions*,
Mathematics in Industry 29, https://doi.org/10.1007/978-3-030-30726-4_19

circuit (IC) with multiple drivers and knowing a required thermal profile. Problems in an IC are that each device has his own electrical and thermal conditions, changing over time and each device has its own required life-time. We have to find the weak spots without the need to define a lot of input information. The chapter covers: (1) Methodology for multi-device electro-migration assessment; (2) Life time model definition for electro-migration. A dedicated model for bond wire interfaces has been described in [3]; (3) Feasibility study to use J (current density), T (temperature) and S (stress) maps for thermal induced failures. For validation of the reliability prediction models we refer to [8].

At first state-of-the-art reliability concepts are reviewed, including concepts like ageing and life-time prediction. Next a new framework is introduced that aims at simplifying and speeding-up the process of assessing the reliability of complex application profiles.

19.2 State-of-the-Art Reliability Concepts

We first define ageing (or aging). This leads to an estimate for the life-time, i.e. the time at which a component has reached its end-of-life. Next we consider several acceleration factors that contribute to ageing: electro-migration, hot-carries-injection, bias-temperature instability and dielectric break-down.

19.2.1 Ageing

Ageing in semiconductor devices is a slow process by which the behaviour of a component X deteriorates over time. This behaviour is usually characterized by series of modeling functions $I_{DCi}(\mathbf{V}, \mathbf{T}|\mathbf{P}_{X0})$ and $Q_i(\mathbf{V}, \mathbf{T}|\mathbf{P}_{X0})$ which relate the static current I_{DCi} and the charges Q_i at terminal i to the vectors $\mathbf{V}$ and $\mathbf{T}$ of externally-applied electric potentials and temperatures, respectively.

The parameter set $\mathbf{P}_{X0}$ allows to adjust the model response to the data collected during the measurements on a "fresh" component X at instant t_0. When a component is subject to ageing, it is standard practice to characterize the ageing effects by applying static stress conditions over a given period of time, then monitor the device behaviour and extract a new parameter set $\mathbf{P}_{Xk}$ at some instant t_k.

Scholten et al. indicated in [14] that the shift in a given parameter P_j during a static stress test can be advantageously expressed as:

$$\Delta P_j(t_k, t_0) = g_{P_j, \mathrm{M}}\big(f_\mathrm{M}(\mathbf{V}_\mathrm{SS}, \mathbf{T}_\mathrm{SS}) \cdot (t_k - t_0)\big), \qquad (19.1)$$

where $\mathbf{V}_{SS}$ and $\mathbf{T}_{SS}$ are the constant stress voltages and temperatures applied during the stress experiment. In equation (19.1), function $f_M(\mathbf{V}_{SS}, \mathbf{T}_{SS})$ defines a stress-dependent acceleration factor, that, multiplied with the stress-time, provides the age $A_{M,SS}$ associated with the degradation mechanism M for the specific stress conditions. The factor $g_{P_j,M}(\cdot)$ is typically a monotone, continuous function of its single argument, and can take particular forms depending on the ageing mechanism and the parameters under study.

The power of formulation (19.1) resides in the fact that it allows to formulate a generic equation for the effect of a variable stress applied to the component, as indicated by Scholten et al. in [14]

$$\Delta P_j(t, t_0) = g_{P_j,M}\left(\int_{t_0}^{t} f_M(\mathbf{V}(\tau), \mathbf{T}(\tau)) \cdot d\tau \right). \tag{19.2}$$

From the above equation, a general definition for the age A_M associated with degradation mechanism M is obtained

$$A_M(t, t_0) = \int_{t_0}^{t} f_M(\mathbf{V}(\tau), \mathbf{T}(\tau)) \cdot d\tau. \tag{19.3}$$

In the case of periodic stress conditions with Δt_{SC} as duration of the cycle, such that for all t $[\mathbf{V}(t), \mathbf{T}(t)] = [\mathbf{V}(t - \Delta t_{SC}), \mathbf{T}(t - \Delta t_{SC})]$, the age function has an interesting scaling property:

$$A_M(t + N \cdot \Delta t_{SC}, t_0) = A_M(t, t_0) + N \cdot A_M(t + \Delta t_{SC}, t). \tag{19.4}$$

This scaling property allows to define an ageing per stress cycle $\Delta A_{M,SC}$ which summarizes the increase of age during such a cycle, characterized by a given periodic evolution of electrical ($\mathbf{V}(\tau)$) and thermal ($\mathbf{T}(\tau)$) excitations

$$\Delta A_{M,SC} = A_M(t + \Delta t_{SC}, t) = \int_{t}^{t + \Delta t_{SC}} f_M(\mathbf{V}(\tau), \mathbf{T}(\tau)) \cdot d\tau. \tag{19.5}$$

In the absence of saturation effects in function $f_M(\mathbf{V}, \mathbf{T})$, every time the same stress conditions will occur on a given device for a time Δt_{SC}, it will incur an additional ageing of $\Delta A_{M,SC}$.

19.2.2 Life-time

For reliability analysis, it is current practice to define a critical parameter shift setting the limit at which a component has reached its end-of-life

$$\Delta P_{EOL} = 0.1 \cdot P(t_0), \tag{19.6}$$

where $P(t_0)$ is the fresh or initial value of the parameter that is being monitored for shifts.

The monotonicity and continuity properties of the $g_{P,M}(\cdot)$ function ensure the existence of its inverse function, such that an end-of-life age can be determined

$$A_{M,EOL} = g_{P,M}^{-1}(\Delta P_{EOL}) \approx f_M(\mathbf{V}_{SS}, \mathbf{T}_{SS}) \cdot TTF_{MSS}(\mathbf{V}(\tau), \mathbf{T}(\tau)), \quad (19.7)$$

where $TTF_{MSS}(\mathbf{V}_{SS}, \mathbf{T}_{SS})$ is the life-time as function of the static stress conditions for the given failure mode M.

19.2.3 Acceleration Factor

The function $f_M(\cdot, \cdot)$ is known in the reliability literature as the acceleration factor AF_M, which is broadly used in the case of static stress. Bernstein, [1], provides an overview of the acceleration factors for the following ageing mechanisms (M)

- **Electro-migration** AF_{EM}

$$AF_{EM} = (J/J_{ref})^n \cdot \exp\left(qE_{A,EM} \cdot \frac{1}{k_B} \cdot \left[\frac{1}{T_{ref}} - \frac{1}{T}\right]\right), \quad (19.8)$$

where J and T are the current density and temperature of some section of interconnect with uniform dimensions, whilst J_{ref} and T_{ref} are the reference conditions used to determine the life-time TTF_{EM} of tracks with similar dimensions. Furthermore, q is the electron charge, $E_{A,EM}$ is the activation energy and k_B is the Boltzmann constant. This model scales well for tracks, which are significantly wider than the typical grain-size. For narrow tracks, the effective life-time tends to increase and it is recommended to consider $TTF_{EM}(W)$ as a function of track width W. The value of the exponent n is typically close to 2, but varies depending on the metal type and processing conditions.

- **Hot-carrier injection** AF_{HCI}

$$AF_{HCI} = (I_B/I_{DS})^m \cdot \frac{I_{DS}}{K_{HCI}W}, \quad (19.9)$$

where I_B and I_{DS} represent the body and drain-source current respectively. $K_{HCI}W$ is a process- or device-dependent sensitivity parameter, W is the transistor width. The model must be completed by an impact ionization model, which allows to predict the body current $I_B(\mathbf{V}, T)$.

- **Bias-temperature instability** AF_{nBTI} and AF_{pBTI} have similar dependencies, but with specific parameters

$$AF_{\text{BTI}} = \sqrt[n]{\exp\left(qE_{A,\text{BTI}} \cdot \frac{1}{k_B} \cdot \left[\frac{1}{T_{\text{ref}}} - \frac{1}{T}\right] + \gamma_{\text{BTI}} \cdot \left(V_G - V_{G,\text{ref}}\right)\right)}.$$
(19.10)

- **Time-dependent dielectric breakdown** AF_{TDDB} consists of a low-field and a high-field contribution

$$AF_{\text{TDDB}} = AF_{\text{TDDB,LF}} + AF_{\text{TDDB,HF}}.$$
(19.11)

The two contributions correspond to separate physical mechanisms that contribute in reducing the life-time of the gate oxides in a complementary fashion

$$AF_{\text{TDDB,HF}} = \left(\frac{A}{A_{\text{ref}}}\right)^{\frac{1}{\beta}} \cdot \left(\frac{E_{\text{ox}}}{E_{\text{HF,ref}}}\right)^2 \cdot \exp\left(1 - \frac{G(T)}{E_{\text{ox}}} + qE_B \cdot k(T)\right),$$
(19.12)

$$G(T) = E_{\text{HF,ref}} \cdot (1 + \delta \cdot k(T)),$$
(19.13)

$$k(T) = \frac{1}{k_B} \cdot \left[\frac{1}{T_{\text{ref}}} - \frac{1}{T}\right],$$
(19.14)

and

$$AF_{\text{TDDB,LF}} = \left(\frac{A}{A_{\text{ref}}}\right)^{\frac{1}{\beta}} \cdot \exp\left(\gamma(T) \cdot \left(E_{\text{ox}} - E_{\text{LF,ref}}\right) + \Delta H_0 \cdot k(T)\right),$$
(19.15)

$$\gamma(T) = \gamma_{\text{ref}} + \gamma_{\text{slope}} \cdot k(T).$$
(19.16)

Most ageing effects described here have a strong temperature-driven acceleration effect typically accounted for by an exponential term with activation energy and thermal voltage $q/(k_B T)$. Only the hot-carrier effect follows a different trend: it is maximal at low temperatures and slows down when the temperature increases mainly due to the reduction of the mobility.

19.2.4 Bond Wire Ageing Model and Calculator

The life-time of the bonds with wires made of gold, gold-palladium and copper where characterized in different high-voltage technologies. Parameters such as metal thickness, bond pad size, temperature and current levels were varied. The parametric dependencies of the life-time extracted from the measurements were complicated, but could be summarized to the formula shown in 19.17 showing the acceleration factor AF_{BWF} for a given bond pad geometry and metal species, a specific bond wire type and mould compound material. The formula allows to compute the acceleration factor as a function of temperature and the level of current through the wire, which is sufficient in order to evaluate the life-time consumption corresponding to a given application profile.

$$AF_{\mathrm{BWF}} = \sqrt{I/I\mathrm{ref}} \cdot \exp\left(qE_{A,\mathrm{BWF}} \cdot \frac{1}{k_B}[\frac{1}{T_{\mathrm{ref}}} - \frac{1}{T}]\right). \tag{19.17}$$

Actually, this model is more a life-time model than an ageing model: indeed the degradation of the contact resistance at the interface of the wire and the bond pad is hardly measurable and the failure of the wire is nearly immediate upon a measurable change of the wire resistance. Contrary to most ageing mechanisms listed before, the degradation of the contact resistance is not a good advance predictor of the ageing of the bond.

The bond wire calculator [5], see Fig. 19.1, is a graphical user interface that helps designers assessing the life-time constraints of a particular bond wire or bond shell design. It contains the data relative to all characterized automotive processes. Based on temperature profiles input for several operating modes ("biased" and "unbiased") it computes the relative amount of life-time consumption with respect to the qualified levels. In Section 19.5 we will describe how bond wires are treated by the PTM-ET simulator of MAGWEL [10].

19.2.5 Mission Profiles

As indicated by Nirmaier and colleagues in [12], mission profiles have become a critical ingredient for the design of reliable and energy-efficient applications. A mission profile is defined as a collection of possible states in which the application will operate for a given amount of time during its life. The states consist in a complete specification of the environmental loads and operating modes of the application. Usually, the mission profile will define distributions of transition times from one state to the other, allowing to construct realistic scenarios and compute aggregates such as the expectation for the amount of time spent in a given state. G. Jerke demonstrated in [9] how mission profiles can be put to use in the design for reliability of an electronic circuit, considering electro-migration effects.

Fig. 19.2 shows a collection of representative junction temperature profiles for various automotive applications, obtained from the MEDEA+ project ELIAS, see [11]. These profiles specify that typically a silicon IC will spend some amount of hours during its life-time at a specified temperature. The profiles shown are quite diverse and from Fig. 19.2 it is not easy to compare the severity of each of them. In order to compare the profiles, the cumulated ageing corresponding to each profile for an hypothetical mechanism with an activation energy of $0.7eV$ was computed and is shown in Fig. 19.3.

As can be seen in Fig. 19.2, the temperature profiles contain typically 5 to 20 points. In the future, with the pressure towards higher energy-efficiencies and a lower ecological impact of systems, one can anticipate that application profiles will become more and more refined, counting more temperature points, but also accounting for more diverse situations specifically instead of merging and averaging out as often performed in the past. As a result, the number of total temperature points at which to evaluate ageing rates can be expected to increase significantly and to become an issue in terms of computational resources and simulation time. Especially if one attempts to predict the lifetime consumption using "the brute force" method, which is to simulate the circuit in operation at each of the specified temperatures and process corners in order to extract the waveforms allowing to evaluate the stress levels and obtain the ageing rates.

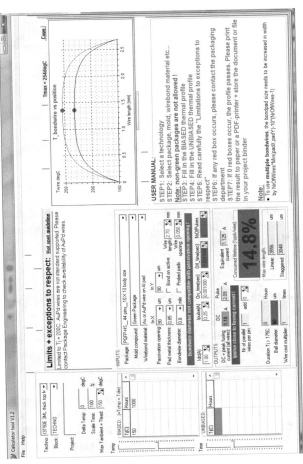

Fig. 19.1: View of the bond wire calculator GUI [5].

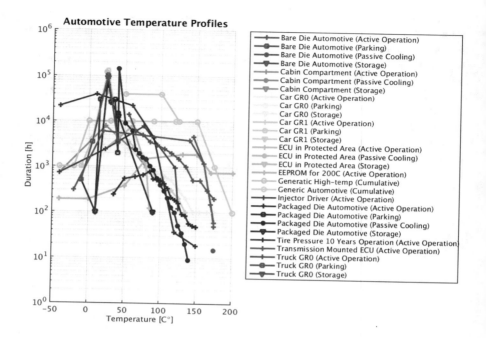

Fig. 19.2: A collection of typical automotive thermal profiles as obtained from MEDEA+ project ELIAS [5, 11].

In Section 19.3, an alternative approach relying on the simulation of a limited number of temperatures along with the interpolation to all specified temperature points will be introduced.

19.3 Framework for Fast Evaluation of Mission Profiles

The time-scales involved in a mission profile are typically several order of magnitudes larger than those related to the evolution of signals and state changes in electronic circuits. Hence, it is appropriate to abstract the combined ageing effect of the signals into an average ageing rate characterizing the life-time consumption of a component with respect to a particular mechanism in a specific operating mode or state defined in the mission profile.

In order to properly capture the ageing effects and the resulting life-time consumption in an application, it is important that temperatures making up the mission profile are specified with sufficient granularity / resolution. In practice, it is typical to encounter application profiles with several tens of specified

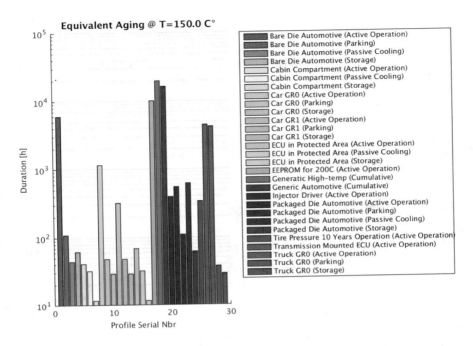

Fig. 19.3: Cumulative equivalent ageing for the automotive profiles, see [5].

temperature levels. This number of temperature levels is a significant challenge for the design engineer, who has to manage a large design-of-experiment to validate his circuit across temperature, bias and process variations.

A typical number of temperature points simulated for an integrated electronic circuit is three to four. Attempting to add more temperature points in the design-of-experiment for the circuit characterization would extend the simulation effort beyond acceptable limits. Hence there is need to find suitable interpolation methods allowing to compute ageing rates at all temperature specified in the mission profile without having to simulate the circuit at all those temperatures.

19.3.1 Operating Modes

For the purpose of reliability analysis, let's define an operating mode as a set of states in which an electronic circuit operates and fulfils a certain function continuously or at least repetitively. The rate at which the circuit cycles through the states is fast with respect to the pace of ageing and to the time-scale of changes in operational states at the system level. From the perspective of the system, the operation of the circuit in an operating mode can best be characterized in terms of averaged parameters over the duration the circuit is in that mode. Typical averaged quantities used to characterize such modes are supply voltage $\overline{V_{\mathrm{bat}}}$, power consumption, operating temperature $\overline{T_{\mathrm{amb}}}$, emitted power-spectral density (mainly for digital blocks), effective ageing rate, etc. The usage of averaged quantities to model a system has been extensively investigated in the field of switched power-converters [2, 13]. We have applied similar concepts to the field of reliability. The set of parameters allowing to characterize a given operating mode will be noted as $\mathbf{S}_{\mathrm{Mode}}$.

19.3.2 Effective Ageing Rate

From the ageing concept introduced in (19.5), one can define the effective ageing rate AF_{eff} as follows

$$AF_{\mathrm{eff}}(\overline{T_{\mathrm{amb}}}, \mathbf{S}_{\mathrm{Mode}}) = 1/\Delta t_{\mathrm{Mode}} \int_{t_0}^{t_0 + \Delta t_{\mathrm{Mode}}} f_{\mathrm{M}}(\mathbf{V}(\tau), \mathbf{T}(\tau)) \cdot d\tau. \quad (19.18)$$

The effective ageing rate for a given mechanism operating on a component X in the circuit under a specific operating mode is computed as the average rate of increase of the age for that particular mechanism over a number of cycles in the given operating mode.

Modelling the effective ageing rate AF_{eff} as a function of the state parameters $\mathbf{S}_{\mathrm{Mode}}$ and the average ambient temperature $\overline{T_{\mathrm{amb}}}$ is an innovation. This procedure formalizes the concept of operating modes as a kind of temporal abstraction, allowing to quickly assess the impact of various application level scenarios without having to re-run all the underlying mode-characterization simulations in the circuit simulator.

Indeed considering an application profile AP defined as a set of pairs combining various ambient temperatures $\overline{T_{a,k}}$ with different operating modes $\mathbf{S}_{\mathrm{OM},k}$ then the total life-time consumption $\Delta A_{\mathrm{AP,M}}$ for a particular failure mode can be obtained as

$$\Delta A_{\mathrm{AP,M}} = \sum_{k} \Delta t_{\mathbf{S}_{\mathrm{OM},k}} \cdot AF_{\mathrm{eff}}(\overline{T_{a,k}}, \mathbf{S}_{\mathrm{OM},k}). \quad (19.19)$$

The innovation realized consists in approximating the effective ageing rate for a given operating mode by interpolating between a limited number ($N \approx 2...4$) of pre-characterized temperature points with an appropriate regressor $\mathfrak{R}$

$$\widehat{AF_{\mathrm{eff}}}(T, \mathbf{S}_{\mathrm{Mode}}) = \mathfrak{R}(T, \{(T_k, AF_{\mathrm{eff}}(\overline{T_k}, \mathbf{S}_{\mathrm{Mode}})) : \forall k \in [1..N]\}). \quad (19.20)$$

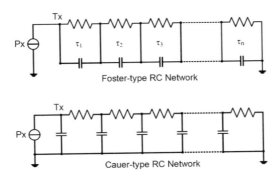

Fig. 19.4: Comparison of Foster- and Cauer-type thermal networks.

19.3.3 Compact Thermal Network

In order to relate the component-level temperatures $\mathbf{T}(\tau)$ to the environmental constraints driven by the mission profile, it is essential to extract a compact thermal network. This network allows to infer component-level temperatures from the effective power dissipation levels in the circuit and the thermal boundary conditions inherited from the mission profile (mainly the T_{amb}).

Using the PTM-ET solver from MAGWEL [10], heating or cooling curves for specific IC and PCB geometries can be extracted. By applying a properly chosen set of excitations it is possible to fully characterize the target thermal system and to model it as a thermal network using identification techniques. Thermal networks of the Cauer type (see Fig. 19.4) are preferred, as, on the contrary to Foster-type networks, they are fully modular and re-usable, [4]. Alternatively, Model Order Reduction techniques may also be used to obtain compact thermal networks from the large mesh built by PTM-ET [10] for the target geometry.

19.3.4 Implementation

In the previous sections we have introduced the notion of operating modes and the associated effective ageing rate. Fig. 19.5 shows the key steps in the flow for evaluating a consumed life-time. The steps in the reddish part serve to characterize the key operating modes. The effective ageing rate at the ambient temperature T_{amb} is computed (average ageing per time-unit). In the blue part the effective ageing rate versus T_{amb} is fitted. We will show in the next sections that the effective ageing rate can be interpolated as a function

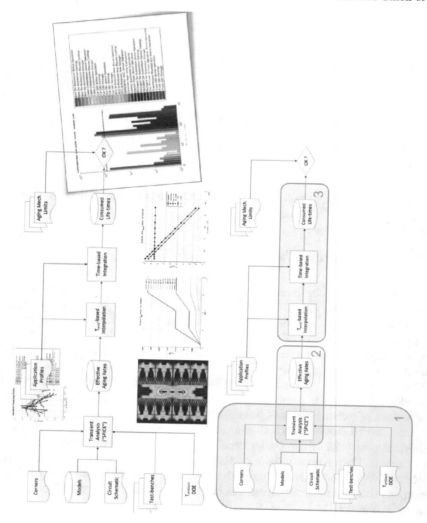

Fig. 19.5: Key steps of the flow in evaluating the consumed life-time. Phase 1 (reddish): Characterize the key operating modes. Phase 2 (blue): Fit the effective ageing rate. Phase 3 (green): Interpolate and integrate the ageing rate and evaluate application profiles.

of temperature from a limited amount of samples. Hence, after fitting, by interpolation and integration of the ageing rate the different application profiles can be determined (green part). By this a very significant speed-up over the vendor recommended approach (to simulate every point of the profile) is achieved.

19.3.4.1 Ageing Framework

We have implemented a series of MATLAB classes and objects allowing to evaluate instantaneous ageing rates as a function of electrical and thermal operating conditions as well as the effective rates in a target operating condition. Theses classes implement the models of Section 19.2.3. Fig. 19.6 shows a typical usage example of these classes.

We also automated the extrapolation of effective ageing rates from a limited set of circuit simulations at few ambient temperatures towards arbitrary mission profile temperatures. Results of this work are shown in Section 19.3.5.

19.3.4.2 Thermal Network Extraction

We extracted compact thermal networks from MAGWEL's PTM-ET [10] simulations using the Foster-Cauer transformation published by Gerstenmaier [4], combined with the method of Habra [6]. Fig. 19.7 shows the quality of the model fit to the simulation data. Fig. 19.8 illustrates that the initially extracted model can be reduced by grouping grid-based time-constants and mapping the groups to a single appropriately chosen time-constant.

Fig. 19.6: Example usage of the MATLAB ageing framework.

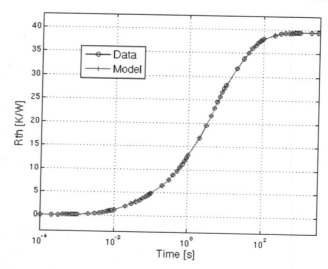

Fig. 19.7: Heating curve showing the fit of the extracted thermal network model.

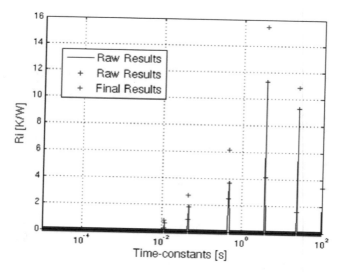

Fig. 19.8: Model compaction though the grouping of time-constants.

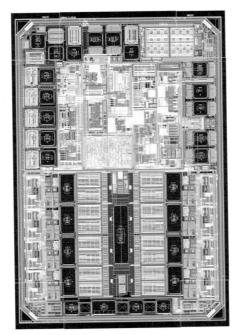

Fig. 19.9: Layout view of Test-Case 3 [7], showing the driver transistors at the bottom.

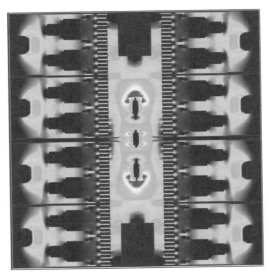

Fig. 19.10: Current-density map of the driver transistors of Test-Case 3 [7], generated with PTM-ET [10].

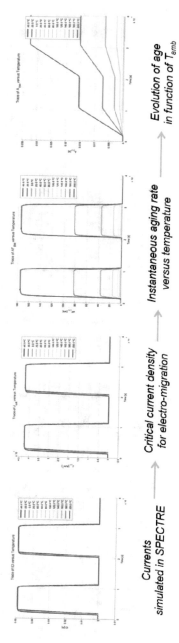

Fig. 19.11: For each operating mode the average ageing rate is determined.

19.3.5 Application to a Test-Case

This section describes the application of the concepts of operating modes and effective ageing rate to the fast computation of life-time consumption in the various application scenarios listed in Section 19.2.5. Fig. 19.11 indicates how for each operating mode the average ageing rate is determined.

19.3.5.1 Description of the Test-Case

The driver IC of Test-Case 3 (see [7]) is used to demonstrate the application of the new ageing framework concepts. Fig. 19.9 and Fig. 19.10 show a view of the layout of the IC as well as a current density map generated with PTM-ET [10] for the driver transistors seen at the bottom of the layout. The current density maps generated with PTM-ET and shown in (19.10) allow to identify the locations with highest current density in the target operating mode. By this the terminal currents are linked to the worst current density. The operating mode was characterized using the Spectre circuit simulator [16]. In the target operating mode, the IC generates current pulses with tightly controlled spectral characteristics in order to meet automotive electromagnetic compatibility constraints. Fig. 19.13 and Fig. 19.12 show the current and gate-voltage waveforms occurring at a series of temperatures covering the

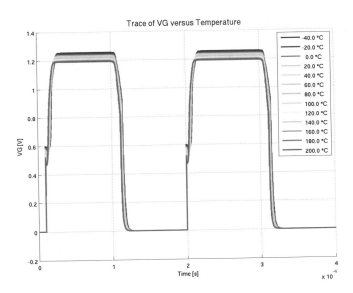

Fig. 19.12: Gate-voltage waveforms in the operating mode of interest.

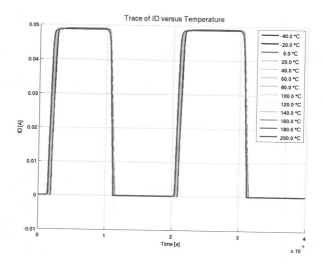

Fig. 19.13: Drain-current waveforms in the operating mode of interest.

range of interest. Fig. 19.11 show the subsequent steps in computing the average ageing rate over each operating mode.

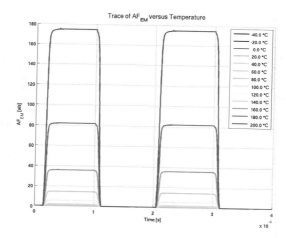

Fig. 19.14: Instantaneous effective ageing-rate for electro-migration.

19.3.5.2 Extraction of the Effective Ageing-rate Model

From the current-density analysis shown in Fig. 19.10, the location for worst-case current density was identified. Using the ageing models described in Section 19.2.3, the instantaneous ageing rates for electro-migration were then computed at all temperature points. Next, the instantaneous ageing rates were averaged over a few cycles of the signals in order to obtain the effective ageing rate.

From Fig. 19.14 and Fig. 19.15 it is immediately apparent that temperature has an overwhelming influence on the ageing rate and the cumulative age.

Ageing rates at low temperature are several orders of magnitude lower than the ageing rates at high temperatures as can be expected from the activation-energy formulation of the temperature dependence of the acceleration factor. In order to further study this effect, the effective ageing rate was split in operating-point-dependent and ambient-temperature-dependent factors. As shown in Fig. 19.16, the operating-point dependent factor (shown as a blue line) has still a weak temperature dependency due to the temperature-sensitivity of the operating point of the circuit. Fig. 19.17 shows the operating-point dependent factor on a linear scale. This smooth behaviour is typical for analogue circuits where minimizing the impact of temperature is a key objective, which is achieved by favoring linear or quadratic temperature behaviours and applying compensation techniques.

The effective ageing rate function $\widehat{AF}_{\text{eff}}(T, \mathbf{S}_{\text{Mode}})$, defined in equation (19.18), was extracted by fitting the bias-dependent factor and the temperature dependent factor from Fig. 19.16 with SPLINE functions in the linear and logarithmic domain of temperatures. Four knots (corresponding to the two extreme and two intermediate temperatures) were used to extract the SPLINE

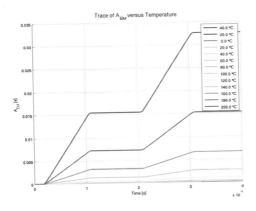

Fig. 19.15: Evolution of the age for electro-migration over two cycles of the target operating mode.

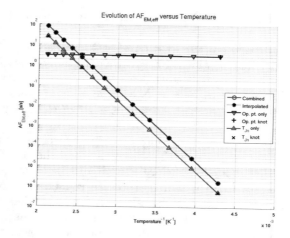

Fig. 19.16: Evolution of effective ageing rate versus temperature for Test-Case 3 [7].

functions. The tight fit obtained between the interpolated points and the originally simulated points in the effective ageing rate Fig. 19.16 and Fig. 19.17 demonstrate the validity of the approach and the fact that a complete temperature profile can be characterized by a limited number of simulation points allowing to realize a substantial reduction of the simulation load.

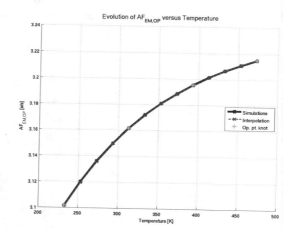

Fig. 19.17: Evolution of the operating-point related factor in the effective ageing rate versus temperature.

19.3.5.3 Quick Evaluation of the Life-Time Consumption

After having computed the effective ageing rate at the target temperatures, we next integrate the effective ageing rate over the target duration. Using the effective ageing rate function $\widehat{AF}_{\text{eff}}(T, \mathbf{S}_{\text{Mode}})$ extracted in the previous section, the evaluation of the life-time consumed in the given operating mode over a collection of temperature profiles such as those shown in Fig. 19.2 becomes rather straight-forward. Indeed the cumulative age is easily computed by summing the product of the effective ageing rates and the time spent in the given mode at a specific temperature

$$\Delta A_{\text{EM,Profile}} = \sum_{k=1}^{N_{\text{pts}}} \widehat{AF}_{\text{EM,eff}}(T_k, \mathbf{S}_{\text{Mode}}) \cdot \Delta t_{\text{Profile}}(T_k, \mathbf{S}_{\text{Mode}}). \quad (19.21)$$

This evaluation does not require circuit simulations any more and can be realized very quickly. Fig. 19.18 shows the life-time consumption obtained in the hypothetical case where the application would remain all the time in the operating mode described in Section 19.3.5.1. The life-time limit for electro-migration is shown as the horizontal black line crossing the figure. One can see, that only a few of the most demanding profiles, would cause the chip to exceed its life-time limit. In reality, the chip of Test-Case 3 [7] is designed to operate in application conditions corresponding to the case labelled "ECU in Protected Area", and the model predicts a margin of two orders of magnitude (meaning that electro-migration is probably not the life-time limiting mechanism for this application).

19.3.5.4 Discussion

The methodology described above shows that it is possible to evaluate the life-time consumption of an IC for a detailed temperature profiles without having to run actual circuit simulations at all temperatures. In order to do so, one must define operating modes, characterize them and extract an effective ageing-rate function $\widehat{AF}_{\text{eff}}(T, \mathbf{S}_{\text{Mode}})$.

The extraction of the effective ageing rate function over a limited set of simulated temperature points will only be valid if the evolution of the internal state of the circuit over the temperature range is appropriately smooth. Today, this assumption must be validated by the designer's insight or by running a sufficient number of simulations. However, one can anticipate that by inspection of the effective ageing rates obtained for various degradation mechanisms at several locations in the circuit for a given operating mode, it should be possible to detect if a given temperature point is behaving significantly different from the others with a very limited extra computational cost. This topic is the subject of future work.

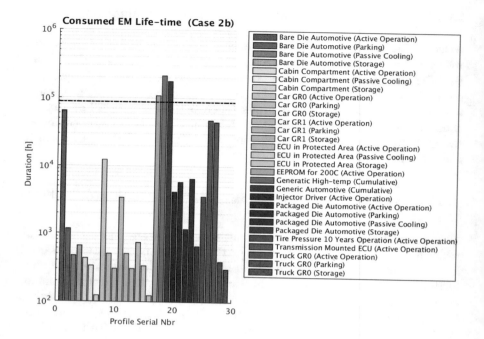

Fig. 19.18: Evaluation of the electro-migration life-time consumed by the circuit of test-case 3 for various application profiles.

The demonstration of the calculation of the consumed life-time was done in the case of a circuit without self-heating. In practice, self-heating in analogue or power circuits is limited to a few tens of degrees Celsius, which is significant from the perspective of accelerating ageing mechanisms, but still limited compared to a typical range of temperature spanned by an application. We therefore anticipate that the methodology developed here will still apply in the case of self-heating. The temperature used as argument in the effective acceleration factor function will then be one of the following possibilities : the ambient temperature, the average temperature at the back of the die in the target operating mode. The self-heating effects will then be incorporated via the operating-point factor of the $\widetilde{AF}_{\text{eff}}(T, \mathbf{S}_{\text{Mode}})$ function.

19.4 Bond Wire Fusing Experiments

An special aspect in reliability is estimating and measuring of bond wire fusing. Bond wires are commonly used to connect the chip and the pins during device assembling. These wires are heated up due to Joule effects and their temperature. Since wire melting is a potential source of failure in IC devices, one would like to estimate, in a expedient manner, the current amplitude and duration that could cause such a failure. Ideally, the sought formula should involve the most important physical parameters that define the package.

In Chapter [3] a transient and fully analytic model for the estimation of the heating of bond wires within a package has been developed. The model resorts to simple mathematical functions, and retains the most important geometric and physical parameters of the package. The model readily permits to estimate the wire melting current and the moulding compound deterioration current. The validation of this model is reported in the Chapters [7] and [8], where its predictions are compared to experimental data.

A number of packages were provided with copper and gold bond wires of varying diameter and length. Then, an experimental set-up was built and time-to-failure data were collected for about two thousand individual wires for four different types and two length categories. Fig. 19.19 shows a snapshot of the Graphical User Interface (GUI) built to control the measurement set-up.

X-ray pictures were also collected on many samples in order to assess the failure mode and ensure that the wires failed by fusion at the expected hot-stop location.

The data collected on the various types of bond wires was analyzed in order to extract the time-to-fail and other representative parameters of the test. Fig. 19.21 shows typical waveforms during the wire-fusing experiments.

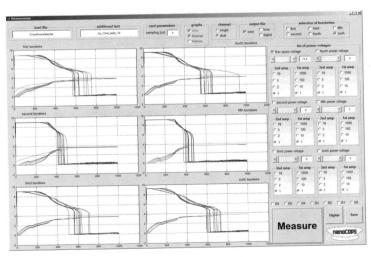

Fig. 19.19: Graphical User Interface built by Brno University of Technology to control the bond wire test-set-up.

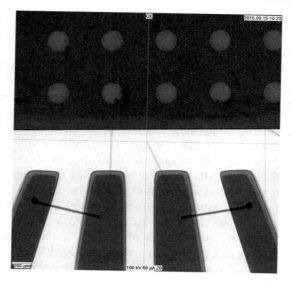

Fig. 19.20: X-ray picture of a failed bond wire (wire leaving third lead-finger from the left).

Key events detected automatically by the post-processing scripts are also marked on the curve.

- The on-set of the current pulse is marked with a triangle pointing forward.
- The end of the current pulse, when the current has fallend to zero, is marked by a triangle pointing backwards.
- The last instant at which the dissipation in the wire is non-zero is marked with a star and defines the "power-death time".
- The instant at which the wire resistance exceeds a value corresponding to a completely molten wire is marked with a ring.

Fig. 19.22 shows the good correlation between the "time to melting point" ($TTMP$) and the "time to power-death" ($TTPD$) indicating that fusion is the dominating failure mode of these experiments.

More details on extensive measurements and simulating bond wire fusing are found in Chapter [7].

19.5 Inclusion of Bond Wires as External Nets in PTM-ET

In this section we shortly describe how bond wires have been included in the PTM-ET simulator of MAGWEL [10]. Bond wires play an important role

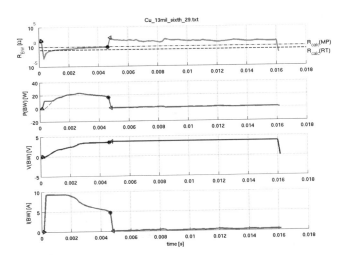

Fig. 19.21: Typical current: $I(BW)$, voltage: $V(BW)$, power: $P(BW)$ and resistance R_{BW} waveforms extracted from the data supplied by Brno University of Technology.

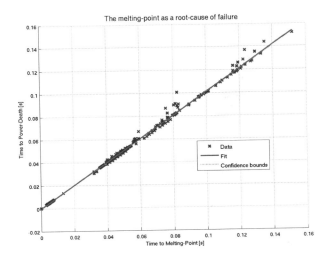

Fig. 19.22: Correlation between the $TTMP$ and $TTPD$, indicating that fusion is the main failure mode.

in setting up the proper electro-thermal boundary conditions for a PTM-ET simulation. In Fig. 19.23 a sketch is given of the mold-package/bond wire/die configuration. The bond wires (black) are located in the mold package (blue) and are connectors between between contacts (red) in the mold and on the die (green). The die is simulated using a detailed field solving. In the mold there is a coarser mesh. The bond wires acquire a thermal profile. They thermally interact with their environment. Inclusion of a field solving approach of the bond wires leads to an prohibitively explosion of the problem size, since a detailed meshing is then required to capture the thin bond wires. As a solution to this each bond wire will be fragmented in sections. We replace a detailed bond wire description by a simplified model of the bond wire while maintaining its thermal profile and exchange of heat with the mold. Some limitations are applied to incorporate the mold in the die simulation but it is not ignored in the computation of the electro-thermal simulation of the die. So far, bond wires are included only as lumped external resistors. We refine the lumped resistor approach by replacing a bond wire by a one-dimensional entity that is fragmented into N sections.

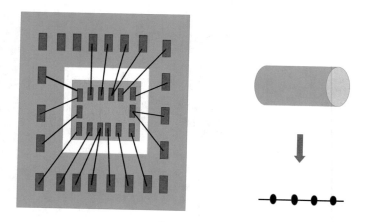

Fig. 19.23: Left: Location of the bond wires (black) between contacts (red) of the mold package (blue) and the die (green). Right: A bond wire fragmented in sections.

The bond wire is a metallic structure for which the regular electro-thermal balance equations apply. We consider the voltage and the temperature as constants over the cross section of the wire. Between the cross sections the balance equations can be taken as one-dimensional. With the bond wire fragmented into N sections this results in a finite-difference formulation in each section for the coupled electro-thermal equations. For more details, also including a more refined modeling, see Chapter [3] in this book.

A number of bond wire parameters must be provided. These parameters are:

- Length
- Diameter
- Material
- Number of fragments
- Die contact
- Ambient thermal contact

The bond wire can be categorized as a specific realization of a more generic entity being the *external net*. An external net is a compact model object attached to a die contact group. The applied excitations are assigned to this die contact, whereas they are physically applied at the mold contact. The external net approach computes the effective voltage and temperatures at the die contact, as well as the ET profiles along the bond wire.

The bond wire data are loaded into PTM-ET using the XML interface in the 'physics section':

```
< physics >

<externalNets>
<externalNet>
<externalNetId>46</externalNetId>
<contact>C1</contact>
<contact>C2</contact>
<type>bondwire</type>
<dofs>10</dofs>
<bondWireLength>150</bondWireLength>
<bondWireDiameter>8</bondWireDiameter>
<bondWireMaterial>METAL1</bondWireMaterial>
<bondWireAmbientMaterial>MOLD1</bondWireAmbientMaterial>
</externalNet>
</externalNets>
<materials>
```

Above $N = 10$, the total bond wire length$=150\mu$ and the diameter equals 8μ. Some general guidelines are

- Rule 1: Every bond wire is declared by adding a section *externalNet* into the XML part

    ```
    <physics>   --   <externalNets> --   ...
    ```

- Rule 2: For each contact C1 in the externalNet, we can apply an electrical voltage or a current-boundary condition. A current boundary conditions leads to a modified DOF counting and a modified assembly of the system of equations. For the same contact we also apply a thermal Dirichlet boundary condition. The value corresponds with the temperature at the other end of the bond wire.

- Rule 3: For each contact C2 in the externalNet, we must apply an electrical current boundary condition with applied current set to zero. For this contact we additionally apply a thermal flux boundary condition with applied flux set to zero. This guarantees that the only role of C2 is to exchange thermal info between the wire and the mold.

The output files 'results_static.csv' and 'results_transient.csv' contain additional rows where one can find the values of the voltage and temperature for each bond wire DOF.

In the mode "etstatic" one finds voltages and temperatures:

- #externalNetVolts_C1 wire attached to C1
- #externalNetTemps_C1 wire attached to C1

In the mode "ettransient" there are $2N$ columns

- dofVolt(C1_k),...,dofTemp(C1_k)...

where k indicates the kth bond wire DOF, and $1 \leq k \leq N = $ dofs.

19.6 Current Density, Temperature and Stress Maps

As last item for addressing reliability we analysed several test-cases in order to understand how a combination of current density, temperature and stress maps would allow to identify the risk areas for isolation cracks without requiring a detailed finite-element simulation of the deformation of the metals over a series of thermal cycles.

A proper combination of mechanical stress and temperature can cause metal to deform as the yield-stress lowers at high temperature, and then freeze as

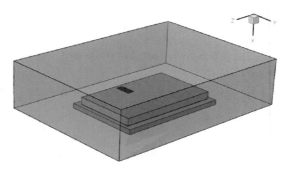

Fig. 19.24: 3D model set up to simulate the distributions of stress on Test-Case 2 [7].

Fig. 19.25: View of the mesh used to simulate stress distributions on Test-Case 2 [7].

the temperature lowers again. The correlation between the stress-state and the ambient temperature ensures that the metal always deforms in the same direction, causing it to move away from the hot-spots and to accumulate in the colder areas, where the accumulation generates a high-level of stress in the surrounding isolation oxide.

Simulations of the distribution of stress on Test-Case 2 [7] were realized in order to identify the high-stress regions. Fig. 19.24 and Fig. 19.25 show the 3D model and the mesh which were built in the Synopsys' Sentaurus Interconnect tool [15] for this purpose.

This gives way to evaluate different metal deformation models in order to

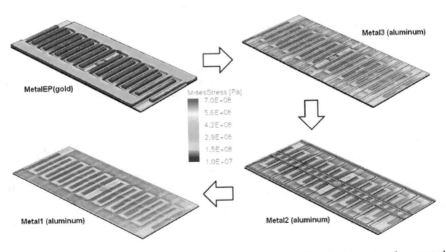

Fig. 19.26: Visualization of the Von Mises stress distribution in the metal layers of Test-Case 2 [7].

assess their ability to predict the zones with risks of creation of cracks in the dielectrics or void in the metals.

Stress distributions such as those shown in Fig. 19.26 were simulated at various ambient temperatures and power-dissipation levels on the die. They were analyzed in order to identify post-processing methods allowing to automatically pin-point risk areas. Details can be found in the companion chapter [8].

19.7 Conclusion

A new generic framework for the fast evaluation of the life-time consumed for a given mission profile was developed and demonstrated. We realized initial analyses of combined temperature, current-density and stress maps produced from 3-dimensional electro-thermal and thermo-mechanical simulations. This gives way to derive a clear criterion allowing to pin-point at the failure locations in designs.

References

1. BERNSTEIN, J.B., GURFINKEL, M., LI, X., WALTERS, J., SHAPIRA, Y., AND TALMOR, M.: *Electronic circuit reliability modeling*. Microelectronics Reliability, 46(12):1957–1979, 2006.
2. CHINIFOROOSH, S., JATSKEVICH, J., YAZDANI, A., SOOD, V., DINAVAHI, V., MARTINEZ, J.A., AND RAMIREZ, A.: *Definitions and applications of dynamic average models for analysis of power systems*. IEEE Transactions on Power Delivery, 25:2655–2669, 2010.
3. DUQUE GUERRA, D.J, CASPER, T., SCHÖPS, S., DE GERSEM, H., RÖMER, U., GILLON, R., WIEERS, A., KRATOCHVIL, T., GOTTHANS, T., AND MEURIS, P.: *Bond Wire Models*. Chapter in this Book, 2019.
4. GERSTENMAIER, Y.C., KIFFE, W., AND WACHUTKA, G.: *Combination of thermal subsystems modeled by rapid circuit transformation*. In: *13th International Workshop on Thermal Investigation of ICs and Systems, 2007. THERMINIC 2007*, pp. 115–120, 2007.
5. GILLON, R., DUQUE, D., GÖTTHANS, T., SCHOENMAKER, W., BRACHTENDORF, H.-G., AND TER MATEN, J.: *D1.6 Public Report (b) on Robustness, Reliability, Ageing*, Project nanoCOPS – nanoelectronic COupled Problems Solutions, FP7-ICT-2013-11/619166, 2016. http://fp7-nanocops.eu/.
6. HABRA, W., TOUNSI, P., MADRID, F., AND DORKEL, J.-M.: *New method of dynamic compact thermal model extraction*. In: *International Conference on Thermal, Mechanical and Multi-Physics Simulation and Experiments in Microelectronics and Micro-Systems, 2008. EuroSimE 2008*, pp. 1–4, 2008.
7. JANSSEN, R., GILLON, R., WIEERS, A., DELEU, F., GUEGNAUD, H., REYNIER, P., SCHOENMAKER, W., AND TER MATEN, E.J.W.: *Test Cases for Power-MOS Devices and RF-Circuitry*. Chapter in this Book, 2019.
8. JANSSEN, R., GILLON, R., WIEERS, A., DELEU, F., GUEGNAUD, H., REYNIER, P., SCHOENMAKER, W., AND TER MATEN, E.J.W.: *Validation of Simulation Results on Coupled Problems*. Chapter in this Book, 2019.

9. JERKE, G., AND KAHNG, A.B.: *Mission profile aware IC design – a case study.* In: *Proc. Design, Automation and Test in Europe Conference and Exhibition (DATE),* http://ieeexplore.ieee.org/document/6800278/, 6p, 2014.

10. MAGWEL NV: *An Electro-thermal module of a power transistor modeler: PTM-ET.* http://www.magwel.com/, 2014.

11. MEDEA+: *2T204 – End of life investigations for automotive systems (ELIAS).* Open access: http://www.catrene.org/web/downloads/profiles_medea/2T204-ELIAS-profile(13-11-07).pdf, 2011.

12. NIRMAIER, T., BURGER, A., HARRANT, M., VIEHL, A., BRINGMANN, O., ROSENSTIEL, W., AND PELZ, G.: *Mission profile aware robustness assessment of automotive power devices.* In: *Proc. Design, Automation and Test in Europe Conference and Exhibition (DATE),* http://ieeexplore.ieee.org/document/6784162/, 6p, 2014.

13. SANDERS, S.R., NOWOROLSKI, J.M., LIU, X.Z., AND VERGHESE, G.C.: *Generalized averaging method for power conversion circuits.* IEEE Transactions on Power Electronics, 6:251–259, 1991.

14. SCHOLTEN, A.J., STEPHENS, D., SMIT, G.D.J., SASSE, G.T., AND BISSCHOP, J.: *The relation between degradation under DC and RF stress conditions.* IEEE Transactions on Electron Devices, 58:2721–2728, 2011.

15. SENTAURUS INTERCONNECT: *Stress related reliability analysis of interconnects.* Synopsys Inc., Mountain View, CA, USA, 2017, urlhttps://www.synopsys.com.

16. SPECTRE: *Spice-class Circuit Simulator.* Cadence Design Systems, Inc., San Jose, CA, USA, 2017, https://www.cadence.com.

Part VI
Test Cases, Measurements, Validation and Best Practices

Part VI covers the following topics.

- **Test Cases for Power-MOS Devices and RF-Circuitry**
  Authors:
  Rick Janssen (NXP Semiconductors, Eindhoven, the Netherlands),
  Renaud Gillon, Aarnout Wieers, Frederik Deleu (ON Semiconductor, Oudenaarde, Belgium),
  Hervé Guegnaud, Pascal Reynier (ACCO Semiconductor, Louveciennes, France),
  Wim Schoenmaker (MAGWEL, Leuven, Belgium),
  E. Jan W. ter Maten (Bergische Universität Wuppertal, Germany).
- **Measurements for RF Amplifiers, Bond Wire Fusing and MOS Power Cells**
  Authors:
  Tomas Kratochvil, Jiri Petrzela, Roman Sotner, Jiri Drinovsky,
  Tomas Gotthans (Brno University of Technology, Czech Republic),
  Aarnout Wieers, Renaud Gillon (ON Semiconductor, Oudenaarde, Belgium),
  Pascal Reynier, Yannick Poupin (ACCO Semiconductor, Louveciennes, France).
- **Validation of Simulation Results on Coupled Problems**
  Authors:
  Rick Janssen (NXP Semiconductors, Eindhoven, the Netherlands),
  Renaud Gillon, Aarnout Wieers, Frederik Deleu (ON Semiconductor, Oudenaarde, Belgium),
  Hervé Guegnaud, Pascal Reynier (ACCO Semiconductor, Louveciennes, France),
  Wim Schoenmaker (MAGWEL, Leuven, Belgium),
  E. Jan W. ter Maten (Bergische Universität Wuppertal, Germany).
- **Methodology and Best-Practice Guidelines for Thermally Optimized Driver Design**
  Authors:
  Renaud Gillon, Aarnout Wieers, Frederik Deleu (ON Semiconductor, Oudenaarde, Belgium),
  Rick Janssen (NXP Semiconductors, Eindhoven, the Netherlands),
  Wim Schoenmaker, Bart De Smedt (MAGWEL, Leuven, Belgium),
  Hervé Guegnaud, Pascal Reynier (ACCO Semiconductor, Louveciennes, France).

Chapter 20
Test Cases for Power-MOS Devices and RF-Circuitry

Rick Janssen, Renaud Gillon, Aarnout Wieers, Frederik Deleu,
Hervé Guegnaud, Pascal Reynier, Wim Schoenmaker, E. Jan W. ter Maten

Abstract This chapter gives an overview of realistic test cases, provided by industrial partners, for validation of the simulation tools, that were developed in the last years. They allowed to compare the results of measurements and simulations for real-life size applications.

20.1 Industrial Uses Cases

Designs in nanoelectronics often lead to large-size simulation problems with different dynamics in sub-systems and include strong feedback couplings [8, 11]. Typical application areas of the industrial partners include power management systems for automobile and LEDs for ON Semiconductor, connecting automotive and secure identification for NXP Semiconductors, and power amplifiers for mobile phones for ACCO Semiconductor. NXP Semiconductors, ON Semiconductor, ACCO Semiconductor, and MAGWEL (tool provider,

Rick Janssen
NXP Semiconductors, Eindhoven, the Netherlands, e-mail: `Rick.Janssen@nxp.com`

Renaud Gillon, Aarnout Wieers, Frederik Deleu
ON Semiconductor Belgium BVBA, Oudenaarde, Belgium, e-mail: {`Renaud.Gillon,`
`Aarnout.Wieers,Frederik.Deleu`}`@onsemi.com`

Hervé Guegnaud, Pascal Reynier
ACCO Semiconductor, Louveciennes, France, e-mail: `Herve.Guegnaud@acco-semi.`
`com,Pascal.Reynier@yahoo.fr`

Wim Schoenmaker
MAGWEL NV, Leuven, Belgium, e-mail: `Wim.Schoenmaker@magwel.com`

E. Jan W. ter Maten
Bergische Universität Wuppertal, Germany, e-mail: `Jan.ter.Maten@math.`
`uni-wuppertal.de`

459
E. J. W. ter Maten et al. (eds.), *Nanoelectronic Coupled Problems Solutions*,
Mathematics in Industry 29, https://doi.org/10.1007/978-3-030-30726-4_20

see [10]) provided test cases that were building blocks from two industrial Use Cases:

- Power-MOS devices, with applications in energy harvesting, that involve couplings between electromagnetics (EM), heat and stress, and
- RF-circuitry in wireless communication, which involves EM-circuit-heat coupling and multirate behaviour, together with analogue-digital signals.

We describe these Use Cases first in some detail. Next we map the Test Cases to these Use Cases. Finally, the Test Cases are described in more detail.

20.1.1 Use Case I: "Power-MOS" - Electro-Thermal-Stress Coupling

With the increased need of smart grids for a sustainable energy infrastructure, **power devices** play an important role in both energy harvesting and distribution and thus in controlling overall energy efficiency. Automotive applications are also an important field that requires the handling of demanding electro-thermal operational constraints to the design of both components and systems. Power devices consist of several thousands of parallel channel devices to deliver high throughput of current and/or current control in both CMOS and bipolar technology. The layout of metal interconnects, bond pads, and wirebonds (or bumps/balls) of large-area power semiconductor devices has a profound effect on metal de-biasing, device ON-resistance (Rdson) and reliability (electro-migration, thermal-induced stress, device & material life times, etc.). Metal-interconnects resistance is especially critical in the context of large-area devices that are designed to have a very low (up to a few tens of milliohms) Rdson value.

Fig. 20.1 shows the complex geometrical finger structure in a typical device design of a **power transistor** using six layers of metal. More details on this device can be found in [6, Section 2]. The source drain contacts are located at the top of the design. A series of metal stripes and via patterns will ultimately transport the current to the drains and away from the sources of the individual channels. Having found the electro-thermal solution, hot spot detectors (design-rule checking) can be activated to post-check the full power transistor array. The current flow is really multi-dimensional.

In order to be able to model and simulate such devices we first need to accurately capture the complicated current-flow patterns in such devices with complicated top metal layouts constrained by the package, wire bonding, or ball array requirements. The design challenge here is further increased by the need to separate out the consequences of the **competing physical effects**, which are **strongly coupled**, and moreover depend on the time-dependent input stimuli. **Variability, reliability, ageing** and **robustness** issues are related to electro-migration, thermal induced stress and thermal hot spot

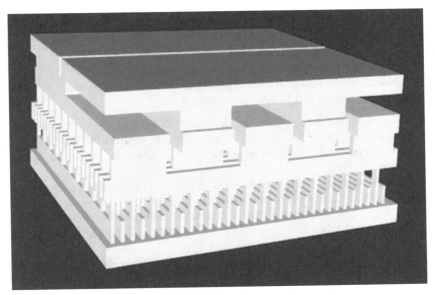

Fig. 20.1: Typical layout of the power transistor (stretched vertical direction) showing its complex geometry

phenomena and, therefore, the current density and temperature distribution has to be computed in the power transistor in operation. The ultimate goal of the design is to provide a **compact model** for the power transistor with wide range of validity that can then be incorporated into the design flow for **yield estimations**, which has to be based on electro-thermal coupling and stress calculations on a chip caused by the heat gradients. These outcomes can then be used to estimate the **power transistor lifetime** (**ageing** problem) by correlating the 3D transient thermal data to electro-migration and stress models, and to generate lifetime estimations for the bond wire interfaces from local currents and temperatures. In order to make possible the evaluation of electro-migration at chip level, where multiple drivers can be combined, a new **methodology** was developed that will convert a product thermal profile into a devices specific requirement, in order to establish the coupled electro-migration constraints. The aspects of **uncertainty** enter into this problem mainly through the closure assumptions: the power transistor is part of a bigger environment (chip) and as such it will be sensitive to third party effects. The impact of the neighboring chip ingredients such as the substrate, the wiring and the loads depend on the electrical activity of the chip and as such are unpredictable. In addition, power transistors typically consist of a large array of small unit devices. Each unit device shows small variations in geometry, and will exhibit correlated variations in device characteristics. Also this type of uncertainties were reviewed and converted

into smart algorithms to efficiently predict the impact of the variations on
the unit device characteristic changes.

Indicative approach and challenges for Use Case I
The solution of the above design challenges requires a multi-disciplinary approach, consisting of the following ingredients:

- **Electro-thermal model:** A more accurate field model for the electrical
  potentials and for the temperature; compact models for the individual
  channels or channel segments and a compact thermal model for the substrate.
- **Analyses:** Accurate determination of the thermal sources and sinks and
  boundaries; high-end processing techniques for finding hot spots and critical segments. Perform stress calculations. Determine efficiently ageing.
  Upgrade the tools to deal with power transistors with 10^8 mesh nodes or
  more.
- **Include variability and reliability constraints** (6σ parameter variations). Derive Model Order Reduction (MOR) methods for capturing the
  I/O relations of the power transistor. Reduced models for the individual field equations require the other field variables as inputs/parameters,
  thus, if they are entered as parameters, parametric MOR methods are
  needed to preserve these quantities as symbolic variables in the reduced
  models.
- **Electro-thermal-stress:** Use the available temperature, current density
  and stress data to make first estimations or predictions of device, bonding
  life times, electro-migration limitations and some thermal induced failures
  like passivation cracks.

20.1.2 Use Case II: "RF-Circuitry" - Circuit-EM-Heat Coupling

The emergence of wireless mass products at centre frequencies in the GHz
range generates a demand from industry for more reliable simulation tools
and design methodologies for RF-circuitry (RF front-end), based on circuit-
EM-heat coupling and mixed analogue-digital simulation, which overcome
the present limitations of lumped circuit models. Some bottlenecks in the
design process will be highlighted by the low Intermediate Frequency (IF)
receiver depicted in Fig. 20.2, which includes a quadrature filter or 90° phase
splitter and a quadrature mixer stage. In this device, the bandpass signal
from the antenna is amplified, filtered, and mixed to the baseband (envelope
or lowpass signal) before digitalization.

The first bottleneck within the design process to be addressed here is
the **multirate** or **multi-scale** problem, i.e., the huge gaps in the spectrum

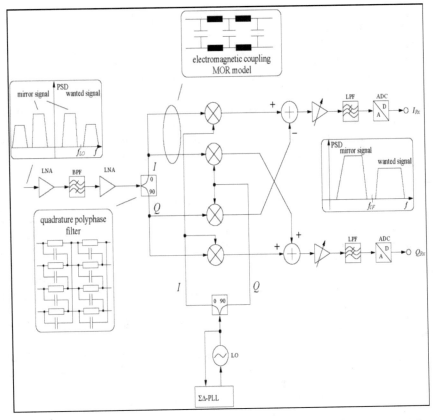

Fig. 20.2: Block diagram of a low-IF or zero-IF receiver architecture with an image reject filter and a quadrature mixer stage. The schematic also displays the power spectral density (PSD) at different stages. Moreover it shows the realization of the image reject filter by a polyphase RC circuit and the symbolic representation of the crosstalk by a lumped LC network (reduced-order model).

of the signals, and to offer a solution to this in the time domain in order to be able to deal with nonlinear couplings. In addition, frequency-domain techniques cannot cope with modulated waveforms such as Orthogonal Frequency Division Multiplexing (OFDM). Though the envelope is of primary interest, the high frequency carrier slows down the time-domain simulation process significantly. A time-domain multirate circuit simulation method [2], also referred to as the Multirate Partial Differential Equation (MPDE) technique, has been enhanced in recent years by the Fachhochschule Oberösterreich (FHO), Hagenberg im Mühlkreiss, Austria and the Bergische Universität Wuppertal (WUP), Germany. This method can handle multi-dynamics problems. Spline/wavelet bases can deal with **discontinuities**, occurring in mixed analogue-digital circuits. Consequently, an efficient and accurate simulation of circuits in the GHz or even THz range can be derived. The same method was enhanced to deal with the implications of **couplings to heat**.

The second bottleneck to be addressed deals with the accuracy of the device models. The phase splitter in Fig. 20.2 is often realized as a polyphase filter with several stages. The on-chip realization of such a polyphase filter structure necessitates device simulation and characterization for taking the **parasitics** and **device variations** and **self-heating** into account. Moreover a severe problem in current receivers is the **crosstalk** between adjacent signal paths over substrate or air. Hence, for highly accurate verification of the various interactions between devices in the RF front-end at these very high frequencies, **EM analysis** is an absolute necessity. A full envelope simulation on the level of detail of an EM analysis tool is out of scope even for the far future. Instead, it is possible to take only a small part of the total design into account in EM analysis, while the rest of the circuit is still simulated at a higher level of abstraction. As in the electro-thermal coupling, **MOR methods** were employed in case the respective other field variables enter as linear inputs in the system, while parametric MOR is required if they enter as parameters. **The coupled system can be solved by achieving co-simulation with dynamic iteration or by holistic/monolithic time-integration** to deal properly with nonlinearities and modulated signals.

The third bottleneck, that needs to be addressed, is the occurrence of **uncertainty in parameters** (Problem 1 in Table 1), which has to be taken into account, to get sufficiently accurate predictions on reliability. The self-mixing of the local oscillator (LO) in Fig. 20.2 is a concern in direct conversion or zero IF (Intermediate Frequency) designs because it directly harms the desired signal through DC offsets [7]. DC offsets reduce the range of operation of the A/D converter, hence increasing the Bit-Error Rate (BER). On the other hand, the output signal of the transmitter arm couples back via the substrate or air to the local oscillator (VCO, Voltage Controlled Oscillator). This is a concern when a Power Amplifier (PA), or at least one stage, is realized on the same IC. This effect is known as **locking** or **pulling**, leading to an unstable oscillator frequency. The operational impacts of the mentioned effects depend strongly on parameter variations. Hence we have to include the effect due to uncertainties in the parameters. Based on this, **yield estimates** should be derived when statistics on the parameter variations are known. **Ageing** causes variations in parameters over a long-term period, which cannot be predicted exactly and thus are typically uncertain. Variability due to ageing was addressed. Problems in an IC are that each device has his own electrical + thermal conditions, changing over time (=ageing); each device has its own required lifetime.

Indicative approach and challenges for Use Case II

Next generation transceiver designs at high carrier frequencies and baseband waveforms such as OFDM require:

- New transient solvers based on the multi-rate envelope method in conjunction with spline/wavelet bases for an optimal signal representation. Enhancements for EM-heat simulations.
- An improved co-simulation or even a holistic/monolithic circuit-EM-heat simulation approach for accurately predicting the signal waveforms in the presence of crosstalk, substrate coupling, mismatch, etc.
- Improved yield analysis based on statistical tools for reliability from uncertainty quantification.
- Efficiently predict ageing effects: lifetime models for electro-migration, thermal induced failures, the construction of accurate probability distributions or probability density functions.

20.1.3 Mapping Test Cases in Industrial Use Cases

The success rate of the newly developed algorithms was defined by demonstrating successful applications of the software developments in several "use cases", provided by the industrial partners (in order to drive these developments). In the previous sections, these Use Cases were defined, as well as the requirements that should be fulfilled. For the industrial partners these are urgent practical real life problems, for which it is essential that they will be solved as soon as possible, in order to keep the competitive edge by reducing cost and time-to-market (design cycle).

The use case description is based on a "top-down" approach. Therefore, the selection of requirement must meet some constraints. The starting point is some product. Therefore, in order to select the use case, it is required to identify an actual product (by name). In this product, some blocks were identified and were submitted to the software developers. Therefore, some design information needed to be provided and the critical parts had to be lifted out for more profound study. The minimum set of use cases is two, which refers to the main coupling cases that were identified in the previous sections. Both use cases address transient issues.

We addressed the problem of coupling from a design point-of-view with the goal of identifying best-practice rules and tool-coupling techniques that can overcome the problems arising in the specific applications in this project, but will also be more widely applicable. In Table 1, the two end-use cases, where the above considerations are clearly present, are given: (1) electro-thermal-stress coupling and (2) circuit-EM-heat coupling, and summarizes the main issues for each Use Case, like the coupling of electromagnetic fields and heat to detailed device semiconductor models.

An inventory was made of available (test cases, preferably with existing experimental results at the end-user sites, and a *test cases repository* was formed. The sample cases from the repository covered all possible types of structures that are suffering from the targeted problems, to be used as working material for the development of simulation software, but also for final validation.

In this chapter we give a description of the selected test cases from the industrial partners. In general, the test cases are quite relevant, since they require a very strong cooperation, making it impossible to simulate when this is not optimally achieved. **Test Cases 1-6 are connected to the specific Use Case I (Power-MOS) and Test Cases 7-12 to Use Case II (RF-circuitry).**

Use Case	Problem 1	Problem 2	Problem 3	Problem 4
I: "Power-MOS": Electro-thermal stress coupling	Electro-Heat coupling. Uncertainty in parameters Ageing.	How to handle the multi-physics.	Strong feedback loop of the thermal effect. Current lack of robustness within co-simulation.	Scale of devices is large: many details lead to 10^8 mesh nodes.
II: "RF-circuitry": Circuit-EM-heat coupling.	Parasitic coupling and cross-talk. Self-heating. Uncertainty.	Multirate baseband-RF signal (envelope + carrier), coupling between analog-digital signals.	Lumped models are not valid anymore. Huge problem size.	How to get around the current long run times.

Table 20.1: Coupled systems: Problems in simulating both targeted Use Cases: Power-MOS (I) and RF-circuitry (II).

20.2 Test Case 1 - A Realistic-Size Power MOS at Constant Temperature

Test Case 1 deals with a design provided by ON Semiconductor. We consider here operation at constant temperature and the focus is on computing Rdson. The structure consists of four layers. The layout of the structure is shown in Fig. 20.3, where we see the top view (top metal).

When launching the PTM (MAGWEL's Power Transistor Modeler [10]) solver, the required mesh consists of approximately 19 million nodes. Tests were done for an applied voltage difference of VDS=0.1V and produce the following results:

The total resistance:

R_total and Power_total

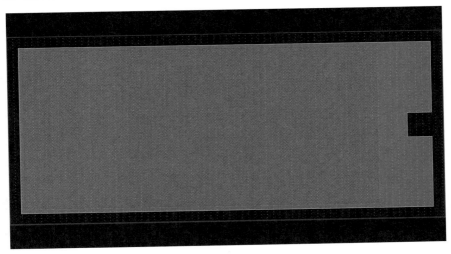

Fig. 20.3: Top view (top metal layer) of Test Case 1

and the R_metal breakdown:

R_Drain_metal, Power_Drain_metal, R_Active, Power_Active, Power_Channel, R_Contact, Power_Contact, R_Metal2, and Power_Metal2.

The Test Case 1 power MOSFET is used in fast switching DC-DC converter modules manufactured by ONN. Those modules can deliver up to 40 A of DC current at full load. The high-side and low-side power MOSFET are hence very large devices with total widths measured in several centimeters. Their on-resistance (Rdson) is typically below 10mΩ. Measuring such low-level resistance is a challenge, even if Kelvin (force-sense) type of contacts are used, as it is essential to ensure a proper repartition of the current over the whole structure.

20.3 Test Case 2 - A Realistic-Size Power MOS in ET Coupling Mode

The next test case represents a realistic industrial Power-MOS device from ON Semiconductor. Measurement data are available and the full strength of the ET (electro-thermal) solver is addressed by both computation speed and memory footprint. Another interesting aspect of this test case is that the impact of packaging and the substrate is also accounted for. Therefore, being successful on this test case means that one goal, e.g., the construction of an electro-thermal solver, both in the strong coupling mode as well as

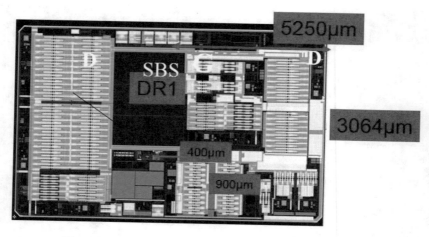

Fig. 20.4: Floorplan of the test chip for Test Case 2.

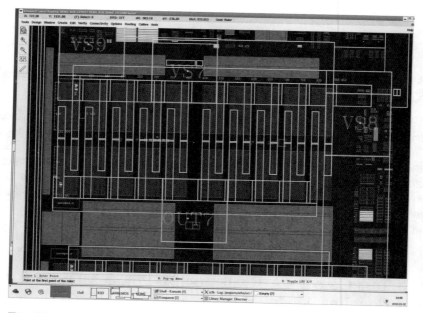

Fig. 20.5: Zoom-in to the DR1 section (see Fig. 20.4) of Test Case 2.

the co-simulation mode, is achieved. The material of the metals and vias is Aluminum. The material of the contacts is Tungsten, which has 174 W/Km as thermal conductivity, and 2.55E+6 J/Km^3 as heat capacity per unit volume. The limiting factor in this test case is the dynamic response, e.g., the static solution is aimed for.

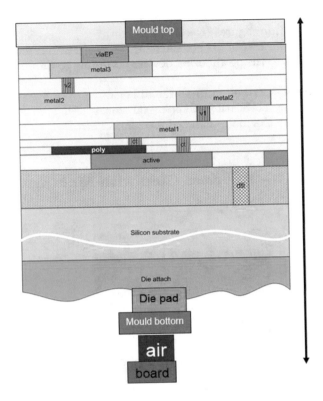

Fig. 20.6: Metal/via stack of the back-end for Test Case 2.

The floorplan of the full test chip is shown in Fig. 20.4. For testing the PTM solver [10], as well as the PTM-ET solver, the part of interest is the one found as DR1 in Fig. 20.4. In Fig. 20.5 a zoom-in to this part is shown. In Fig. 20.6 the metallization stack is illustrated.

The top view is showing the poly fingers and the width of one finger is 152.7 μm. Every pocket contains 8 fingers, so 16 channels. The total width of one pocket equals 2443 μm. Moreover, there are 20 pockets and the effective area is $(152.7 * 2) * (68 * 10) = 0.208$ mm^2.

The test conditions for the device of interest DR1 are: During the short circuit test the driver in question sees the following biases: 0V on source and 14V on drain. The gate is controlled by an external circuit, VGS is 1.63V. The current ID will be limited to approximately 3A. The power is expected to be 42 Watts at the beginning, but then due to self-heating the power decreases.

The timing of the pulses on the drain is: ON time 118μs, period 500μs, duty cycle 23%. Stop these repetitive pulses at 300ms (first cycle). The desired output of the test is:

- Simulation of total Rdson of the driver + interconnect,
- Mean time to failure figures (electromigration),
- Transient dynamic 2D (or 3D) of the temperature distribution on OUT7, and the gradient to the neighboring devices according to the specified bias and power condition,
- Current Density distribution in all metal layers,
- Simulation of the self-heating of the metal tracks,
- Simulation of the thermally induced mechanical stress.

20.4 Test Case 3 - A Driver Chip with Multiple Heat Sources

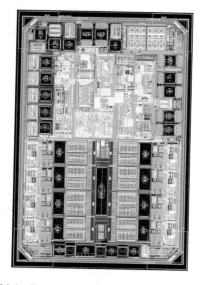

Fig. 20.7: Driver chip layout of Test Case 3.

This test case is a LED driver chip (for lighting, see Fig. 20.7), that contains several power transistors that can each drive one LED string. The drivers can operate in direct current or pulsed-current mode. The chip is very compact and can dissipate a significant amount of heat. It is packaged in an exposed-diepad SSOP (Shrink Small-Outline Package) package. It operates in an automotive environment where it must meet specific reliability requirements.

This chip was selected as main test case for the activity on a methodology for the estimation of ageing and lifetime in the presence of multiple heat-sources and for complex operating profiles with many possible temperature points and operating modes.

20.5 Test Case 4 - Smart Power Driver Test Chip with Thermal Sensor

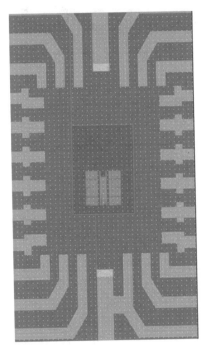

Fig. 20.8: Top view of PTM structure of Test Case 4.

The next test case (see 20.8-20.10) represents a realistic industrial smart power driver with a build-in thermal sensor. This test case allows to dynamically measure the temperature of the driver and also takes the impact of a package and the substrate into account. It is a relevant test case to evaluate the dynamic responses of the electro-thermal solver in strong coupling mode as well as in co-simulation mode.

The smart power driver was developed as a stand-alone driver in a testchip. It consists of a 45V NLDMOS (n-channel lateral, double diffused, MOS device) of which drain, source and gate terminals are made externally accessible through bondpads. Additionally, a thermal sensor is embedded into the

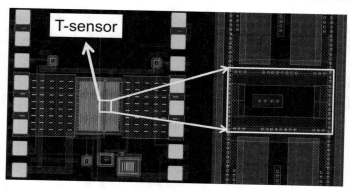

Fig. 20.9: Smart power driver testchip, using a 45V NLDMOS, for Test Case 4.

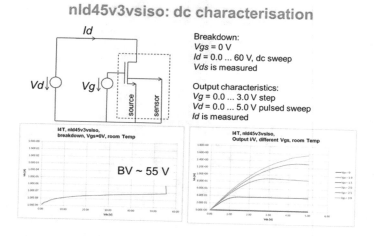

Fig. 20.10: DC BV and I/V characteristic of the Smart power driver for Test Case 4.

testchip, to allow measuring the dynamic temperature over time. This sensor is also made accessible through bondpads.

Ceramic packaged devices have been used at ON Semiconductor to perform first measurements on the power driver. The driver was confirmed to be functional and shows the expected I/V characteristics. Fig. 20.10 shows the typical device characteristic measured at room temperature.

20.6 Test Case 5 - A Fast and Reliable Model for Bond Wire Heating

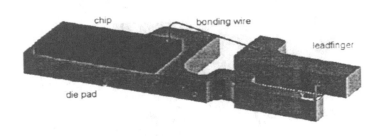

Fig. 20.11: Bond wire test case configuration showing its main components (Test Case 5, Fig. is taken from [13]).

This test case deals with the development and validation of a fast and reliable model for the calculation of the temperature distribution in bond wires (see Fig. 20.11), and thus the determination of the maximum allowable current for its posterior dimensioning as requested by ON Semiconductor. The model permits to take into account the parameters (i.e., moulding compound material and dimensions, bond wire characteristics, etc.) that geometrically define a package [3–5]. To validate our model, results were compared with those obtained through high-fidelity computer simulations of a realistic configuration, as illustrated in Fig. 20.11 taken from [13], and measurement data.

In the smart power electronics industry, bond wires play an essential role in predictive simulation of electro-thermal coupled problems. The reasons are that bond wires often create reliability limitations and depending on its length and diameter also can determine the boundary conditions of the ET solver. Without assessing for reliability issues and without properly setting the boundary conditions the simulation results are "void".

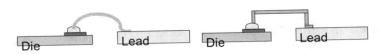

Fig. 20.12: Left panel: real bond wire. Right panel: idealized shape bond wire to deal with meshing issues or to convert to 1D characterization (Test Case 5).

In general, the bond wire is found between a die and lead finger (see Fig. 20.12).

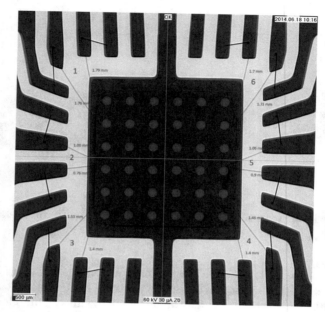

Fig. 20.13: X-ray picture of several samples of the bond wire test-vehicle provided by ON Semiconductor and tested by Brno University of Technology (Test Case 5).

To extract reliability data, a test plan was defined between the Technical University of Brno (BUT) and ON Semiconductor. ON Semiconductor focused on steady-state measurements, while BUT focused on pulsed measurements [9].

The Technical University of Darmstadt developed an analytical electro-thermal model for the bond wire [3–5]. ON Semiconductor specified the input and output requirements for the model. Actually, a data format reduction algorithm was agreed upon, to strongly reduce the data file sizes and measured and calculated data was exchanged.

ON Semiconductor provided a large number of samples of the bond wire test-vehicle shown in Fig. 20.13. The set comprises samples with gold or copper wires of different thicknesses and two lead-frame variants. The fusing tests were performed on the bond wires by the Brno University of Technology [9, section 3]. During the tests, a large constant current was forced into the wires until it would stop to conduct current.

20.7 Test Case 6 - Inductor Test Case

Also test cases were foreseen to simulate device/circuit structures. As a particular choice for a device, an inductor was selected by ACCO Semiconductor, see Fig. 20.14. This test case has been added to the Validation Test Set. In the next stage, this inductor was used by Fachhochschule Oberösterreich (FHO) to set up combined circuit/device simulations. Moreover, it was foreseen that the time integration methods of FHO are to be applied to the field state space description of the inductor. As a first step, we needed to simulate the inductor in stand-alone mode. This already required much of the work that was needed anyway at later stages. For example, the inductor had to be uploaded in the MAGWEL software [10], starting from the technology file and the gds file.

Some extensions are mentioned as well. Among all structures, it was decided to use in a first phase one inductor on silicon (Test Case 6), one Balun (Test Case 11) and one 50Ω transmission line (Test Case 12) on BT (Bismaleimide-Triazine) substrate. All BT cells have been cut for stack analysis and physical data were reinjected in the simulation. Simulations with Momentum (Keysight) [12] were done for comparison.

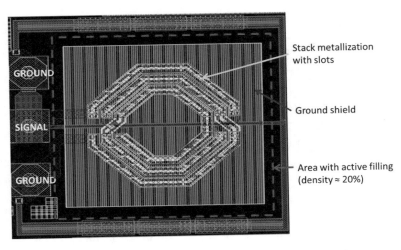

Fig. 20.14: Layout of the Si inductor (Test Case 6).

The inductor is a custom inductance design on silicon technology for an application at 1GHz. The inductance characteristics are:

- Inductance: L = 3.5nH @ 1GHz
- Quality factor: Q = 13 @ 1GHz

- Max quality factor: Qmax = 16 @ 1.7GHz

As can be seen in Fig. 20.14 (layout inductor), this structure can be measured by probing (with a ground-signal-ground, GSG, probe).

20.8 Test Case 7 - RF and Electro-Thermal Simulations

This test case of ACCO Semiconductor (Fig. 20.15) represents a realistic stand-alone power stage used in an actual Power Amplifier (PA) product. The driver stage has been removed to facilitate electro-thermal simulation. In order to measure the transistor junction and the board ground pad temperature, two thermal sensors (diodes) have been integrated. This allows to extract the junction to case (Rjc) and case to ambient (Rca) thermal resistances. All input and output matching networks are integrated on the PCB board (cf. Fig. 20.15).

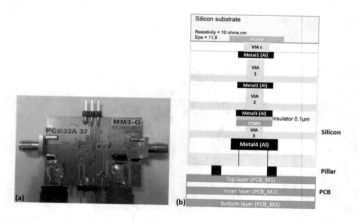

Fig. 20.15: Power stage test boards: (a) picture of the board and (b) stack-up description (Test Case 7).

The stand-alone RF power stage is realized on a $0.18\mu m$ standard CMOS technology and directly flip-chipped on the PCB. Designed for band II (@ 1.9GHz) it delivers on board 26dBm linear output power (@ -40dBc) with 44% efficiency.

20.9 Test Case 8 - Reliable RFIC Isolation

In order to minimize interference issues and coupling effects in RF products, it is essential to apply proper floorplanning and grounding strategies. The interaction of the IC with its physical environment needs to be accounted for, to certify that the final packaged and mounted product meets the specifications. This test case has been provided by NXP Semiconductors. The first focus was on the key requirements to address physical design issues in the early design phases of complex RF designs. Typical physical design issues encountered, such as on-chip coupling effects, chip-package interaction, substrate coupling and co-habitation, have been investigated. The main challenges were the first order prediction of cross-domain coupling. Therefore, we applied a floorplan methodology to quantify the impact of floorplanning choices and isolation grounding strategies (Figs. 20.16-20.17). This methodology is based on a very high-level floorplan EM/circuit simulation model, including the most important interference contributors and including on-chip, package and PCB elements, to be applied in the very early design phases (initial floor planning).

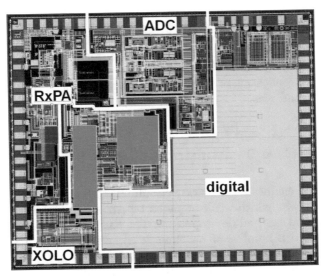

Fig. 20.16: Integrated RF-CMOS automotive transceiver design, with digital aggressor and analog victims (XOLO, RxPA, ADC) (Test Case 8).

The overall model of a complete RF product contains the following parts (see Fig. 20.17):

- On-chip: domain-regions, padring, sealring, splittercells, substrate effects.
- Package: ground and power pins, bond wires/downbonds, exposed diepad.
- PCB: ground plane and exposed diepad connections.

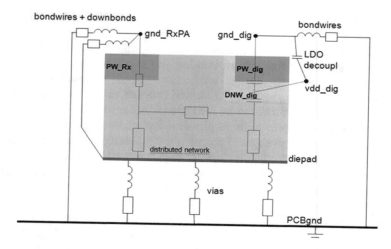

Fig. 20.17: Floorplan testbench model for isolation and grounding strategies (Test Case 8).

The effect of the number of parameter variations on the impact of noise from digital parts on the isolation sensitive RF domains is investigated, i.e., the number of downbonds, the number of ground pins, the domain spacing and shape, the application of deep-Nwell and exposed diepad, and the number of exposed diepad vias.

20.10 Test Case 9 - Multirate Circuit Examples

The simulation engines for multirate circuit simulation were tested against a benchmark library, which was completed by circuit examples from partners, third parties and internal developments. The Fachhochschule Oberösterreich (FHO) provided a number of benchmark circuits. The circuits were available to other partners both as a schematic, as well as a C++ class model. The device models used include standard Spice models, BSIM, VBIC, models as well as selected models from the partner NXP Semiconductors (MOS9, MEXTRAM). The circuit classes focused on oscillators with high quality factors, mixers in the up- and downlink and amplifiers with multitone excitation.

LNA test example
The transAmp is a wide-band low noise amplifier (LNA) in trans-impedance topology in 0.25μm RF-CMOS design. For simulation, the transAmp amplifier is embedded in a testing environment using an external bias tree and input and output modelled as PORT sources, i.e., a voltage source in series

with a resistor (or a current source parallel to a resistor).

Mixer example

The c9linmix mixer is a simple single-ended version of a Gilbert mixer (enhanced by a cascade stage to yield better linearity) for GSM in 0.25μm standard CMOS. For GSM usually the RF input frequencies lie in a frequency band from 925 MHz up to 960 MHz. To down-convert an input signal to the band of a low-IF receiver, the LO frequency must lie very close to the RF input frequency ($|f_{lo} - f_{rf}| < 1MHz$). Furthermore, the bandwidth of the GSM channel is about 200 kHz, consequently, for intermodulation distortion computations, a second input frequency must lay within this distance. Note that the RF input source is a source of type PORT, i.e., a voltage source in series with a resistor. The parameters of such a source should be chosen according to the purpose of the simulation.

Voltage controlled oscillator (VCO)

The vcoBi oscillator is a simplified version of a fully integrated 1.3 GHz LC-tank VCO for GSM in 0.25μm standard CMOS. The VCO is tunable from about 1.2 GHz up to 1.4 GHz and has a very low phase noise of less than -130 dB at 1 MHz offset from the carrier frequency. For simulation, the oscillator is embedded in a testing environment using a virtual output buffer load and tuning voltage as well as core current modelled as independent DC sources.

Quartz crystal oscillator (Colpitts)

This is a standard Colpitts oscillator circuit operating at 20 MHz in bipolar technology. The crystal is replaced by a series of RLC elements and a capacitor in parallel representing the package.

PLL Design

The Phase-Locked Loop (PLL) design was designed for the industrial, scientific, and medical radio band (ISM band) at 5.8 GHz with a tuning range between 5.6 and 6 GHz. 85 channels can be allocated with a channel bandwidth of 150 MHz. This design is described in [1] in detail. The frequency divider chain of the design divides the frequency in 5 stages of ratio 2 from approximately 1MHz to approximately 31.25 kHz. Fig. 20.18 shows the input signals for different stages in the divider chain. Especially the huge spread of the frequencies in the divider chain is highly challenging for multirate methods.

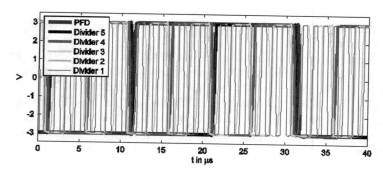

Fig. 20.18: Several signals in a frequency divider chain as part of a PLL (part of Test Case "set" 9).

20.11 Test Case 10 - Silicon Test Chips for Step-by-Step Testing and Validation: Driver and Power Stage

For large-scale electro-thermal co-simulation of power circuits, ACCO Semiconductor has prepared three different Silicon test chips (see Fig. 20.19). They were used for a step-by-step study, measurements and validation of the MAGWEL software [10].

This test case represents a realistic stand-alone Power Amplifier (PA) used in an actual PA product at 3 different steps of integration. It is realized in a $0.18\mu m$ standard CMOS technology. Designed for band II (@$1.9GHz$), the PA delivers on board plane 26dBm linear output power ($@ - 40dBc$) with 44% PAE (Power Added Efficiency).

This test case is composed of three different dies:

1. Stand-alone power stage (PS) mounted directly on the testing board (PCB): The driver stage has been removed to facilitate electro-thermal retro simulation. In order to measure the transistor junction and the board ground pad temperature, two thermal sensors (diodes) have been integrated. This allows to extract the junction temperature (Tj) and case temperature (Tc). This part reuses Test Case 6 (see section 20.8).
2. Stand-alone driver stage (DS) mounted directly on the testing board (PCB): The simulation process is the same as for the stand-alone power stage. All input and output matching networks are integrated on the PCB board.
3. Stand-alone driver and power stage cascaded.
4. Full PA (driver + power stage integrated on the same die) mounted on Bismaleimide-Triazine (BT) + PCB. Input and output matching networks are integrated on the PCB. Inter-matching network is integrated on Silicon (capacitances) and BT (inductances).

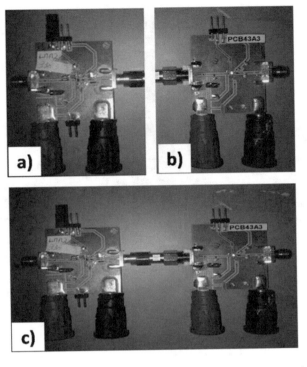

Fig. 20.19: Stand-alone (a) driver stage, (b) power stage and (c) DS + PS cascaded (Test Case 10).

Performances of the two stand-alone driver and power stages in series can be compared to the full PA. These structures allowed to evaluate the package and substrate thermal impact on RF performances. To optimize the heat flow, the two first stand-alone power and driver stages are directly mounted on the PCB. Then, the third one, the BT substrate (4 copper layers) is insert between the die and PCB (see Fig. 20.21). All the junctions between PCB, BT and the test chip are made by bump pads.

20.12 Test Case 11 - Transmission Line

The transmission line is a 50 Ω coplanar line integrated on the top layer (metal 1). The physical characteristics (length, width, gap), determined by cross section analysis, are given in Fig. 20.22, together with the view (MAG-WEL).

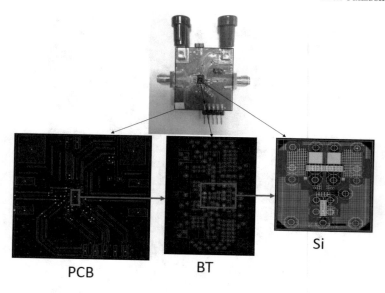

Fig. 20.20: Full power amplifier integrated on silicon + BT + PCB (Test Case 10).

20.13 Test Case 12 - Balun

A balun is an electrical device that converts between a balanced signal (two signals working against each other, where ground is irrelevant) and an unbalanced signal (a single signal working against ground or pseudo-ground). The balun used is a 3-port component integrated on 4 layers BT technology (Fig. 20.23). It has two main functions:

- Convert balanced (ports 1 and 3) input signals to an unbalanced output signal (port 2)
- Make an impedance transformation from 50 Ω single to 20 Ω differential

The primary winding (on the differential side) is integrated in the second metal layer (metal 2) and the secondary (on the single ended side) in the first metal layer (metal 1). As for the coplanar line, the cross section analysis has been done and the measured data are used for simulation.

References

1. BRACHTENDORF, H.-G., AND BITTNER, K.: *Fast Algorithms for Grid Adaptation Using Non-uniform Biorthogonal Spline Wavelets*, SIAM Journal on Scientific Computing, vol. 37, no. 2, pp. 283–304, 2015.

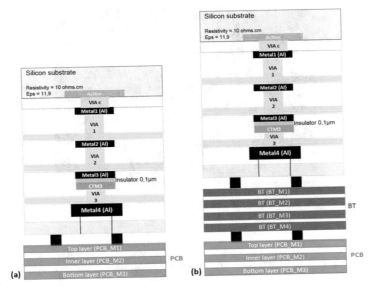

Fig. 20.21: Assembly (a) silicon + PCB and (b) silicon + BT + PCB (Test Case 10).

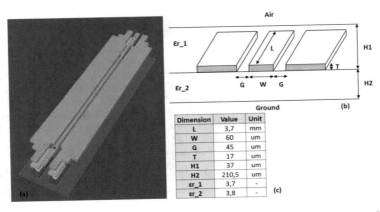

Fig. 20.22: 50 Ω line. (a) DevEM 3D view, (b) model and (c) physical characteristics (Test Case 11).

2. BRACHTENDORF, H.-G., WELSCH, G., LAUR, R., AND BUNSE-GERSTNER, A.: *Numerical steady state analysis of electronic circuits driven by multi-tone signals*, Electrical Engineering, vol. 79, Issue 2, pp. 103–112, 1996
3. DUQUE, D., GOTTHANS, T., GILLON, R., AND SCHÖPS, S.: *An Extended Analytic Model for the Heating of Bondwires.* https://arxiv.org/pdf/1709.10069.pdf, 2017.
4. DUQUE, D., AND SCHÖPS, S.: *Fast and Reliable Simulations of the Heating of*

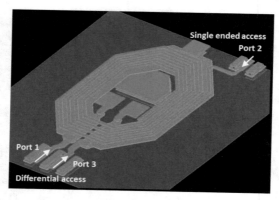

Fig. 20.23: Balun structure (Test Case 12).

Bond Wires. In: RUSSO, G., CAPASSO, V., NICOSIA, G., AND ROMANO, V. (EDS.): *Progress in Industrial Mathematics at ECMI 2014*, Series Mathematics in Industry Vol. 22, Springer International Publishing, pp. 819–827, 2016. https://doi.org/10.1007/978-3-319-23413-7_114.

5. DUQUE, D., SCHÖPS, S., DE GERSEM, H., AND WIEERS, A.: *nanoCOPS: Analytical Approach for Estimating the Heating of Bond-wires*, ECMI Newsletter 56, pp. 70–71, 2014. Open access: http://www.mafy.lut.fi/EcmiNL/issues.php?action=viewart&ID=353.

6. FENG, L., AND BENNER, P.: *Parametric Model Order Reduction for Electro-Thermal Coupled Problems*. Chapter in this Book, 2019.

7. ONLINE: https://en.wikipedia.org/wiki/Intermediate_frequency, 2018.

8. JANSSEN, H.H.J.M., BENNER, P., BITTNER, K., BRACHTENDORF, H.-G., FENG, L., TER MATEN, E.J.W., PULCH, R., SCHOENMAKER W., SCHÖPS, S., AND TISCHENDORF, C.: *The European Project nanoCOPS for Nanoelectronic Coupled Problems Solutions*. In: RUSSO, G., CAPASSO, V., NICOSIA, G., AND ROMANO, V. (EDS.): *Progress in Industrial Mathematics at ECMI 2014*, Series Mathematics in Industry Vol. 22, Springer International Publishing, pp. 835–842, 2016. http://dx.doi.org/10.1007/978-3-319-23413-7_116

9. KRATOCHVIL, T., PETRZELA, J., SOTNER, R., DRINOVSKY, J., GOTTHANS, T., WIEERS, A., GILLON, R., REYNIER, P., AND POUPIN, Y.: *Measurements for RF Amplifiers, Bond-Wire Fusing and MOS Power Cells*. Chapter in this Book, 2019.

10. MAGWEL NV: *Device-Electro-Magnetic Modeler (devEM)* and *Power Transistor Modeler: PTM*, MAGWEL NV, Leuven, Belgium, 2018, http://www.magwel.com/.

11. TER MATEN, E.J.W., PUTEK, P.A., GÜNTHER, M., PULCH, R., TISCHENDORF, C., STROHM, C., SCHOENMAKER, W., MEURIS, P., DE SMEDT, B., BENNER, P., FENG, L., BANAGAAYA, N., YUE, Y., JANSSEN, R., DOHMEN, J.J., TASIĆ, B., DELE, F., GILLON, R., WIEERS, A., BRACHTENDORF, H.-G., BITTNER, K., KRATOCHVÍL, T., PETRZELA, J., SOTNER, R., GÖTTHANS, T., DŘÍNOVSKÝ, J., SCHÖPS, S., DUQUE GUERRA, D.J., CASPER, T., DE GERSEM, H., RÖMER, U., REYNIER, P., BARROUL, P., MASLIAH, D., AND ROUSSEAU, B.: *Nanoelectronic COupled Problems Solutions - nanoCOPS: Modelling, Multirate, Model Order Reduction, Uncertainty Quantification, Fast Fault Simulation*, Journal for Mathematics in Industry 7:2, 2016. Open access: http://dx.doi.org/10.1186/s13362-016-0025-5

12. MOMENTUM: *3D planar electromagnetic (EM) simulator used for passive circuit analysis*, Keysight Technologies, Santa Rosa, CA, USA, 2018, https://www.keysight.com.

13. NÖBAUER, G.T., AND MOSER, H.: *Analytical Approach to Temperature Evaluation in Bonding Wires and Calculation of Allowable Current*, IEEE Transactions on Advanced Packaging, vol. 23-3, pp. 426–435, 2000.

Chapter 21
Measurements for RF Amplifiers, Bond Wire Fusing and MOS Power Cells

Tomas Kratochvil, Jiri Petrzela, Roman Sotner, Jiri Drinovsky,
Tomas Gotthans, Aarnout Wieers, Renaud Gillon, Pascal Reynier,
Yannick Poupin

Abstract This chapter focuses on several areas regarding measurements (including methodology, test-bed development and measurement of results itself) of experimental results in the framework of an established cooperation with several partners from industry. We give attention to measurements for RF amplifiers, bond wire fusing and MOS power cells. We also include on-chip measurements of RF passive components and pay attention to heat, stress and reliability measurements. Measurements set-ups have been made in close cooperation with ON Semiconductor Belgium (Oudenaarde, Belgium) and with ACCO Semiconductor (Louveciennes, France).

21.1 RF Power Amplifiers Measurements and Verifications

In order to create an efficient automated and flexible test bench for measuring the Power Amplifiers (PA), the possibility of Software Defined radio (SDR) needs to be considered. The standard methods, nowadays, are using expensive and not much flexible devices in order to test the parameters of

Tomas Kratochvil (Tomáš Kratochvíl), Jiri Petrzela (Jiří Petržela),
Roman Sotner (Roman Šotner), Jiri Drinovsky (Jiří Dřínovský),
Tomas Gotthans (Tomáš Götthans)
Brno University of Technology, Brno, Czech Republic, e-mail: {KratoT,PetrzelJ,
Sotner,Drino,Gotthans}@feec.vutbr.cz

Aarnout Wieers, Renaud Gillon
ON Semiconductor Belgium BVBA, Oudenaarde, Belgium, e-mail: {Aarnout.Wieers,
Renaud.Gillon}@onsemi.com

Pascal Reynier, Yannick Poupin
ACCO Semiconductor, Louveciennes, France, e-mail: Pascal.Reynier@yahoo.fr,
Yannick.Poupin@acco-semi.com

© Springer Nature Switzerland AG 2019
E. J. W. ter Maten et al. (eds.), *Nanoelectronic Coupled Problems Solutions*,
Mathematics in Industry 29, https://doi.org/10.1007/978-3-030-30726-4_21

PAs. One of the biggest benefits of SDRs is their flexibility. They can evaluate GSM, EDGE and 3G networks, as well as LTE networks (we collected several abbreviations used in Communication Technology in Table 21.1). For creating an automated test bench, reconfigurability is necessary that allows for preparation, transition, reception and evaluation of the test signals. Due to multiband properties of the PA, an external control device has to be used as well. The PA under consideration required several control signals. We used a development board with the microcontroller STM32 controlled via USB. Among many other benefits, synchronization is relatively easy with the transmitted signal. First, we tested three different SDRs (ETTUS USRP, NI FLEXRio and BladeRF). After the testing stage, we chose the one with the best performance in terms of precision. After receiving the test signal in the feedback path, the signal post-processing needed to be done. The SDR BladeRF is used with the GPIO board in order to control additional devices (such as the PA). The capabilities of SDRs have been tested in the operating systems Windows and Linux. We tested the possible usage of GNU Radio in combination with Python. The next item was the control of power sources. Here, the voltage sources (V_{ramp} and V_{batt}) have to be controlled and monitored. Such switching needs to be evaluated in order to have precise timing and to restrict voltage overshoot (current respectively). Another fact that was taken into account is measuring the power of the RF signal that is transmitted in time-divided buffers. Such measurements have to be triggered with power detection. To obtain the same test results, a set of test signals had to be delivered. In fact, using different test signals could possibly lead to differences in results. Another approach that has been evaluated is the usage of an RF generator (Agilent MXG[1]) and a real-time spectrum analyzer.

21.1.1 Universal Test Bench for RF Amplifiers

The test bench was made using a power supply, a personal computer with MATLAB [11], a 12 GHz Agilent direct sampling generator, an Agilent digital oscilloscope and a Rohde & Schwarz[2] spectrum analyzer. The first tests were performed with the PA on the test-bench with an Agilent MXG generator and with an Agilent DSO[3] oscilloscope with 8-bit resolution and sampling frequency 4 GHz. Preliminary measurements have been acquired by the test-bench that is in Fig. 21.1. For measuring the power cell, we have as-

---

[1] MXG – X-Series Microwave Analog Signal Generator (9 kHz–40 kHz), or RF Vector Signal Generator (9 kHz–6 GHz). https://www.keysight.com/en/pcx-x205194/x-series-signal-generators-mxg-exg?cc=NL&lc=dut.

[2] Rohde & Schwarz, GmbH & Co. KG, Munich, Germany, https://www.rohde-schwarz.com.

[3] DSO – Digital Storage Oscillator. https://www.keysight.com/en/pcx-x2015004/oscilloscopes?cc=NL&lc=dut.

Table 21.1: Abbreviations used in communication technology (signals, Software Defined Radio).

GSM	Global System for Mobile Communications, https://en.wikipedia.org/wiki/GSM.
EDGE	Enhanced Data rates for GSM Evolution, https://en.wikipedia.org/wiki/Enhanced_Data_Rates_for_GSM_Evolution.
3G	Third Generation of wireless mobile telecommunication technology, https://en.wikipedia.org/wiki/3G.
3GPP	The 3rd Generation Partnership Project is a collaboration between groups of telecommunications standards associations, known as the Organizational Partners. http://www.3gpp.org/about-3gpp.
LTE	Long-Term Evolution, standard for high-speed wireless communication for mobile devices and data terminals, https://en.wikipedia.org/wiki/LTE_(telecommunication).
ETTUS USRP	Universal Software Radio Peripheral (USRP) is a range of software-defined radio designed and sold by Ettus Research, https://en.wikipedia.org/wiki/Universal_Software_Radio_Peripheral.
NI FLEXRio	Offers the flexibility of custom hardware without the cost of custom design by combining large, user-programmable FPGAs and high-performance analog, digital, and RF I/O (National Instruments), http://www.ni.com/nl-nl/shop/electronic-test-instrumentation/flexrio/what-is-flexrio.html.
BladeRF	A Software Defined Radio (SDR) platform designed to enable a community of hobbyists, and professionals to explore and experiment with the multi-disciplinary facets of RF communication, https://www.nuand.com/.
GPIO	General-Purpose Input/Output is a generic pin on an integrated circuit or computer board whose behavior (including whether it is an input or output pin) is controllable by the user at run time. GPIO pins have no predefined purpose, and go unused by default. https://en.wikipedia.org/wiki/General-purpose_input/output.
GNU Radio	A free software development toolkit that provides signal processing blocks to implement software-defined radios and signal-processing systems. https://en.wikipedia.org/wiki/GNU_Radio.
WCDMA	Wideband Code Division Multiple Access is a designated air interface for one of the International Telecommunications Union's (ITU's) family of 3G mobile communications systems. WCDMA is used in the radio leg of both UMTS and HSPA networks. http://www.3gpp.org/technologies/keywords-acronyms/104-w-cdma.
HSUPA	High Speed Packet Access (HSPA) is an amalgamation of two mobile protocols, High Speed Downlink Packet Access (HSDPA) and High Speed Uplink Packet Access (HSUPA), that extends and improves the performance of existing 3G mobile telecommunication networks using the WCDMA protocols. https://en.wikipedia.org/wiki/High_Speed_Packet_Access.

sembled a test-bench made of an R&S SMU 200A Vector Signal Generator, R&S FSVR real-time spectrum analyzer, and a precise multi-meter Agilent 34401A. The power stage (active part) is integrated on silicon inside the flip chip and is assembled on a 3-layer test board. The biasing was externally driven. The input and output matching network was integrated on a printed circuit board (PCB) with a loss of about 0.5 dB, see Fig. 21.2(left). An example of the measured ultra-wideband RF spectrum in the EDGE mode with

$V_{\text{batt}} = 4.5\,\text{V}$, $V_{\text{ramp}} = 1.2\,\text{V}$ is shown in Fig. 21.2(right). High-Speed Uplink Packet Access (HSUPA) with enhanced uplink using a new transport channel to WCDMA was used. More specifically: 3GPP - WCDMA, Uplink, and the Release 6 standard was employed. The carrier frequency was 1900 MHz.

Fig. 21.1: Experimental test-bench for PA measurement.

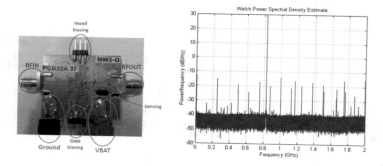

Fig. 21.2: Experimental PCB of PA and initial test.

The dependency of the output power with respect to the input power in the linear regime can be seen in Fig. 21.3(Top-left). The frequency-dependent gain of the characterized PA was calculated and is presented in Fig. 21.3(Bottom-left). Another key property that was measured is Adjacent Channel Power (ACP), measured in the 5 MHz and 10 MHz offset side bands. The results of maximal ACP are presented in Fig. 21.3(Bottom-right). The Power Added Efficiency (PAE) with respect to output power is depicted in Fig. 21.3(Top-right) as well.

Scattering parameters (S-parameters)[4] describe the electrical transfer (output as function of input) behavior of linear electrical networks when

---

[4] https://en.wikipedia.org/wiki/Scattering_parameters

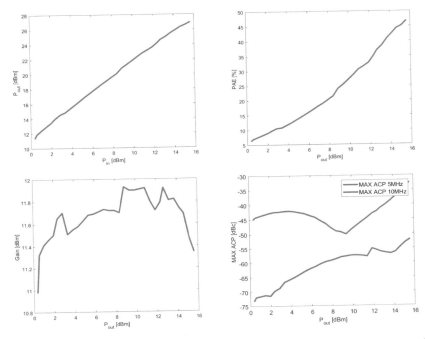

Fig. 21.3: Exemplary results of measurement. **Top-left**: Output power be-haves as a linear function of the input power. **Top-right**: PAE (Power Added Efficiency) as function of the output power. **Bottom-left**: Frequency-dependent gain as function of the output power. **Bottom-right**: The max-imum Adjacent Channel Power (ACP), measured in the 5MHz and 10MHz offset side bands, as depending on the output power.

undergoing various steady-state stimuli by electrical signals. S-parameters change with the measurement frequency, so the frequency must be specified for any S-parameter measurement stated, in addition to the characteristic impedance or system impedance. For measuring the S-parameters, an Ag-ilent E5071C ENA Series Network Analyzer was used. In the S-parameter approach, an electrical network is regarded as a 'black box' containing var-ious interconnected basic electrical circuit components or lumped elements such as resistors, capacitors, inductors and transistors, which interacts with other circuits through ports. The network is characterized by a square ma-trix of complex numbers, called the S-parameter matrix, which can be used to calculate its response to signals applied to the ports. The scattering param-eters have been measured in the frequency range from 300 kHz up to 6 GHz. Fig. 21.4 depicts the results of S_{11}, S_{12}, S_{21} and S_{22}.

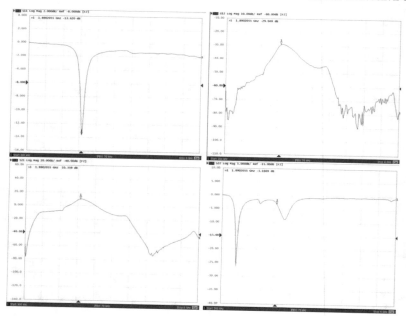

Fig. 21.4: S-parameters as depending on the input frequency in the range 300 kHz to 6 GHz (all four horizontal axes). **Top-left**: The screen shot shows the S_{11} parameter (input matching) as a function of frequency. **Top-right**: shows the S_{12} parameter. **Bottom-left**: S_{21}, known as gain. **Bottom-tight**: the S_{22} parameter (output matching).

21.2 RF Passive Components On-Chip Measurements

Our activities in this area of research were focused on tests of integrated passive elements (RF coils and capacitors - technology $0.18\,\mu$m with five metal layers. Components are integrated on the first two layers, seen from the top: first a thick aluminium ($4\,\mu$m) layer and just below, the thick copper ($2\,\mu$m) metal layer). For the measurements, we used a GSG[5] probing system (Fig. 21.5) with a Cascade table and a ZVA Rohde & Schwarz Vector Network Analyzer. On silicon, the technology uses $0.18\,\mu$m with five metal layers.

The tested silicon wafer, as can be seen in Fig. 21.6, consists of several components. We performed measurements on the identified parts S1 (inductor $L = 1.78\,$nH, $Q_{\max} = 15$ at $3.45\,$GHz), S3 (inductor $L = 1.2\,$nH, $Q_{\max} = 18$ at $3\,$GHz), S4 (inductor $L = 4.2\,$nH, $Q_{\max} = 16$ at $1.65\,$GHz) and S5 (capacitor

---

[5]   https://www.lakeshore.com/products/Cryogenic-Probe-Stations/Pages/GSG-Microwave-Probes.aspx

Fig. 21.5: Probing station and measuring equipment.

$C = 17 \, \mathrm{pF}$). The details of probing ground-signal-ground (GSG[6] Picoprobe on silicon wafer) can be seen in Fig. 21.7(left). The crucial parameter is the probing resistance. In order to minimize this issue, important caution had to be taken, see the detailed view on the probing contact (Fig. 21.7(right)).

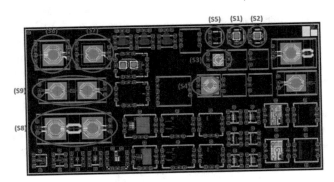

Fig. 21.6: Layout of the test chip. Components are integrated on the first two layers as seen from the top: first a thick aluminium (4 μm) layer and just below a thick copper (2 μm) metal layer.

---

[6] http://www.ggb.com/40a.html

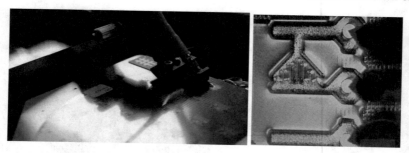

Fig. 21.7: Probing of the silicon wafer including a detailed view on the contacts.

21.3 Fusing Tests and Bond Wires Testing

The mathematical approaches to the problem of predicting and estimating the fusing current were defined many years ago [12, 13]. However, the majority of published works focusing on bonding problems provide a prediction of bond wire behavior only if the DC current is conducting [8]. Particular attention is paid to the destructive static tests [15]. However, practical experiments were not provided in many cases and only a limited number of papers were supported by real experiments. The measurement set-up proposed in [9] is dedicated for high-power IC chips and assumes test wires with large diameters. Such a network solution requires a bank of high quality capacitors, which is able to provide constant current up to hundreds of Amperes over 100 μs time scale. Nevertheless, the transient behavior is more closely related to a practical operation regime of real integrated devices; a promising example was introduced in [18]. The next work [1] focused on experimental testing of the dynamical performances of bond wires, including a statistical study of the time required for wire bond fusing by a specific amplitude of current. However, details about the used measurement set-up were not given. The quality of a wire bond is very important for the reliability of the integrated circuit and, of course, the final application itself. A significant group of research papers dealing with this topic focused on mechanical aspects of the bond wire fabrication process, such as [3] and references cited therein. Approaches like these are mostly based on destructive and nondestructive mechanical and electrical testing. Electrical tests regarding material ageing due to current density flowing through a wire are closely related to mechanical changes such as expansion or retraction [7, 16] caused by temperature effects of flowing current leading to fusing of the material. The mechanical and electrical quality of galvanic interconnections depends on more aspects than only on wire bonds. For example, the bond pad degradation process is also very important for the reliability of the IC.

The general concept of the test set-up is shown in Fig. 21.8. It contains four important blocks of the measurement set-up. The measurement procedure is controlled by MATLAB [11] software with developed programs and user interfaces for easily setting up of test conditions and evaluating results. After summing-up all necessary requirements and parameters to be achieved, we decided to utilize the National Instruments' internal PCI card 6251 (abbreviated as NI card in further text), supplemented by a professional interface. The NI card directly controls the developed tester (testing board) of bonding wires.

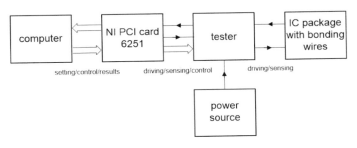

Fig. 21.8: General block definition of experimental set-up.

21.3.1 Hardware Platform for Bond Wire Testing

The tester (block diagram in Fig. 21.9) consists of six power channels (for six bonding wires in the SOIC (Small Outline Integrated Circuit) package provided by ON Semiconductor). The channel is addressed by a sophisticated control system based on demultiplexing of a driving signal to a specific bonding wire. Then, the current flowing through the wire and the voltage drop across this wire are sensed (and amplified by an available bank of amplifiers using six channels) and multiplexed to outputs digitized by the NI card and returned back to the computer. This NI card comprises two analog outputs; only a single channel is used at the moment for arbitrary pulse driving signal generation required by the testing board. The eight-bit digital port of the NI card is used as follows: three bits are dedicated for multiplexing (addressing) the individual bond wires connected to the tester, two bits directly control the gain of the first Programmable Gain Amplifier (PGA), two bits adjust the gain of the second PGA (parts of the bank of amplifiers) in a cascade and the remaining bit signalizes if the measurement sequence is forwarded to the test board. For capturing the necessary analog signals from the bonding wire and tester, two channels with 10-bit resolution A/D converters are used. Referring to this concept, the experimental set-up can be divided into two parts, namely the software, which directly controls this NI card (and tester consequently), and the PCB of the tester, with practical implementation of the multi-channel switching power source.

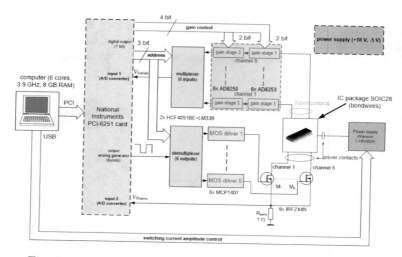

Fig. 21.9: Principal block diagram of the bonding wire tester.

The individual bonding wires are addressed by the HCF4051BE analog multiplexer, i.e., three bits are used to select one power channel from six possibilities. In order to avoid failing states, all addresses set on L or H means that no power channel is selected. This conception simply means that the PCB contains six drivers, six power switches (MOSFET transistors IRFZ044) and that the same amount of amplifiers are connected to bond wires via Kelvin probes. These integrated amplifiers have differential voltage inputs and non-symmetrical output with digitally controlled gain. Since a very large final gain is required, a two-stage cascade connection is utilized. Up to four bits can be used to set a final gain factor; these bits can be directly set via a MATLAB script. For amplification of the voltage at Kelvin probes, a high-gain stage AD8253 PGA, with a gain taken from the set 1, 10, 100 or 1000, is combined with an integrated circuit AD8250, having a gain taken from the set 1, 2, 5 or 10. All power transistors have sources connected to a single sensing $1\,\Omega/50\,\text{W}$ resistor. It is advantageous since the dominant current will always come from the power branch of the active bond wire, while contributions from other bonding wires are negligible. The detailed principle of the specified single channel test is indicated in Fig. 21.10. The hardware solution of the tester is shown in Fig. 21.11. The connection of the IC package (including bond wires) to the testing board was provided by sockets. The standard socket OTS-28-1.27 suffers from long-term exposure of high DC current levels. Therefore, its life-time was highly reduced to several tens of tests. Better thermal and mechanical features are available using the CLIPS-SOIC28 "probe". This probe sustains many times longer exposure of continuing DC tests using units of Ampere due to larger diameters of contacts and better mechanical arrangement.

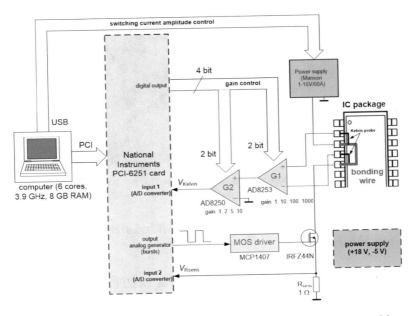

Fig. 21.10: Detailed power and signal path if a specific bond wire is addressed.

Note that the tester should be compatible with another source of the TTL signal (for example a waveform generator controlled by VEE[7] or LabVIEW[8]) and different types of sensing devices (oscilloscope, external A/D converters, etc.). The universality of the tester for future compatibility with different types of measuring equipment has been taken into account. Only different or highly modified communication/control software development will be required because the measuring set-up is designed for communication with the NI card.

---

[7] VEE – Visual Engineering Environment, a graphical dataflow programming software development environment for automated test, measurement, data analysis and reporting. https://en.wikipedia.org/wiki/Keysight_VEE.

[8] LabVIEW – Laboratory Virtual Instrument Engineering Workbench is a system-design platform and development environment for a visual programming language from National Instruments. https://en.wikipedia.org/wiki/LabVIEW.

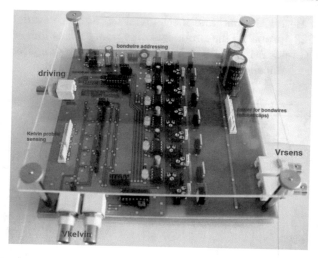

Fig. 21.11: Final bond wire tester cooperating with the NI card.

21.3.2 Software for Controlling Measurement Procedures

The software part consists of two executable MATLAB [11] scripts with a Graphical User Interface (GUI). The first one creates a text file with the analog driving signal. The driving force can be derived for non-inverting as well as for inverting MOSFET drivers. Up to nine bursts, with an arbitrary time window and duty cycle, can be created with cooling time delays placed between these bursts, see Fig. 21.12, for example. The test sequence has no time scale; this parameter is variable and is included into the test data just before the procedure of the measurement itself. Nevertheless, to easily visualize the test sequence in some suitable program (MATLAB, Mathcad, Microsoft Excel, etc.), the time vector can be included as a part of the saved file.

The generated file can then be directly loaded by the second program, which is capable to do the rest, i.e., choose a bond wire to be measured, and specify the power voltages for the individual measurements and gains of amplifiers in the V_{Kelvin} sensing path. The sampling rate of the NI card allows for a single pulse width scaled down to $1\,\mu s$, while the upper boundary is arbitrary; in fact, it can be boosted up to DC. The program is designed for two basic measurement regimes. For the purpose of quantifying the fusing time, a single channel measurement should be activated. Doing so, only the necessary amount of data is stored in single-column text format with reduced number precision. For a bonding wire resistivity change calculation, a dual channel method should be activated because it is necessary to precisely measure and store the current through the bonding wire as well as the voltage drop across this wire via Kelvin probes. As is evident from Fig. 21.13, it is possible to measure up to six bonding wires and for each bonding wire, up to six consequent measurement scenarios can be applied. The power voltage is adjusted automatically on a MANSON 16 V/60 A switching supply via a serial port. Before the start of the particular measurement, the total gain applied on the voltage across the Kelvin contacts should be guessed and set (so far manually). A measurement set-up is essential in order to correctly repeat the same measurement plan for the different wire materials (represented by the different integrated circuits with bonding wires). That is the reason why a "saving" icon has been added to the GUI (Fig. 21.13); the set-up is comprised in the text file and loaded before the start of the script measurement.m. The "digital" icon (Fig. 21.13) is used only if the hardware part of the fusing system is debugged and its individual parts are tested. It allows for sending L or H levels towards the digital output immediately . By sending appropriate addressing, it can also serve as an immediate cut-off driving signal from the power source in the test board. For graphical visualization, different colors are used for different natures of measurements.

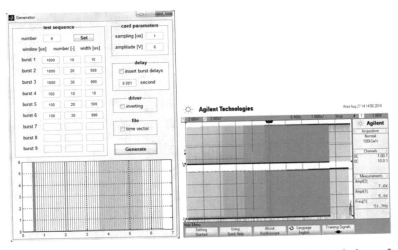

Fig. 21.12: Generation of test sequences in MATLAB GUI of the software generator and an example of a generated sequence with a stepwise increasing duty cycle.

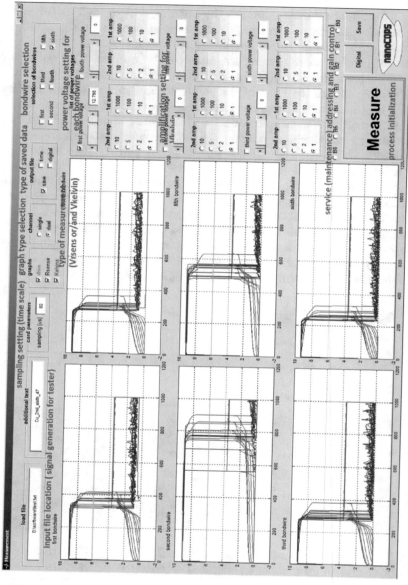

Fig. 21.13: MATLAB GUI for setting and controlling bond wire fusing measurement (example of records for all six channels/bond wires for DC/continual fusing test for Au bond wire with a diameter of $1\,\mathrm{mil}=10^{-3}\,\mathrm{in}=25.40\,\mu\mathrm{m}$).

21.3.3 Exemplary Experimental Results of Bond Wire Fusing

The proposed tester was designed to work in two regimes (DC current continual fusing and dynamic fusing by current pulses). The majority of results focused on DC fusing due to the requirements to fit the mathematical model of bonding wire to real behavior and conditions.

Tests of continual fusing use generated sequences of long-time duration of high logic level H (5 V) in coordination with different values of voltage produced by the power supply Manson to obtain several values of current flowing to the bond wire (in dependence on specific type/material/diameter of tested bonding wire). A typical example of resulting traces for $V_{R_{sens}}$ and V_{Kelvin} is shown in Fig. 21.14(left). The most interesting results of these tests are important for mathematical modeling of bonding wire behavior [2, 10, 14]. The time required to break the bond wire was measured for each applied current and the resulting characteristics obtained from these tests were provided for all available types (material and diameter) of bonding wires (Au 1 mil, Cu 1 mil, Cu 1.3 mil, Cu 2 mil; here a diameter of 1 mil=10^{-3} in=25.40 μm). As example for Au 1 mil, see Fig. 21.14(right).

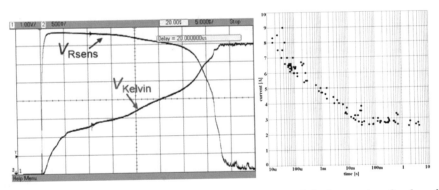

Fig. 21.14: Typical example of destructive continual fusing tests of a bond wire (supervising oscilloscope) and dependence of maximal fusing current on time to break for Au 1 mil (bond wire #1).

The dynamic fusing tests supposed square wave bursts of different duty cycles generated by the NI card. This operation of the tester in the dynamic/pulse regime is directly given by the NI card (test sequences are available from the generator GUI shown in Fig. 21.12). The maximum frequency of pulse excitation for controlling the bonding wire switch load is approximately 500 kHz for a 50% duty cycle. Otherwise, ringing caused by switching of very high loads (currents in units of amperes) with inductive character damages the shape of received/recorded responses significantly. A signal with 10%

duty cycle and 1 μs active level duration can be generated by the NI card as the shortest impulse that also limits the frequency range of expected tests. A typical example of a destructive scenario for switching by current pulses with 90% duty cycle is indicated in Fig. 21.15. This test was provided for stepwise increments of the duty cycle in bursts from 10% to 90%. The bonding wire was broken by the fusing process when the duty cycle reached 90%.

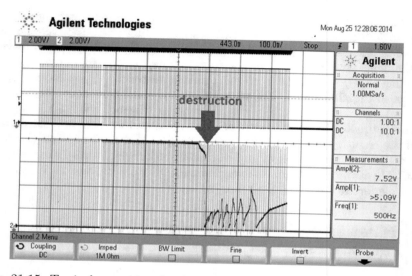

Fig. 21.15: Typical example of a destructive dynamic fusing test of a bond wire for stepwise increment of duty cycle ($V_{R_{\text{sens}}}$ observed at supervising oscilloscope, pulsed $I_{\text{sens}} = 7.5\,\text{A}$ at $1\,\Omega$).

Many integrated circuits with encapsulated bonding wires were tested until the moment of destruction. High currents also created drastic damage of the plastic package surface, as shown in Fig. 21.16(left). It seems that a fused bonding wire (especially when using lower currents) does not mean an immediate open loop; various degradation processes were intercepted by the delayed time base feature of the oscilloscopes (Fig. 21.15). An example of an X-ray photo of a tested and destroyed bond wire is shown in Fig. 21.16(right).

21.4 MOS Power-Cell Testing

This investigation focused on the current stress causing cracks on the surface of the MOS cell [4,5,17]. ON Semiconductor prepared test-chips in the DIP28 package (with an open lid - window) to see the surface under a microscope. These test chips consist of three types of MOS power cells.

Fig. 21.16: An example of destructive fusing by DC continual current (left: destroyed IC bond wire packages, right: X-ray screening after successful fusing).

The tester cooperates with a standard waveform generator and an oscilloscope from a common laboratory equipment. This equipment was supplemented with a counter and microscope with a camera to observe the surface of the uncovered die of MOS cells. The measuring set-up was controlled by LabVIEW software. The general measuring block set-up is indicated in Fig. 21.17. The records of all important signals (if a square wave excitation is used) are provided by this experimental set-up in the form of digitized transient responses of the gate-source (V_{gs}) and the drain-source voltage (V_{ds}) together with the drain current (I_d) or its equivalent voltage drop at the external sensing resistance and information about the internal temperature in the form of the voltage (V_{temp}). These results are also accompanied by screening (microphotograph) of the cell surface to see the effect of the transient stress caused by a long series of current pulses.

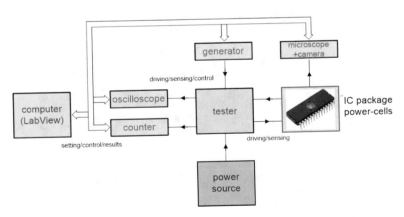

Fig. 21.17: General block concept of the experimental set-up for power-cell tests.

21.4.1 Hardware of the Power-Cell Tester

The hardware solution of the tester is shown in Fig. 21.18. In fact, the power-cell is the power MOS transistor. There are three types of power MOS transistors in the DIP28 package. We selected type NLD45V3 for our initial tests. The power transistor serves as the switching device working to a defined load. Therefore, the testing arrangement is very simple. The package of the MOS cell contains an auxiliary thermal sensor and a current mirror for measuring the drain current itself. Our tester offers a choice of selecting the method of drain current measurement based on external sensing resistance ($0.22\,\Omega$) or internal current mirror (proportional part of drain current). The tester operates with an inductive load of $1\,\mathrm{mH}$. The tester provides all significant electrical signals V_{gs} (or voltage between gate and GND in the case of external R_{sens} utilization), V_{ds} (or voltage between drain and GND in the case of external R_{sens} utilization), $V_{R_{sens}}$ (I_d), V_{temp} for further processing in accordance to Fig. 21.17. This tester consists of protection (input diodes), a connector for attaching a resistive load and auxiliary circuits for measuring and sensing required signals. It supposes a direct connection of these terminals to high-impedance inputs of sensing devices (oscilloscope, counter, etc.).

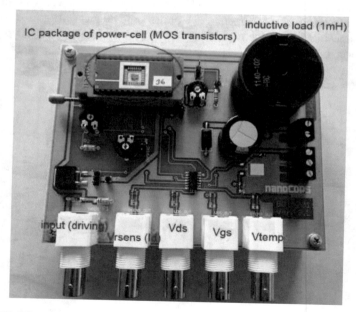

Fig. 21.18: Photograph of the power-cell tester prototype. For further details, see also Fig. 21.20.

21.4.2 *Software for Controlling Measurement Procedures*

The GUI (Fig.21.19) for easily controlling and setting of the tests was programmed in the software LabVIEW. This software is responsible for pulse generation, recording (from used equipment, camera, oscilloscope, and counter) of all important signals (V_{gs}, V_{ds}, $V_{R_{sens}}$, V_{temp}) and screening of the power-cell surface, all fully automated. The user is able to set the frequency, the number of pulses (bursts), the amplitude of voltage signal from generator and its DC offset. The four-channel oscilloscope returns all sensed transient responses to be displayed in the GUI and the software produces records of all mentioned transient responses to a text file as data points for simple post-processing in Excel.

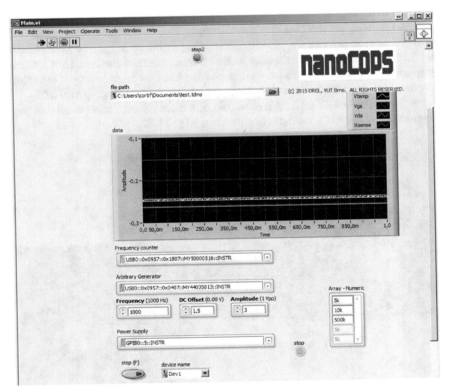

Fig. 21.19: GUI software in LabVIEW for controlling and recording measured signals.

21.4.3 Exemplary Experimental Results of Power-Cell Testing

The measurement set-up established in our laboratory is shown in Fig. 21.20. The developed software controls the Agilent (Keysight) generator to produce an arbitrarily defined number of bursts of rectangular (square wave) impulses and their period as well as duty cycle. This signal is used for driving the MOS power-cell to observe its long-term abrasion and ageing by a substantially accelerated procedure of routine operation (switching). These effects are monitored by a camera in a microscope as a physical demonstration of long-term exposure to accelerated (or unsuitable) operational conditions (standard or large switching currents). Additional monitoring of all significant signals (all voltages by oscilloscope and number of impulses by counter) of this MOS transistor based switch is also important. Exemplary results are shown in Fig. 21.21 for these conditions: current from supply source $I_{dd} = 0.4\,A$, $V_{inp} = 2.1\,V(@1\,kHz)$, and supply $V_{DD} = 6\,V$. The integrated temperature sensor provides a slowly fluctuating value of V_{temp} between $0.10\,V$ and $0.25\,V$. This parameter seems to be almost constant for constant internal temperature if the MOS power-cell is not stressed by high levels of switching current I_d. The measured value of I_d (calculated from $V_{R_{sens}}$ and R_{sens}) in active level reaches $330\,mA$. An example of a microphotograph available from the microscope with a camera is shown in Fig. 21.22.

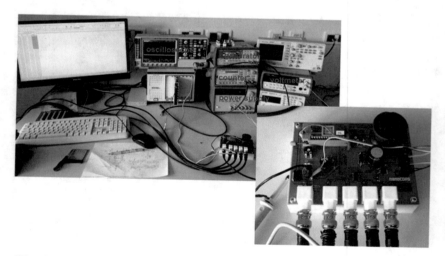

Fig. 21.20: The experimental power-cell test set-up. See also Fig 21.18.

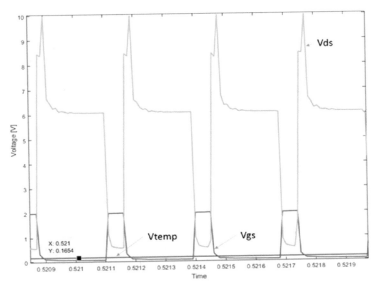

Fig. 21.21: Exemplary data recorded by software to Excel and edited in MAT-LAB.

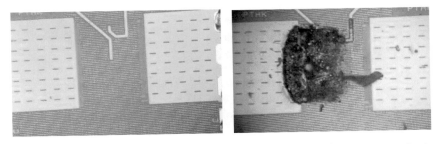

Fig. 21.22: Microphotograph of power-cell NLD45V3 (left: virgin device, right: junction failure).

21.5 Active Load and Pull-Function for RF Measurements

The testing of the Power Amplifier (PA) consists of several different steps and produces different data, which in turn yields a complex characterization of the amplifier. One of the several tests is the testing of the output stage of the amplifiers. This test is commonly performed by passive load tuners. In principle, the tuner sets the desired value of the output impedance, by which the PA is terminated. The characterization could be done by sweeping the impedance from $0\,\Omega$ to several hundred Ω. The value of terminated impedance is usually given as the Voltage Standing Wave Ratio (VSWR). Therefore, the absolute value of the impedance is changed; moreover, the phase shift of the terminating impedance is also required to be changed. The passive tuners are usually very slow due to the mechanics, which they very often contain. Passive tuners are able to terminate or attenuate high load power. The main disadvantages are the tuning speed, cost and operational signal (they reflect/produce only the same signal as the incident one).

21.5.1 Experimental Set-Up Definition

The motivation for developing an active tuner approach is lower cost, higher tuning speed, generating higher order signals/harmonics, extension of the impedance range over the border of the Smith chart[9], etc. The basic idea of the active load pull is obvious from the configuration, which is given in Fig. 21.23. The test set-up consists of the test signal generator (G), a Directional Coupler (Mini-Circuits ZABDC20-252H-S+), a Vector Voltmeter (Hewlett-Packard 8508A) and an RF circulator (Pasternack PE8400). The key equipment of the set-up is the Voltage Signal Transceiver (consists of NI 5791 as a Front Adapter Module and NI 7975R), which is able to receive signals, change the properties of the received signals and transmit these modified signals back to the amplifier through the RF circulator. The Voltage Signal Transceiver (VST) is placed in the feedback loop to increase the speed and accuracy of the reflected signals. For controlling and monitoring the incident and reflected waves' properties, the Directional Coupler (DC) is used. The output signals are measured by the Vector Voltmeter (VV), which is able to measure both amplitudes and the phase shift between these two signals (incident and reflected waves). As is obvious, this approach requires calibration of VV. Adjustment techniques for the calibration of VV have to be developed and a calibration procedure has to be performed before each measurement.

21.5.2 Active Load Pull Function Description

The function of the whole set-up is as follows and is illustrated in Fig. 21.23. The generator (G) produces the incident wave as the input signal for the PA. The output signal from the amplifier is led through the directional coupler (DC). The input coupling signal is connected from DC to port A on the VV for measuring the incident wave properties. From the directional coupler, the incident wave goes to the RF circulator at port 1, where it is shifted to

---

[9] Smith chart, see https://en.wikipedia.org/wiki/Smith_chart

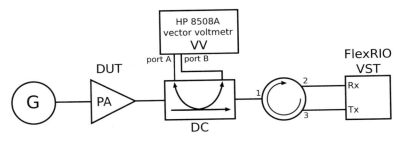

Fig. 21.23: The active load pull basic set-up with the PA as Device Under Test (DUT). VV is a Vector Voltmeter.

the input of the VST, which is connected to port 2 of the RF circulator. The receiver part of the transceiver obtains the data from the incident wave signal. The core of the VST, which consists of a powerful Field-Programmable Gate Array (FPGA) system, makes the required signal changes, like gain, signal attenuation or signal delay, by which the phase shift of the signal is performed. The amended signal is transmitted afterwards by the output stage of the VST to port 3 of the RF circulator. The RF circulator shifts this signal from port 3 to port 1. Doing this, the property of the reflected wave is changed. The reflected wave smoothly continues through the directional coupler to the output of the PA. The incident and reflected waves are combined there and they represent the virtual impedance for the PA. The contribution of the reflected wave is measured by the VV at the output coupling terminal of the directional coupler. By this approach, the virtual impedance could be changed within the dynamic range limits of the test set-up. These limits are induced by the used components of the test set-up. Significant problems emerge from insufficient isolation of the direction coupler and the RF circulator, which causes leakage of the incident wave signal to the reflected signal path. These effects are similar to the crosstalk effects. A minor effect, but similar to the previous one, is produced by the VST itself.

21.5.3 Calibration

In order for the proposed set-up to function properly, calibration before measurement needs to be performed, similar to the classic Vector Network Analyzer (VNA) calibration. Therefore, we introduce a calibration method based on the classic calibration procedure of the VNAs by using short, open and match loads. The calibration procedure can be divided into several steps. Firstly, the VNA has to be calibrated for the measured frequency range and for the power of the output signal, which will be at the output terminal of the tested amplifier. In fact, this is the normal calibration of the VNA. In the next step, the VNA is plugged in according to Fig. 21.24 instead of the tested power amplifier. Now we are able to measure the response of the VST in the feedback loop. In this step, we have to calibrate the whole test set-up including the VV. The VNA has to be switched to a single carrier frequency mode, if possible, because the VST and VV require a stable signal to be able to lock to the carrier signal. If this mode is not present in the VNA, the sweep time has to be set to more than 1 s for one frequency. The ZVL6 from Rohde & Schwarz has been used for calibration. This VNA does not support a single frequency mode without sweeping. Therefore, it is necessary to set the parameter sweep time manually to 1 s or more. This setting yields sufficient time to the VV to get a lock on the carried signal and provide the necessary measurements. The VV is able to get up to 6 measurements in one second, but it has to lock to the carrier frequency (incident wave). Other VNA settings have been set as follows. The sweep points setting is fixed to the minimal value of two and three points, respectively. The sweeping frequency range has been set to the minimum, which is a 2 Hz frequency bandwidth for the ZVL6.

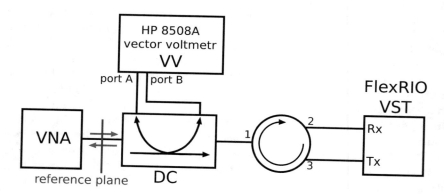

Fig. 21.24: The active load pull calibration set-up.

During this calibration procedure, we record the measured data from the VV, e.g., the reflection coefficient including the amplitudes and phase shift. The VNA in this set-up generates the incident wave and also analyses the reflected wave, which is produced by the VST after appropriate processing. The VNA measures the quantity of the virtual load, generated by VST. This data is sent to the LabVIEW control software, which modifies the reflected signal to tune the active load into the required state. Therefore, by calibration, the VST tunes the virtual impedance to the desired values (match, open and short). The data from VV, VST and VNA are recorded and the calibration error matrix is calculated.

Fig. 21.25 shows the error network for the proposed set-up in Fig. 21.24. The calibration is suggested to be done for the Open (Γ_O), Short (Γ_S) and

Match load (Γ_M) [6, 19]. The error network, according to their scattering parameters, satisfies:

$$\begin{bmatrix} a_\text{VST} \\ b_\text{DUT} \end{bmatrix} = \mathbf{S}_\text{G} \begin{bmatrix} b_\text{VST} \\ a_\text{DUT} \end{bmatrix}, \text{ with } \mathbf{S}_\text{G} = \begin{bmatrix} e_{11} & e_{10} \\ e_{01} & e_{00} \end{bmatrix}. \tag{21.1}$$

Here a_DUT and b_DUT are values for the Device Under Test (DUT) and a_VST and b_VST can be replaced by values measured with the VV

$$a_\text{VV} = L_{\text{DC}_\text{in}} a_\text{VST}, \tag{21.2}$$
$$b_\text{VV} = L_{\text{DC}_\text{out}} b_\text{VST}, \tag{21.3}$$

where L_{DC_in} and L_{DC_out} represent the insertion losses caused by the directional coupler. The VNA provides measurements for the open, short and match load. There are only three measured impedances, but the error network consists of four coefficients. Here, we can introduce the presumption that the error coefficient e_{01} satisfies $e_{01} = 1$. With this assumption, we can make a little mistake or error, which will be covered by the error coefficient e_{10}. Afterwards, the measured value M can be specified by the following equation

$$M = \frac{b_\text{DUT}}{a_\text{DUT}} = e_{00} + \frac{e_{10} \, \Gamma_\text{DUT}}{1 + e_{11} \, \Gamma_\text{DUT}}. \tag{21.4}$$

Now we can provide measurements for the three proposed impedance terminations: open ($\Gamma_\text{DUT} = \Gamma_\text{O}$), short ($\Gamma_\text{DUT} = \Gamma_\text{S}$) and match load ($\Gamma_\text{DUT} = \Gamma_\text{M} = 0$). After this step we get three measured values $M = M_\text{O}$, $M = M_\text{S}$, and $M = M_\text{M}$. These measurements are done by the VV. If we substitute the measured values for the open, short and match by the VNA (Γ_VNAO, Γ_VNAS and Γ_VNAM, resp.) into equation (21.4), we get the following equations:

$$M_\text{O} = e_{00} + \frac{e_{10} \, \Gamma_\text{VNAO}}{1 + e_{11} \, \Gamma_\text{VNAO}}, \tag{21.5}$$

$$M_\text{S} = e_{00} + \frac{e_{10} \, \Gamma_\text{VNAS}}{1 + e_{11} \, \Gamma_\text{VNAS}}, \tag{21.6}$$

$$M_\text{M} = e_{00}. \tag{21.7}$$

From previous equations (21.6) and (21.7), the error coefficients (e_{00}, e_{10}, e_{11}) can be derived (as defined above $e_{01} = 1$) and each error coefficient can be quantified by the following equations:

$$e_{00} = M_\text{M}, \tag{21.8}$$
$$e_{10} = \frac{(\Gamma_\text{VNAO} - \Gamma_\text{VNAS}) \, (M_\text{O} - M_\text{M}) \, (M_\text{S} - M_\text{M})}{\Gamma_\text{VNAO} \, \Gamma_\text{VNAS} \, (M_\text{O} - M_\text{S})}, \tag{21.9}$$
$$M_\text{M} = \frac{\Gamma_\text{VNAS} \, (M_\text{O} - M_\text{M}) - \Gamma_\text{VNAO} \, (M_\text{S} - M_\text{M})}{\Gamma_\text{VNAO} \, \Gamma_\text{VNAS} \, (M_\text{O} - M_\text{S})}. \tag{21.10}$$

Γ_{VNAO}, Γ_{VNAS} and Γ_{VNAM} are measured by the VNA and the M values are acquired by the VV. After this calibration procedure and calculation of the error coefficients, the value of the virtual impedance can be computed as follows

$$\Gamma_{\text{DUT}} = \frac{M_{\text{VV}} - e_{00}}{e_{10} - e_{11}\left(M_{\text{VV}} - e_{00}\right)}, \tag{21.11}$$

where M_{VV} is data measured by the VV. After that, the system is calibrated and the VNA should be replaced by the DUT, and fed by the generator.

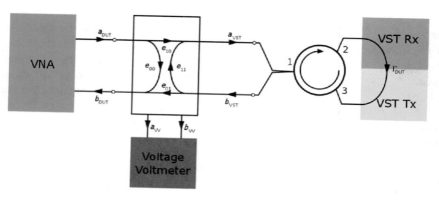

Fig. 21.25: The error port network with e coefficients.

The error coefficients could be characterized as follows: e_{00} represents the directivity of the system, e_{10} is the total reflection tracking (loss), and finally, e_{11} is the test port match. Here, it should also be mentioned that all the written reflection coefficients and all error coefficients are represented as complex numbers. Consequently, it is necessary to measure amplitudes and phases for each coefficient.

21.5.4 The LabVIEW Software Description

The LabVIEW program environment has been chosen for the measurement control and FPGA design for the VST control. The designed control software package consists of a controlling program based on the FPGA VI file, which has been made in the LabVIEW FPGA. This environment contains the design environment, simulation tools for basic function checking and a compiler based on the Xilinx Vivado[10] core. The compiler built the binary files

---

[10] Vivado Design Suite is a software suite produced by Xilinx for synthesis and analysis of HDL designs, https://www.xilinx.com/products/design-tools/vivado.html.

for the FPGA unit and assures proper configuration of the FPGA core. The Top level VI is dedicated for setting the proper carrier frequency for accurately configuring the front end of the VST. In addition, it is necessary to set the reference level of the incident wave and the proper nominal level of the reflected wave. The maximum possible VSWR setting could be infinity in an ideal case, if we neglect insertion losses of the test system. In fact, for compensation of set-up losses, the setting of the output power of the VST output signal must be over two or three dBs higher than the reference signal (incident wave). It is necessary to have enough power for insertion loss compensation. There is also the possibility of setting the phase shift of the reflected wave in comparison to the incident wave, if necessary.

LabVIEW also includes the possibility of controlling the VV, which measures the parameters of the incident and reflected wave as well. These measurements are executed in the while loop, to have full control from the main program by which it is possible to change the VSWR value continuously.

21.6 Summary

We tested several test set-up modifications of the active load pull based on generating the reflected wave. As the reference test device, the NI FlexRIO platform was used with the front adapter model NI 5971. This module was configured as an RF transceiver with a frequency range from 200 MHz up to 4.4 GHz. A vector voltmeter was used for measuring the value of the VSWR. The vector voltmeter was used as the reference measuring equipment in the feedback loop. The measured values were used for directly controlling the vector signal transceiver. The big disadvantage of the used Hewlett-Packard 8508A is its measuring speed. The VV was able to take up to six measurements per second. For some applications, the measuring speed of the vector signal transceiver was not enough. For this reason, we also proposed some test set-up modifications, where the VV was replaced by a pair of USRP radios or by USRP RIO (includes LabVIEW reconfigurable I/O). Additionally, we also proposed a calibration procedure for the vector voltmeter and for the vector signal transceiver. For proper calibration, we used standard calibration impedances, the same as the impedances used for the VNA calibration. The calibration procedure determines the error coefficients, which are used for correcting the measured values by the vector voltmeter. Only by this calibration and correction, we are able to obtain correct information about the virtual impedance, which is produced by active load pull. The calibration requires using a VNA as the reference device. This VNA has to be calibrated by standard procedure before use.

Tomas Kratochvil et al.

Tomas Kratochvil et al.

Acknowledgements Research described in this chapter was financed by Czech Ministry of Education in frame of National Sustainability Program under grant LO1401. For research, infrastructure of the SIX Center was used.

References

1. ANTONY, J.: *Improving the wire bonding process quality using statistically designed experiments.* Microelectronics Journal, 30 (2), pp. 161–168, 1999.
2. CASPER, T., DE GERSEM, H., GILLON, R., GOTTHANS, T., KRATOCHVIL, T., MEURIS, P., AND SCHÖPS, S.: *Electrothermal simulation of bonding wire degradation under uncertain geometries.* Proceedings of the 2016 Design, Automation & Test in Europe Conference & Exhibition (DATE), Dresden, pp. 1297–1302, 2016.
3. COXON, M., KERSHNER, CH., AND MCELIGOT, D.M.: *Transient current capacities of bond wires in hybrid microcircuits.* IEEE Transactions on Components, Hybrids and Manufacturing Technology, 9 (3), pp. 279–285, 1986.
4. GOEHRE, J., GEISSLER, M., SCHNEIDER-RAMELOW, M., AND LANG, K.D.: *Influence of Bonding Parameters on the Reliability of Heavy Wire Bonds on Power Semiconductors.* 7th International Conference on Integrated Power Electronics Systems (CIPS), Nuremberg, pp. 1–6, 2012.
5. GOEHRE, J., SCHNEIDER-RAMELOW, M., GEISSLER, U., AND LANG, K.D.: *Interface degradation of Al heavy wire bonds on power semiconductors during active power cycling measured by the shear test.* 6th International Conference on Integrated Power Electronics Systems, Nuremberg, pp. 1–6, 2010.
6. HIEBEL, M.: *Fundamentals of Vector Network Analysis.* Rohde & Schwarz GmbH & Co.KG, 2005. 419 p. ISBN: 978-3-939837-06-0.
7. ISHIKO, M., USUI, M., OHUCHI, T., AND SHAI, M.: *Design concept for wire-bonding reliability improvement by optimizing position in power devices.* Microelectronics Journal, 37 (3), pp. 262–268, 2006.
8. KRABBENBORG, B.: *High current bond design rules based on bond pad degradation and fusing of the wire.* Microelectronic Reliability, 39, pp. 77–88, 1999.
9. MALLIK, A., AND STOUT, R.: *Simulation methods for predicting fusing current and time for encapsulated wire bonds.* IEEE Transactions on Electronic Packaging Manufacturing, 33 (4), pp. 255–264, 2010.
10. TER MATEN, E.J.W., PUTEK, P., GÜNTHER, M., PULCH, R., TISCHENDORF, C., STROHM, C., SCHOENMAKER, W., MEURIS, P., DE SMEDT, B., BENNER, P., FENG, L., BANGAAYA, N., YUE, Y., JANSSEN, R., DOHMEN, J., TASIC, B., DELEU, F., GILLON, R., WIEERS, A., BRACHTENDORF, H., BITTNER, K., KRATOCHVIL, T., PETRZELA, J., SOTNER, R., GOTTHANS, T., DRINOVSKY, J., SCHÖPS, S., DUQUE GUERRA, D., CASPER, T., DE GERSEM, H., RÖMER, U., REYNIER, P., BARROUL, P., MASLIAH, D., AND ROUSSEAU, B.: *Nanoelectronic coupled problems solutions - nanoCOPS: Modelling, multirate, model order reduction, uncertainty quantification, fast fault simulation.* Journal of Mathematics in Industry, 7 (2), 2016. Open access: http://dx.doi.org/10.1186/s13362-016-0025-5
11. MATLAB: Release R2016a. MathWorks, Natick, MA, USA, 2016. https://www.mathworks.com/products/matlab.html.
12. MERTOL, A.: *Estimation of aluminum and gold bond wire fusing current and fusing time.* IEEE Transactions on Components, Hybrids and Manufacturing Technology, 18 (1), pp. 210–214, 1995.
13. NÖBAUER, G.T., AND MOSER, H.: *Analytical approach to temperature evaluation in bonding wires and calculation of allowable current.* IEEE Transactions on Advanced Packaging, 23 (3), pp. 426–435, 2000.

14. PETRZELA, J., SOTNER, R., GOTTHANS, T., DRINOVSKY, J., KRATOCHVIL, T., AND WIEERS, A.: *Measurement setup for identifying parameters of the encapsulated bond wires.* IFAC-PapersOnLine, 48 (1), pp. 936–937, 2015.
15. RAMMINGER, S., TURKES, P., AND WACHUTKA, G.: *Crack mechanism in wire bonding joints.* Microelectronic Reliability, 38, pp. 1301–1305, 1998.
16. SAIKI, H., NISHITAKE, H., YOTSUMOTO, T., AND MARUMO, Y.: *Deformation characteristics of Au wire bonding.* Journal of Materials Processing Technology, 191, pp. 16–19, 2007.
17. STECHER, M., NELLE, P., BUSCH, J., AND ALPERN, P.: *Interconnect technologies for SmartPower integrated circuits in the area of automotive power applications.* 2011 IEEE International Interconnect Technology Conference, Dresden, pp. 1–3, 2011.
18. WANG, C., AND SUN, R.: *The quality test of wire bonding.* Modern Applied Science, 3 (12), pp. 50–56, 2009.
19. *The Essentials of the Vector Network Analysis (From α to Z0).* 2009 Anritsu Company, Printed in United States 2009-01, PN: 11410-00476A.

Chapter 22
Validation of Simulation Results on Coupled Problems

Rick Janssen, Renaud Gillon, Aarnout Wieers, Frederik Deleu,
Hervé Guegnaud, Pascal Reynier, Wim Schoenmaker, E. Jan W. ter Maten

Abstract The Test Set for the Power-MOS Devices and RF-Circuitry, as described in Chapter [11], and the Measurements for RF-Amplifiers, Bond Wire Fusing and MOS Power Cells (see Chapter [18]) were used to validate simulation results of coupled problems by comparing with measurements, as well as by comparing to outcomes with other simulation tools (when possible, mostly without full coupling).

22.1 Introduction and Validation Plan Description

The validation plan was based on the test cases, as described in chapter [11] of this book, which gives coverage of the complete functionality of the tools developed. One of the objectives was to facilitate validation of new algorithms/tools and modelling methodologies developed during the project, by:

Rick Janssen
NXP Semiconductors, Eindhoven, the Netherlands, e-mail: `Rick.Janssen@nxp.com`

Renaud Gillon, Aarnout Wieers, Frederik Deleu
ON Semiconductor Belgium BVBA, Oudenaarde, Belgium, e-mail: `{Renaud.Gillon, Aarnout.Wieers,Frederik.Deleu}@onsemi.com`

Hervé Guegnaud, Pascal Reynier
ACCO Semiconductor, Louveciennes, France, e-mail: `Herve.Guegnaud@acco-semi. com`,`Pascal.Reynier@yahoo.fr`

Wim Schoenmaker
MAGWEL NV, Leuven, Belgium, e-mail: `Wim.Schoenmaker@magwel.com`

E. Jan W. ter Maten
Bergische Universität Wuppertal, Germany, e-mail: `Jan.ter.Maten@math. uni-wuppertal.de`

- Comparison with available designs and experimental results. An inventory was made to set up a number of relevant test cases from the repositories of the end-user partners. The outcome of this inventory was the Test Set as described in chapter [11].
- Comparison with simulation results of the tools/algorithms developed against the results obtained with the existing commercial and/or public domain tools.

The simulations described in this Chapter include:

- Coupled EM/heat/circuit simulations,
- Exploitation of multirate dynamics in case of RF circuitry,
- Use of parametric MOR techniques for coupled problems,
- Techniques from Uncertainty Quantification (UQ).

In Table 22.1 an overview is given of all test cases with an indication of the main focus of validation. Not all Test Cases are treated here or treated in that full detail.

Uncertainty Quantification and optimisation for Test Case 1 are described in detail in [29–31] and in the chapter [28] in this book. Here [29] gives a brief open access overview.

For the MOR aspects for the Test Cases 2 and 5 we refer to the chapters in this book [5, 9, 39]. Here [39] also combines with Uncertainty Quantification. Chapter [5] in this book discusses multiple inputs (Test Case 3). Furthermore, [10] provides open access.

Test Case 8 is described in [25, 27], in an open access overview in [26] and in the chapter [28] in this book.

Test Case 9 is partly combined with Test Case 6 in Section 22.7. More results on multirate simulations can be found in [12, 22] and in the chapter [6] in this book. Test Cases 11, 12 are extensively described in [13] (open access). For the multirate aspects we refer to the book chapter [6].

In addition to pure validation activities, for some test cases, such as Test Cases 2 and 3, new analyses could be made, which are also described here. In fact the work described here is the extension of a methodology for finding failures in chip design and ageing predictions.

Test Cases 1-6 are connected to the specific industrial Use Case I and Test Cases 7-12 to industrial Use Case II, as described in [11].

22.2 PowerMOS with EM/heat and Uncertainty Quantification (Test Case 1)

Results for Uncertainity Quantification for Test Case 1 are described in full detail in [28–31]. Reference [29] (open access) covers quantifying variations due to uncertainty in geometry (like thicknesses, lengths and widths) that affect material parameters like conductances. The references [28, 30, 31] deal

Table 22.1: Overview of validation activities.

Test Case	Main focus of validation
Test Case 1 - A realistic size powerMOS at constant temperature	EM/heat and UQ
Test Case 2 - A realistic size powerMOS in ET coupling mode	EM/heat and MOR
Test Case 3 - A Driver Chip with Multiple Heat Sources	EM/heat and UQ
Test Case 4 - Smart Power Driver Test Chip with Thermal Sensor	EM/heat and UQ
Test Case 5 - A Fast and Reliable Model for Bond Wire Heating	EM/heat and MOR
Test Case 6 - Inductor test case	EM/circuit and Multirate
Test Case 7 - RF and Electro-Thermal Simulations	EM/heat/circuit
Test Case 8 - Reliable RFIC Isolation	EM/circuit and UQ
Test Case 9 - Multirate Circuit Examples	Multirate
Test Case 10 - Silicon Test Chips for Step-by-Step Testing and Validation: Driver and Power Stage	EM/heat
Test Case 11 - Transmission Line	EM/circuit
Test Case 12 - Balun	EM/circuit

with shape and topology optimization. The book chapter [28] introduces concepts like material and shape derivatives and combines these with Uncertainty Quantification (UQ) for robust shape optimization – including results for Test Case 1. Reference [30] will become available in open access. In reference [31] one minimizes the current density overshoots, since the change of the shape and topology of a device layout is the proven technique for the reduction of a hotspot area. The gradient of a stochastic cost functional is evaluated using the topological asymptotic expansion and the continuous design sensitivity analysis with the Stochastic Collocation Method.

22.3 Validation of the Usage of J, T and S Maps (Test Case 2)

22.3.1 Data Collected and Analysis

For this test case the new simulation options gave rise to study the cause of possible failures. During repetitive short-circuit tests, performed according to AEC-Q100-12 [1], on the packaged die of Test Case 2, we were confronted with a series of failures, where the output driver was destroyed. AEQ-Q100-12 is issued by the Component Technical Committee of the Automotive Electronics Council and is targeted towards short circuit reliability characterization of smart power devices for 12V systems.

The repetitive short-circuit test defined in AEC-Q100-12 aims at determining the reliability of smart-power drivers that operate in "short circuit condition". This is directly connected to the supplies with most of the energy dissipated in the driver itself (due to the accidental absence of load in the circuit). Fig. 22.1 shows the principle of the measurement set-up. The smart-power driver is activated in a "short-circuit set-up" with load impedances set

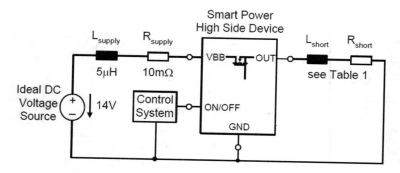

Fig. 22.1: Schematic diagram of the set-up for a repetitive short-circuit test (Test Case 2).

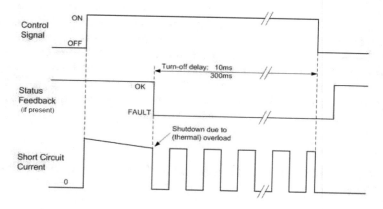

Fig. 22.2: Timing diagram showing the sequence of events during a repetitive "cold" short-circuit test (Test Case 2).

in order to obtain a target current of several amperes (typ. 10 – 20 A, on 12V). After activation, it will take a short time (a few milliseconds) for the driver to detect the short-circuit condition. It will then enter in pulsing mode, until it is de-activated by the controller.

The standardized test procedure specifies a turn-off delay ranging from 10 to 300 ms. Once turned-off, the device-under-test is left to cool down. After this, a new test-cycle is started (see Fig. 22.2). The test is performed in "worst-case" condition with the ambient temperature at -40 °C.

In order to analyze the failures and to find the root-cause, samples were de-capsulated and de-layered. Fig. 22.3 shows a picture of the top-level metal just after de-capsulation (removal of the molding compound of the package). It is slightly blurred by overhanging bond wires, but a failure site can clearly

Fig. 22.3: Test Case 2: die after de-capsulation showing failure site at top-metal level.

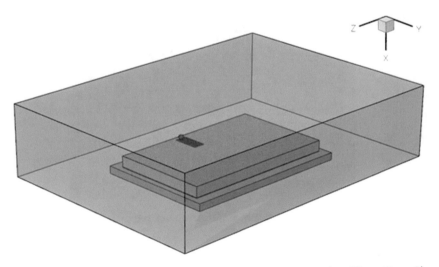

Fig. 22.4: Location of the failure site marked with a red dot (Test Case 2).

be seen as a dark area interrupting the regular fingered layout. Fig. 22.4 shows an overview of the test case 2 device, with the failure site indicated with a red dot. The body of the package is represented as a transparent grey-tinted volume. The silicon substrate is shown in pink, the interconnection back-end is shown in light brown, and the darker rectangle marks the position of the failing driver.

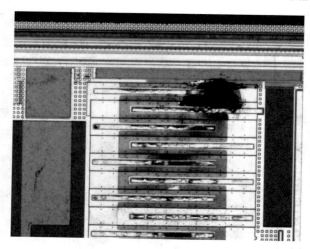

Fig. 22.5: Test Case 2: failure site after removal of the top-metal layer.

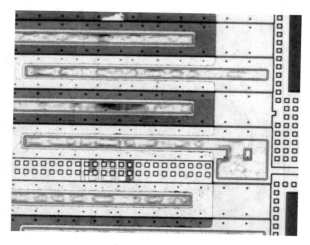

Fig. 22.6: Zoom showing metal deformation patterns (Test Case 2).

Pictures taken after removal of the top-level metal, show clear signs of metal deformation appearing as the contrasted "bubbles" in Fig. 22.5 and Fig. 22.6 below. It is known that under heavy temperature cycling, aluminium metallization has a tendency to deform: its yield strength decreases at temperatures above 120 ^oC, and the mechanical stress, resulting from the CTE mismatches in the package and board, may cause it to flow away from the high-temperature regions and accumulate in lower temperature areas.

The movement of the aluminium metal and especially its accumulation in some areas can significantly increase the stress in the surrounding dielectric layers and initiate or grow cracks. Under the effect of the local stress distribution, the cracks get filled with metal, and during the cycling, the metal-filled crack grows until it reaches another conductor, potentially creating a short-circuit condition.

The effect of the short-circuit depends on the voltage differences between the two networks involved, and the available energy. Typically, the high current flowing through the short will cause the metal to melt, the pressure will increase, the damage in the dielectric will expand, eventually broadening the conductive path of the short-circuit. In the end, the local dissipation is so intense that the molding compound may start to decompose, freeing up gases which can generate enough pressure such that a small-scale explosion occurs, shattering the body of the package. In order to avoid the expulsion of package fragments during a short-circuit test (which can be dangerous for the operator or may perturb the operation of neighboring devices-under-test), usually a safety fuse is inserted in the circuit that allows to limit the damage (as is the case in the figures Fig. 22.5 and Fig. 22.6.

22.3.2 Simulated Maps of Temperature, Current-Density and Stress

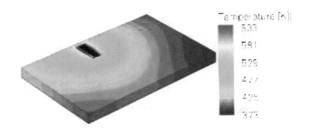

Fig. 22.7: Distribution of temperature in Test Case 2 during the short-circuit test.

In order to understand the chain of events leading to premature failure of the Test Case 2 devices under repetitive short-circuit tests, a series of 3D FEM simulations was set up.

- First PTM-ET [20] simulations where run, in order to investigate the heating of the power transistor and its interconnections during a short-circuit test. It was found that most of the heating is generated in the

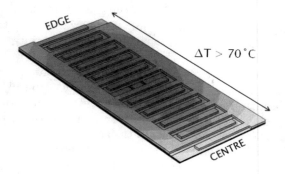

Fig. 22.8: Zoom showing the distribution of temperature across the driver transistor (Test Case 2).

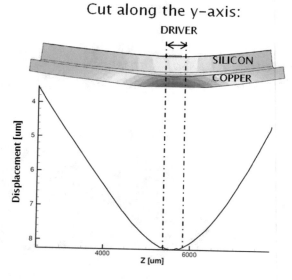

Fig. 22.9: Deformation of the die and paddle during the active portion of the short-circuit test cycle (XY plane) (Test Case 2).

transistor itself, and that the self-heating caused by the test current in the interconnect metal was negligible.

- A 3D model of the Test Case 2 die and its package was created in Sentaurus Interconnect [32], a 3D FEM thermo-mechanical simulator commercialized by Synopsys, Inc.
- In a first series of simulations in Sentaurus Interconnect, the time-averaged power dissipation pattern obtained in PTM-ET was introduced

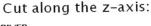
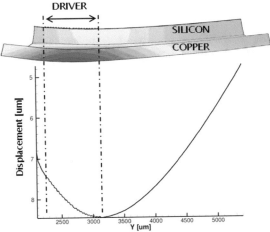

Fig. 22.10: Deformation of the die and paddle during the active portion of the short-circuit test cycle (XZ plane) (Test Case 2).

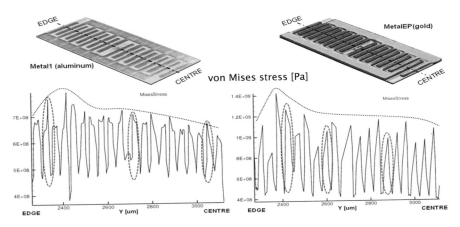

Fig. 22.11: Profile of the von Mises stress level at the lowest metal level (left) and top metal level (right) along the median of the driver (Test Case 2).

in order to obtain the distribution of temperature in the whole structure during the active part of the short-circuit test cycle.

- In a second series of simulations, the distribution of mechanical stress and deformation was computed in the whole structure and analyzed.

Fig. 22.7 and Fig. 22.8 (zoom-in) show the distribution of temperature reached at the end of the active portion of the short-circuit test cycle. A temperature difference of 70 $^{\circ}C$ occurs across the median axis of the driver transistor.

The consequence of the extremely high power dissipation and the high temperatures reached in the structure is a substantial level of deformation as shown in Fig. 22.9 and Fig. 22.10.

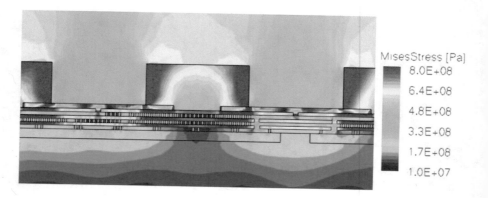

Fig. 22.12: Close-up of the von Mises stress distribution around a top-level metal finger. The yield-stress is attained in the aluminium layers (Test Case 2).

The high levels of deformation go along with high levels of mechanical stress. In order to simplify the analysis, the levels of the von Mises stress are displayed in Fig. 22.11. The von Mises stress is the magnitude of the deviatoric stress sensor (the difference from hydrostatic pressure), which allows to estimate the on-set of plastic deformation in ductile materials by comparison to the yield strength value. The most severe deformation is likely to occur close to the edge of the die, at the location where the peak temperature and the overall temperature excursion is the largest, and where the von Mises stress peak are attaining the highest levels above the yield strength.

As can be seen in the close-up picture of the von Mises stress distribution in Fig. 22.12, the stress levels attained during the active portion of the repetitive short-circuit test are significantly higher than the yield stress levels mentioned in the literature for thin aluminium layers. As a result, plastic deformation of the aluminium metallization can be anticipated during each stress cycle.

22.3.3 Discussion and Benchmarking

The analysis of temperature, current density, and stress-maps allowed to identify the location where the deformation of the aluminium layers is the strongest. The location of the maximum deformation rate within a thermal cycle was found to match well with the collected evidence from the analyzed experimental samples.

The field of modelling of metal migration under forces from electrical or mechanical origins is still a field of active fundamental research [35, 40, 41]. State-of-the-art engineering studies were reported by the teams of Smorodin [33] and Alpern [2–4].

The new coupled EM-heat simulations bring analysis from [33], [2–4, 33] closer to the design engineer, by demonstrating that chip-scale electro-thermal simulations can be derived in a semi-automatic fashion from a standard chip layout. However, further research is necessary in order to derive efficient numerical criteria that will allow to pin-point locations with high risk of failures under specific stress conditions.

Finally, we recall that for the MOR aspects for Test Case 2 we refer to the chapters [5, 9, 39] in this book. Here [39] also combines with Uncertainty Quantification (UQ). Chapter [5] discusses multiple inputs (Test Case 3). Furthermore, [10] provides open access.

22.4 Validation of the Accelerated Evaluation of Life-time Consumption for Arbitrary Application Profiles (Electro-migration case, Test Case 3)

22.4.1 Data Collected and Analysis

Collection of statistical samples of devices having aged according to specific application profiles is a very large undertaking. However, as the purpose of the new method is to accelerate a simulation procedure by interpolating from a limited sample, it is still meaningful to validate the interpolation principle using a simulated design of experiment. This was realized on a small scale and is briefly reported here for the sake of completeness.

22.4.2 Validation of the Ageing-Rate Interpolation Technique

This new technique is applied to assess the life-time consumption relative to electro-migration in the case of the driver from Test Case 3. This driver is

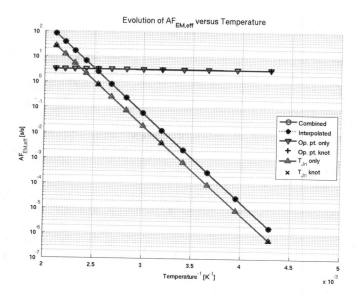

Fig. 22.13: Evolution of the effective ageing rate as a function of the temperature for Test Case 3.

operated in pulsed mode. The operating mode was characterized using PTM-ET [20] in order to locate the points of highest current density. Then, Spectre simulations [34] were run in order to extract the effective acceleration rates for electro-migration in the driver at 13 different temperatures in the range from -40 ^oC to 200 ^oC.

Fig. 22.13 and Fig. 22.14 show how the evolution of the effective ageing rate versus temperature can be very precisely reconstructed over the whole range by considering only 4 extraction points for an "ambient temperature factor" and a "bias-point factor" and applying appropriate interpolation functions. The resulting scheme allows to quickly evaluate the life-time consumption of any circuit using the operating mode and effective ageing rate concepts for arbitrary temperature profiles in an efficient fashion, by avoiding the repetition of circuit simulations at each temperature point of the application profile (which is the standard approach advocated by the commercial tools from Cadence and Mentor Graphics today).

22.4.3 Discussion and Benchmarking

The state-of-the-art concepts relative to mission profiles and operating state-based reliability estimation were outlined in [14] and [23]. However these

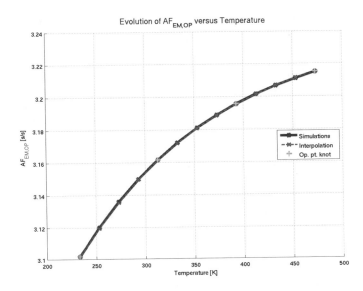

Fig. 22.14: Evolution of the operating-point related factor in the effective ageing rate versus temperature (Test Case 3).

concepts were introduced in a top-down approach (translating application level requirements in development constraints) and lacked to formalize the link from transistor-level simulations upward to life-time consumption rate in a given operating mode. Recent publications from the EDA vendor side attempted to remedy to the situation [36], [37]. They proposed to discretize the temperature profiles in a set of bins. However, they failed to provide a solution to efficiently link the underlying transistor-level simulations for the estimation of ageing rates. Hence, our methodology advances the state-of-the-art today, proposing an efficient method to compute the life-time consumption by characterizing ageing-rates for each operating mode at a limited set of temperatures and interpolating to all target temperatures of the application profile. Our innovation relies in making the link between ageing models and dynamic average modelling as well as splitting the interpolation of the acceleration factor in an ambient-temperature-driven and an operating-point-driven factor, which can both easily be interpolated.

Finally, we recall that Chapter [5] in this book covers Model Order Reduction for multiple inputs (Test Case 3). Furthermore, [10] provides open access.

22.5 Validation of a Smart Power Driver Test Chip with Thermal Sensor (Test Case 4)

Test Case 4 represents a realistic industrial smart power driver with a build-in thermal sensor. This test case allows to dynamically measure the temperature of the driver and also takes the impact of a package and the substrate into account. It is a good test case to evaluate the dynamic responses of the electro-thermal solver in strong coupling mode as well as in co-simulation mode.

The Test Case 4 devices were used for the validation of both PTM (the electro-static FVM (Finite Volume Method) and PTM-ET (dynamic electro-thermal solver), both from MAGWEL [20].

Model Order Reduction with Uncertainty Quantification aspects have been discussed in the book chapter [39].

22.5.1 Validation of the Electrostatic Solver PTM

The isothermal R_{DSON} of six different power-cell layouts was simulated using PTM-ET for 3 temperatures. The layouts covered two n-type devices and two p-type devices, with maximum operating voltages at 30 V and 45V. The target on-resistance (R_{DSON}, the "drain-source on resistance", or the total resistance between the drain and source when the device is "on") for these devices was below 1.0 Ω. Typical layouts for a P-type and an N-type power-cell are shown in Fig. 22.15 and Fig. 22.16.

Table 22.2 shows a comparison of the measured and simulated R_{DSON} for a series of N- and P-type power-cell devices with different maximum operating voltages. The measurements were repeated at three temperatures: 25 ^oC, 85 ^oC and 150 ^oC. On these eighteen test cases, PTM-ET [20] simulates the R_{DSON} with very good accuracy: the error is on average below 5% over the whole temperature range.

	Measured			PTM			Error		
Temperature	25	85	150	25	85	150	25	85	150
30V n-type DMOS	0.304	0.396	0.518	0.291	0.389	0.505	4%	2%	3%
45V n-type DMOS (variant 1)	0.363	0.479	0.634	0.345	0.467	0.602	5%	3%	5%
45V n-type DMOS (variant 2)	0.363	0.479	0.632	0.347	0.470	0.606	5%	2%	4%
45V n-type DMOS (variant 3)	0.371	0.484	0.644	0.353	0.477	0.615	5%	1%	5%
30V p-type DMOS	0.823	1.053	1.325	0.834	1.069	1.305	-1%	-2%	2%
45V p-type DMOS	1.069	1.403	1.802	1.117	1.481	1.942	-4%	-5%	-7%

Table 22.2: Comparison of measured and simulated R_{DSON} (Test Case 4).

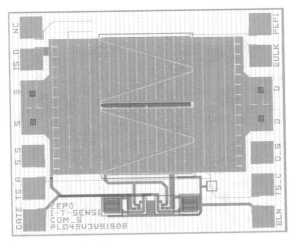

Fig. 22.15: Top-view of the PTM-ET [20] model for a P-type power-cell device (Test Case 4).

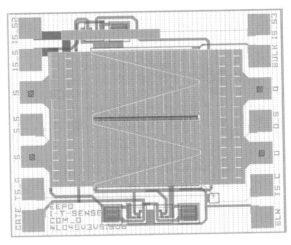

Fig. 22.16: Top-view of the PTM-ET model for an N-type power-cell device (Test Case 4).

22.5.2 Validation of the Dynamic Electro-thermal Solver PTM-ET

The dynamic response of the Test Case 4 device was characterized using on-chip temperature and current sensors whilst the device was activated for a short period of time. Fig. 22.17 shows an overview of the device layout, with the source bondpads located along the left-most edge, the drain bond-

I-sensor + T-sensor

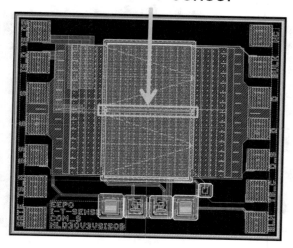

Fig. 22.17: Geometry of the power-cell device with location of the built-in temperature and current sensors (Test Case 4).

pads located along the right-most edge and the location of the current and temperature sensors indicated by the orange arrow and rectangle.

At first, the temperature and current sensors were calibrated. Fig. 22.18 shows the response of the thermal sensor for a few typical biasing current levels and temperatures.

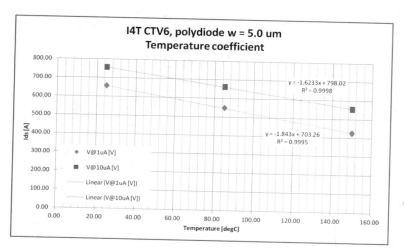

Fig. 22.18: Characterization of the thermal sensor (Test Case 4).

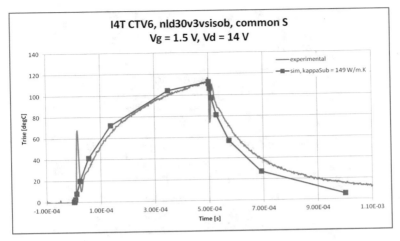

Fig. 22.19: Comparison of measured and simulated responses (Test Case 4).

Next, the device was biased in off-state with the drain at 14V. For the test, a short pulse was applied on the gate in order to activate the transistor. Because of the high-level of current flowing in the device and its operation in saturation, the power dissipation is fairly high and a substantial amount of self-heating is occurring. Fig. 22.19 compares the measured and simulated temperature curves for this test. The good match confirms the validity of the 3D model set-up in PTM-ET for this type of simulations.

22.6 Validation of the Bond Wire Fusing Models (Test Case 5)

Test Case 5 has been the basis for the book chapter [8], the paper [7] and for several measurements described in the book chapter [18]. This test case deals with the development and validation of a fast and reliable model for the calculation of the temperature distribution in bond wires, and thus the determination of the maximum allowable current.

We studied a large number of samples of the bond wire test-vehicle shown in Fig. 22.20. The set comprises samples with gold or copper wires of different thicknesses and two lead-frame variants. The fusing tests are described in Chapter [18]. During the tests, a large constant current was forced into the wires until it would stop to conduct current.

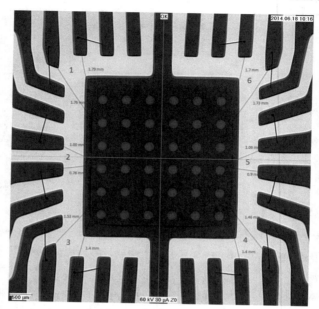

Fig. 22.20: X-ray picture of several samples of the bond wire test-vehicle
(measured and tested as described in Chapter [18], for Test Case 5).

22.6.1 Data Collected and Analysis

We collected voltage and current waveforms during fusing tests on close to a
thousand samples from 3 variants of a bond wire test vehicle: (1) one with
1.3 mil[1] copper wires, (2) one with 2 mil copper wires, (3) one with 1 mil
gold wires. Each test vehicle had 6 test sites, consisting of a pair of force
wires carrying the stress current and a pair of sense wires for accurately measuring the voltage drop in the force wires. The wire length varied from test
site to test site, but these could be grouped in a shorter wires category (approximately 1.9 mm) and a longer wires category (3.2 mm). Fig. 22.21 shows
typical waveforms measured on a 2 mil copper bond wire. The measurements
are the current and voltage traces (bottom), while the upper traces represent
the power dissipated in the wire and the apparent wire resistance. Symbols
were added to the curves in order to mark particular events detected during the post-processing phase. Triangles pointing right and left indicate the
detected starting and end points of the current pulse. The triangle pointing upward indicates the instant at which the dissipated power rises above
10% of its maximum value. The asterix indicates the moment just before
the power dissipation abruptly ceases. This moment will be further referred

---

[1] A circular mil is a unit of area, equal to the area of a circle with a diameter of one mil
(one thousandth of an inch). It corresponds to $5.067 * 10^{-4}$ mm^2.

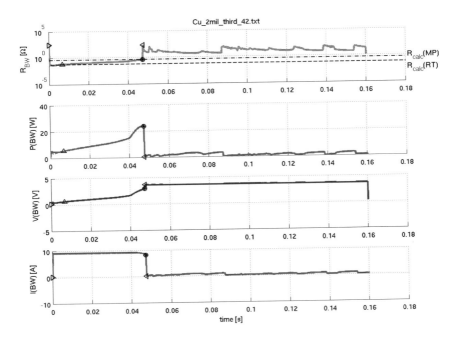

Fig. 22.21: Typical waveforms measured on a copper bond wire of 2 mil diameter (Test Case 5).

to as "power-death time". The resistance plot also displays two limit lines indicating the value of the calculated room-temperature resistance of a wire having the average length at that site $R_{calc}(RT)$ as well as the resistance of this wire at the melting point temperature $R_{calc}(MT)$. The instant at which the apparent resistance value in the measurements crosses the melting-point resistance value is defined as the "melting point".

Fig. 22.22 compares the time until the power dissipation ceases ("time-to-power-death") with the time until the measured resistance reaches above the melting point resistance. The good correlation between the two measures confirms that the underlying failure mechanism is fusion of the copper material.

A further analysis of the measured data, illustrated in Fig. 22.23, shows that there is a significant scatter in the measured values of the apparent resistance of the wire $R_{meas}(MT)$ just before crossing the theoretical melting point value $R_{calc}(MT)$. This might be related to the steep slope of the measurement resistance curve in that region combined with the finite resolution of the time-base. It also indicates a level of variation present in the measured

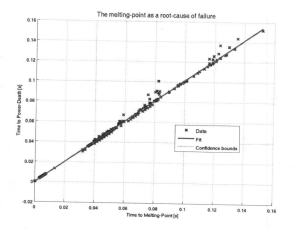

Fig. 22.22: Correlation between the time-to-power-death and the time-to-melting-point for 1.3 mil copper wires (Test Case 5).

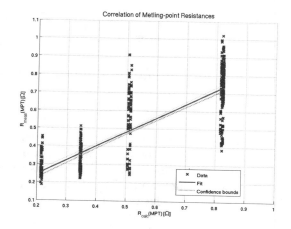

Fig. 22.23: Correlation between the calculated wire resistance at the melting point $R_{calc}(MT)$ and the apparent resistance value measured at the time-point preceding the crossing of that value $R_{meas}(MT)$ (Test Case 5).

data, which contributed to complicate the validation of the bond wire fusing models.

22.6.2 Analytic and Simulated Bond Wire Model Results

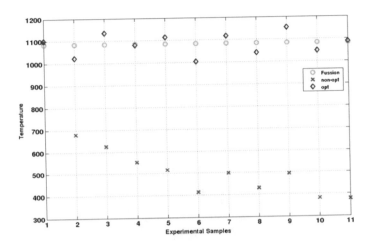

Fig. 22.24: The developed bond wire model results before ($\times$) and after ($\diamond$) joint optimization of empirical model parameters (Test Case 5).

The validation of the bond wire model developed at TU Darmstadt (see [7] and the book chapter [8]) was complicated by the inherent variability in the data collected when measuring. In order to mitigate this issue, it was decided to perform the validation as follows:

- Compute the median time-to-failure per bond wire type and current level, with detection of outliers;
- Set up simulations for a selection of bond wire lengths and current levels;
- Evaluate the time-to-failure of the bond wires in simulation as the time elapses from the start of the current pulse until the moment where the maximum temperature reaches the melting-point temperature of the wire.
- Compare the time-to-failure in simulation to the median of the corresponding measurements.

Fig. 22.24 compares the measured and simulated time-to-failures. Initial results without optimization of the empirical model parameters deviated significantly from the measurements. After optimization of the model parameters across a range of bond wire lengths and current levels, a reasonable match was obtained across all cases.

22.6.3 Simplified 3D Bond Wire Model Results

Fig. 22.25: Simplified geometry for a pair of copper wires (Test Case 5).

In order to obtain better insight in the phenomena occurring during a bond wire fusing test, we also set up simulation experiments using PTM-ET [20]. A first series of experiments was set up, where the bond wire is straightened in the vertical direction and wrapped in a box of mold-compound of about 2 mm in lateral dimension (see Fig. 22.25). In order to accurately predict the time-to-failure of bond wires by simulating the time it takes to reach the melting point at the hottest place, and to provide the latent heat of fusion, a first series of simulations has shown that it was absolutely necessary to use accurate temperature-dependent models of the electrical and thermal conductivity of the wire material. A second learning point was that within several milliseconds, the heating of the mold compound is always confined in a tube of less than 500 μm diameter around the bond wire (see Fig. 22.26). The third learning point was the actual temperature profile reached within a specific amount of time for a given stress current is very strongly dependent on boundary conditions and material parameters. After optimizing the mesh and solver settings, we were able to simulate the wire fusing experiments in less than a day.

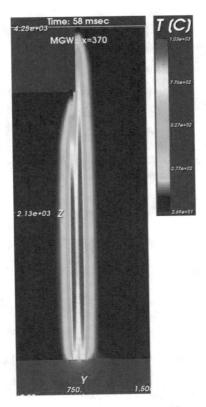

Fig. 22.26: Simulated temperature profile at the instant of failure (Test Case 5). Red color indicates a temperature close to the melting point. Blue is room temperature.

In order to understand the sensitivity of the temperature profiles reached within a given time during wire-fusion tests, several experiments were run with different models for the wire / pad interface (wire with bond-ball, no bond-ball, presence or absence of oxide and nitride materials linking the pad to the substrate), as well as different models of the wire to molding compound interface (uniform conductivity, localized non-uniformity, distributed non-uniformity).

Fig. 22.27 and Fig. 22.28 show the temperature profiles obtained the cases corresponding to the geometries shown in Fig. 22.29. When assuming an ideally uniform molding compound material, the simulator predicts a time-to-failure (TTF) of more than 8 milliseconds. When adding a muffle of perturbed molding compound material over the whole length of the wire, the TTF drops to about 3 milliseconds. When taking only a local non-uniformity into account the simulated TTF is on the order of 6ms.

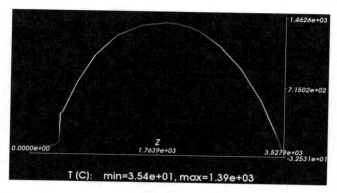

Fig. 22.27: Temperature profile along the wire at the melting-point for the case with localized non-uniformity (Test Case 5).

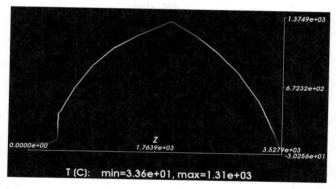

Fig. 22.28: Temperature along the wire at the melting-point for the case with a non-uniform muffle (Test Case 5).

As shown in Table 22.3 below, the presence of a local non-uniformity as spatial initiator for the melting point seems to be the most realistic explanation of the measured data. It also helps to understand why the distribution of time-to-failures and especially the distribution of measured resistance at melting point as shown in Fig. 22.23 is so wide.

Wire	Measured TTF	Simulated TTF (uniform mold)	Simulated TTF(local non-unif.)	Simulated TTF (muffle)
1.3 mil Cu 'third' pair 6.7 A	7.2 ± 0.1 ms	8.16 ms	6.64 ms	3.13 ms

Table 22.3: Comparing measured and simulated TTF's for a 1.3 mil Cu wire (Test Case 5).

Fig. 22.29: Wire geometry showing the location of a pure-epoxy region (left) and an epoxy-muffle (right), where, by absence of silica fillers, the thermal conductivity is 5x lower (Test Case 5).

22.6.4 Embedded 1D Bond Wire Model Results

We implemented a new type of bond wire model in the PTM-ET tool [20]. In this new model, the bond wire is discretized along its length and represented as a 1D string of elementary sections, interacting with their neighbours for the conservation of voltage and currents and with a single "port" embedded in the 3D PTM structure of the package in order to model thermal interactions with the molding compound.

We used a structure out of the bond wire test-chip in order to validate the new model. A view of a typical wire-pair is shown in Figure 22.30. It shows the standard 3D model and the new "1D" model side by side. The 1D model was validated by comparison to the full 3D model. The distribution of

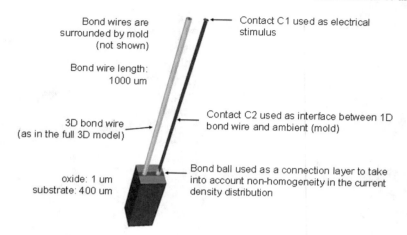

Fig. 22.30: PTM-ET structure [20] showing a pair of bond wires (Test Case 5). The left-hand side is a standard full-3D model, the right-hand-side is the new 1D bond wire model.

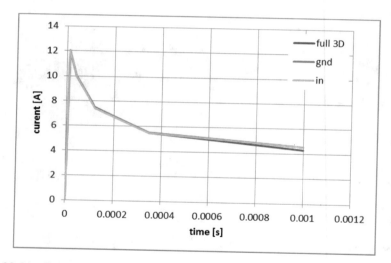

Fig. 22.31: Comparison of the currents in the full 3D model and at both terminals of the 1D model (Test Case 5).

temperature along the wire, as well as the evolution of the wire current and the peak temperature over time were compared.

Fig. 22.31, Fig. 22.32, and Fig. 22.33 show that the 1D model provides a good approximation of the full 3D model. The main advantage of the 1D model is that it has virtually zero impact on the complexity of the mesh

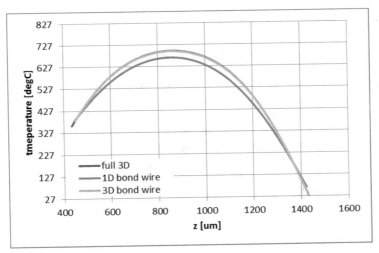

Fig. 22.32: Comparison of temperature profile in the case of both wires as 3D models ("full 3D"), the 3D bond wire model and the 1D wire model for the mixed case (Test Case 5).

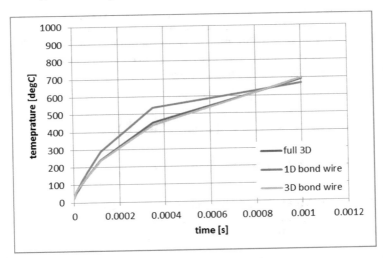

Fig. 22.33: Comparison of the peak temperature over time in the case of both wires as 3D models ("full 3D"), the 3D bond wire model and the 1D wire model for the mixed case (Test Case 5).

required to simulate the structure and therefore allows to incorporate many bond wires in the simulation at a very limited computational cost.

22.6.5 Discussion and Benchmarking

The validation activities related to the fusing of bond wires have produced good results. The developed bondwire model improved over the external state-of-the-art at the beginning of the project [24], and compares very well with recently published state-of-the-art result from the same team [15], [16]. Whereas the model from Jung [15] sticks to the cylindrical geometry for the wire model, the developed model accounts for the finite and asymmetric extent of the mold compound, which provides an advantage for predicting the temperature rise of the wire in more advanced (very thin) packages. The PTM-ET [20] simulations of the 3D model of the bondwire fusing tests also allowed to refine the understanding of the sources of variations dominating the time-to-failure of the wires during the tests. Publically available information on the variability of the TTF of bond wires was devoted to more energetic pulses and much shorter failure times [21]. The impact of the wire interface to the molding compound on the TTF had already been noticed [19], however it is a merit of the nanoCOPS project to pin-point local variations of thermal conductance as main source of variation for TTF's at the millisecond time-scale, corresponding to "realistic" stress current levels. The 1D model published by the nanoCOPS team [7] at DATE 2016 and integrated in PTM-ET allowed to perform uncertainty quantification during electro-thermal simulation of a complete package including the bond wires. This capability is state-of-the-art, and has already been noticed and acclaimed by package experts of the industrial partners.

22.7 Inductor Test Case (Test Case 6)

22.7.1 DevEM Simulation and Set-up

The inductor chosen as one of the test cases is a custom inductance design on silicon technology for an application at 1GHz. It was used to validate the DevEM tool [20]. The inductance characteristics are:

- Inductance: $L = 3.5nH$ @ 1GHz
- Quality factor: $Q = 13$ @ 1GHz
- Max quality factor: $Qmax = 16$ @ 1.7GHz

As can be seen in Fig. 22.34 (layout inductance), this structure can be measured by probing (1 GSG probe). After measurement, the first step was to simulate, with the 2D solver used by ACC (Momentum from Keysight [17]), the inductance with the launcher without any simplification (except via merging). In that case, the complexity is high (high number of mesh nodes), so

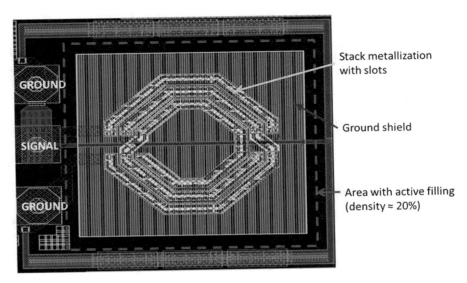

Fig. 22.34: Layout of the Si inductor (Test Case 6).

that the EM simulation needs a lot of time and memory. Nevertheless, without that, the comparison with measurement is not valid. Fig. 22.35 shows:

- The comparison between the measurement and the simulation (blue and red curves)
- As an example, the impact of the active filling on performances knowing that active filling has a second order impact on performances (green curve).

As we can see in Fig. 22.35, if we take the layout without making simplifications, we have a very good correlation between measurement and the simulation (blue and red curves). The only difference on the quality factor between 0 and 1GHz can be explained by calibration inaccuracy and the contact resistance of the probe. Despite the ground shield, if we do not take into account the active filling (higher eddy current effect), inductance and quality factor performances are impacted (inductance resonance is shifted to a higher frequency and the quality factor curve changed). Knowing the time simulation and the memory needed for the simulation, we decided to simplify the structure and make the comparison between Momentum and DevEM results. The simplifications are listed below:

- We removed active filling.
- We merged vias.

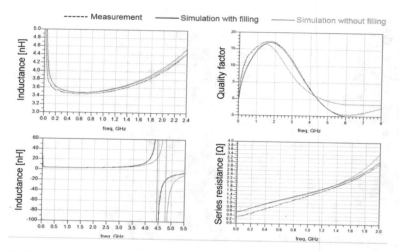

Fig. 22.35: Comparison between measurements and simulation (Momentum) (Test Case 6).

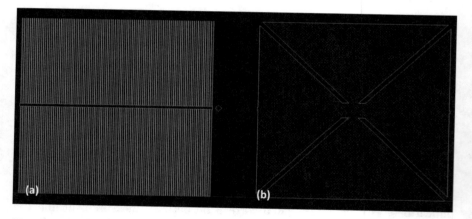

Fig. 22.36: Inductance ground shield (a) from the DK and (b) simplified (Test Case 6).

- We removed metal slots on the inductance.
- We simplify the DK ground shield: the structure with multi thin bars in parallel is replaced by a square metal plane split in 4 parts (see Fig. 22.36).

The layout of the simplified inductance structure used for the validation is represented in Fig. 22.37. In Fig. 22.38 the corresponding DevEM model is shown.

The comparison between Momentum (Keysight) [17] and DevEM (MAG-WEL) [20] simulations is given in Fig. 22.39.

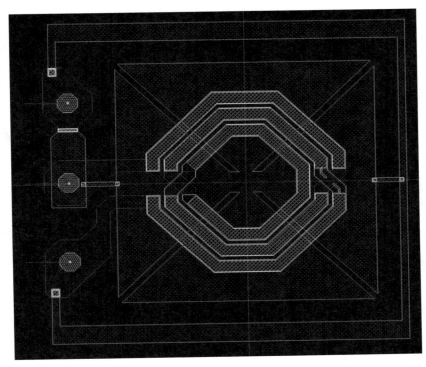

Fig. 22.37: Layout of the simplified inductance structure (Test Case 6).

As we can see in Fig. 22.39, until 2.0 GHZ, DevEM results are really close to Momentum. To match at higher frequency, the meshing needs to be increased.

22.7.2 Coupled Simulations - Colpitts Oscillator

In Fig. 22.40, the inductor in a Colpitts oscillator circuit has been replaced by the on-chip ACCO inductor (Test Case 6), modeled and simulated as 3D component by the DevEM simulator [20]. The simulation of the transient response of this circuit is shown in Fig. 22.41. It shows a multirate effect: a high frequent carrier signal with a superimposed low frequency envelope signal (as part of Test Case 9). The simulation was performed with the coupled LinzFrame/devEM tool, employing trigonometric BDF integration schemes. The settling time is about 50ns and the oscillation frequency 0.42GHz. For further details we refer to the chapter [6] in this book.

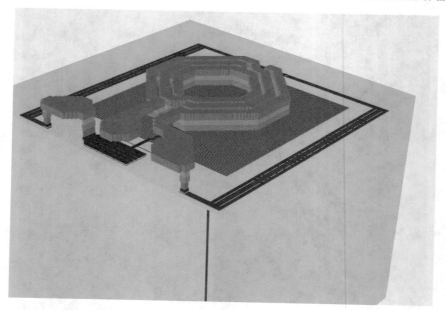

Fig. 22.38: 3D DevEM model [20] of the simplified inductor (Test Case 6).

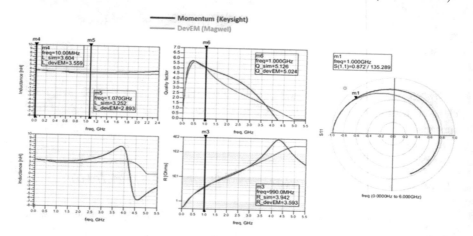

Fig. 22.39: Comparison between Momentum (Keysight) [17] and DevEM (MAGWEL) [20] on the simplified inductance structure (Test Case 6).

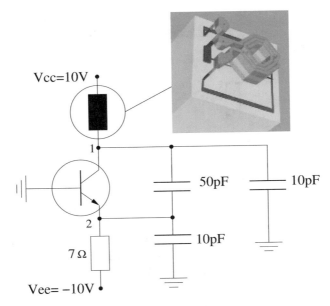

Fig. 22.40: Colpitts oscillator circuit with inductor (Test Case 6).

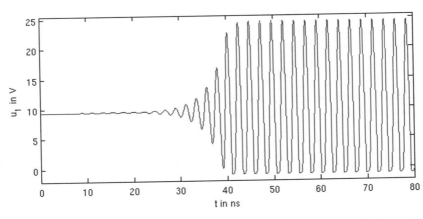

Fig. 22.41: Transient simulation of the oscillator from Fig. 22.40 (Test Case 6, combined with multirate simulations for Test Case 9).

22.7.3 Transient Coupled Circuit/EM Simulation in PyCEM

The on-chip inductor has been used for the validation of the holistic coupled circuit/EM simulation with the PyCEM (Python Circuit EM) simulator, de-

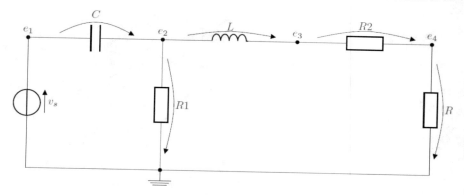

Fig. 22.42: Band-pass filter circuit with lumped modeling (Test Case 6).

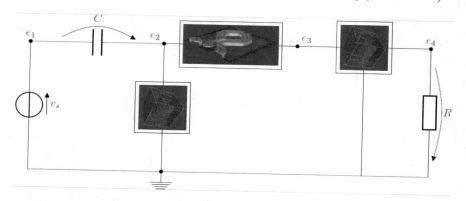

Fig. 22.43: Band-pass filter circuit with 3D EM models for resistances and the on-chip inductor (Test Case 6).

veloped at the Humboldt Universtät zu Berlin. This simulator uses the EM package DevEM for the generation of the spatially discretized EM models and the circuit package MECS for the generation of the network structure and lumped circuit device models.

The on-chip inductor was tested within a band-pass filter, see Fig. 22.43. For validation, the band-pass filter was additionally simulated with pure lumped device models, see Fig. 22.42.

The following scenario for the band-pass-filter has been chosen:

- low stop at $f_L = 10^8$ Hz
- high stop at $f_H = 3 \cdot 10^8$ Hz

All frequencies between f_L and f_H shall pass the device. The resistance values are fixed by R1 = 0.03Ω, R2 = 0.03Ω and R=1Ω. The capacitance and inductance are chosen such that C = $5.3 \cdot 10^{-8}$ F and L = $2.4 \cdot 10^{-9}$ H.

We tested the band-pass filter with the input signal

$$v_s(t) = \sin(2\pi \cdot 1.5 \cdot 10^7 t) + \sin(2\pi \cdot 10^8 t) + \sin(2\pi \cdot 0.8 \cdot 10^9 t). \qquad (22.1)$$

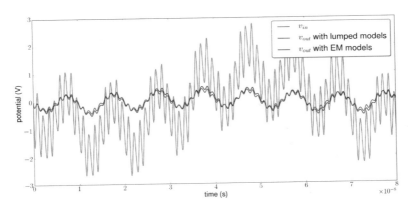

Fig. 22.44: Band-pass filter simulation results (Test Case 6): green line: input signal, blue line: output voltage with pure lumped circuit modeling as shown in Fig. 22.42, red line: output voltage with 3D-EM models as shown in Fig. 22.43 (with on-chip inductor).

The output signal vout (voltage over R) is shown in Fig. 22.44. We see that the resulting signal filters the low frequency f=$1.5 \cdot 10^7$ Hz very well and the high frequency f=$0.8 \cdot 10^9$ Hz quite well for both model variants: pure lumped modeling Fig. 22.42 and circuit modeling coupled with 3D EM field models Fig. 22.43. The latter one yields a little smaller amplitude in comparison with the first one. This seems to be reasonable since the EM model of the inductor includes additional resistive behavior.

22.8 RF and Electro-Thermal Simulations (Combining Test Cases 7 & 10)

The Test Case 7 of ACCO Semiconductor represents a realistic stand-alone power stage used in an actual Power Amplifier (PA) product. The driver stage has been removed to facilitate electro-thermal simulation. In order to measure the transistor junction and the board ground pad temperature, two thermal sensors (diodes) have been integrated. This allows to extract the junction to case (Rjc) and case to ambient (Rca) thermal resistances. All input and output matching networks are integrated on the PCB board (cf. Fig. 22.45). The stand-alone RF power stage is realized on a 0.18 m standard CMOS

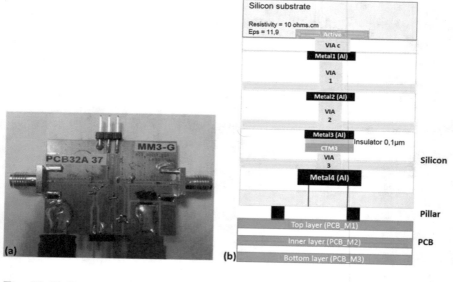

Fig. 22.45: Power stage test boards: (a) picture of the board testing and (b) stack-up description. (Test Case 7)

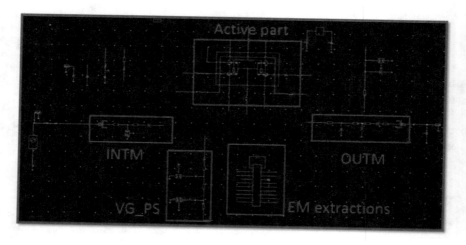

Fig. 22.46: Test bench for design and retro simulation (Test Case 7).

technology and directly flip-chipped on the PCB. Designed for band II (@ 1.9GHz) it delivers on board 26dBm linear output power (@ -40dBc) with 44% efficiency.

Test validations are made in 3 main steps:

- RF and thermal measurements.
- RF small signal validation in order to validate the PCB, Si die and input and output matching networks modeling, as shown in Fig. 22.46.
- Electro-thermal simulation with HeatWave (Keysight software) and the DevEM (MAGWEL software).

22.8.1 RF and Thermal Measurement (Test Case 10)

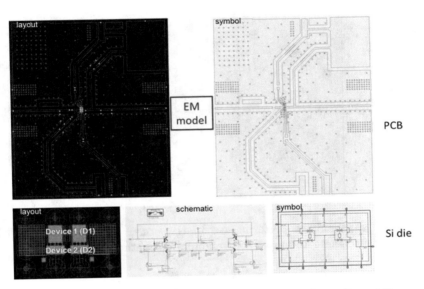

Fig. 22.47: Test bench with full EM extraction (Test Case 10).

The silicon die and the PCB were developed and fabricated. Measurements are available and have been performed at room temperature with different Si dies to evaluate the impact of the process variation on RF performances. RF measurements confirmed the test case functionality, the target specifications and performance reproducibility.

In parallel to that, PCB (Tc) and junction (Tj) temperatures have been measured for each output power value.

22.8.2 Small Signal Validation

For small signal validation, 2 different methods have been used:

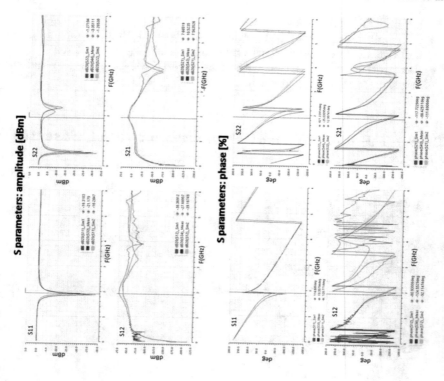

Fig. 22.48: S-parameter performances of the power stage (Test Case 10) (blue=measurements and green/red=simulation method 1/method 2).

- The first one has been used for the design and optimization. In order to reduce simulation time and to bring more flexibility into the design, instead of simulating the complete layout electromagnetically, only the part under the silicon (ground connections and input/output connections) is simulated. Line models model the remaining PCB, taking into account technological data. This approach is described in Fig. 22.46.
- In the second method, after optimization, the PCB is fully extracted with Momentum, like shown in Fig. 22.47.

In order to validate the 2 previous test benches, S-parameter performances are compared to measurements. The S-parameter performances (Fig. 22.48) have been done at 30 ^{o}C temperature. The quiescent current is 44 mA.

Even if some discrepancy appears in the previous figure (in S11 and S12), knowing that SMD (Surface Mounted Device) components have a precision of ±0.1 pF for capacitances and ±0.3nH for inductances, the modeling can be considered as acceptable.

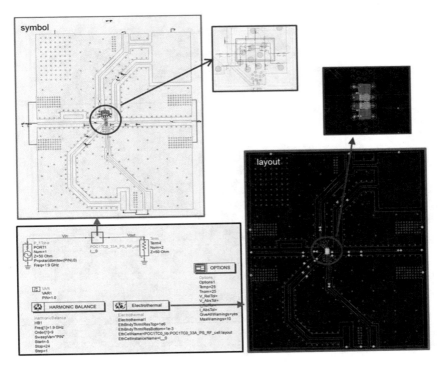

Fig. 22.49: Electro-thermal testbench (Test Case 10).

22.8.3 Electro-Thermal Simulation

For electro-thermal simulation, method 2 (PCB fully extracted with Momentum) is used as reference. The electro-thermal testbench is presented in Fig. 22.49.

For electro-thermal simulation, heat sources have been added to the two cascaded power devices (D1 and D2, cf. Fig. 22.47), and the temperature of the PCB bottom plate was fixed at the ambient temperature ($27\ {}^oC$).

Thermal results on the junction plane are presented in Fig. 22.50 and 22.51 at the maximum linear output power of 29 dBm.

As we can see in Fig. 22.51:

- The temperature is "uniform" for each device.
- The temperature of device 1 is 20 ^oC higher than device 2. This can be explained by bump localization. The bumps of the power device source are inside the device 2, whereas the power device drain bump is outside the device 1 (cf. Fig. 22.47).

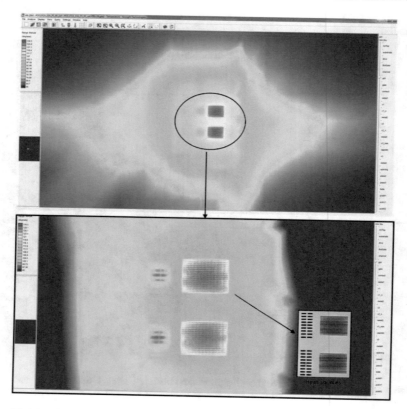

Fig. 22.50: Temperature results on junction plane and heat sources @ Pout=29 dBm (Test Case 10).

In the next phase, we compare the RF electro-thermal simulations with measurements. Except for the gain at low output power, simulations are in phase with measurements.

Thermal sensor measurement allows to extract the junction to case thermal resistance $(Rth_{Jc}) \approx 47\ ^{\circ}C/W$. Knowing that the case temperature is fixed at $27\ ^{\circ}C$, Fig. 22.53 compares the measurements and simulations of the junction temperature of device 2 (D2).

Despite of the difference at lower powers, the simulated performances are acceptable, especially at the maximum output power. The temperature curve is directly linked to the output power curve (cf. Fig. 22.52).

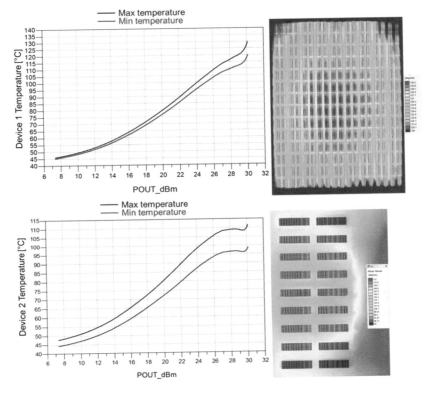

Fig. 22.51: Min and max junction temperature of the devices D1 and D2 vs Pout (Test Case 10).

22.9 Reliable RFIC Isolation and Uncertainty Quantification (Test Case 8)

The increasing integration process of various systems on a single die force a circuit designer to make some trade-offs in preventing interference issues and in compensating coupling effects. This has led to the Radio Frequency Integrated Circuit (RFIC) isolation problem defined by NXP Semiconductors. Unintended RF coupling, which can occur both as a result of industrial imperfections and as a consequence of the integration process, might additionally downgrade the quality of products and their performance or even be dangerous for safety of both environment and the end users.

A specific floorplan/domain set-up was considered that combined digital (aggressor) parts and analog (victim) blocks. Around this set-up several grounding strategies were taken into account (involving bond wires as used in Test Case 5 and inductive groundings).

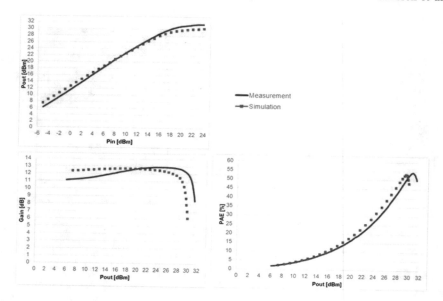

Fig. 22.52: RF performances of the power stage - comparison between measurements and electro-thermal simulation (Test Case 10).

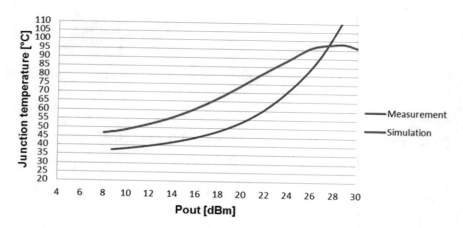

Fig. 22.53: Junction temperature of device 2 (D2) measurement vs simulation (Test Case 10).

The electro magnetic coupling from aggressor to victim part was determined and studied under the variation in bond wires and in grounding inductors. Uncertainty Quantification also provided a sensitivity impression, which was of help in reducing the coupling as well as the variations due to uncertainty. Actually this was a kind of discrete optimization.

The outcomes for Test Case 8 are detailedly described in [25,27], with an extensive final overview in [26] (open access). Related work has been reported in [38].

22.10 Multirate Circuit Examples (Test Case 9), Transmission Line (Test Case 11) and Balun (Test Case 12)

Test Case 9 was partly combined with Test Case 6 in Section 22.7. Results on multirate simulations can be found in [12], in [22] (open access) and in the chapter [6] in this book. The simulations included circuits involving a Transmission Line (Test Case 11) and a Balun (Test Case 12).

22.11 Silicon Test Chips for Step-by-Step Testing and Validation: Driver and Power Stage (Test Case 10)

The simulations have been combined with Test Case 7, see Section 22.8.

22.12 Conclusion

The main theme of this chapter is the confrontation of measurement results with simulation results, with dedicated tester prototypes as in Fig. 22.54. Such a comparison requires a complete understanding of the measurement set up, the simulation tools, including their limitations and the physical assumptions that underlie the construction of the simulation tools. Last but not least the numerical techniques require that the reality can often be captured only approximately. Being aware of these limitations it is impressive to find good correlation between the experimental data and the simulation data. Whereas finding good correlation or even a good match between experimental and simulation data is seen as a justification of the series of underlying assumptions and approximations, one must always be alert that some errors have cancelled. If this is the case, the simulation tool looses predictive power and therefore, it is recommended to perform variations of the validation tests. That has been done also for the cases studied in this chapter. In this way we ruled out accidental match of data and thereby enhance the trust in the simulation tools. Finally, it should be stressed that a negative results of a validation, i.e. the experimental results to not correlate with the simulation results, also constitutes valuable information. In this case it implies that an essential ingredient for the understanding of the test results is still missing.

Fig. 22.54: Photograph of the power-cell tester prototype. For further details and usage, see Fig. 21.18 and Fig. 21.20.

References

1. AEC-Q100-12: *Short Circuit Reliability Characterization of Smart Power Devices for 12V Systems*, Document Automotive Electronics Council, Revised version, 2006.
2. ALPERN, P., NELLE, P., BARTI, E., GUNTHER, H., KESSLER, A., TILGNER, R., AND STECHER, M.: *On the Way to Zero Defect of Plastic-Encapsulated Electronic Power Devices – Part I: Metallization*, IEEE Transactions on Device and Materials Reliability, vol. 9, pp. 269–278, 2009.
3. ALPERN, P., NELLE, P., BARTI, E., GUNTHER, H., KESSLER, A., TILGNER, R., AND STECHER, M.: *On the Way to Zero Defect of Plastic-Encapsulated Electronic Power Devices – Part II: Molding Compound*, IEEE Transactions on Device and Materials Reliability, vol. 9, pp. 279–287, 2009.
4. ALPERN, P., NELLE, P., BARTI, E., GUNTHER, H., KESSLER, A., TILGNER, R., AND STECHER, M.: *On the Way to Zero Defect of Plastic-Encapsulated Electronic Power Devices – Part III: Chip Coating, Passivation, and Design*, IEEE Transactions on Device and Materials Reliability, vol. 9, pp. 288–295, 2009.

5. BANAGAAYA, N., FENG, L., AND BENNER, P.: *Sparse (P)MOR for Electro-Thermal Coupled Problems with Many Inputs.* Chapter in this Book, 2019.
6. BITTNER, K., AND BRACHTENDORF, H.-G.: *Multirate Circuit - EM - Device Simulation.* Chapter in this Book, 2019.
7. CASPER, T., DE GERSEM, H., GILLON, R., GOTTHANS, T., KRATOCHVIL, T., MEURIS, P., AND SCHÖPS, S.: *Electrothermal simulation of bonding wire degradation under uncertain geometries.* Proceedings of the 2016 Design, Automation & Test in Europe Conference & Exhibition (DATE), Dresden, pp. 1297–1302, 2016.
8. DUQUE GUERRA, D.J., CASPER, T., SCHÖPS, S., DE GERSEM, H., RÖMER, U., GILLON, R., WIEERS, A., KRATOCHVIL, T., GOTTHANS, T., AND MEURIS, P.: *Bond Wire Models.* Chapter in this Book, 2019.
9. FENG, L., AND BENNER, P.: *Parametric Model Order Reduction for Electro-Thermal Coupled Problems.* Chapter in this Book, 2019.
10. FENG, L., YUE, Y., BANAGAAYA, N., MEURIS, P., SCHOENMAKER, W., AND BENNER, P.: *Parametric Modeling and Model Order Reduction for (Electro-)Thermal Analysis of Nanoelectronic Structures.* Journal of Mathematics in Industry, 6 (10):1–16, 2016. Open access: https://dx.doi.org/10.1186/s13362-016-0030-8.
11. JANSSEN, R., GILLON, R., WIEERS, A., DELEU, F., GUEGNAUD, H., REYNIER, P., SCHOENMAKER, W., AND TER MATEN, E.J.W.: *Test Cases for Power-MOS Devices and RF-Circuitry.* Chapter in this Book, 2019.
12. JANSSEN, H.H.J.M., BENNER, P., BITTNER, K., BRACHTENDORF, H.-G., FENG, L., TER MATEN, E.J.W., PULCH, R., SCHOENMAKER W., SCHÖPS, S., AND TISCHENDORF, C.: *The European Project nanoCOPS for Nanoelectronic Coupled Problems Solutions.* In: RUSSO, G., CAPASSO, V., NICOSIA, G., AND ROMANO, V. (EDS.): *Progress in Industrial Mathematics at ECMI 2014,* Series Mathematics in Industry Vol. 22, Springer International Publishing, pp. 835–842, 2016. http://dx.doi.org/10.1007/978-3-319-23413-7_116.
13. JANSSEN, R., TER MATEN, J., BRACHTENDORF, H.-G., BITTNER, K., SCHOENMAKER, W., MEURIS, P., DE SMEDT, B., KRATOCHVÍL, T., GÖTTHANS, T., GILLON, R., WIEERS, A., DELEU, F., REYNIER, P., PULCH, R., TISCHENDORF, C., STROHM, C., AND PUTEK, P.: *nanoCOPS Deliverable D3.3 - Final Report on Validation and Measurements,* 2016, http://www.fp7-nanocops.eu/.
14. JERKE, G., AND KAHNG, A.B.: *Mission profile aware IC design – a case study.* In: Proc. Design, Automation and Test in Europe Conference and Exhibition (DATE), https://ieeexplore.ieee.org/document/6800278/, 6p, 2014.
15. JUNG, C.C., SILBER, C., AND SCHEIBLE, J.: *Heat Generation in Bond Wires,* IEEE Transactions on Components, Packaging and Manufacturing Technology, vol. 5, pp. 1465–1476, 2015.
16. JUNG, C.C., SILBER, C., AND SCHEIBLE, J.: *Temperature profiles along bonding wires, revealed by the bond calculator, a new thermo-electrical simulation tool,* 16th International Conference on Thermal, Mechanical and Multi-Physics Simulation and Experiments in Microelectronics and Microsystems (EuroSimE), 2015, http://ieeexplore.ieee.org/abstract/document/7103148.
17. MOMENTUM/EMPRO: *3D planar electromagnetic (EM) simulator used for passive circuit analysis,* and HEATWAVE: *Electro-Thermal Analysis Software* (also available in *ADS Electro-Thermal Simulator Element*), Keysight Technologies, Santa Rosa, CA, USA, 2018, https://www.keysight.com.
18. KRATOCHVIL, T., PETRZELA, J., SOTNER, R., DRINOVSKY, J., GOTTHANS, T., WIEERS, A., GILLON, R., REYNIER, P., AND POUPIN, Y.: *Measurements for RF Amplifiers, Bond-Wire Fusing and MOS Power Cells.* Chapter in this Book, 2019.
19. LIU, K., FRYE, R., KIM, H., LEE, Y., KIM, G., PARK, S., AND AHN, B.: *Electrical-thermal characterization of wires in packages.* In: 2014 IEEE 64th Electronic Components and Technology Conference (ECTC), 2014.
20. MAGWEL NV: *Device-Electro-Magnetic Modeler (devEM)* and *Power Transistor Modeler: PTM,* MAGWEL NV, Leuven, Belgium, 2018, http://www.magwel.com/.

21. MALLIK, A., AND STOUT, R.: *Simulation methods for predicting fusing current and time for encapsulated wire bonds*. IEEE Transactions on Electronic Packaging Manufacturing, 33 (4), pp. 255–264, 2010.

22. TER MATEN, E.J.W., PUTEK, P.A., GÜNTHER, M., PULCH, R., TISCHENDORF, C., STROHM, C., SCHOENMAKER, W., MEURIS, P., DE SMEDT, B., BENNER, P., FENG, L., BANAGAAYA, N., YUE, Y., JANSSEN, R., DOHMEN, J.J., TASIĆ, B., DELE, F., GILLON, R., WIEERS, A., BRACHTENDORF, H.-G., BITTNER, K., KRATOCHVÍL, T., PETŘZELA, J., SOTNER, R., GÖTTHANS, T., DŘÍNOVSKÝ, J., SCHÖPS, S., DUQUE GUERRA, D.J., CASPER, T., DE GERSEM, H., RÖMER, U., REYNIER, P., BARROUL, P., MASLIAH, D., AND ROUSSEAU, B.: *Nanoelectronic COupled Problems Solutions - nanoCOPS: Modelling, Multirate, Model Order Reduction, Uncertainty Quantification, Fast Fault Simulation*, Journal for Mathematics in Industry 7:2, 2016. Open access: http://dx.doi.org/10.1186/s13362-016-0025-5

23. NIRMAIER, T., BURGER, A., HARRANT, M., VIEHL, A., BRINGMANN, O., ROSENSTIEL, W., AND PELZ, G.: *Mission profile aware robustness assessment of automotive power devices*. In: *Proc. Design, Automation and Test in Europe Conference and Exhibition (DATE)*, http://ieeexplore.ieee.org/document/6784162/, 6p, 2014.

24. NÖBAUER, G.T., AND MOSER, H.: *Analytical Approach to Temperature Evaluation in Bonding Wires and Calculation of Allowable Current*, IEEE Transactions on Advanced Packaging, vol. 23-3, pp. 426–435, 2000.

25. PUTEK, P., JANSSEN, R., NIEHOF J., TER MATEN, E.J.W., PULCH, R., GÜNTHER, M., TASIĆ, B.: *Robust optimization of a RFIC isolation problem under uncertainties*. In: LANGER, U., AMRHEIN, W., AND ZULEHNER, W. (EDS): *Scientific Computing in Electrical Engineering – SCEE 2016*. Springer International Publishing AG, Series Mathematics in Industry, Vol. 28, pp. 177–186, 2018. http://dx.doi.org/10.1007/978-3-319-75538-0_17.

26. PUTEK, P., JANSSEN, R., NIEHOF J., TER MATEN, E.J.W., PULCH, R., TASIĆ, B., AND GÜNTHER, M.: *Nanoelectronic Coupled Problems Solutions: Uncertainty Quantification for Analysis and Optimization of an RFIC Interference Problem*. Submitted to Journal for Mathematics in Industry (Open Access).

27. PUTEK, P., JANSSEN, R., NIEHOF J., TER MATEN, E.J.W., PULCH, R., TASIĆ, B., AND GÜNTHER, M.: *Nanoelectronic coupled problem solutions: Uncertainty Quantification of RFIC Interference*. In: QUINTELA, P., BARRAL, P., GÓMEZ, D., PENA, F.J., RODRIGUEZ, J., SALGADO, P., AND VÁZQUEZ-MÉNDEZ, M.E. (EDS): *Progress in Industrial Mathematics at ECMI 2016*. Springer International Publishing AG, Series Mathematics in Industry, Vol. 26, pp. 271-279, 2017. http://dx.doi.org/10.1007/978-3-319-63082-3_42.

28. PUTEK, P., TER MATEN, E.J.W., GÜNTHER, M., BARTEL, A., PULCH, R., MEURIS, P., AND SCHOENMAKER, W.: *Robust Shape Optimization under Uncertainties in Device Materials, Geometry and Boundary Conditions*. Chapter in this Book, 2019.

29. PUTEK, P., MEURIS, P., GÜNTHER, M., TER MATEN, J., PULCH, R., WIEERS, A., AND SCHOENMAKER, W.: *Uncertainty Quantification in Electro-Thermal Coupled Problems based on a Power Transistor Device*, IFAC-PapersOnLine, Vol. 48, No. 1, pp. 938–939, 2015, http://dx.doi.org/10.1016/j.ifacol.2015.05.206.

30. PUTEK, P., MEURIS, P., PULCH, R., TER MATEN, E.J.W., GÜNTHER, M., SCHOENMAKER, W., DELEU, F., AND WIEERS, A.: *Shape optimization of a power MOS device under uncertainties*, Proceedings DATE-2016, Design, Automation & Test in Europe, March 14-18, 2016, Dresden, Germany, pp. 319–324, 2016, http://dx.doi.org/10.3850/9783981537079_0994.

31. PUTEK, P., MEURIS, P., PULCH, R., TER MATEN, J., SCHOENMAKER, W., AND GÜNTHER, M.: *Uncertainty quantification for a robust topology optimization of power transistor devices*, IEEE Transactions on Magnetics 52-3, paper: 1700104, 2016, http://dx.doi.org/10.1109/TMAG.2015.2479361.

32. SENTAURUS INTERCONNECT: *Stress related reliability analysis of interconnects*. Synopsys Inc., Mountain View, CA, USA, 2017, urlhttps://www.synopsys.com.

33. SMORODIN, T., WILDE, J., ALPERN, P., AND STECHER, M.: *A Temperature-Gradient-Induced Failure Mechanism in Metallization Under Fast Thermal Cycling*, IEEE Transactions on Device and Materials Reliability, vol. 8, pp. 590–599, 2008.
34. SPECTRE: *Spice-class Circuit Simulator*. Cadence Design Systems, Inc., San Jose, CA, USA, 2017, `https://www.cadence.com`, `https://www.cadence.com/content/cadence-www/global/en_US/home/tools/custom-ic-analog-rf-design/circuit-simulation/spectre-circuit-simulator.html`.
35. SUKHAREV, V., KTEYA, A., AND ZSCHECH, E.: *Physics-Based Models for EM and SM Simulation in Three-Dimensional IC Structures*, IEEE Transactions on Device and Materials Reliability, vol. 12, pp. 272–284, 2012.
36. SZEL, A., SARKANY, Z., BEIN, M., BORNOFF, R., VASS-VARNAI, A., AND RENCZ, M.: *Lifetime estimation of power electronics modules considering the target application*. In: Thermal Measurement, Modeling Management Symposium (SEMI-THERM), 2015 31st, 2015.
37. SZEL, A., SARKANY, Z., BEIN, M., BORNOFF, R., VASS-VARNAI, A., AND RENCZ, M.: *Mission profile driven component design for adjusting product lifetime on system level*. In: 2015 International Conference on Electronics Packaging and iMAPS All Asia Conference (ICEP-IACC), 2015.
38. YILDIZ, Ö.F.: *Analysis of Electromagnetic Interference Variability on RF Integrated Circuits*, MSc-Thesis, Technische Universität Hamburg, 2016.
39. YUE, Y., FENG, L., BENNER, P., PULCH, R., AND SCHÖPS, S.: *Reduced Models and Uncertainty Quantification*. Chapter in this Book, 2019.
40. ZARBAKHSH, J., KARUNAMURTHY, B., TREJO-CABALLERO, C.O., BARTI, E., AND DETZEL, T.: *Microscopic stress simulation of non-planar chip technologies' Micro-electronics Reliability*, vol. 50, no. 9 - 11, pp. 1666-1671, 2010.
41. ZHU, X., LU, H., AND BAILEY, C.: *Modelling the stress effect during metal migration in electronic interconnects*. In: 14th IEEE International Conference on Nanotechnology, 2014.

Chapter 23
Methodology and Best-Practice Guidelines for Thermally Optimized Driver Design

Renaud Gillon, Aarnout Wieers, Frederik Deleu, Rick Janssen,
Wim Schoenmaker, Bart De Smedt, Hervé Guegnaud, Pascal Reynier

Abstract An important outcome of the validation activities is the formulation of methodologies to address coupled problems in EDA. This chapter describes typical coupled problems arising in the field of activities of the industrial partners. It also highlights the challenges and makes the link to some test cases. In order to address these challenges and solve the design problems illustrated by the test cases, several design flows were put in place based on outcomes of the project activities. We describes these flows as well as the tools set-up in order to support them. Finally, the recommended usage of these flows is proposed.

23.1 Design of a Power Transistor for Cost and Robustness

Designing a power transistor, see Fig. 23.1, for use in automotive applications involves finding a solution that meets many types of constraints. There are functional constraints such as the on-resistance and the off-state current. There are reliability constraints stating that the transistor should sustain

Renaud Gillon, Aarnout Wieers, Frederik Deleu
ON Semiconductor Belgium BVBA, Oudenaarde, Belgium,
e-mail: {Renaud.Gillon,Aarnout.Wieers,Frederik.Deleu}@onsemi.com

Rick Janssen
NXP Semiconductors, Eindhoven, the Netherlands, e-mail: Rick.Janssen@nxp.com

Wim Schoenmaker, Bart De Smedt
MAGWEL NV, Leuven, Belgium,
e-mail: {Wim.Schoenmaker,Bart.DeSmedt}@magwel.com

Hervé Guegnaud, Pascal Reynier
ACCO Semiconductor, Louveciennes, France, e-mail: Herve.Guegnaud@acco-semi.com,
Pascal.Reynier@yahoo.fr

© Springer Nature Switzerland AG 2019
E. J. W. ter Maten et al. (eds.), *Nanoelectronic Coupled Problems Solutions*,
Mathematics in Industry 29, https://doi.org/10.1007/978-3-030-30726-4_23

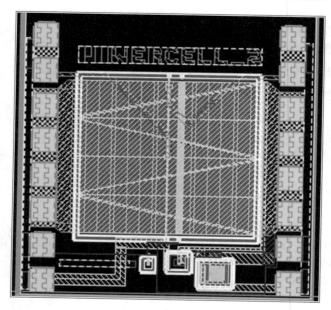

Fig. 23.1: Top-level view of a typical power transistor design.

operation in various modes for a certain amount of time. There are also cost constraints stemming from profitability objectives and often imposing limits on the area of silicon consumed for the realization of the power transistor.

Besides the fact that functional and reliability constraints are often driving towards solutions involving higher costs, the designer's job is complicated by the sheer size and complexity of a power transistor design, as its overall size is often 3 or 4 orders of magnitude larger than the size of its basic constitutive elements like a contact, a gate finger, a minimally-sized metal track. The complexity of a power transistor design is a very big challenge for typical simulation models, which are most easily formulated in terms of a combination of elementary constructs, which are present in very large numbers (varying from thousands to billions) in the power transistor structure. Simulating with models consisting of such large number of elements implies significant simulation times and computational resources which might not be compatible with the tight planning and budgets imposed by the competitive landscape on those markets. In order to minimize the risk of spending time and computational resources analyzing designs or layout that would not result in feasible solutions, it is important to establish a stepwise methodology that allows to identify issues and take appropriate measures as early as possible in the product definition, refinement, implementation and validation cycles.

23.1.1 Extrapolate the Performance of Power Transistors

Extrapolation allows to infer quantitative prediction outside of the range of an available set of measured basis points. Being able to extrapolate transistor performance from small unit cells towards large designs is critical for applications of power transistors. Indeed full-scale testing of power transistor is expensive, as most of the time the currents and voltages handled by these devices exceed the capabilities of the standardly available commercial equipment. Dedicated setup have to be custom-built, tested and debugged in order to enable full-scale testing of the power FETs (Field-Effect Transistors). On the other hand, small unit cells are more easily amenable to testing with standardly available commercial instruments. Hence the ability to extrapolate power transistor performance from a small unit cell towards a full-scale application level transistor is very valuable. It allows to reduce product design costs and speed up developments by removing the lead time required to validate the dedicated set-ups.

23.1.2 Use Realistic Models of Power Transistors at Application Level

Architectural design is a critical stage of product development, where application level requirements must be translated into lower level requirements defining design objectives for an IC and its constitutive blocks. The task is even more critical as it is often linked to the quotation phase, when the IC supplier has to come up with an estimate of its costs in order to make an offer to its customer. Taking a wrong decision during the architectural design phase, may turn out to be very costly, especially if the issue is only detected during a late stage of the development (typically during the validation and test phase). Hence it is important that architectural design decision are based on accurate and representative data or assumptions. In particular, for applications involving smart power ICs, a proper estimation of the size of the power transistors is critical, as they usually occupy a significant fraction of the total die area. A second critical aspect is a correct assessment of impact of the application level temperature profile and its impact on the product lifetime.

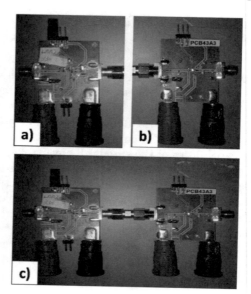

Fig. 23.2: Stand-alone (a) driver stage, (b) power stage and (c) DS + PS cascaded.

23.1.3 Power Amplifier Test Case

This test case represents a realistic stand-alone power amplifier used in an actual Power Amplifier (PA) product at 3 different integration steps. It is realized on a 0.18 μm standard CMOS technology. Designed for band II (@ 1.9GHz), the PA delivers on board plan 26dBm linear output power (@-40dBc) with 44

This user case is composed of 3 different dies (see Figs. 23.2 and 23.3):

1. Stand-alone power stage (PS) mounted directly on board testing (PCB): Driver stage has been removed to facilitate electro thermal simulation. In order to measure the transistor junction and the board ground pad temperature, two thermal sensors (diodes) have been used. This allows to extract junction to case (Rjc) and case to ambient (Rca) thermal resistances. All input and output matching networks are integrated on PCB board.
2. Stand-alone driver stage (DS) mounted directly on board testing (PCB): Simulation process is the same than for the stand-alone power stage. All input and output matching networks are integrated on PCB board.
3. Stand-alone driver and power stage cascaded.
4. Full PA (driver + power stage integrated on the same die) mounted on BT + PCB. Input and output matching networks are integrated on PCB.

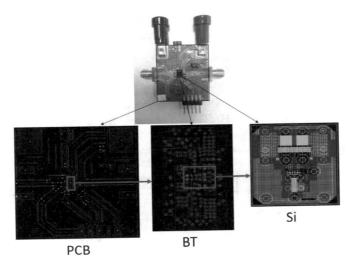

Fig. 23.3: Full power amplifier integrated on silicon + BT + PCB.

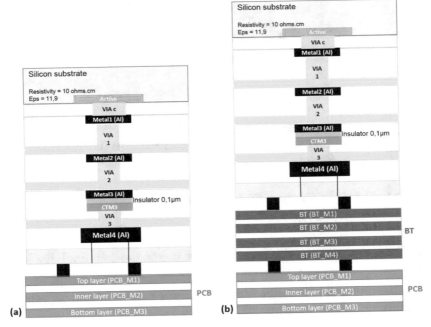

Fig. 23.4: Assembly (a) silicon + PCB and (b) silicon + BT + PCB.

Inter matching network is integrated on Silicon (capacitances) and BT (inductances).

These structures allow to evaluate package and substrate thermal impact on RF performances. To optimize the heat flow, the two stand-alone power and driver stages are directly mounted on PCB. On the full PA, the BT substrate (4 copper layers) is inserted between die and PCB (see Figure 23.4). All the junctions between PCB, BT and the test chip are made by bump pads.

Product name:	AC28180 stand-alone
IC	Power amplifier for WCDMA application for band II
Sub block design subtleties:	Step by step integration
	Electro-thermal issues, thermal runaway
	Thermal coupling effect between power stage and driver
Desired info from simulation:	Local temperature + joule heating-aware current density simulation
	Mean temperature + hot spot
Simulation inputs:	GDS + technology description
	Power sources localization
	Ambient temperature
	Frequeny
	Input power (sweep)
	Type of signal (CW pulsed)

Table 23.1: Power amplifier design and simulation details.

23.2 Integration of Tool Developments in Industrial Design Flows

23.2.1 Design Flows for Power FETs and Smart Power ICs

The main steps for the design of a smart power product involving strong electro-thermal coupling effects are listed below:

- Initial estimate of the transistor size, current-carrying capability, thermal impedance.

- Detailed implementation of the layout. Identification of the critical areas for current density. Extraction of dissipation levels in various operating modes.
- Life-time analysis considering the application-level profiles, the various operating modes, the effective dissipation levels, the device thermal impedance.
- Layout corrections and some iterations.
- Extraction of a reduced-order model allowing to represent the power transistor in circuit level simulations for validating the driver and sensor chains design, as well as the control algorithms.

RDSON and Current-density Maps Extraction Flow

A typical RDSON extraction flow is shown in Figure 23.5. This flow allows to validate the electrical performance of a power transistor in terms of its on-resistance (RDSON), but also visualize current density maps, or check for violations of current-density constraints. This flow is used in order to optimize the layout of power transistors during the implementation phase of the design. It allows to check if the layout meets the on-resistance target, but also to highlight locations where current density constraints are eventually violated.

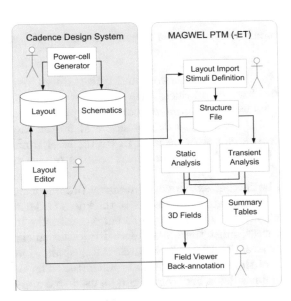

Fig. 23.5: Flowchart describing the typical RDSON extraction flow.

Thermal Network and Uncertainty Quantification Flow

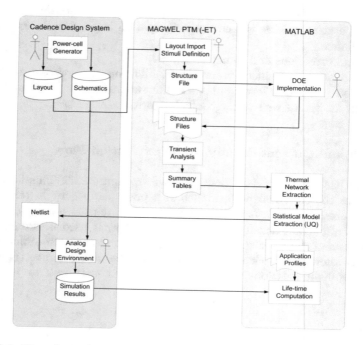

Fig. 23.6: Flowchart describing the thermal network / UQ extraction flow.

Fig. 23.6 shows the thermal network and Uncertainty Quantification (UQ) flow put in place. This flow produces a thermal network linking power transistors as heat sources with the boundary conditions prevailing at the interfaces of the package: pins, exposed pad, top side and lateral faces of the body. The thermal network can then be used in circuit level simulations in order to evaluate the impact of the various heat sources on the operation of the system.

The flow can be further extended to quantify the impact of variations occurring during the packaging process (e.g., thickness variations). The initial Design of Experiment (DOE), required to characterize the impact of the individual heat-sources, must then be extended in order to capture variations in material parameters or layer thicknesses. After simulation of the DOE, a statistical model of the thermal network is generated. The thermal networks generated by the flows described above, can be used to compute the distributions of temperatures, currents and voltages occurring in all critical devices and to finally assess the life-time consumption of the product in various operating modes and for various temperature profiles.

Model-order Reduction Flow

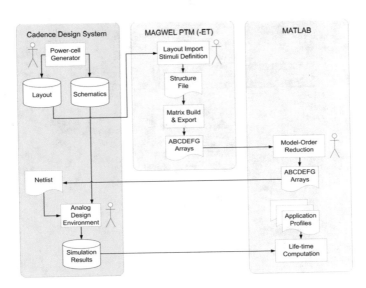

Fig. 23.7: Flowchart describing the model-order reduction flow.

Fig 23.7 describes a flow based on Model Order Reduction (MOR) which was tested in the frame of the nanoCOPS project for generating circuit-level electro-thermal models for power transistors from a 3D geometry. The advantage of this flow is that it generates a compact representation of the thermal and electrical interconnection network of the power transistor that can be easily interfaced with existing transistor models that described the behavior of the core transistor. These models may be characterized on smaller scale devices. The combination of the core transistor models (e.g., BSIM or EKV-like) with an extracted electrical and thermal network is likely to be more accurate than the combination of a "empirically scalable" transistor model and a thermal network as described in the previous section. However, this increase in accuracy may come at the cost of longer simulation times in the circuit simulator as the netlist incorporating the ABCDE matrices produced by the MOR procedure is still significantly larger than the one produced by the procedure from the previous section.

23.2.2 Electro-thermal Design Flow

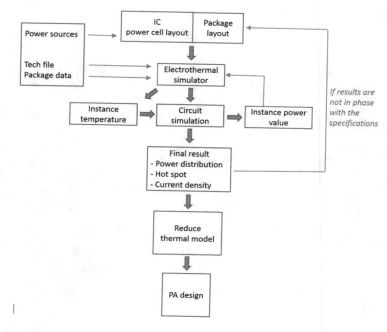

Fig. 23.8: Flowchart describing the electro-thermal design flow.

The design flow for electro-thermal simulations is described in Fig. 23.8. For power amplifier designs, thermal aspects are crucial. We used the electro-thermal simulator at different steps / phases of the design of a PA product.

In the design phase, the electro-thermal simulator is used to optimize power cell and package drawing in order to:

- ensure a uniform power distribution,
- optimize the heat flow,
- estimate the hot spot and the mean temperature at maximum power,
- estimate the current into the IC interconnections and compare to electro migration rules.

If power cell results are not in line with the expectations in terms of RF performances and/or electro-migration, IC and/or package layout are modified. To save design time, we create after that realistic thermal model (in phase with electro-thermal results) that we used for the full PA design.

Before fabrication we used the electro-thermal simulator for the product validation.

23.2.3 Power Transistor Electro-thermal Design Flow

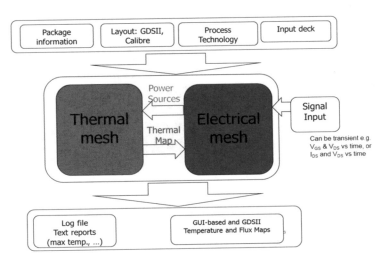

Fig. 23.9: Flowchart describing the power transistor electro-thermal simulation flow.

Challenges/Requirements for Power Transistor Applications (PTA)

- Hot Swap applications require both low RDSON and high level of linear mode performance.
- We need to design out (eliminate) the hot spots, achieving a better temperature uniformity across the die.
- We need to optimize the SOA (Safe Operating Area) performance of the PTA, while minimizing the impact on RDSON.

How Can We Use Simulation To Improve The Design?

- It is known that the linear mode performance can be improved by modifying the gain of the MOSFET.
- The formation of hotspots that lead to thermal runaway and destruction also depends on the
  - Current density,
  - Thermal impedance.
- The approach is to modify the die design by varying the MOSFET gain factor in areas susceptible to thermal runaway.
- Therefore, to achieve better current distribution and temperature uniformity across the die, and eliminate hotspot formation.
- The electro-thermal simulation flow is described in Fig. 23.9.

23.2.4 Back-Annotation

The design flow of power management ICs can be supported further by back-annotating the outcome of detailed electro-thermal simulation results into the layout editing environment. Using a graphical interface, violations of peak current densities in the interconnect can be browsed through. The corresponding layout structure can be highlighted in the design framework. Such a feature points the layout engineer directly to the problematic regions and simplifies the task to correct the layout.

23.3 Recommended Design Methodologies for Electro-thermal Problems

23.3.1 Full 3D electro-thermal simulation

PTM is a forerunner of PTM-ET [5] (both tools developed by MAGWEL). The latter has electro-thermal coupling simulation facilities.

The PTM (no ET) Tool

PTM's integrated environment combines a state of the art 3D field solver, an easy-to-use results viewer featuring cross linked reports and field view. Incorporating a proprietary unique edge-based 3D mesher and 64-bit numerical solvers, PTM extracts RDSON in large power transistor arrays by computing detailed non-uniform current distributions in metal & poly interconnect and vias while including bond wires in the simulation. PTM's Field View allows users to easily create current, voltage, power dissipation and other types of graphs through any slice of the layout. The tool offers extremely fast simulation on standard Linux workstations with RDSON results matching silicon typically within 2%-5%. PTM reports and displays current densities and electro-migration rule violations on the layout. Current density and voltage distributions can be viewed in 2D or cross-cut formats. Designers can focus in on potential trouble spots by locally increasing mesh density for better accuracy. Designers also have full control of terminal placement and excitation, including test benches with Voltage Controlled Current Source (VCCS) to help with high precision sense device design.

The PTM-ET Tool

PTM-ET builds forth on the PTM tool by adding coupled static and dynamic electro-thermal simulation capabilities. In static mode, the tool evaluates voltage, current density, temperature and heat flux density fields across a power transistor or smart power IC in steady state, taking the electro-thermal coupling into account arising mainly from the temperature dependency of the electric and thermal conductivity of the materials. In transient mode, the tool allows to compute the evolution of the above fields under time-varying excitations applied at the ports. The tool allows to visualize the distribution of these fields at selected time points of the simulation run. Tables summarizing the evolution of port quantities (voltage, current, temperature and heat flux) at all time points are generated in the run directory.

Tool Usage patterns

Table 23.2 gives an overview of the tool usage patterns that have been tested.

23.3.2 3D Electro-Thermal Simulation With Effective Parameters

When the structure to be analyzed is really too big to be handled with all details in the PTM field solver (despite its very large capacity), it is still possible to abstract details from the original structure by running simulations on a subdomain that is repeated regularly (see Fig. 23.10). This procedure was successfully tested in the FP7-SMAC [6] project on a case similar to the test case 1, which is described in [3,4]. Yet, the recent further progress in the PTM solver has made it possible to handle the test case 1 structure directly

Design Phase	Objective	PTM Outcomes	PTM-ET Outcomes
Architectural	Initial estimates based on approximate layout	RDSON; Current-density;	TRISE[1] in regime and pulsed condition; ABCDE[2] matrices (electro-thermal); Thermal network; Temperature maps; Current density
Implementation	Layout refinement and optimization integrating floor-planning constraints	RDSON; Current density;	
	Check life time constraints (adjust layout)		TRISE in regime and pulsed condition; Thermal network; Temperature maps; Current density
Verification	Validate interaction of drivers and sensors with power FET	ABCDE matrices (electrical only);	ABCDE matrices (electro-thermal); Thermal network.

[1] TRISE: Abbreviation and parameter name often used in circuit simulators to denote a local rise of temperature.
[2] ABCDE matrices: A state-space formulation of differential equations describing the behaviour of a dynamic system.

Table 23.2: Tool usage patterns.

in PTM without having to run the special procedure tested in the earlier FP7-SMAC Project [6].

23.3.3 Equivalent Thermal and Electrical Network Extraction

Running electro-thermal simulation on large structures is necessary in order to obtain accurate performance predictions and to validate the reliability of a given design. However, as these simulations tend to consume a significant amount of computational resources and time, they cannot be used at any time a designer or system architects needs information about the electro-thermal behavior of a power transistor. The main limitation is here the mismatch between the allotted time frame for the job and the set-up and simulation time. In order to bridge this gap, it is important to have methods that allow to capture the result of the detailed simulation in a compact and reusable form. Methods for generating thermal and electro-thermal equivalent networks were tested and validated (see Chapters [1, 2, 7] and [4]). These methods opened new perspectives in the validation of product lifetimes in function of arbitrary

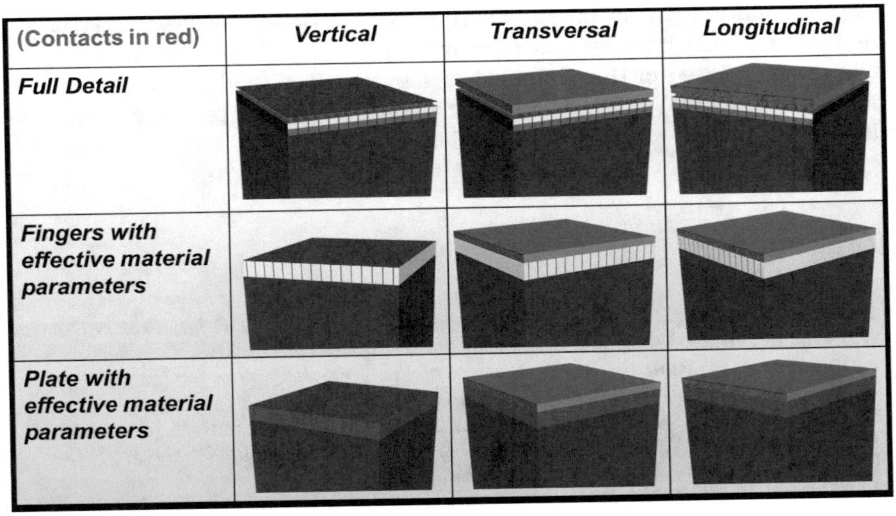

Fig. 23.10: Design of experiment allowing to extract the effective mesoscopic parameters [6].

application level profiles, as well as for the validation of complete sensor chains using virtual prototyping techniques.

23.4 Conclusion

In this chapter, design flows, which were developed and tested for the design of thermally optimized power drivers, were described. The characterization of the electro-thermal coupling is performed using the PTM and PTM-ET tools from MAGWEL and subsequently "packaged" in different forms in order to facilitate interaction with the standard design tools. The feedback from the electro-thermal tool was made either by back-annotation or by highlighting of warnings in the layout, or the creation of netlist components representing the thermal and electrical networks of the power transistor in compact form. These compact forms where either obtained algebraically using model-order reduction procedures or using system identification techniques.

Simulating coupled problems in order to address issues that may affect the design of products implies a risk of spending a lot of time and computational resources for the simulations. The new tools brought by MAGWEL allow to simulate complex coupled problems starting from the "physical level" (material level) in a reasonable but still significant time-frame. From the user's perspective it is important to be able to get the most out of a set of "physical level" simulations by extracting abstract models or representations that can be used at higher level for quick evaluations in very short time frames.

The work reported in this chapter has shown that several flows can be built on top of the MAGWEL tools in order to support a more precise and efficient design of products affected by cross-domain couplings.

References

1. BANAGAAYA, N., FENG, L., AND BENNER, P.: *Sparse (P)MOR for Electro-Thermal Coupled Problems with Many Inputs.* Chapter in this book, 2018.
2. FENG, L., AND BENNER, P.: *Parametric Model Order Reduction for Electro-Thermal Coupled Problems.* Chapter in this Book, 2018.
3. JANSSEN, R., GILLON, R., WIEERS, A., DELEU, F., GUEGNAUD, H., REYNIER, P., SCHOENMAKER, W., AND TER MATEN, E.J.W.: *Test Cases for Power-MOS Devices and RF-Circuitry.* Chapter in this book, 2018.
4. JANSSEN, R., GILLON, R., WIEERS, A., DELEU, F., GUEGNAUD, H., REYNIER, P., SCHOENMAKER, W., AND TER MATEN, E.J.W.: *Validation of Simulation Results on Coupled Problems.* Chapter in this book, 2018.
5. MAGWEL NV, Leuven, Belgium. ONLINE: http://www.magwel.com/products/, 2018.
6. *SMAC: SMArt systems Co-design.* Smart Components & Smart Systems Integration. Project FP7-ICT-2011-7 – 288827 – CP IP, 2011-2015. ONLINE: http://www.fp7-smac.org/, 2018.
7. YUE, Y., FENG, L., BENNER, P., PULCH, R., AND SCHÖPS, S.: *Reduced Models and Uncertainty Quantification.* Chapter in this book, 2018.

Index

© Springer Nature Switzerland AG 2019
E. J. W. ter Maten et al. (eds.), *Nanoelectronic Coupled Problems Solutions*,
Mathematics in Industry 29, https://doi.org/10.1007/978-3-030-30726-4

Printed in the United States
By Bookmasters